AN OUTLINE OF MATHEMATICAL LOGIC

SYNTHESE LIBRARY

MONOGRAPHS ON EPISTEMOLOGY,

LOGIC, METHODOLOGY, PHILOSOPHY OF SCIENCE,

SOCIOLOGY OF SCIENCE AND OF KNOWLEDGE,

AND ON THE MATHEMATICAL METHODS OF

SOCIAL AND BEHAVIORAL SCIENCES

Editors:

DONALD DAVIDSON, *Rockefeller University and Princeton University*

JAAKKO HINTIKKA, *Academy of Finland and Stanford University*

GABRIËL NUCHELMANS, *University of Leyden*

WESLEY C. SALMON, *University of Arizona*

VOLUME 70

AN OUTLINE OF MATHEMATICAL LOGIC

Fundamental Results and Notions
Explained with All Details

by

ANDRZEJ GRZEGORCZYK

Warsaw University

D. REIDEL PUBLISHING COMPANY
DORDRECHT-HOLLAND/BOSTON-U.S.A.

PWN—POLISH SCIENTIFIC PUBLISHERS
WARSZAWA

Translated from the original Polish
Zarys Logiki Matematycznej, Warszawa 1969

by

Olgierd Wojtasiewicz *and* Wacław Zawadowski

Library of Congress Catalog Card Number 72-97956
ISBN-13: 978-90-277-0447-4   e-ISBN-13: 978-94-010-2204-0
DOI: 10.1007/978-94-010-2204-0

Co-publication with Państwowe Wydawnictwo Naukowe, Warszawa, Poland

Distributors for the socialistic countries
ARS POLONA — RUCH
Krakowskie Przedmieście 7, 00-068 Warszawa 1, Poland

Distributors for the U.S.A., Canada and Mexico
D. REIDEL PUBLISHING COMPANY, INC.
306 Dartmouth Street, Boston, Mass. 02116, U.S.A.

Distributors for all other countries
D. REIDEL PUBLISHING COMPANY
P.O. Box 17, Dordrecht, Holland

TO MY PARENTS

PREFACE

Recent years have seen the appearance of many English-language hand-books of logic and numerous monographs on topical discoveries in the foundations of mathematics. These publications on the foundations of mathematics as a whole are rather difficult for the beginners or refer the reader to other handbooks and various piecemeal contributions and also sometimes to largely conceived "mathematical folklore" of unpublished results. As distinct from these, the present book is as easy as possible systematic exposition of the now classical results in the foundations of mathematics. Hence the book may be useful especially for those readers who want to have all the proofs carried out in full and all the concepts explained in detail. In this sense the book is self-contained. The reader's ability to guess is not assumed, and the author's ambition was to reduce the use of such words as *evident* and *obvious* in proofs to a minimum. This is why the book, it is believed, may be helpful in teaching or learning the foundation of mathematics in those situations in which the student cannot refer to a parallel lecture on the subject. This is also the reason that I do not insert in the book the last results and the most modern and fashionable approaches to the subject, which does not enrich the essential knowledge in foundations but can discourage the beginner by their abstract form.

A. G.

CONTENTS

INTRODUCTION TO THE PROBLEMS OF THE FOUNDATIONS OF MATHEMATICS

The foundations of mathematics consist of two basic theories, the logical calculus and set theory, together with methodological research concerning these and other mathematical theories. The logical and the set theoretical concepts arise from an analysis of the subject matter of mathematics as well as from an analysis of the construction of mathematical concepts. Let us first consider the subject matter of mathematics.

Beginning in antiquity numbers and geometrical figures were the subject matter of mathematics, but in modern times the emphasis was more on numerical functions than on the numbers themselves. Recently this abstraction has gone still further. Mathematicians have begun to investigate arbitrary functions defined on sets of arbitrary objects and limited only by very general conditions central to each given branch of mathematics.

The most convenient way to describe the subject matter of present-day mathematics is to use the concept of mathematical domain. By a mathematical domain, or simply a domain, we mean what is often called a relational system, or an algebra. We do not find these terms satisfactory. So we will say that a mathematical domain is a set of objects connected by some relations.

1. MATHEMATICAL DOMAINS

A *mathematical domain* is any sequence of objects $\langle Z, f_1, f_2, ..., f_n, a_1, a_2, ..., a_k, R_1, R_2, ..., R_m \rangle$ satisfying the following conditions:

a) the first of these objects, denoted here by Z, is a set;

b) the next n objects: $f_1, f_2, ..., f_n$ are functions with arguments in Z, and values also in Z;

c) the objects: $a_1, a_2, ..., a_k$ are some elements chosen from the set Z;

d) the last m objects: $R_1, R_2, ..., R_m$ are some relations which interest us. These relations hold between some elements of the set Z.

Before giving examples of various mathematical domains we shall explain the meanings of the concepts used: *set, element, function,* and *relation.* We shall soon see that even a cursory analysis of these concepts leads to elements of set theory and the logical calculus. These concepts are partly familiar from high school. A set may, for instance, consist of all the chairs that are in this room. There are four of them. This set has, therefore, four elements. A set may also consist of all natural numbers less than the number 4. They are the numbers: 0, 1, 2, 3. We say that these numbers are the elements of the set of natural numbers less than the number 4.

We shall adopt the following notation. If we know all the elements of a finite set and if we write their names separated by commas between the brackets $\{\ ,\ \}$, then this composite sign will be considered the name of that set. The names of its elements may be written in an arbitrary order. For instance, the set of natural numbers less than 4 may be denoted by $\{0, 1, 2, 3\}$, as well as by $\{1, 2, 0, 3\}$.

Sometimes instead of saying that an object a is an element of a set X, we shall use the usual abbreviation

$$a \in X.$$

Accordingly, a sentence stating that the number 1 belongs to the set $\{0, 1, 2, 3\}$ is symbolically written

$$1 \in \{0, 1, 2, 3\}.$$

The following sentences are true: $0 \in \{0, 3\}$, $0 \in \{0\}$, $1 \in \{1\}$, $3 \in \{0, 5, 3, 7\}$, while the following are false: $0 \in \{1, 2\}$, $2 \in \{0, 3, 1\}$. To show that an object does not belong to a set we shall use the same symbol with a bar, for instance: $0 \notin \{1, 2\}$ and $2 \notin \{0, 3, 1\}$. In this notation the following general sentences are true:

For every x, $x \in \{0, 3\}$ if and only if $x = 0$ or $x = 3$.

For every x, $x \in \{0, 1, 2\}$ if and only if $x = 0$ or $x = 1$ or $x = 2$.

All the above sets were finite. In contrast the following sets are infinite:

The set of natural numbers: 0, 1, 2, 3, 4, 5, ..., which will be denoted by $\mathcal{N}$.

The set of even natural numbers: 0, 2, 4, ...

The set of odd natural numbers: 1, 3, 5, ...

The set of rational numbers, which we denote by $\mathcal{W}$.

The set of real numbers, denoted by $\mathcal{R}$.

1. MATHEMATICAL DOMAINS

A set is called *empty*, if it contains no elements. For instance, the set of even numbers which are also odd is empty. It is empty, because there are no such numbers. The set of whales living in the Parliament Building is also empty. If the elements of a set X are sets, then we often call this set X a *family of sets*. We do so to avoid the unpleasent repetition which occurs in the phrase: a set of sets.

We say that a set X is *included* in a set Y, if every element of X is also an element of Y. In writing this proposition down, we often use the abbreviation: $X \subset Y$. Using the symbol $\in$ we can formulate a rigorous definition of inclusion for sets:

(1) $X \subset Y$ *if and only if, for every* x, *if* $x \in X$ *then* $x \in Y$.

In plain English formula (1) could be read: *a set X is included in a set Y if and only if every element of X is also an element of Y.*

For instance, the set of even natural numbers is included in the set of natural numbers, since every even natural number is natural. The set of rational numbers is included in the set of real numbers, since every rational number is real. We also have $\{0, 1, 2\} \subset \{0, 1, 2, 3, 4\}$.

A *unit set* is a set which has only one element. The sets $\{0\}$, $\{1\}$, $\{3\}$, $\{7\}$, etc. are unit sets. If $x \in X$, then the unit set $\{x\}$ is included in the set X: $\{x\} \subset X$, and conversely, if $\{x\} \subset X$, then $x \in X$. The following obvious statement can be made using the identity symbol:

$x \in \{y\}$ *if and only if* $x = y$.

We shall also write concisely: $x, y, z, \ldots \in A$ instead of $x \in A$, $y \in A$, $z \in A$, etc., for instance, $0, 1 \in \{0, 1\}$, $0, 1, 2 \in \{0, 1, 2, 3\}$.

Before discussing the concept of a function, we shall characterize the concept of a relation, which is simpler. Any connection between two objects is a *relation*. Let us illustrate the meaning of the concept. For instance, the connection between real numbers based on the property of one number being smaller than another is a relation. The sentence: *a number x is less than a number y* is written concisely as

$x < y$.

Among the numbers of the set $\{0, 1, 2\}$ we have the following instances of one number being smaller than another: $0 < 1$, $0 < 2$, $1 < 2$. On the other hand the following connections do not hold: $0 < 0$, $1 < 1$, $2 < 2$, $2 < 1$, $2 < 0$, $1 < 0$.

The following relations are very important:
the relation of identity among numbers, denoted by $=$,
the relation of being not greater than, denoted by $\leqslant$,
the relation of divisibility for integers, denoted by $\mid$ (the sentence: *an integer x is divisible by an integer y* is denoted symbolically by $y|x$).

The seniority relation between persons can be expressed by the phrase: *x is older than y*. Many other relations also hold among persons.

If R stands for a given relation, then the statement that an object x is related by R to an object y may written concisely as:

$$xRy,$$

For instance, $x = y$, $x \leqslant y$, $z \leqslant 3$, etc.

A function is a special case of a relation. Let us consider for instance the relation: y is the square of x:

$$y = x^2,$$

which is a relation between the numbers x and y. This relation has the property that 1) for every number x, there is a y such that $y = x^2$, and 2) the number y which is the square of a given number x is unique. Every relation that satisfies conditions parallel to these is called a *function*. The relation $y = x^2$ is accordingly a function.

We say that a relation R, (xRy), is a *function* defined on the elements of a set X with values in a set Y if and only if

(2)   *for every element x of the set X there exists an element y of the set Y such that xRy,*

and

(3)   *for every element x of the set X $(x \in X)$ there exists only one element y which satisfies the condition xRy.*

For instance, the relation xR_1y, interpreted as $x^2 = y$, is a function defined on the elements of the set $\mathscr{R}$ of real numbers, with values also belonging to the set $\mathscr{R}$. The relation: y is the natural number which succeeds x is a function defined on the set $\mathscr{N}$ of natural numbers, with values in the set $\mathscr{N}$. This function is called the *successor* function. The relation $2x = y$ may be regarded as a function defined on the set $\mathscr{N}$, with values in the set of even natural numbers. The relation $\sqrt{x} = y$ is a function defined on the set of non-negative real numbers, with values in the set of non-negative real numbers.

4

2. EXAMPLES

If a relation R is a function defined on a set X with values in a set Y, then we often say that this relation R establishes a *correspondence*: to every element x of the set X there corresponds at least one element (condition (2)) and at most one element (condition (3)) of the set Y. We also say that this relation (function) *maps* X into Y, or that it *transforms* X into Y. For instance, the function $y = 2x$ maps the set of integers into the set of even integers.

All these functions: x^2, $2x$, $\sqrt{x}$ are functions of one argument. There are also functions of two or more arguments. For instance, the operation of addition is a function of two arguments. For any two numbers x and y there is a number z which is their sum, $z = x + y$. Thus, addition maps the set of pairs of real numbers into the set of real numbers, since it satisfies the conditions: 1) for every pair of real numbers x, y there is a real number $x + y$; and 2) for every such pair the sum is unique. Likewise, multiplication maps the set of pairs of numbers into the set of their products.

If a relation is a function, then we often denote it by a formula which indicates the element which is the value for given argument(s) $x^2 = y$ $\sqrt{x} = y$, $x+y = z$. Hence, if we have an arbitrary function we give up the notation used for relations, and instead of xRy we use the functional notation: $f(x) = y$. When we use functional symbols: $y = f(x)$, $y = F(x)$, etc., it always means that we tacitly assume that the relation $y = f(x)$ is a function. To denote functions, we most often use the letters $f_1, f_2, ..., F_1, F_2, ..., \varphi_1, \varphi_2, ...$, and for relations, we use $R_1, R_2, ..., \varrho_1, \varrho_2, ...$ If f is a function of two arguments, then we say that it is defined on a set X and has its values in a set Y, if f maps every pair of elements of X to one and only one element of Y.

Functions of two arguments, defined on X and with values in X, are called *operations*. Addition and multiplication are thus operations on the set of real numbers.

2. EXAMPLES OF MATHEMATICAL DOMAINS

An example of a mathematical domain is provided by the system $\langle \mathscr{C}, +, 0 \rangle$, which consists of the set of integers, of the operation of addition, and of zero, which is a designated element of $\mathscr{C}$. This domain is called

5

the *additive group of integers*. If we are also interested in the relation "less than", $<$, between integers, then we must consider another domain, $\langle \mathscr{C}, +, 0, < \rangle$. If we are interested in the operation of multiplication, we consider the domain $\langle \mathscr{C}, \cdot, + \rangle$, which is called the *ring of integers*. The system $\langle \mathscr{R}, \cdot, +, 0, 1, < \rangle$ consisting of the set of real numbers, the operation of addition and the operation of multiplication between these numbers, two designated elements 0 and 1, and the relation "less than" $<$, is called the *ordered field of real numbers*. A similar system $\langle \mathscr{W}, +, \cdot, 0, 1, < \rangle$ is called the *ordered field of rational numbers*. These systems are important examples of mathematical domains which are studied in arithmetic and algebra. One can easily define a number of mathematical domains which are less important, for instance, the domain $\langle \{0, 1, 2\}, \oplus \rangle$, where the operation $\oplus$ is defined by the following table:

$\oplus$	0	1	2	y
0	0	1	2	
1	1	2	0	
2	2	0	1	
x				

The first column on the left indicates the values of the first argument. The top row indicates the values of the second argument. The value of the operation $\oplus$ for given arguments x, y is found at the intersection of the corresponding row and column. This is the usual way of defining operations by means of tables. For instance, $2 \oplus 0 = 2$, $1 \oplus 2 = 0$, $1 \oplus 1 = 2$.

Every branch of mathematics is in constant growth. It is difficult to describe exactly what the subject matter of any given branch of mathematics is. Still, the concept of a mathematical domain helps us to describe the interests of mathematicians to a good approximation. It may be noted in this connection that there are fields such that the experts in these fields are interested most in some fixed mathematical domains. For instance, people interested in arithmetic are mainly concerned with the ring $\langle \mathscr{C}, +, \cdot \rangle$ of integers, and those studying calculus focus their attention on the field $\langle \mathscr{R}, +, \cdot, < \rangle$ of real numbers. However, a mathematician interested in calculus does not limit himself to the objects $\mathscr{R}, +, \cdot, <$, specified in the description of the domain of real numbers.

He is interested in many other objects: functions, sets, etc., which can be defined by the concepts $\mathscr{R}$, $+$, $\cdot$, $<$ occurring in the description of that domain. Hence, if a person is interested in the domain $\langle \mathscr{R}, +, \cdot \rangle$, he is also interested in the domain $\langle \mathscr{R}, +, \cdot, < \rangle$, because the relation "less than" can easily be defined by means of the operations $+$ and $\cdot$.

People interested in other fields, which are called more abstract, are interested in all domains satisfying some general conditions. For instance, group theory covers the investigation of the properties common to all domains that are called groups. A *group* is any domain $\langle G, \oplus \rangle$ such that G is a set of any objects, and $\oplus$ is an operation satisfying the following conditions:

1) For every x, y, and z in the set G,
$$x \oplus (y \oplus z) = (x \oplus y) \oplus z.$$

2) For every x and y in the set G, there are elements z and u in G such that
$$x \oplus z = y \quad \text{and} \quad u \oplus x = y.$$

Thus, both the additive group of integers $\langle \mathscr{C}, + \rangle$ and the domain $\langle \{0, 1, 2\}, \oplus \rangle$ with the operation $\oplus$ defined on the preceding page, prove to be groups. Likewise, general topology covers every domain $\langle X, \cup, \cap, 0, 1, C \rangle$, where X is a set, 0 and 1 are designated elements of that set, $\cup$ and $\cap$ are operations, and C is a function of one argument, and where the operations $\cup$ and $\cap$ satisfy the axioms of Boolean algebras and C satisfies the axioms of closure algebras (cf. K. Kuratowski [40], p. 20).

The number of various groups and of closure algebras is infinite. Group theory is concerned mainly with general properties of groups. Likewise, general topology is concerned mainly with general properties of all closure algebras. Elements of a group or a closure algebra may be numbers or points or any other objects as long as they satisfy those general conditions which define a group or a closure algebra. The subject matter of present-day mathematics thus goes far beyond what the subject matter of mathematics was previously.

We have already mentioned that when speaking about a fixed domain, for instance, the domain of real numbers, we may say that a person is interested in the domain $\langle \mathscr{R}, +, \cdot \rangle$ or we may as well say that he is interested in the domain $\langle \mathscr{R}, +, \cdot, < \rangle$, meaning thereby the same thing. On the other hand, if we speak about a class of domains, then the addition of a new relation can essentially restrict the subject of interest. For instance, if we say that a person is interested in rings, which are the domains $\langle X, +, \times \rangle$ satisfying the axioms of ring theory (cf. p. 208), then it does not mean the same as if we said that he is interested in the domains $\langle X, +, \times, < \rangle$, which are rings with the relation "less than". It is so because not all rings admit of the relation "less than" satisfying some additional natural axioms.

3. SELECTED KINDS OF RELATIONS AND FUNCTIONS

Before outlining a rigorous mathematical theory of the concepts of function, relation, and set, singled out of the above analysis of the subject matter of mathematics, we shall discuss some special kinds of relations, functions, and sets, which are important in mathematics. This discussion will help us to understand the meaning of these concepts.

Even in the most elementary mathematical considerations an important role is played by some simple properties of mathematical domains, which are mainly properties of relations and functions. Let us consider a domain $\langle X, R \rangle$, which consists of a set X and a relation R. We say that the relation R is *symmetric* in the set X, if for all elements x and y of a set X (for all $x, y \in X$) the following condition holds:

(4)   If xRy, then yRx.

For instance, the following relations are symmetric. The relation of equality of numbers, the congruence relation for figures, the relation between straight lines of being parallel, and also for straight lines, the relation of being perpendicular. It is true that if a figure x is congruent with a figure y, then conversely the figure y is congruent with the figure x. Likewise, the relations of equality, of being parallel, and of being perpendicular satisfy schema (4).

A relation R is *reflexive*, if for every element x of a set X ($x \in X$) we have

(5) xRx.

Every number x is equal to itself, every line is parallel to itself. The relations of equality, of being parallel, and of congruence are reflexive.

The relation of being perpendicular is not; no line is perpendicular to itself.

A relation is *transitive*, if for all elements x, y, and z of a set X $(x, y, z \in X)$ we have

(6)   *If xRy and yRz, then xRz.*

Except for the relation of being perpendicular, all the relations mentioned above are transitive. It is common knowledge that two lines which are parallel to a third line are parallel to one another. Likewise if a triangle x is congruent with a triangle y, and the triangle y is congruent with a triangle z, then the triangle x is congruent with the triangle z.

A relation which is symmetric, reflexive, and transitive in a set X is called an *equivalence relation* in X. The relations of equality for numbers, being parallel for lines, and being congruent for triangles are therefore equivalences.

The equality of numbers is a special case of the *identity* relation, the relation of being the same object. The identity relation defined on an arbitrary set of objects is an equivalence, since every object is identical with itself; and, if x is identical with y, then y is identical with x; and two objects identical with a third object are identical with one another. Identity has one more property to which we often refer. If an object a is identical with an object b, then b has every property that a has. This means that whatever holds for a, also holds for b.

There are relations which have one or two of the properties, described above, but do not have the others. For instance, the relation for lines of being perpendicular is symmetric, but it is neither reflexive, nor transitive. The relation of being less than is transitive, since if a number x is less than y and y is less than z, then it follows that x is less than z. On the other hand, that relation is neither symmetric, nor reflexive.

Let us consider the relation "less than". Beside being transitive, that relation has the property which is exactly converse to the symmetry condition:

if $x < y$, then it is not true that $y < x$.

This condition is a particular case of the following scheme:

(7)   *if xRy, then it is not true that yRx.*

Any relation R which satisfies condition (7) is called *antisymmetric*.

The relation "less than" has also the property that for any two distinct numbers one is always less than the other. This property may be formulated as follows:

for any real numbers x and y, x = y, or x < y, or y < x.

This is a special case of the general scheme:

(8)   *for any x and y in a set X, x = y, or x R y, or y R x.*

Any relation satisfying condition (8) is called *connected* in the set X. A relation R which is antisymmetric and transitive and connected in a set X is an *ordering* of the set X. Hence the relation "less than" establishes an ordering of real numbers. In this sense ordering relations are often called *linear order relations*, since they as it were arrange all objects in one row. Every element of the set X precedes some other elements and is preceded by some. This corresponds to the usual concept of order.

Ordering relations are sometimes called *orders*. A given set may be ordered by various kinds of orders. The set $\mathcal{N}$ of natural numbers can be ordered in many ways. There are infinitely many ways in which that set can be ordered. The simplest order is the ordering of the elements according to their size: $<$.

We may also define the following order:

$x R y$ if and only if at least one of the following conditions is satisfied:
1) x and y are even numbers and $x < y$,
2) x is even and y is odd,
3) x and y are odd and $x < y$.

The order thus defined divides the whole set of natural numbers into two parts: even numbers and odd numbers. In each of these parts the order according to size holds, and even numbers precede odd numbers. It is easy to demonstrate formally, by considering in turn conditions 1)–3), that the relation so defined satisfies conditions (6), (7), (8) which characterize order relations.

More will be said about various kinds of orderings and equivalences throughout the book.

One-to-one functions are of special importance. A function is a relation satisfying conditions (2) and (3) of p. 4. Condition (2) guarantees the existence of an element which is designated as the value:

(9)   *For every x ∈ X there is a y ∈ Y such that x R y.*

Condition (3) states that that element is unique. It is customary to formulate condition (3) by referring to the identity relation:

(10)   *If* xRy *and* xRy', *then* $y = y'$.

For instance, the relation: y is the square of x, is a function, since every number has only one square. To put it otherwise, if y is the square of x and so is y', then it follows that $y = y'$.

If in addition to condition (10) a relation R satisfies a condition which is very much like a converse condition to (10), namely

(11)   *If* xRy *and* $x'Ry$, *then* $x = x'$,

then we say that the relation R *establishes a one-to-one correspondence*, or that it *is a one-to-one correspondence*, or *one-to-one function*. The function $x^2 = y$ is not one-to-one in the set of real numbers, since it fails to satisfy condition (11), e.g. $2^2 = 4$ and $(-2)^2 = 4$, whereas $2 \neq -2$. The following function is one-to-one in the set of real numbers: $y = \frac{1}{2}x$. If in addition to conditions (9), (10), (11) a relation R satisfies the following condition:

(12)   *for every* $y \in Y$ *there is an* $x \in X$ *such that* xRy,

then we say that R determines a *one-to-one mapping of the set X onto the whole set Y*. We also say that the relation R determines (or is) a *one-to-one correspondence between the elements of X and Y*.

Let us consider some examples. The relation $y = x/(1+|x|)$ maps the set $\mathscr{R}$ of all real numbers one-to-one onto the set of real numbers contained in the open interval $(-1, 1)$, which is the set of numbers that are less than 1 and greater than -1. The relation $y = 2x$ determines a one-to-one mapping of the set $\mathscr{N}$ of all natural numbers onto the set of even natural numbers. Every increasing function defined on a certain set of numbers determines a one-to-one correspondence between the set of its arguments and the set of its values. Given the graph of the function on the plane, the vertical lines establish a one-to-one correspondence of each point of the graph to a point on the horizontal axis. The relation of lying on the same vertical line passing through a point x of the graph and a point y of the horizontal axis is one-to-one.

If R is a relation, then it is customary to denote by $\breve{R}$ the relation converse to R. This relation is

(13) $x\breve{R}y$ *if and only if* yRx.

11

E.g. the relation converse to $x < y$ is $x > y$. The converse relation $x \lesssim y$ is therefore identical with $x > y$.

It may thus be said that a relation R is one-to-one, if both R and its converse $\breve{R}$ are single-valued, i.e., are functions. In other words, a relation R establishes a one-to-one mapping of the set X onto the whole set Y, if R determines a function which maps X into Y and $\breve{R}$ determines a function which maps Y into X.

If a relation R establishes a one-to-one mapping of a set X onto a set Y, and if this fact is symbolized by a formula for the values of the function

$$y = f(x),$$

then the function given by the relation $\breve{R}$ converse to R is called the *inverse function* of the function f and is often denoted by f^{-1}. Hence, if $y = f(x)$, then it follows that $x = f^{-1}(y)$. If we consider the operation of squaring a number and that of taking the square root of a number, both for positive arguments only, then these functions are one-to-one, and one is the inverse function of the other. The inverse of the exponential function $y = e^x$ is the logarithmic function $x = \log_e y$. The inverse of the function $y = 2x$ is the function $x = \frac{1}{2}y$.

In place of the relational formula $x R y$ we shall now use almost exclusively the functional notation: $y = f(x)$. By using it, we imply that it is assumed or that it has been proved that f is a function. In the functional notation the condition that f be a one-to-one function defined on a set X may be briefly formulated as follows:

(14)   *For every x, $x' \in X$, if $f(x) = f(x')$, then $x = x'$.*

In fact, since the formula $y = f(x)$ means the same as $x R y$, then condition (11) shall be rewritten as follows:

if $y = f(x)$ and $y = f(x')$, then $x = x'$.

But in this expression we can replace the formula: $y = f(x)$ and $y = f(x')$ by the formula $f(x) = f(x')$ and thus obtain (14). This replacement is legitimate, since the relation of identity is symmetric and transitive and $y = f(x)$ and $y = f(x')$ implies $f(x) = f(x')$; conversely, if $f(x) = f(x')$, then it follows that for some y, $y = f(x)$ and $y = f(x')$.

By condition (14), the function $f(x) = 2x$ is one-to-one, since for every two numbers x, x'

if $2x = 2x'$, then $x = x'$.

The function $\tan x$ is one-to-one for angles in the interval $0 < x < 90°$, because

if $0 < x, x' < 90°$ and $\tan x = \tan x'$, then $x = x'$.

On the other hand, if we admit arbitrary angles, then the trigonometric functions, being periodic functions, are not one-to-one.

Two more operations on sets are related closely with the concept of function, the image operation and the counter-image operation. If f is a function defined on a set X with values in a set Y (for $x \in X$, the value of the function $y = f(x)$ belongs to Y, $f(x) \in Y$), and if a set A is contained in the set X, $A \subset X$, then, by (1), for every element of A, $x \in A$, this element belongs to the X, $x \in X$. Hence, by condition (2), there is an $y = f(x)$, such that $f(x) \in Y$. The set of all those elements $f(x)$ for which $x \in A$ is called the *image* of the set A. The image of the set A yielded by the function f is denoted by $O_f(A)$. The set $O_f(A)$ can be described by reference to its elements as follows:

(15) $y \in O_f(A)$ *if and only if there is an x such that $x \in A$ and $y = f(x)$.*

It follows from this formula that

(16)   *if $x \in A$, then $f(x) \in O_f(A)$.*

In fact, if $x \in A$, then obviously $f(x) = f(x)$, and thus the condition: there is an x such that $x \in A$ and $f(x) = f(x)$ is satisfied, which by definition (15), on replacement in it of y by $f(x)$, means that $f(x) \in O_f(A)$.

If a set A is finite: $A = \{x_1, x_2, \ldots, x_k\}$, then its image consists of the elements $f(x_1), \ldots, f(x_k)$:

$$O_f(\{x_1, \ldots, x_k\}) = \{f(x_1), \ldots, f(x_k)\}.$$

Let us consider some examples.

The function $y = 2x$, which maps the set of natural numbers into the set of even numbers, maps for instance the set $\{1, 2, 3\}$ exactly into the set $\{2, 4, 6\}$. The set $\{2, 4, 6\}$ is therefore the image of set $\{1, 2, 3\}$ formed by the function $f(x) = 2x$, or $O_f(\{1, 2, 3\}) = \{2, 4, 6\}$. Likewise, if $f(x) = 3x$, then $O_f(\{1, 2, 3\}) = \{3, 6, 9\}$, and $O_f(\{3, 2, 5, 7\}) = \{9, 6, 15, 21\}$. If $f(x) = \sqrt{x}$, then $O_f(\{4, 9, 25\}) = \{2, 3, 5\}$.

Geometric projections are a special case of this concept. If we project a plane figure on a horizontal axis, then to every point p of that figure A we assign a point $r(p)$ of that horizontal axis which lies exactly under the point p (Fig. 1). The set of the values of the function r for the arguments from the set A is the image $O_r(A)$. That image is usually called an *orthogonal projection* of the set A onto a horizontal axis.

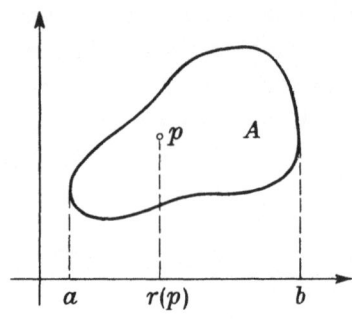

Fig. 1

If f is a function defined on a set X with Y as its set of values, and if a set B is included in Y, $B \subset Y$, then the *counter-image of the set B* is a set A contained in X, such that its elements are all those elements $x \in X$ for which $f(x) \in B$. The counter-image operation is denoted by $O_f^{-1}(B)$. We can symbolically define the counter-image in terms of its elements:

(17) $x \in O_f^{-1}(B)$ *if and only if* $f(x) \in B$.

If $f(x) = 2x$, then the counter-image of the set $\{2, 4, 6\}$ is the set $\{1, 2, 3\}$,

$$\{1, 2, 3\} = O_f^{-1}(\{2, 4, 6\}).$$

If a function f is one-to-one, $A \subset X$ and $B \subset Y$, then B is the image of A if and only if A is the counter-image of B. $B = O_f(A)$ if and only if $A = O_f^{-1}(B)$. The counter-image of $O_f(A)$ is again A, $A = O_f^{-1}(O_f(A))$, which we shall soon prove. It is so, for instance, for the function $2x$, since $\{1, 2, 3\} = O_f^{-1}(\{2, 4, 6\})$ and at the same time $\{2, 4, 6\} = O_f(\{1, 2, 3\})$. On the other hand, if a function is not one-to-one, then the counter-image of a set $O_f(A)$ need not be identical with A.

The function $r(p)$ (projection on the horizontal axis), being defined on the whole plane, is not one-to-one. The point $r(p)$ on the horizontal axis is the projection not only of p, but also of every point lying on the vertical line passing through p and $r(p)$ (Fig. 2). Hence the counter-image of $r(p)$ is the whole line passing through p and $r(p)$. The counter-image of an interval (a, b) is the whole vertical band above and below that interval. The interval (a, b) is the projection of the set A (Fig. 1). The set A is therefore contained in the counter-image of its projection:

(18) $\qquad A \subset O_f^{-1}(O_f(A))$,

but it need not be identical with it. This formula is true for all functions defined on a set A. Indeed, if $x \in A$, then by (16), $f(x) \in O_f(A)$. Next, replacing B by $O_f(A)$ in formula (17), we obtain that if $f(x) \in O_f(A)$, then $x \in O_f^{-1}(O_f(A))$. This proves that, for any x, if $x \in A$, then $x \in O_f^{-1}(O_f(A))$. This, by (1), yields $A \subset O_f^{-1}(O_f(A))$.

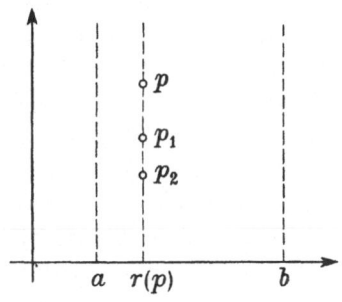

Fig. 2

Another example of a function which is not one-to-one is $f(x) = |x|$, the absolute value of the number x, which is defined for all real numbers. Since $|2| = 2$ and $|-2| = 2$, $|3| = 3$ and $|-3| = 3$, etc., it follows that the counter-image of the set $\{2, 3\}$ is the set $\{2, -2, 3, -3\}$. Likewise, $O_f^{-1}(\{1, 2, 5\}) = \{1, -1, 2, -2, 5, -5\}$.

The *counter-image of an element* x is the counter-image of the unit set $\{x\}$. Hence, if $f(x) = |x|$, then $O_f^{-1}(\{1\}) = \{1, -1\}$. The counter-image of a single number other than 0 is a set of two elements.

The following function is not one-to-one either:

$\qquad f(x) =$ the remainder x leaves when divided by 10.

15

By this definition: $f(15) = 5$, $f(16) = 6$, $f(27) = 7$, $f(147) = 7$ etc. The counter-image of the set $\{5, 7\}$ is the infinite set $\{5, 7, 15, 17, 25, 27, 35, 37, ...\}$; likewise $O_f^{-1}(\{5\}) = \{5, 15, 25, 35, ...\}$.

We shall now state the formula we mentioned earlier:

If a function f defined on a set X is one-to-one, then for every set A contained in X the following identity holds:

(19) $A = O_f^{-1}(O_f(A))$.

To prove that two sets A and B are identical one has to prove that every element of A is also an element of B, or that $A \subset B$, and conversely, that $B \subset A$. The first inclusion, $A \subset O_f^{-1}(O_f(A))$, holds for any function, as follows from formula (18). It therefore suffices to prove that the converse inclusion $O_f^{-1}(O_f(A)) \subset A$ holds for one-to-one functions.

Suppose that $x' \in O_f^{-1}(O_f(A))$. This means, by formula (17), that $f(x') \in O_f(A)$. To verify it we have only to replace B in formula (17) by $O_f(A)$. Next, we make use of the definition of an image, (15). If we replace y in formula (15) by $f(x')$, then it follows that if $f(x') \in O_f(A)$, then there is an $x \in A$ such that $f(x') = f(x)$. If the function is one-to-one, then it follows by (14) that $x = x'$. Since x is identical with x' and $x \in A$, it follows that $x' \in A$. Thus, assuming that $x' \in O_f^{-1}(O_f(A))$, we have proved that $x' \in A$ for any x'. We have therefore proved that, for every x', if $x' \in O_f^{-1}(O_f(A))$, then $x' \in A$; by (1), this means that $O_f^{-1}(O_f(A)) \subset A$. This inclusion combined with formula (18) yields the desired identity (19).

We have proved so far that if a function is one-to-one, then that is a sufficient condition for formula (19) to hold for any A. This condition is also necessary. We have the following converse formula:

(20)   *If a function f is defined on a set X and if formula (19) holds for any $A \subset X$, then the function f is one-to-one.*

We leave the proof to the reader as an exercise.

Concerning the symbolic abbreviation for image and counter-image it is worth mentioning that most authors use a somewhat different notation. If objects of a universe are denoted by small letters, $x, y, z, ...$, and sets by capitals, $X, Y, Z, ...$ and $f, g, h, ...$ denote functions defined on these objects, then the image of a set X is usually denoted by $f(X)$, and the counter-image, by $f^{-1}(X)$. There is no danger of misunderstanding provided the function f is defined not on the sets X, Y, Z, but on the elements of those sets, since then the values of the function f are $f(x)$, $f(y)$, ... and not $f(X)$, $f(Y)$, ... This notation is not used here, because we do not wish to restrict the shape of the variables.

Superposition is an important operation on functions. A function h is the *superposition* of the functions f and g if and only if for every x

$$h(x) = f(g(x)).$$

The superposition of f and g is denoted by fg.

For instance, if $f(x) = 2x$ and $g(x) = 3x+1$, then $fg(x) = 2(3x+1) = 6x+2$.

The superposition of a one-to-one function and its inverse function is the identity function, because

$$f(f^{-1}(x)) = x$$

and also

$$f^{-1}(f(x)) = x.$$

For example, for the angles x in the interval $-90° < x < 90°$, $\arcsin x = \sin^{-1}x$, it follows that $\arcsin (\sin x) = x$ and also $\sin (\arcsin x) = x$.

Sequences are a special kind of functions. In arithmetic, we consider infinite sequences $\{a_n\}$ of real numbers. The infinite sequence of fractions of the form $\dfrac{1}{n}$ is, for instance, such a sequence. It consists of the numbers: $\frac{1}{1}, \frac{1}{2}, \frac{1}{3}, \frac{1}{4}, \frac{1}{5}, \frac{1}{6}, \ldots$ There is as many of them as natural numbers, since with each natural number n we can associate the fraction $\dfrac{1}{n}$. This correspondence is single-valued, and therefore it is a function. In general, a *countable sequence of real numbers* is a function defined on the set $\mathcal{N}$ of natural numbers with values in the set $\mathcal{R}$ of real numbers. It is usual to denote a sequence by a formula giving its n-th term taken in brackets. For instance $\left\{\dfrac{1}{n^2}\right\}$ is the sequence of fractions: $\frac{1}{1}, \frac{1}{4}, \frac{1}{9}, \frac{1}{16}, \frac{1}{25}, \frac{1}{36}, \ldots$, or a function which associates the fraction $\dfrac{1}{n^2}$ with a natural number n.

1. For each of the three of the following properties: symmetry, reflexivity, and transitivity, find an example of a relation which has one property and has not the remaining two. Find also the converse examples, i.e., such relations which have two of these properties, but not the remaining one.

2. Define a relation R between pairs of numbers as follows: $\langle x, y \rangle R \langle u, w \rangle$ if and only if 1) $x < u$ or 2) $x = u$ and $y < w$. Prove that the relation R orders the set of pairs of numbers.

3. Prove that a necessary and sufficient condition for a function f to be one-to-one is that the counter-image of every unit set be a unit set, so that $\{x\} = O_f^{-1}(\{f(x)\})$.

4. Prove condition (20) formulated in the text above.
H i n t: it suffices to consider unit sets $A = \{x\}$.

5. Define arithmetically a function which maps one-to-one the set of numbers from the interval $\langle 0, 2 \rangle$ onto the whole set of numbers from the interval $\langle 0, 5 \rangle$.

4. LOGICAL ANALYSIS OF MATHEMATICAL CONCEPTS

After this brief survey of selected mathematical concepts we shall go further by analysing their structure and the structure of some other sample mathematical concepts. Most representative are the concepts taken from the calculus. The calculus is a method of investigating properties] of real numbers, sequences of real numbers, and functions defined on sets included in the set $\mathscr{R}$ of real numbers, with values also in the set $\mathscr{R}$. We use there the general concepts of function, sequence, and set on which a function is defined. The most important properties of functions considered in the calculus are continuity and differentiability. The most important property of sequences is convergence. The concept of a continuous or differentiable function is usually introduced by reference to the concept of convergence of sequences. We therefore conclude that the concept most representative for the calculus is that of the limit of a sequence. That concept was understood intuitively rather early, but it was given an exact definition much later, in the middle of the 19th century. The intuitive background is the following:

A sequence $\{a_n\}$ converges to a limit a ($\{a_n\} \Rightarrow a$) if and only if the terms which are sufficiently remote, i.e. which have large enough subscripts, differ arbitrarily little from the number a.

18

For instance the sequence $\left\{\dfrac{1}{n}\right\}$ converges to 0 $\left(\left\{\dfrac{1}{n}\right\} \Rightarrow 0\right)$, and so does $\left\{\dfrac{1}{n^2}\right\}$ $\left(\left\{\dfrac{1}{n^2}\right\} \Rightarrow 0\right)$.

More penetrating analysis led mathematicians to the opinion that the above definition is not quite correct. Definitions using phrases *arbitrarily small, sufficiently large*, etc., are too vague. Eventually, the following definition came to be considered a rigorous version of that given above:

(21) $\{a_n\} \Rightarrow a$ *if and only if for every natural number k there is a natural number m such that, for every natural number n, if $n > m$, then it follows that $|a_n - a| < \dfrac{1}{k}$.*

For instance, $\left\{\dfrac{1}{n^2}\right\} \Rightarrow 0$, since for the number 1 there is a number $m = 1$ such that for every n, if $n > 1$, then $n^2 > 1$ and $\dfrac{1}{n^2} < \dfrac{1}{1}$, hence $\left|\dfrac{1}{n^2} - 0\right| < \dfrac{1}{1}$. For the number 2 there is an $m = 1$, the same as before, such that for every $n > 1$ we have $n^2 = 4, 9, 16, \dots$ etc., hence $n^2 > 2$, $\dfrac{1}{n^2} < \dfrac{1}{2}$, and $\left|\dfrac{1}{n^2} - 0\right| < \dfrac{1}{2}$. Likewise, for 3 we may set $m = 1$. For the number 4, $m = 2$ will do: for every $n > 2$, $n^2 > 4$ and $\left|\dfrac{1}{n^2} - 0\right| < \dfrac{1}{4}$. Similarly $m = 2$ will do for 5, 6, and 7. For the number 10 there is again an m, $m = 3$, such that for every $n > 3$ we have $\left|\dfrac{1}{n^2} - 0\right| < \dfrac{1}{10}$ and so on. We can prove in general that for every $k \in \mathcal{N}$ there is an m, which of course must be large enough, such that for $n > m$ we have $\left|\dfrac{1}{n^2} - 0\right| < \dfrac{1}{k}$.

Now, let us consider definition (21). First, there is the definiendum: $\{a_n\} \Rightarrow a$. Next, there is *if and only if* which is used in many kinds of definitions, and which is termed the equivalence connective. It belongs to logic. The rest of the definition which follows the connective serves to describe the phrase which is being defined, and is called the *definiens*. The majority of rigorous definitions describing some concepts have this form. In definition (21), the defining phrase, definiens, begins with the phrase *for every natural number k*. Logical analysis turns this phrase into *for every k, if k is a natural number, then*. Likewise, the phrase *there is a natural number m such that* is equivalent to *there is an m such that m is a natural number and* .... This way the phrases used in (21) break up into the following phrases:

(22)   *For every k, ...,*

          There is a k such, that ...

and the phrases of the form *If ..., then ..., ... and ..., m is a natural number*, or *m belongs to the set $\mathcal{N}$, $m \in \mathcal{N}$*. Such a phrase is followed again by a phrase of the form (22), but with a different variable letter, and once more a phrase which fits the general scheme *if ..., then ...* with two arithmetical expressions, the inequalities $n > m$ and $|a_n - a| < \dfrac{1}{k}$.

We have already met the phrases used in definition (21) in previous definitions, e.g. they occurred in (7), (8), (9). We can find similar expressions, phrases, and concepts in any other mathematical definition.

In the second half of the last century, logicians examining the apparatus of classical mathematics found three kinds of concepts occurring in mathematics:

1. logical concepts,
2. set-theoretical concepts,
3. arithmetical concepts.

Let us review the concepts which we have met so far.

L o g i c a l n o t i o n s. They are, first of all, the following phrases: *if ..., then ...*, which occurs e.g. in (4), (6), (7), (10), (14), (21);

... or ..., which occurs in (8);

... and ..., which occurs in (6), (10), (11);

... if and only if ..., occurring in (13), (15), (17), (21);

it is not true that ..., occurring in (7).

Those phrases are called *sentential connectives*, or briefly *connectives*, since they connect two sentences into one, except for the last phrase, which can only be used with one sentence. If every occurrence of dots is replaced by a sentence, we obtain compound sentences, for instance: *if $x > 3$, then $x > 2$*, where the expressions $x > 3$ and $x > 2$ are sentences in the grammatical sense of the term, $x > 0$ *or* $x = 0$, *or* $0 < x$, *it is not true that* $2 > 2$, etc.

20

4. ANALYSIS OF CONCEPTS

The way of handling the connectives is described in detail in the part of logic called the *sentential calculus*, or the *logic of sentences*.

There are two other important phrases we have met: *for every x ...*, occurring in (8), (9), (21), and *there is an x such that ...*, occurring in (9), (21).

They are called *quantifiers* and they belong to the part of logic called the *calculus of quantifiers*, or else the *logic of quantifiers*. The name of these phrases arose from their meaning; they indicate whether the sentence which follows them is true for all objects or for at least some of them. For instance, the sentence *for every x, x is even* is false. On the other hand, the sentence *there is an x such that x is even* is true.

Logical concepts have abbreviational symbols which we shall use. There are several notations which are used with equal frequency. These symbols are listed in the table below. The table also makes use of sentential variables.

Sentential variables are variables which, when replaced by sentences, yield a formula which is a sentence. For instance, in the formula

if p, then q

the variables p and q are sentential variables, since if we replace them by sentences, the whole formula turns into a sentence. It may be said that sentential variables play more or less the same role as the three dots used in the above list of sentential connectives.

Phrases of common usage	Names of phrases	Abbreviational symbols in use
if p, then q	implication	$p \to q$, $p \supset q$, Cpq
p or q	disjunction	$p \lor q$, $p + q$, Apq
p and q	conjunction	$p \land q$, $p \cdot q$,
		$p \& q$, Kpq
p if and only if q	equivalence	$p \equiv q$, $p \leftrightarrow q$, Epq
it is not true that p	negation	$\sim p$, p', $\neg p$, Np
(not p)		
for every x p	universal quantifier	$\bigwedge xp$, $\Pi_x p$,
		$\forall_x p$, $(x)p$
there is an x such that p	existential quantifier	$\bigvee xp$, $\sum_x p$, $\exists_x p$,
		$(Ex)p$

We shall use those symbols which are listed first in each group. The concept of identity, as expressed in the formula $x = y$ (x is the same object as y), also is included in the logical concepts.

Some of the properties of these concepts were already used in the brief proofs we gave above and more properties will appear in the text below; e.g., we used the following property: if assuming P we can prove Q, then we consider that we can prove the implication: if P, then Q (cf. proofs of formula (19)). We also reasoned the following way: assuming P if and only if Q, we considered that the following implication holds: if P, then Q, as well as the converse implication: if Q then P (cf. proofs of formulas (16) and (18)). In addition, we shall use the following rule: if it is true that if P, then Q, and if it is true that if Q, then P, then it is also true that P if and only if Q. The equivalence $P \equiv Q$ is in a sense the conjunction of two implications: $P \to Q$ and $Q \to P$.

Let us write these relationships provisionally as follows:

(23)   *Accepting* ($p \equiv q$) *we accept* ($p \to q$) *and we accept* ($q \to p$).
        Accepting ($p \to q$) *and* ($q \to p$) *we accept* ($p \equiv q$).

In this very chapter we shall use the following additional rules of logic:

        Accepting ($p \equiv q$) *we accept* ($q \equiv p$).
        Accepting ($p \equiv q$) *and* ($q \equiv r$) *we accept* ($p \equiv r$).
(23′)   *Accepting* ($p \to q$) *and* ($q \to r$) *we accept* ($p \to r$).
        Accepting ($p \vee q$) *we accept* ($q \vee p$).
        Accepting ($p \wedge q$) *we accept* ($q \wedge p$).

S e t   t h e o r e t i c a l   c o n c e p t s. These include the following: the set, the concept of an element of a set with its symbol $\in$, the general concept of relation, the general concept of function. So far of all these concepts the term $\in$ occurred most explicitly (cf. (8), (9), (14), (15), (17)). The symbol $\in$ is the fundamental term in set theory, and it can be used to define all other concepts. The concepts of set, relation, and function did not occur in formulas. They did occur, but only in the accompanying explanatory text. If our analysis had been more precise, their role would have been emphasized. We shall formalize them soon.

T h e   a r i t h m e t i c a l   c o n c e p t s. These include: the set of natural numbers $\mathcal{N}$ (cf. (21) and definition of a sequence), the set of real numbers $\mathcal{R}$, and arithmetic operations: addition, multiplication, and other operations which can be defined by these two, e.g. subtraction, absolute value, inverse, <, which occur in definition (21).

The role the concepts of these three groups play in classical mathematics may be formulated as follows. Logic supplies mainly general methods for the construction of new concepts and proofs; set theory supplies the most general concepts; while arithmetical concepts supply the most specific content for mathematics.

EXERCISES

1. Write down, using the abbreviations given in the table, the following formulas: (1), (13), (14), (15), (16), and (17), and the definition of the limit (21).

H i n t: e.g., the definition of an image (15) should run as follows:

$$y \in O_f(A) \equiv \bigvee x(x \in A \wedge y = f(x)).$$

2. Recall some other mathematical definition containing quantifiers and write it down in symbols.

5. ZERMELO'S SET THEORY

In Sections 5–11 an analysis of mathematical concepts from the set-theoretical standpoint will be made.

The concepts of set, relation, and function were investigated in their whole generality by mathematicians beginning only in the second half of the 19th century. The theory of the most general properties of these concepts developed into an independent and self-contained discipline, which is called *set theory*. That discipline was developed first on an intuitive basis, but in the beginning of the 20th century it became an axiomatic system, as was the case with most mathematical theories at that time. There are, however, several rival approaches to that discipline. We shall now outline here the system of set theory due to E. Zermelo ([105]), which is presently the most popular among mathematicians. Another equally sufficient system is the *theory of types*.

The concepts of set theory are so general that human intuition is not capable of deciding many problems which arise there. This is why there are so many different views on how to interpret that part of mathematics, and why we must be cautious in formulating axioms.

23

Zermelo's set theory strives to characterize two fundamental notions:

1. The concept of elementhood: $x \in y$, x is an element of y;

2. The concept of set: X is a set, which is concisely and symbolically written: $Z(X)$.

All definitions and axioms will be written in symbols, and in place of the frequent phrase *if and only if* we shall use the symbol of equivalence. Beside partly symbolic versions we shall also give completely symbolic versions. We do so to make the reader familiar with symbolic formulas which will often be used below.

The concepts $\in$ and $Z(X)$ will be characterized by axioms. The axioms express those properties of sets which seem to be most obvious and fundamental. Other properties, which are less obvious, are considered to be hypotheses.

E. *Zermelo's axioms for set theory* follow with discussion.

1. *Axiom of Extensionality*

(24)   *If $Z(X)$ and $Z(Y)$, and if for every x $(x \in X \equiv x \in Y)$, then $X = Y$.*

Symbolically:

$$\{Z(X) \land Z(Y) \land \bigwedge x(x \in X \equiv x \in Y)\} \to (X = Y).$$

In plain English, two sets X and Y which consist of the same elements are identical. The idea that two sets X and Y consist of the same elements has been expressed by reference to the equivalence relation: for every x, $x \in X$ if and only if $x \in Y$. This axiom lays down a very important property of sets, namely: it depends only on their elements whether two sets are the same or different — if their elements are the same, the sets are the same, too. To put it another way: sets of the same extension are identical, hence the name of the axiom.

This axiom has been used to derive formula (19): to prove that two sets were the same, we proved there that $A \subset B$ and $B \subset A$. Next, by definition (1) and the law of logic stated in (23), we have

$$\bigwedge x(x \in A \equiv x \in B)$$

which according to the extensionality axiom yields the desired identity $A = B$.

The next four axioms have a somewhat different structure. They state that there are some new sets, provided some objects already are sets.

2. *Axiom of the Pair*

This axiom says that for arbitrary objects x and y there exists a set W which consists exactly of these two objects, $W = \{x, y\}$. Strictly speaking,

(25)   *For every x, y there exists a W such that $Z(W)$ and for every z, $z \in W \equiv z = x$ or $z = y$.*

In symbols,

$$\bigwedge x, y \bigvee W[Z(W) \wedge \bigwedge z\{z \in W \equiv (z = x \vee z = y)\}].$$

For instance, the sentence stating that there is a set W such that

$$\bigwedge z\{z \in W \equiv (z = 0 \text{ or } z = 1)\}$$

is a consequence of the axiom of the pair. The set W is a two-element set containing the numbers 0 and 1, and according to notation adopted in Section 1:

$$W = \{0, 1\}.$$

We point out the special case of (25) when x and y are the same object. In that case, we obtain the theorem that *for every object x there is a unit set containing exactly that object and nothing else* (the singleton of x):

$$\bigwedge x \bigvee W[Z(W) \wedge \bigwedge z(z \in W \equiv z = x)].$$

Axiom (25) alone is not enough to prove that there exists a set which contains exactly three elements.

3. *Axiom of the Union*

This axiom says that if we have a family of sets X, then there exists a set Y each of whose elements is an element of some set of the family X. Such a set is called the *union of the sets* of the family X. We formulate this axiom as follows:

(26)   *If $Z(X)$ and [for every x, if $x \in X$, then $Z(x)$], then there exists a Y such that $Z(Y)$ and [for every x, $x \in Y \equiv$ (there exists a z such that $z \in X$ and $x \in z$)].*

25

In symbols:

$$\left(Z(X) \wedge \bigwedge x(x \in X \to Z(x))\right) \to$$
$$\bigvee Y[Z(Y) \wedge \bigwedge x\{x \in Y \equiv \bigvee z(z \in X \wedge x \in z)\}].$$

By Axiom (26), if the elements of a set X are themselves sets, i.e. if X is a family of sets, then there is a set Y which is the union of all sets that belong to the family X. Therefore, an arbitrary object x belongs to the set Y if and only if it belongs to at least one set z which belongs to the family X.

Axioms (25) and (26) combined make it possible to prove that there exist sets which consist of 3, 4, or an arbitrary finite number of elements. The axiom of the pair, as we have already seen, makes it possible to prove the existence of sets W_1 and W_2 such that

$$W_1 = \{0, 1\}, \quad W_2 = \{2\}.$$

Next, by referring to that axiom again, we prove that there is a set X whose elements are exactly two objects, namely W_1 and W_2: $X = \{W_1, W_2\}$. X is therefore a family of sets, and it follows from the axiom of the union (26) that there is a set Y, which is the union of the sets of the family X. That family consists of W_1 and W_2:

$$x \in Y \equiv \bigvee z(z \in X \wedge x \in z) \equiv (x \in W_1 \vee x \in W_2) \equiv$$
$$(x = 0 \vee x = 1 \vee x = 2).$$

4. *Axiom of the Power Set* (*Axiom of the family of subsets*)

(27)   *If $Z(X)$, then there exists a Y such that $Z(Y)$ and [for every $x\{x \in Y \equiv (Z(x)$ and every element of the set x is an element of $X)\}$].*

In symbols:

$$Z(X) \to \bigvee Y[Z(Y) \wedge \bigwedge x\{x \in Y \equiv$$
$$(Z(x) \wedge \bigwedge u(u \in x \to u \in X))\}].$$

By (27), for every set X there exists a family Y whose elements are all the subsets x of the set X. A *subset* of a set X is a set which is included in X, i.e., a set whose elements belong to X. To denote the fact that a set Y is included in a set X we have adopted the symbolic abbreviation $Y \subset X$, defined by formula (1).

Symbolically:

(28) $Y \subset X \equiv \bigwedge u(u \in Y \rightarrow u \in X)$.

Using abbreviation (28) we can reformulate Axiom (27) as follows:

$$Z(X) \rightarrow \bigvee Y\{Z(Y) \wedge \bigwedge x(x \in Y \equiv Z(x) \wedge x \subset X)\}.$$

By axiom (27), for the set $\mathcal{N}$ of all natural numbers there exists the family Y of all subsets of $\mathcal{N}$ which consists of all sets of natural numbers, i.e., of all sets x for which $x \subset \mathcal{N}$.

5. *Axioms of Set Construction* (*of comprehension*)

In this section we are concerned with an infinity of axioms of set construction, but they are all alike. They fall under the same schema. Their role is as follows: using mathematical symbols one may describe many different properties. For instance, using multiplication and the whole set of natural numbers $\mathcal{N}$ we can define divisibility without remainder for natural numbers:

$x|y \equiv x \in \mathcal{N}$ *and* $y \in \mathcal{N}$ *and there exists a* $z \in \mathcal{N}$ *such that* $x \cdot z = y$,

where $x|y$ means: x is a divisor of y.
Symbolically:

$$x|y \equiv x \in \mathcal{N} \wedge y \in \mathcal{N} \wedge \bigvee z(z \in \mathcal{N} \wedge x \cdot z = y).$$

Using the relation of divisibility we may easily define the property of being an even number, or the property of being divisible by 3:

x *is an even number* $\equiv 2|x$,
x *is divisible by* $3 \equiv 3|x$.

Likewise, we can easily describe the property of being a prime. A number is prime, if it is greater than 1, $x > 1$, and if it is divisible without remainder by 1 and by itself only:

x *is a prime number* $\equiv$
$$x \in \mathcal{N} \wedge x > 1 \wedge \bigwedge u\{u|x \rightarrow (u = 1 \vee u = x)\}.$$

For each of these properties there is a corresponding set of natural numbers: the set of even numbers, the set of numbers divisible by 3, the set of prime numbers. It might be suspected that every property which can be described in words defines some set, namely the set of all the objects that have that property. This assumption, however, results in a contradiction. There are examples of reasonings based on this assumption which result in contradictions. Many of them were discovered by the end of the 19th century. This is why today we make a weaker assumption. Instead of saying that every property defines a set of objects that have that property, we say that, given a set X, each property which can be expressed in mathematical language defines in X a subset Y which consists of these elements of X which have that property, and of these elements only. For instance, since we assume that the natural numbers form a set, the property of being an even number singles out in the set $\mathcal{N}$ a subset Y which consists of those, and only those, elements of $\mathcal{N}$ which are even. Likewise, the property of being divisible by 3 singles out in the set $\mathcal{N}$ a subset Y of the numbers divisible by 3 without remainder. Let $\Phi(x)$ stand for any mathematical expression which describes a property. The axioms of set construction can be written in the form of the following schema:

(29)   If $Z(X)$, then there is a Y such that $Z(Y)$ and [for every $x\big(x \in Y \equiv x \in X$ and $\Phi(x))\big]$.

In symbols:

$$Z(X) \to \bigvee Y\{Z(Y) \wedge \bigwedge x(x \in Y \equiv [x \in X \wedge \Phi(x)])\}.$$

For instance, if $Z(\mathcal{N})$, then there is an Y such that $Z(Y)$ and $\big[$for every $x\big(x \in Y \equiv x \in \mathcal{N} \wedge x > 1 \wedge$ for every $u(u|x \to (u = 1 \vee u = x))\big)\big]$. Since we indeed assume that the natural numbers form a set, $\mathcal{N}$, it follows that there is a set Y which, by the above definition, is the set of prime numbers. Almost every definition of any set which occurs in this book makes use of this general axiom schema, even though we do not emphasize it every time. Given some sets, we shall define new sets by the formation of the union or power sets or pairs, using axioms (25), (26), (27), or else using (29), the axiom of set construction, we shall define new sets by reference to sets already defined and to some properties which single out these new sets.

The axioms formulated so far make it possible to construct new sets. It can easily be seen that the axioms of the union, of the power set, and of the pair are the only axioms that make it possible to construct larger and larger sets, containing more and more elements. The family of all subsets of a set X always contains more such subsets than X has elements. The union of a family of sets X often has more elements than X has. On the other hand, a subset singled out in a set X cannot have more elements than the set X itself. The axiom of set construction cannot be used to construct ever larger sets. The axioms of set theory provide only for a limited possibility to construct ever larger sets. This limitation is not accidental, it is imposed on purpose.

6. *Axiom of Infinity*

All axioms formulated so far have been in the conditional form. They state that there exist sets with certain properties, provided some other objects already are sets. When set theory is being formulated apart from other branches of mathematics, then one more axiom is usually included which says that there exists at least one set A having infinitely many elements. The property of having infinitely many elements can be expressed in many ways. One of them is that there exists an infinity of subsets of set A, which are increasingly large, hence that there is a family X of sets such that for each member x of that family there is another member of this same family, y, which is greater than x, i.e. such that $x \subset y$ and $x \neq y$:

(30)   *There is an X such that $Z(X)$ and for some x, $x \in X$ and [for every x, if $x \in X$, then $Z(X)$] and [for every x, if $x \in X$, then there is an y such that $y \in X$, $x \subset y$, and $x \neq y$].*

In symbols:

$$\bigvee X[Z(X) \wedge \bigvee x(x \in X) \wedge \bigwedge x(x \in X \to Z(x)) \wedge$$
$$\bigwedge x \{x \in X \to \bigvee y(y \in X \wedge x \subset y \wedge x \neq y)\}].$$

Intuitively, it is clear that the family X consists of infinitely many different sets, and of course the union $\bigcup X$ of the sets of the family X, too, is a set A, that contains infinitely many elements.

There are some more axioms which are usually adopted in set theory, e.g., the axiom of substitution, the axiom of foundation and the axiom of choice, but axioms 1–6 are the most characteristic. We will discuss these axioms when we formalize set theory (Chapter I, Section 7). The theory based on them is for mathematicians a kind of minimal set theory. We shall also mention other approaches to set theory, but for the time being we shall concentrate on this particular approach. Axioms 1–6 characterize well enough the two principal concepts: that of set and that of elementhood. All axioms except the axiom of extensionality and the axiom of infinity assert that some operations on sets can be performed. We say that an operation is performable, if there is an object which is the result of that operation. The operations mentioned in Axioms 2, 3, 4 have their names and symbols. The union of the sets belonging to a family X is denoted by $\bigcup X$. The set of all subsets of X is denoted by 2^X. Axiom (25) states that the operation of pairing is performable. Axioms (26) and (27) state that the operations $\bigcup X$ and 2^X are performable. We shall characterize informally these operations by indicating under wnaht coditions an element belongs to a given set:

$$x \in \{a, b\} \equiv (x = a \vee x = b),$$
$$(31) \qquad x \in \bigcup X \equiv \bigvee U(U \in X \wedge x \in U),$$
$$x \in 2^X \equiv Z(x) \wedge x \subset X.$$

When introducing a new name into the set theory we shall always follow the rule, generally accepted in mathematics, that a new name is introduced when it can be proved that there is a unique object we want to bear that new name, i.e. when it can be proved that there exists exactly one object having a characteristic property. Such an object may be, of course, a set containing many objects, but it must be unique. If the new name includes a variable parameter, i.e., if it is of the form $\alpha(X)$ or $\alpha(X, Y)$, for instance 2^X or $\{x, y\}$, then we must prove that for every X or for every pair X, Y there exists a unique object bearing that name. The equivalences of (31) are just such characteristic properties of $\{a, b\}$, $\bigcup X$ and of 2^X. Axioms (25), (26), and (27) just state that there exist sets satisfying conditions (31). These conditions together with the axiom of extensionality imply that the sets in question are unique. For instance, by (31), if both Y_1 and Y_2 were the sets of all subsets of X, then they both would satisfy the following equivalences:

$$x \in Y_1 \equiv Z(x) \wedge x \subset X,$$
$$x \in Y_2 \equiv Z(x) \wedge x \subset X.$$

Hence it follows from the rules of logic (23) that for every x, x belongs to Y_1 if and only if x belongs to Y_2. By the axiom of extensionality, Y_1

and Y_2 are then one and the same set, $Y_1 = Y_2$. Thus the set of all subsets of a given set exists and is unique. We might denote that set by a special name. The same holds for the union. Later on, the principle of introducing new names will be analysed in greater detail.

Definitions (31) can be used to formulate the axioms of pairs, of union, and of power in the following way:

$$Z(\{x, y\}),$$

(32) $\quad [Z(X) \wedge \bigwedge x(x \in X \rightarrow Z(x))] \rightarrow Z(\bigcup X).$

$$Z(X) \rightarrow Z(2^X).$$

This formulation of these axioms shows clearly that the issue in question is that these three operations can always be performed.

We shall introduce some more set-theoretical operations which are commonly used in mathematics. For instance the meet, or the *intersection* of a family of sets. We denote it by $\bigcap X$. The set $\bigcap X$ consists of those elements which belong to every set of the family X.

Symbolically:

(33) $\quad x \in \bigcap X \equiv \bigwedge u \, (u \in X \rightarrow x \in u).$

The intuitions connected with the concept of set suggest that for every non-empty family of sets there exists a set which is their intersection. (The assumption of non-emptiness is necessary, since should X be an empty family, any x would satisfy formula (33) in accordance with the truth-table for implication.)

An axiom expressing this idea might be as follows:

(34) $\quad$ *If $\bigwedge u(u \in X)$ and $Z(X)$ then $Z(\bigcap X)$.*

Such an axiom, however, would be superfluous, since it follows from axioms (26) and (29). The intersection of all sets of a family X is by definition included in every set of that family, and hence it is also included in the union of the family X. By axiom (26), the union of the family X is a set, and using the property specified in formula (33) we can define $\bigcap X$ as a subset of $\bigcup X$ satisfying the condition (33). We have

$$x \in \bigcap X \equiv x \in \bigcup X \wedge \bigwedge u(u \in X \rightarrow x \in u).$$

The formula $\bigwedge u(u \in X \rightarrow x \in u)$ thus describes the property $\Phi(x)$ which makes it possible to designate among the elements of $\bigcup X$ those

31

which are contained in $\bigcap X$. The axiom of set construction (29) states that if $\bigcup X$ is a set, then there is also a subset of it which consists of those elements of $\bigcup X$ which have the property $\Phi(x)$, specified earlier. That subset is exactly the intersection of all the sets of the family X. The statement that for every family of sets there is a set which is the intersection of all the sets of that family thus follows from the axioms already adopted. If the sets of the family X have no common elements, then their common part exists, too, but it is the empty set. If the family X consists of only two sets, Y and Z, $X = \{Y, Z\}$, then their intersection is denoted by $Y \cap Z$: $Y \cap Z = \bigcap \{Y, Z\}$.

Likewise the union of a family $\{X, Y\}$, consisting of two sets X, Y, is denoted by $X \cup Y$: $X \cup Y = \bigcup \{X, Y\}$.

The proof that the set $X \cup Y$, which is the union of the sets X and Y, exists is based on the axiom of the pair (25) and the axiom of the union (26). If $Z(X)$ and $Z(Y)$, then the pair $W = \{X, Y\}$ is a family of sets, and hence, by (26), there exists a set Z which is the union of the family W. That set has just been denoted by $X \cup Y$. Likewise it can be proved that the set $X \cap Y$ exists too.

These sets can be characterized by the following equivalences:

$$\bigwedge x\{x \in X \cup Y \equiv (x \in X \vee x \in Y)\},$$
$$\bigwedge x\{x \in X \cap Y \equiv (x \in X \wedge x \in Y)\}.$$

The empty set will be denoted by Λ. It can easily be proved that the empty set exists and is unique. The elements of the empty set can be characterized by any contradictory property, that is, one described by a sentence and its negation, for instance

(35) $x \in \Lambda \equiv (x \in x \wedge \sim (x \in x))$.

Of course, no object can at the same time have and have not a given property. Hence there is no x such that $x \in \Lambda$. The set Λ, if it exists, has no elements, and hence it is empty. The existence of the empty set follows directly from the axiom of infinity. Infinity itself is not referred to for that purpose. We only refer to the fact that that axiom ensures the existence of a set. We thus consider the first part of Axiom (30), which states that there is an X such that $Z(X)$. By (29), we designate in X, its empty part, i.e., the set Λ_X characterized by

$$x \in \Lambda_X \equiv x \in X \wedge x \in x \wedge \sim (x \in x),$$

which consists of those elements of X which have a contradictory property. Obviously there are no such elements. The set Λ_X, which exists on the strength of the axiom of set construction (29), is thus empty, i.e. has the property that for every x it is not true that $x \in \Lambda_X$. It can easily be seen that there is at most one set which is empty. If Λ_X and Λ_Y are both empty, then for every x, x is in Λ_X if and only if x is in Λ_Y, since no x is either in Λ_X or in Λ_Y. By the axiom of extensionality, $\Lambda_X = \Lambda_Y$. Hence there is exactly one empty set. The empty set does not depend on the set X, because using any set in place of X we always obtain one and the same empty set that satisfies condition (35). In this way, starting from the existence of any set, we have proved the existence of a unique empty set. On the other hand, it is not possible to prove that there exists a set which would satisfy the condition: for every x, $x \in X$, i.e. a set which would consist of all objects. On the contrary it can be proved that there is no set X that has that property.

Two sets are called *disjoint*, if they have no common elements, i.e. if the set of their common elements is empty: $X \cap Y = \Lambda$.

Using the union of sets A and B, we can define a set which is called the *difference* of A and B, and which is denoted by $A - B$:

$$x \in (A - B) \equiv x \in A \ \text{and} \ \sim (x \in B).$$

Using the terms $\cup, \cap, -, \Lambda$ we can formulate many theorems which have the form of equations between sets. This part of set theory is usually called the *calculus of sets*. For instance

$$(A \cup B) = (B \cup A), \quad (A \cap B) = (B \cap A),$$

$$A \cap (B \cup C) = (A \cap B) \cup (A \cap C), \quad (A - B) \cap (B - A) = \Lambda,$$

$$A \cup (B \cap C) = (A \cup B) \cap (A \cup C), \quad A \cap (B \cap C) = (A \cap B) \cap C.$$

For modern approach to set theory see: Cohen [10], Felgner [14], Kuratowski and Mostowski [42], Morse [51].

EXERCISE

Prove the formulas written above using the definitions of set-theoretical operations and the set-theoretical axioms.

H i n t: To prove each equality it is necessary to use the axiom of extensionality.

6. SET-THEORETICAL APPROACH TO RELATIONS AND FUNCTIONS

Relations and functions can be defined in set-theoretical language by the axiom of pairs. There are two kinds of pairs: *unordered pairs* and *ordered pairs*. Unordered pairs are simply sets consisting of two elements. According to the notation we have assumed here, we denote them by curled brackets: $\{x, y\}$. The equivalence which defines unordered pairs is

$$z \in \{x, y\} \equiv (z = x \text{ or } z = y).$$

If X is a fixed set given in advance, we may consider all those subsets $\{x, y\}$ of X which consist of two elements each. The family of those subsets of X forms part of the family of all subsets of X. By the axiom of the power set (27), all the subsets of X form the set 2^X. Using the axiom of set construction we may distinguish in the set 2^X the subset of those sets which consist of two elements. But unordered pairs are less important than ordered pairs. To define a relation we need ordered pairs. A *relation* is a connection between x and y. When a relation R between x and y holds, that fact is recorded as: xRy. If a relation is not symmetric, then the sentences xRy and yRx are not equivalent. The fact that a relation R holds of the pair x, y may not, therefore, be identified with its holding of the pair y, x. Now the symbol $\{x, y\}$ is symmetric, $\{x, y\}$ and $\{y, x\}$ are one and the same set of two elements: $\{x, y\} = \{y, x\}$. To obtain a non-symmetric pair $\langle x, y \rangle$ it suffices to assume that:

(36) $\qquad \langle x, y \rangle = \{\{x\}, \{x, y\}\}.$

If $x \neq y$, then $\langle x, y \rangle \neq \langle y, x \rangle$, for we have

(37) $\qquad \langle y, x \rangle = \{\{y\}, \{y, x\}\}.$

Since $\{x, y\} = \{y, x\}$, hence

(38) $\qquad \langle y, x \rangle = \{\{y\}, \{x, y\}\}.$

By comparing (36) and (38) we can see that the sets $\langle x, y \rangle$ and $\langle y, x \rangle$ have one element in common, namely $\{x, y\}$, but the remaining element of $\langle x, y \rangle$ is different from the remaining element of $\langle y, x \rangle$, provided $x \neq y$. For the set $\langle x, y \rangle$ that element is $\{x\}$, and for $\langle y, x \rangle$ it is $\{y\}$.

The sets $\langle x, y \rangle$ and $\langle y, x \rangle$ thus consist of different elements, and hence they are different sets. The pairs which have the scheme (36) are asymmetric. The order in which x, y occur in the scheme (38) is essential. This is why such pairs are called *ordered pairs*.

If a relation R holds between x and y, which is usually written xRy, then we can say that the ordered pair $\langle x, y \rangle$ has the property that the relation R holds between the two elements. As we shall see, all ordered pairs whose members belong to a certain set form a set, in which we can single out a subset R' which consists of those pairs $\langle x, y \rangle$ which have the property that their first members are in relation R to their second members, xRy. The set R' can be considered to stand for the relation R. That set has the following property:

$$
(39) \quad \begin{aligned} & xRy \equiv \langle x, y \rangle \in R', \\ & \sim(xRy) \equiv \sim(\langle x, y \rangle \in R'). \end{aligned}
$$

Whether the relation R holds between x and y or not one can tell simply by seeing whether $\langle x, y \rangle$ belongs to R' or not. The study of relations can therefore be entirely replaced by the study of sets of pairs. To every relation R there corresponds a set R' of pairs $\langle x, y \rangle$ such that R holds between the elements of those pairs: xRy and every set R' of pairs whose members are in X defines by (39) a relation R. For instance to the relation $<$ for natural members there corresponds a set R' such that

$$
\langle x, y \rangle \in R' \equiv x < y.
$$

Thus $\langle 1, 2 \rangle \in R'$, $\langle 3, 7 \rangle \in R'$, but $\langle 6, 3 \rangle \notin R'$.

Let us now define more rigorously the set of all pairs with members in a set X, and the set of all relations that hold in the set X. By (36), the pairs $\langle x, y \rangle$ are two-element subsets of 2^X and hence belong to the family of all subsets of 2^X, i.e., to 2^{2^X}. The set of all pairs (denoted by X^2) can be singled out in the set 2^{2^X} on the strength of the axiom of set construction (29).

We have

(40) $\quad z \in X^2 \equiv z \in 2^{2^X}$ *and there exist* $x, y \in X$ *such that* z *is the pair* $\langle x, y \rangle$.

In symbols:

$$
z \in X^2 \equiv z \in 2^{2^X} \wedge \bigvee x \{ \bigvee y (x, y \in X \wedge z = \langle x, y \rangle) \}.
$$

The set X^2 of all pairs from the set X is also called the *Cartesian product* of X. This concept is very important for mathematics. For instance, complex numbers may be considered to be pairs of real numbers. The set of complex numbers in theoretical arithmetic is thus identified with the product $\mathscr{R}^2$ or real numbers.

Relations may be identified with subsets of the set X^2, i.e., with elements of the power set 2^{X^2}:

$$R \text{ is a relation in the set } X \equiv R \subset X^2.$$

Functions may be considered to be single-valued relations. The set Y^X of all functions defined on the set of arguments X and taking on values from the set Y can be characterized as follows:

$f \in Y^X \equiv \{f \in 2^{(X \cup Y)^2}$ and

    1. for every x, if $x \in X$, then there is an y such that $y \in Y$ and $\langle x, y \rangle \in f$,

    2. for any x, y, z, if $\langle x, y \rangle \in f$ and $\langle x, z \rangle \in f$, then $y = z\}$.

In symbols:

(41) $f \in Y^X \equiv f \in 2^{(X \cup Y)^2} \wedge \bigwedge x\{x \in X \to \bigvee y(y \in Y \wedge$

    $\langle x, y \rangle \in f)\} \wedge \bigwedge x, y, z\{(\langle x, y \rangle \in f \wedge \langle x, z \rangle \in f) \to y = z\}.$

Obviously, conditions 1 and 2 are the conditions (2) and (3) of existence and uniqueness of the original definition of function, cf. p. 4.

An abbreviation of the quantifier notation is used in formula (41). In place of

$$\bigwedge x[\bigwedge y\{\bigwedge z(...)\}]$$

we used a shorter expression

$$\bigwedge x, y, z\{...\},$$

which is read: for all $x, y, z\{...\}$ holds. Likewise, a succession of existential quantifiers will be written

$$\bigvee x, y, z(...)$$

in place of

$$\bigvee x[\bigvee y\{\bigvee z(...)\}]$$

and shall be read: there exist x, y, z such that...

7. CONSTRUCTION OF NATURAL NUMBERS

In the above way, the concepts of relation and function are reduced in set theory to the concept of set. This procedure does not remove all the difficulties and problems concerning these concepts. They are only focused on the interpretation of the concept of set. They are studied by researchers in the foundations of set theory, forming a separate branch of mathematics, with which we shall not be concerned here.

EXERCISES

1. Give another definition of the ordered pair using an equivalence relation of the form

$$u \in \langle x, y \rangle \equiv \ldots$$

where the right hand member of this equivalence contains only the constants $\bigwedge$, $\bigvee$, $\wedge$, $\vee$, $\rightarrow$, $\equiv$, $=$, $\in$, and variables.

2. Reformulate definition (40) without using the pair symbol $\langle \; \rangle$ on the right side of the equivalence symbol. Use the result of exercise 1.

3. How many different subsets are there of a set containing n elements? (The answer is 2^n.) Check it for 2, 3, 4 and give a general proof.

4. How many different ordered pairs are there in an n-element set?

5. How many different functions are there defined on an n-element set and taking on values in a k-element set?

7. THE GENETIC CONSTRUCTION OF NATURAL NUMBERS

The third group of concepts discussed in classical mathematics, namely the arithmetical concepts, seems to come closest to the specific nature of mathematics. The origin of arithmetical concepts has for the last 100 years been studied by researchers working on the foundations of mathematics. There are two main viewpoints. One of them, called *intuitionistic*, is that the arithmetical concepts are created by a specifically mathematical intuition which is closely related to the succession of moments in time. That intuition suggests, first of all, that the concept of natural number is a result of successive accumulation of identical units. All other kinds of numbers can then be reduced to natural numbers. That philosophical standpoint suggests that the arithmetic of

37

natural numbers be treated as a separate, fundamental discipline. Another viewpoint, called *logistic*, is that even the arithmetic of natural numbers can be derived genetically from more general intuitions of a set-theoretical and logical nature. Apart from philosophical considerations, the genetic theory of natural numbers is an interesting construction, which goes back to G. Cantor [7] and G. Frege [15]. It will be outlined below in a cursory manner, presenting only the main ideas. A more precise exposition of the subject can be found e.g. in Kuratowski and Mostowski [42] or other textbooks in set theory.

The natural numbers 0, 1, 2, 3, ... are obtained, so it seems, by natural abstraction from everyday experience of dealing with many finite sets, some of which contain equally many objects. Every normal man has two legs, two arms, two eyes, etc. The set of arms of a man consists of two elements, and the set of his eyes of two elements, too. It can be supposed that these observations lead to the general concept of the number of elements in a set, as a property which a given set shares with all other sets which contain as many elements as the set in question. Thus, the number 2 is the property of having two elements. The set of arms of a man has the property, and so has the set of eyes. The number 5 is the quantity of the fingers of one hand, and in general the property of every set which has as many elements as one hand has fingers.

We have just used the concept of a property. In set theory that concept is replaced by the concept of a set. Instead of a property, reference is made to the set of all elements having that property. Thus, the number 5 is the set of all sets containing five elements, or the set of all sets containing as many elements as a right hand has fingers. The number 2 is the set of all those sets which have as many elements as a normal man has ears, etc. Thus, in set theory the concept of equinumerosity is defined independently of the concept of number.

By making this reasoning more precise we arrive at the strict concept of a natural number as a *finite cardinal number*. We shall now write down in a precise manner all definitions which occur in that construction. To make that construction possible, we must have a positive proof that at least one infinite set exists. Let us denote that set by A, and let us assume once and for all that the symbol A will always stand for a certain fixed infinite set. We shall consider subsets of A, i.e., elements of the set 2^A.

7. CONSTRUCTION OF NATURAL NUMBERS

As the first step we shall distinguish in the set 2^A a subset Fin, which consists of all finite subsets of A. Finite sets are such that, 1) the empty set is a finite set, and 2) if one element x is added to a finite set Y, then the set $Y \cup \{x\}$ is again a finite set. All finite sets can be "created" from the empty set by a consecutively adding single elements. This is the property which distinguishes the finite sets from all others. The set Fin may be distinguished in the set 2^A as follows:

$$(42) \quad X \in \text{Fin} \equiv X \in 2^A \wedge \bigwedge U[(\Lambda \in U \wedge \bigwedge Y, W\{(Y \in U \wedge x \in A) \to Y \cup \{x\} \in U\}) \to X \in U].$$

In words: X is a finite subset of the set A, if $X \in 2^A$ and X belongs to every such family of sets U, which satisfies the following two conditions:

1. The empty set Λ belongs to U,

2. If a set Y belongs to U, then the set $Y \cup \{x\}$, which is the union of Y and any single element x of A, belongs to U, too.

By (42), if a set X is finite, then it belongs to every family U satisfying conditions 1 and 2. The family Fin of finite sets satisfies, as we have seen, conditions 1 and 2, hence the family Fin is contained in every family satisfying conditions 1 and 2, i.e., as is often said, it is the *least family* satisfying conditions 1 and 2. Definitions of this type will be encountered further in this book. The term Fin, like all other terms in this section, ought to be restricted to the set A, i.e., it ought to have a subscript A. We shall omit it, though, for simplicity's sake.

Next we consider the concept of equinumerosity. Two sets are *equinumerous* when they have the same quantity of elements. Since, however, our task is to define the concept of quantity, equinumerosity must be defined in a different, more direct way. We refer to the following intuitions. The set of right-hand fingers is equinumerous with the set of left-hand fingers because we can put both hands together so that each finger of one hand touches one and only one finger of the other hand. This establishes a one-to-one correspondence between the elements of both sets. In general we shall say that two sets X and Y are *equinumerous*, X eqnum Y, if there is a one-to-one mapping f of the set X onto the set Y, that is, such that the whole set Y is exactly the image of the set X. Each element of the set X is then associated by the mapping f with exactly one element of the set Y, and conversely, each element of the set Y is associated by the mapping f with exactly one element of

39

the set X. As a result of this association we may say that the set X has as many elements as the set Y has, or that X and Y are equinumerous. This definition may be written in symbolic notation as follows (cf. formula (41)):

$$(43) \quad X \text{eqnum} Y \equiv \bigvee f[1. \, f \in Y^X \wedge$$

$$2. \bigwedge x, y, z(\langle x, y \rangle \in f \wedge \langle z, y \rangle \in f \to$$
$$x = z) \wedge$$

$$3. \bigwedge y\{y \in Y \to \bigvee x(x \in X \wedge \langle x, y \rangle \in f)\}].$$

A procedure suggested by this definition is often used, for instance, by two boys who want to divide, without counting, a number of peanuts between themselves, and who want to make it a square deal. They sit beside the peanuts, one facing the other and they simultaneously take one peanut each, put the peanuts aside, and then reach for another pair of peanuts, and so on. If in the end there are no peanuts left, they may be sure their shares are equal. The one-to-one correspondence between their shares is established by the succession of time, and the notion of the natural number is eliminated entirely. If there is a great heap of peanuts this method is even more efficient than counting them all and then counting one half again.

Mappings will usually be denoted by functional symbols. To denote which function establishes the one-to-one correspondence we shall use the abbreviation $X \text{eqnum}_f Y$. The exact definition of this phrase is:

$$(44) \quad X \text{eqnum}_f Y \equiv 1. \, f \in Y^X \wedge$$

$$2. \bigwedge x, y(f(x) = f(y) \to x = y) \wedge$$
$$3. \bigwedge y\{y \in Y \to \bigvee x(x \in X \wedge y = f(x))\}.$$

It follows from (43) and (44) that

$$X \text{eqnum} Y \equiv \bigvee f(X \text{eqnum}_f Y).$$

EXAMPLES

The set $\{1, 3, 7\}$ is equinumerous with $\{2, 6, 14\}$. A function establishing this equinumerosity is, for instance, the function $y = 2x$ restricted to the set $\{1, 3, 7\}$. Likewise, $\{2, 5, 8, 6\}$ eqnum$_f$ $\{6, 15, 24, 18\}$, where $f(x) = 3x$ for $x \in \{2, 5, 8, 6\}$. The set of natural numbers is equinumerous with the set of even numbers and a function establishing such equinumerosity is $f(x) = 2x$. The set of natural numbers is also equinumerous with the set of all primes, and with the set of numbers divisible by 10, etc., and more generally with any of its infinite subsets. The fact that this set $\mathcal{N}$ is equinumerous with its various subsets may seem strange and paradoxical. The set of even numbers is only one half of that of natural numbers. A part is equinumerous with the whole set. It turns out, however, that this is a characteristic property of infinite sets. Under some additional assumptions that property may be used as a definition of infinite set.

If we consider the equinumerosity of subsets of a given set, then it can be distinguished as a set of pairs of subsets, i.e., a relation in the set-theoretic sense.

Using the relation of equinumerosity we can define the general concept of the *power of a set* or of the *cardinal number of a set*.

As mentioned in the introductory, intuitively formulated part of this section, the number of elements of a set X is a property of that set X which is also a property of every other set that is equinumerous with X. In place of talking about properties of sets we have decided to consider families of sets, i.e., sets of sets, because this device enables us to stay within the scope of set theory. The *power of a set* X, symbolically $\overline{X}$, is the family of sets which contains X and any other set that is equinumerous with it.

If these considerations are restricted to subsets of the set A, the family $\overline{X}$ is then defined in symbols as follows:

(45) $Y \in \overline{X} \equiv Y \in 2^A \wedge Y \,\mathrm{eqnum}\, X.$

The relation eqnum is, of course, reflexive, which means that every set is equinumerous with itself, and this equinumerosity is represented by the identity function. Hence by (45), $X \in \overline{X}$. The relation eqnum is symmetric and transitive. It is symmetric, because if $X \,\mathrm{eqnum}_f\, Y$, then $Y \,\mathrm{eqnum}_{f^{-1}}\, X$, and it is transitive, since if $X \,\mathrm{eqnum}_f\, Y$ and $Y \,\mathrm{eqnum}_g\, Z$, then $X \,\mathrm{eqnum}_h\, Z$, where h is the superposition of g and f: $h(x) = g(f(x))$ for $x \in X$. Equinumerosity is thus an equivalence relation. Hence by (45), if $X \in \overline{Y}$ and $Z \in \overline{Y}$, then $X \,\mathrm{eqnum}\, Z$.

The *cardinal numbers* (CN) or *powers of sets*, are the cardinal numbers, or powers of some sets. If we are confined to subsets of a set A, then the cardinal numbers can be easily specified as follows:

(46) $a \in \mathrm{CN} \equiv a \in 2^{2^A} \wedge \bigvee Y(Y \in 2^A \wedge a = \overline{\overline{Y}}).$

This means that a is a cardinal number if and only if a is identical with a family of sets which are equinumerous with a certain subset Y of A.

Using (45) and the axiom of extensionality we can formulate the definition of the family of cardinal numbers as follows:

(47) $a \in \mathrm{CN} \equiv a \in 2^{2^A} \wedge \bigvee Y \{Y \in 2^A \wedge \bigwedge X (X \in a \equiv X \mathrm{eqnum} Y)\}.$

The natural numbers are the cardinal numbers of finite sets. We must insert in the definition the proviso that $Y \in \mathrm{Fin}$:

(48) $a \in \mathcal{N} \equiv a \in 2^{2^A} \wedge \bigvee Y (Y \in \mathrm{Fin} \wedge a = \overline{\overline{Y}}).$

This is the definition of the *set of natural numbers*. Operations on natural numbers can also be defined by reference to sets. If we form the union of two disjoint sets X and Y, such that $\overline{\overline{X}} = a$ and $\overline{\overline{Y}} = b$, then, of course, $\overline{\overline{X \cup Y}} = a+b$. The family $a+b$ can thus be defined as follows:

(49) $Z \in (a+b) \equiv Z \in 2^A \wedge \bigvee X, Y \{a = \overline{\overline{X}} \wedge b = \overline{\overline{Y}} \wedge$
$X \cap Y = \Lambda \wedge Z \mathrm{eqnum} (X \cup Y)\}.$

The *operation of addition* defined in this way leads from natural numbers to natural numbers. It can easily be proved that

(50) *If $a, b \in \mathcal{N}$, then $(a+b) \in \mathcal{N}$.*

In fact, suppose that $a, b \in \mathcal{N}$. Then the sets X and Y from formula (49) are finite and so is their union $X \cup Y$. By (49), $a+b$ is the family of all sets which are equinumerous with the finite set $X \cup Y$. Hence, by (48), $(a+b) \in \mathcal{N}$.

The *multiplication of natural numbers* can be described as follows. If we want to multiply 3 by 5, then we may take a family X consisting of 3 disjoint sets containing five elements each. Then the union of the family X has $3 \cdot 5$ elements. Generally speaking, the number $a \cdot b$ is the number of the elements of the union of every such family X of disjoint sets for which $\overline{\overline{X}} = a$ and for every U, if $U \in X$, then $\overline{\overline{U}} = b$. Hence the product can be defined by the following formula:

(51) $Z \in (a \cdot b) \equiv Z \in 2^A \wedge \bigvee X \{\overline{\overline{X}} = a \wedge \bigwedge U (U \in X \to \overline{\overline{U}} = b) \wedge$
$Z \mathrm{eqnum} \bigcup X \wedge \bigwedge u, v ((u, v \in X \wedge u \neq v) \to u \cap v = \Lambda)\}.$

Strictly speaking, the cardinality function was defined only on sets $X \in 2^A$, whereas the family X belongs to 2^{2^A}. However, instead of

42

speaking about the cardinality of the family X, one can speak about the cardinality of the set of representatives of the family. We leave the correction of definition (51) in the way indicated as an exercise.

Since the union of a finite number of finite sets is again a finite set, it follows that the set $a \cdot b$ satisfying condition (51) is the family of all sets that are equinumerous with the finite set $\bigcup X$, and hence this set is a natural number in the sense of formula (48). The operation of multiplication leads from natural numbers to natural numbers (that is, the set of natural numbers is closed under multiplication):

(52) $a, b \in \mathcal{N} \rightarrow (a \cdot b) \in \mathcal{N}$.

The system $\langle \mathcal{N}, +, \cdot \rangle$ is a mathematical domain which is called the *semi-ring of natural numbers*. It can easily be verified, which will not be done here, that all fundamental laws of the arithmetic of natural numbers can be deduced from the axioms of set theory.

EXERCISES

1. Define arithmetically a function which demonstrates that the set of numbers divisible by 10 and the set of numbers divisible by 3 are equinumerous.

2. Define a function which demonstrates that the set of integers and the set of natural numbers are equinumerous.

3. How many different one-to-one functions are there which map a set A of n elements into another equinumerous set?

4. Prove that for the natural numbers defined in the text the following theorems are true:

$$a+b = b+a,$$
$$a \cdot b = b \cdot a,$$
$$a+(b+c) = (a+b)+c,$$
$$a \cdot (b \cdot c) = (a \cdot b) \cdot c.$$

5. Write down symbolically the following definition of equinumerosity: sets A and B are equinumerous if there is a family C such that:

1. Elements of C are two-element sets $\{x, y\}$,
2. The sets from the family C are disjoint,
3. Each $z \in C$ has a non-empty intersection with the set $A - B$ and a non-empty intersection with $B - A$,
4. $(A \cup B) \subset (C \cup (A \cap B))$.

Prove that this definition is equivalent to the original one given in the text.

6. Correct Definition (51) as was indicated.

8. EXPANSION OF THE CONCEPT OF NUMBER

It appears that the first concept of number, historically, was precisely that of the very natural and indispensable concept in the counting of objects. The need to determine the number of elements in a set was, we may assume, the first mathematical need of man. The necessity to measure length with a fixed measure gave rise to rational numbers, i.e., fractions

$$a = \frac{p}{q},$$

where p, q are integers. Investigations in geometry next showed that rational numbers do not suffice to express the lengths of arbitrary line segments which occur in geometric constructions. For instance, no fraction p/q can express the length of the diagonal of a square whose side has one unit of length. Hence, to be able to express every geometric value by a number, it was necessary to introduce the concept of real number. It was useful, too, to extend the concept of number still further so as to cover the complex numbers.

These concepts of number were growing spontaneously more and more general before the logical foundations of mathematics began to be investigated. Today, beginning with the arithmetic of natural numbers we can construct, by using logical concepts, analogues of what we may call rational, real, and complex numbers, and we can do so in a way reminiscent of their historical development. These constructions shall be outlined below.

First, consider the concept of integer. *Integers* include natural numbers and negative integers. It is taught in the elementary school that integers are natural numbers preceded by the plus (+) or minus (−) sign. Hence, now that we have at our disposal the concepts of set theory, it is most convenient to define integers as pairs of the form $\langle +, k \rangle$ or $\langle -, k \rangle$, where k is a natural number other than zero and the signs "+" and "−" may be considered names of some fixed symbols of the form + and −. There is nothing against choosing for "+" and "−" names of any two different objects, for instance of zero and one. In this way we do not have to use any other objects than natural numbers. We can assume, for instance, that the negative numbers are the pairs $\langle 0, k \rangle$ and positive numbers are the pairs $\langle 1, k \rangle$. Furthermore, 0 is considered to be an integer. We shall hereafter use the signs $+$ and $-$ as corresponding very well to our intuitions. The set of integers shall be denoted by $\mathscr{C}$. It can be defined as follows:

$$(53) \qquad x \in \mathscr{C} \equiv x = 0 \vee \bigvee y, z \in \mathscr{N}(x = \langle z, y \rangle \wedge y \neq 0 \wedge$$
$$(z = + \vee z = -)).$$

We have used in (53) an abbreviation we shall often use in the future. Instead of writing $\bigvee y(y \in X \wedge \ldots)$ we shall use a shorter symbol $\bigvee y \in X(\ldots)$, which is to be read: there exists an y from the set X such that $(\ldots)$. Likewise, in place of $\bigwedge y(y \in X \to \ldots)$, we shall use the symbol $\bigwedge y \in X(\ldots)$, which is to mean: for every element y from the set X $(\ldots)$. These abbreviated symbols are called *quantifiers restricted to the set X*.

Operations on integers are defined much in the same way as it is done in elementary school, i.e., by referring to operations on natural numbers:

$$\langle +, k \rangle + \langle +, l \rangle = \langle +, k+l \rangle,$$

$$\langle +, k \rangle + \langle -, l \rangle = \begin{cases} \langle +, k-l \rangle & \text{for} \quad k > l, \\ 0 & \text{for} \quad k = l, \\ \langle -, l-k \rangle & \text{for} \quad l > k, \end{cases}$$

$$\langle -, k \rangle + \langle +, l \rangle = \langle +, l \rangle + \langle -, k \rangle,$$

$$\langle -, k \rangle + \langle -, l \rangle = \langle -, k+l \rangle,$$

$$\langle i, k \rangle + 0 = 0 + \langle i, k \rangle = \langle i, k \rangle \quad \text{for} \quad i = + \text{ or } i = -,$$

$$0 + 0 = 0.$$

These formulae define *addition* for all possible pairs of integers. To be strict we must use another symbol to denote the addition of integers so defined, since they are entirely different from natural numbers. This is why this symbol is set in bold type $+$. Other operations, too, must be redefined for the integers so defined. A good definition ought to have the form of an equivalence which defines a three-place relation, e.g., $+(x, y, z)$, for any triplets chosen from the set $\mathscr{C}$. Then it should be proved that such a relation is a function of two arguments. Another correct form of definition could be modelled on definitions (49) or (51) and would define the pair representing the sum as a set by indicating the elements of that set.

Among the integers we can designate those which correspond to natural numbers, i.e., 0, and the pairs $\langle +, k \rangle$, i.e., positive integers. There is a one-to-one correspondence between natural numbers, in the old sense of the term and positive integers (including zero). This correspondence maps zero into zero and each $k > 0$ into the pair $\langle +, k \rangle$:

$$0 \leftrightarrow 0, \quad k \leftrightarrow \langle +, k \rangle.$$

This correspondence has a very important property. It maps sums of natural numbers into corresponding sums of integers:

(54)
$$
\langle +, k \rangle + \langle +, l \rangle = \langle +, k+l \rangle,
$$
$$
k \quad + \quad l \quad = \quad k+l.
$$

The new addition of integers fully coincides with the former addition of natural numbers. This guarantees that the same symbol used to denote both operations cannot result in errors in calculation.

Multiplication is defined unambiguously by the following formulae:

1. If $z \in \mathscr{C}$, then $z \cdot 0 = 0 \cdot z = 0$,

2. $\left. \begin{array}{l} \langle +, k \rangle \cdot \langle +, l \rangle \\ \langle -, k \rangle \cdot \langle -, l \rangle \end{array} \right\} = \langle +, k \cdot l \rangle,$

3. $\left. \begin{array}{l} \langle -, k \rangle \cdot \langle +, l \rangle \\ \langle +, k \rangle \cdot \langle -, l \rangle \end{array} \right\} = \langle -, k \cdot l \rangle.$

By this definition the product of two positive integers is associated with a product of natural numbers by the correspondence symbol $\leftrightarrow$:

(55)
$$
\langle +, k \rangle \cdot \langle +, l \rangle = \langle +, k \cdot l \rangle,
$$
$$
k \quad \cdot \quad l \quad = \quad k \cdot l.
$$

It can easily be verified that the operations of addition and multiplication so defined do not lead outside the set of integers $\mathscr{C}$. Both of these operations are associative and commutative. Multiplication is distributive with respect to addition. Because of these special properties the mathematical domain $\langle \mathscr{C}, +, \cdot \rangle$ is often called the *ring of integers*.

In the set $\mathscr{C}$ we shall distinguish a set $\mathscr{N}^*$ which consists of analogues of natural numbers, i.e. zero and integers with the sign $+$:

(56) $\quad x \in \mathscr{N}^* \equiv x \in \mathscr{C} \wedge \left(x = 0 \vee \bigwedge u, n(x = \langle u, n \rangle \wedge u = +) \right).$

The relation $\leftrightarrow$ just defined is a one-to-one mapping of the set $\mathcal{N}^*$ onto the whole set $\mathcal{N}$. By (54) and (55), this mapping satisfies the following conditions:

(57)   If $x, y \in \mathcal{N}^*, n, m \in \mathcal{N}$ and $x \leftrightarrow n, y \leftrightarrow m$,
        then $x + y \leftrightarrow n + m$,

(58)   If $x, y \in \mathcal{N}^*, n, m \in \mathcal{N}$ and $x \leftrightarrow n,\ y \leftrightarrow m$, then $x \cdot y \leftrightarrow n \cdot m$.

We say that the map $\leftrightarrow$ induces an *isomorphism* between the domain $\langle \mathcal{N}^*, +, \cdot \rangle$, and the domain $\langle \mathcal{N}, +, \cdot \rangle$ (cf. Section 10, p. 58). The concept of isomorphism will soon be defined in full generality, but right now let us consider, as another example, the construction of rational numbers, i. ., fractions.

Every *fraction* is determined by two integers p and q and it is customarily denoted by the symbol p/q. This symbol suggests that fractions may be considered to be pairs $\langle p, q \rangle$ such that the first element is the numerator and the second is the denominator. Thus fractions would be elements of the set $\mathscr{C}^2$ and the set of fractions $\mathscr{U}$, included in $\mathscr{C}^2$ could be defined as follows.

(59) $x \in \mathscr{U} \equiv x \in \mathscr{C}^2 \wedge \bigvee p, q(x = \langle p, q \rangle \wedge q \neq 0)$,

with the special provision that the denominator q must be other than 0 ($q \neq 0$).

Operations on fractions considered as pairs can easily be defined by the ordinary rules from elementary school arithmetic:

(60) $\langle p, q \rangle \oplus \langle r, s \rangle = \langle (p \cdot s) + (q \cdot r), q \cdot s \rangle$,
        $\langle p, q \rangle \otimes \langle r, s \rangle = \langle p \cdot r, q \cdot s \rangle$.

But the arithmetic of fractions so defined does not satisfy strictly our intuitions about the arithmetic of rational numbers. Intuition tells us that the fractions 2/5, 4/10, 6/15, 14/35, ... stand for one and the same rational number. Yet each of the pairs $\langle 2, 5 \rangle, \langle 4, 10 \rangle, \langle 6, 15 \rangle, \langle 14, 35 \rangle$, ... is an entirely different set. Of course, each of the labels: "2/5", "4/10", "6/15", "14/35" ... is different from others. None of them is considered a rational number. The same applies to the pairs $\langle 2, 5 \rangle, \langle 4, 10 \rangle$, ... Accordingly we shall not say that these pairs, i.e., our fractions (elements of $\mathscr{U}$) are rational numbers. We shall say that they stand for some rational numbers. Infinitely many different fractions can stand for one rational number. We can easily tell, though, when two pairs stand for the same rational number. We use the symbol sfs to denote this relation:

(61) $\langle p, q \rangle \, \text{sfs} \, \langle r, s \rangle \equiv p \cdot s = q \cdot r.$

More precisely, this relation should be defined as follows:

(62) $x \, \text{sfs} \, y \equiv \bigvee p, q, r, s \in \mathscr{C} (x = \langle p, q \rangle \wedge y = \langle r, s \rangle \wedge$
         $p \cdot s = q \cdot r).$

This relation corresponds to arithmetical identity of two fractions,

(63) $p \cdot s = q \cdot r \equiv \dfrac{p \cdot s}{q \cdot s} = \dfrac{q \cdot r}{q \cdot s} \equiv \dfrac{p}{q} = \dfrac{r}{s}.$

As can easily be verified the relation sfs is an equivalence relation. For instance, we shall verify its transitivity. If $\langle p, q \rangle \, \text{sfs} \, \langle r, s \rangle$ and $\langle r, s \rangle$ sfs $\langle u, v \rangle$, then, by (61), $p \cdot s = q \cdot r$ and $r \cdot v = s \cdot u$. Hence, by multiplying the first identity by v and the second by q, we obtain $p \cdot s \cdot v = q \cdot r \cdot v$, $q \cdot r \cdot v = s \cdot u \cdot q$. Since ordinary identity is transitive, it follows from the above that $p \cdot s \cdot v = s \cdot u \cdot q$. On cancelling s, we obtain $p \cdot v = u \cdot q$, which, by (61), means that $\langle p, q \rangle \, \text{sfs} \, \langle u, v \rangle$.

We are now approaching the crucial point in the construction of rational numbers. We want a rational number to be a single mathematical entity, whereas in the procedure used so far one rational number was represented by an infinite set of pairs, any two of them connected by the relation sfs. For instance, $\langle 2, 5 \rangle$, $\langle 4, 10 \rangle$, $\langle 6, 15 \rangle$, ... are such pairs. There are two solutions. We may choose one pair, for example the pair with the least denominator, $\langle 2, 5 \rangle$, and call this pair the rational number represented by $\langle 2, 5 \rangle$, $\langle 4, 10 \rangle$, $\langle 6, 15 \rangle$ ... Another possibility is to say that the rational number we want to define is the set of all those pairs which are connected with $\langle 2, 5 \rangle$ by the relation sfs. Both methods yield the same arithmetic. Usually the second method is chosen, so that the definition of a rational number is:

(64) $x \in \mathscr{W} \equiv x \in 2^{\mathscr{U}} \wedge \bigvee y \{ y \in x \wedge \bigwedge z (z \in x \equiv z \, \text{sfs} \, y) \}.$

This method resembles that of defining cardinal numbers. Definitions of this type will also be encountered in the text below.

Operations on elements of $\mathscr{W}$ can be defined by operations on fractions, i.e., on elements of $\mathscr{U}$:

48

(65) $\quad x \in (a \mp b) \equiv a, b \in \mathcal{W} \wedge \bigvee p, q, r, s \in \mathcal{C}(\langle p, q \rangle \in a \wedge$
$\langle r, s \rangle \in b \wedge x \text{ sfs } (\langle p, q \rangle \oplus \langle r, s \rangle)).$

(66) $\quad x \in (a \bar{\times} b) \equiv a, b \in \mathcal{W} \wedge \bigvee p, q, r, s \in \mathcal{C}(\langle p, q \rangle \in a \wedge$
$\langle r, s \rangle \in b \wedge x \text{ sfs } (\langle p, q \rangle \otimes \langle r, s \rangle)).$

The operations determined by formulae (65) and (66) do not, as can easily be verified, lead outside the set $\mathcal{W}$, so that if $a, b \in \mathcal{W}$, then $(a \mp b) \in \mathcal{W}$ and $(a \bar{\times} b) \in \mathcal{W}$. It can also easily be seen that these operations are associative and commutative, that multiplication is distributive with respect to addition, and that the other laws of the arithmetic of rational numbers hold. The system $(\mathcal{W}, \mp, \bar{\times})$ is thus the mathematical domain called the *field of rational numbers*.

EXERCISES

1. Define the addition of integers using equivalence relations of the form

$$x \in (a + b) \equiv \ldots,$$

much like we did for the addition of natural numbers. This can be done, because the sum of two integers, being an integer, is a set and can be defined by indicating what its elements are.

2. Define the multiplication of integers in the same way.

3. Using equivalences of the form

$$x \leftrightarrow y \equiv \ldots$$

define the relation $\leftrightarrow$ introduced on p. 46.

4. Define the addition and the multiplication of fractions (60) using equivalences of the form

$$x \in (p \oplus q) \equiv \ldots,$$
$$x \in (p \otimes q) \equiv \ldots$$

9. CONSTRUCTION OF NEW MATHEMATICAL DOMAINS

The content of the last three Sections, 9, 10, and 11, of this Introduction is somewhat more difficult than that of the previous sections. They are, however, not necessary for the comprehension of the next Chapter. A non-mathematician may omit them when reading this book for the first time. These sections discuss concepts to be referred to only in Chapter II.

In this section we shall consider certain general methods of constructing new mathematical domains, namely those which have as their special cases the constructions of the concepts of cardinal number, of integer, and of rational number, as outlined above.

The formation of a product X^2, interpreted as the set of pairs in the sense of Definition (40), played an important role in the construction of integers and rational numbers. The sets $\mathscr{C}$ and $\mathscr{U}$ were subsets of the products $\mathscr{N}^2$ and $\mathscr{C}^2$, respectively, to be precise, $\mathscr{C} \subset \mathscr{N}^2 \cup \{0\}$. The construction of cardinal numbers and of rational numbers was done by the formation of a new domain consisting of abstraction classes for some relation. This operation shall be discussed in greater detail.

If $\langle Z, F, S \rangle$ is a mathematical domain, and if there is an equivalence relation R defined on the set Z, then that relation naturally divides the set Z into disjoint subsets called *abstraction classes* of the relation R. For instance, let $Z = 2^A$, where A is a fixed infinite set, and R is the equinumerosity relation, eqnum, holding between the subsets of A, then all subsets of A, i.e., elements of the set Z, can be classed as follows:

K_1, one-element sets contained in A,

K_2, two-element sets contained in A,

K_3, three-element sets contained in A, etc.

These classes are mutually disjoint, since no one-element set is a two-element set etc. Each of these classes can be characterized by the fact that it consists of equinumerous sets. For instance, if $X \in K_3$ and $Y \in K_3$, then X eqnum Y.

The family of all those sets which are equinumerous with a set X was called the *cardinal number*, or the *power*, of X, and was denoted by $\overline{X}$. The sets $K_1, K_2, K_3, \ldots$ are different cardinal numbers. Cardinal numbers are thus certain disjoint sets. Two cardinal numbers either are the same cardinal number, or else they have no common elements. Cardinal numbers are abstraction classes of the relation of equinumerosity.

The abstraction class determined by an element y with respect to a relation R will be denoted by $[y]_R$. The symbol R will be omitted if it is obvious which equivalence relation is being considered. The general definition of an *abstraction class* is:

(67) $z \in [y]_R \equiv z, y \in Z \land zRy.$

The set of abstraction classes of a relation R in a set Z will be denoted by Abs_R:

(68) $\qquad a \in \mathrm{Abs}_R \equiv a \in 2^Z \wedge \bigvee y \in Z(a = [y]_R).$

When comparing Definitions (45) and (46) of cardinal number with Definitions (67) and (68) we see that in fact $\overline{\overline{X}} = [X]_{\mathrm{eqnum}}$ and $\mathrm{CN} = \mathrm{Abs}_{\mathrm{eqnum}}$, so that cardinal numbers are abstraction classes of the relation eqnum in agreement with the above definitions. If we assume that a relation R is an equivalence relation, then, by (67) and (68),

(69) $\qquad a \in \mathrm{Abs}_R \equiv a \in 2^Z \wedge \bigvee y \in Z(z \bigwedge (z \in a \equiv zRy)),$

which also may be considered a definition of the set of abstraction classes of the relation R. On comparing Definition (64) of rational numbers with Definition (69), we can see that $\mathscr{W} = \mathrm{Abs}_{\mathrm{sfs}}$ for $Z = \mathscr{U}$, so that the rational numbers are abstraction classes of the relation sfs between fractions.

If R is an equivalence relation, then it can be proved in general that two abstraction classes either are the same class or are disjoint. This theorem can be written down as follows:

(70) $\qquad \bigwedge a, b((a \in \mathrm{Abs}_R \wedge b \in \mathrm{Abs}_R) \to (a \cap b = \Lambda \vee a = b)).$

Formula (70) holds under the assumption that R is an equivalence relation. To prove that the classes a and b are either disjoint ($a \cap b = \Lambda$), or identical ($a = b$), it suffices to demonstrate that if they are not disjoint, then they are the same class. If a and b are not disjoint, then, by the definitions on p. 33, they must have a common element. Hence there exists an x such that $x \in a$ and $x \in b$. But if $a, b \in \mathrm{Abs}_R$ then, by (68), for some y and z, $a = [y]_R$ and $b = [z]_R$, so that $x \in [y]_R$ and $x \in [z]_R$, which by (67), means that xRy and xRz. This yields yRx, since R is symmetric, and yRz, since R is transitive, too.

Suppose now that there is an element $u \in a$, hence $u \in [y]_R$. Thus, by (67), uRy. Since yRz, then it follows from the transitivity of R that uRz. Consequently, $u \in [z]_R$, i.e., $u \in b$. Conversely, too, if for some u, $u \in b$, hence $u \in [z]_R$, then, by (67), uRz, and thus zRu by symmetry, and yRu by transitivity, since yRz. Since yRu, then uRy by symmetry, i.e., $u \in [y]_R$, by (67), and hence $u \in a$. We have thus proved that, for any element u, if $u \in a$, then $u \in b$, and conversely, if $u \in b$, then $u \in a$.

Hence, $u \in a$ if and only if $u \in b$ (the rule of logic (23) has been used here). The conclusion just arrived at is formulated in symbols

$$\bigwedge u(u \in a \equiv u \in b).$$

By the axiom of extensionality this formula implies $a = b$. We have thus proved formula (70).

The following formulae are easy to prove:

(71) $x \in [x]_R$ for any $x \in Z$,

(72) $x R y \equiv [x]_R = [y]_R$ for any $x, y \in Z$,

and will be used below.

The first follows from the reflexivity of the relation R. Since $x R x$, then, by (67), $x \in [x]_R$.

The second formula can be proved in the following way: If $x R y$, then, by (67), $x \in [y]_R$, and by (71) $x \in [x]_R$. The sets $[y]_R$ and $[x]_R$ thus have a common element x. Hence they are not disjoint and, accordingly, by (70), they must be identical.

Conversely, if $[x]_R = [y]_R$, then, by (71), $x \in [x]_R$, hence $x \in [y]_R$ which, by (67), means that $x R y$. These two implications, by the rule of logic (23), yield (72).

Since the relation eqnum is an equivalence relation, by (70) two cardinal numbers are either disjoint or identical. The relation sfs, defined by (62), is an equivalence relation, too, hence two rational numbers, defined as abstraction classes of the relation sfs, are either disjoint or identical.

In defining rational numbers we started with the domain $\langle \mathscr{C}, +, \cdot \rangle$ and first we defined the domain of fractions $\langle \mathscr{U}, \oplus, \otimes \rangle$. Only then, using that domain did we define a new domain $\langle \mathscr{W}, \mp, \bar{\times} \rangle$ of rational numbers by forming abstraction classes of the relation sfs. In this transition from the domain $\langle \mathscr{U}, \oplus, \otimes \rangle$ to the domain $\langle \mathscr{W}, \mp, \bar{\times} \rangle$ not only does the construction of the set $\mathscr{W}$ fall under a general schema, but the definition of the operations $\mp$ and $\bar{\times}$ on rational numbers, too, is a special case of a widely used general construction, which we shall now describe in detail.

Let $\langle Z, F, G, S \rangle$ be any domain, where Z is a set, F is a one-argument function, G is a two-argument function, and S is a relation. We say that a relation R, which is an equivalence relation and which is defined on a set Z, *preserves* the operations F and G if and only if, for every $x, y, z, v \in Z$, the following conditions hold:

(73)
$$xRy \to F(x)\,RF(y),$$
$$(xRy \wedge zRv) \to G(x, z)\,RG(y, v).$$

We say that an equivalence relation R preserves a relation S, if and only if, for every $x, y, z, v \in Z$,

(74) $(xRy \wedge zRv) \to (xSz \equiv ySv)$.

An equivalence relation which preserves all functions and relations that are essential for a certain domain is called a *congruence* in that domain. For instance, the equivalence relation sfs is a congruence in the domain $\langle \mathcal{U}, \oplus, \otimes \rangle$, since it satisfies the condition

$$(\langle p, q \rangle \text{ sfs } \langle r, t \rangle \wedge \langle s, u \rangle \text{ sfs } \langle w, z \rangle) \to$$
$$(\langle p, q \rangle \oplus \langle s, u \rangle) \text{ sfs } (\langle r, t \rangle \oplus \langle w, z \rangle),$$

and also an analogous condition for multiplication.

The identity relation is a congruence on any domain. It is such that all abstraction classes consist of one element each, and conditions (73) and (74) are obviously satisfied. Every congruence is very much like identity. This similarity is especially striking when a given congruence is denoted by a symbol resembling equality, for instance $\simeq$. Conditions (73) and (74) then turn into

$$x \simeq y \to F(x) \simeq F(y),$$
$$(x \simeq y \wedge z \simeq v) \to G(x, z) \simeq G(y, v),$$
$$(x \simeq y \wedge z \simeq v) \to (xSz \equiv ySv).$$

Conditions of this type are characteristic of the identity relation, for if x is the same thing as y, then every operation peformed on x is identical with the same operation performed on y. Congruent elements may be treated as if they were identical, if we are only interested in operations the congruence preserves. Such an identification of congruent elements is made when abstraction classes with respect to a congruence relation R are being discussed. We can define operations and relations on abstraction classes that correspond to basic operations and relations in the original domain $\langle Z, F, G, S \rangle$. Let us denote these operations and relations by $\bar{F}^R, \bar{G}^R, \bar{S}^R$, respectively. These operations are uniquely determined for all elements of the set Abs_R by the following formulae:

$$\bar{F}^R([x]_R) = [F(x)]_R,$$

(75) $\quad \bar{G}^R([x]_R, [y]_R) = [G(x, y)]_R,$

$$[x]_R \bar{S}^R [y]_R \equiv x S y.$$

Each element of the set Abs_R is of the form $[x]_R$, where $x \in Z$. Formulae (75) thus define the functions $\bar{F}^R$ and $\bar{G}^R$, and the relation $\bar{S}^R$ for all elements of Abs_R.

Definitions (75) are not constructed in agreement with the method adopted here of defining functions and relations. But in the case of a congruence such abbreviated definitions of the type (75) may be used. To see this, we will define, for instance, the function $\bar{F}^R$ in the usual way, using the axiom of set construction, and we will prove the first of formulae (75) as a theorem. The usual definition should run as follows:

(76) $\quad u \in \bar{F}^R(a) \equiv u \in Z \wedge \bigvee x_1 \in Z(a = [x_1]_R \wedge u \in [F(x_1)]_R).$

Suppose now that $u \in \bar{F}^R([x]_R)$. By (76), this means that there is an $x_1 \in Z$ such that $[x]_R = [x_1]_R$ and $u \in [F(x_1)]_R$. Since $[x]_R = [x_1]_R$, it follows from (72) that $x R x_1$. Hence, by (73), $F(x) R F(x_1)$, and also, by (72), $[F(x_1)]_R = [F(x)]_R$. Therefore, if $u \in [F(x_1)]_R$, then $u \in [F(x)]_R$.

Conversely, if $u \in [F(x)]_R$, and if it is true that $[x]_R = [x_1]_R$, then the right hand member of equivalence (76) with $a = [x]_R$ and x in place of x_1 is true. Hence the left side of (76) must be true, too, which proves that $u \in \bar{F}^R([x]_R)$.

By the rule of logic (23), these two implications just proved yield the equivalence: $u \in \bar{F}^R([x]_R)$ if and only if $u \in [\bar{F}^R(x)]_R$ for every u. It follows then from the axiom of extensionality that $\bar{F}^R([x]_R) = [F(x)]_R$. In this way the first of the formulae (75) has been established. Likewise, using proper set-theoretic definitions of $\bar{G}^R$ and $\bar{S}^R$, the second and the third formula (75) can be proved.

In place of the lengthy definitions (65) and (66) of addition and multiplication of rational numbers, this shorter method could have been used due to the fact that relation sfs is a congruence in the domain $\langle \mathcal{U}, \oplus, \otimes \rangle$. This is simply done:

(77) $[\langle p, q \rangle]_{\text{sfs}} \mp [\langle r, s \rangle]_{\text{sfs}} = [\langle p, q \rangle \oplus \langle r, s \rangle]_{\text{sfs}}$

and analogically for multiplication.

If $\mathfrak{A} = \langle Z, F, G, S \rangle$ is a mathematical domain and R is a congruence in that domain $\mathfrak{A}$, then the new domain $\mathfrak{B} = \langle \text{Abs}_R, \overline{F}^R, \overline{G}^R, \overline{S}^R \rangle$, with operations defined by formulae (75), is called the *quotient domain* obtained from $\mathfrak{A}$ on division by R. This new domain is denoted by $\mathfrak{A}/R$. Since $\mathcal{W} = \text{Abs}_{\text{sfs}}$, $\mp = \oplus^{\text{sfs}}$, $\overline{\times} = \otimes^{\text{sfs}}$, it follows that the field of rational numbers is the quotient domain

(78) $\langle \mathcal{W}, \mp, \overline{\times} \rangle = \langle \mathcal{U}, \oplus, \otimes \rangle / \text{sfs}.$

Formation of quotient domains is one of the fundamental constructions leading to new mathematical domains.

By using (77) it is easier to derive various laws of the arithmetic of rational numbers directly from the corresponding laws for fractions. For instance, the fact that $\mp$ is commutative is a direct consequence of the commutativity of $\oplus$:

$$[\langle p, q \rangle]_{\text{sfs}} \mp [\langle r, s \rangle]_{\text{sfs}} = [\langle p, q \rangle \oplus \langle r, s \rangle]_{\text{sfs}} =$$
$$[\langle r, s \rangle \oplus \langle p, q \rangle]_{\text{sfs}} =$$
$$[\langle r, s \rangle]_{\text{sfs}} + [\langle p, q \rangle]_{\text{sfs}}.$$

Likewise, all other laws involving addition and multiplication which can be expressed by using the identity sign automatically cover every quotient domain. We shall soon see that this is a general phenomenon.

EXERCISES

1. Define the relation "less than" for fractions:

$$\langle p, q \rangle < \langle r, s \rangle \equiv \ldots$$

2. Prove that the relation sfs is a congruence also with respect to the relation just defined, cf. formula (74).

3. Using Definition (76) as a pattern, define in the language of set theory the sets $\overline{G}^R$ and $\overline{S}^R$ and prove that they satisfy the second and third formula of (75).

4. Using the results of this section derive for fractions and for rational numbers some laws of addition and multiplication, for instance $a + b = b + a$, $a + (b + c) = (a + b) + c$; do the same for multiplication, and for the distributivity of multiplication with respect to addition: $a(b + c) = ab + ac$.

10. SUBDOMAINS, HOMOMORPHISMS, ISOMORPHISMS

To gain a broader view we shall introduce some more notions from universal algebra. Universal algebra is a fairly distinct part of set theory and is concerned with arbitrary mathematical domains. The following concepts from universal algebra are important: subdomain, homomorphism, and isomorphism.

A domain $\mathfrak{A} = \langle Z', F', G', S' \rangle$ is a *subdomain* of $\mathfrak{B} = \langle Z, F, G, S \rangle$ if and only if

1. $Z' \subset Z$,
2. For any $x, y \in Z'$

$$F'(x) = F(x), \quad G'(x, y) = G(x, y), \quad xS'y \equiv xSy.$$

To put it in words: $\mathfrak{A}$ is a subdomain of $\mathfrak{B}$ if the set Z' of elements of the domain $\mathfrak{A}$ is included in the set Z of elements of the domain $\mathfrak{B}$, and the operations and relations in the domain $\mathfrak{A}$ are analogues of respective operations and relations in $\mathfrak{B}$.

EXAMPLES

1. If $\mathscr{P}$ denotes the set of even natural numbers, then the system $\langle \mathscr{P}, +, \cdot \rangle$ is a mathematical domain, since the sum of two even natural numbers and the product of two even natural numbers are again even natural numbers. The domain $\langle \mathscr{P}, +, \cdot \rangle$ is a subdomain of the semiring of natural numbers $\langle \mathscr{N}, +, \cdot \rangle$.

2. The domain $\langle \mathscr{N}^*, +, \cdot \rangle$, where $\mathscr{N}^*$ is the set of nonnegative integers defined by formula (56), is a subdomain of the ring of integers $\langle \mathscr{C}, +, \cdot \rangle$.

3. If we distinguish, in the set $\mathscr{W}$ of rational numbers, the numbers that correspond to integers $\mathscr{C}^*$:

(79) $\qquad a \in \mathscr{C}^* \equiv a \in \mathscr{W} \wedge \bigvee p \in \mathscr{C} \ (a = [\langle p, 1 \rangle]_{\text{sts}}),$

then the domain $\langle \mathscr{C}^*, \mp, \bar{\times} \rangle$ is a subdomain of the field of rational numbers $\langle \mathscr{W}, \mp, \bar{\times} \rangle$.

Homomorphism and isomorphism correspond to the intuitive concept of similarity.

We say that a function h is a *homomorphism* mapping a domain $\mathfrak{A} = \langle Z, F, G, S \rangle$ into a domain $\mathfrak{B} = \langle Z', F', G', S' \rangle$ (in symbols: $\mathfrak{A} \, \text{hom}_h \, \mathfrak{B}$), if and only if

1. h is a function defined on the set Z with values in Z',

2. h satisfies the following conditions:

(80)
$$\text{(i) } h(F(x)) = F'(h(x)) \text{ for any } x \in Z,$$
$$\text{(ii) } h(G(x, y)) = G'(h(x), h(y)) \text{ for any } x, y \in Z,$$
$$\text{(iii) } x'S'y' \equiv \bigvee x, y \in Z(x' = h(x) \wedge y' = h(y) \wedge xSy)$$
for any $x', y' \in Z'$.

If the values of the function h cover the whole set Z', then we say that h is a *homomorphism* of the domain $\mathfrak{A}$ onto the domain $\mathfrak{B}$.

If h is a homomorphism of a domain $\mathfrak{A}$ onto a domain $\mathfrak{B}$ and if h is a one-to-one function, then we say that it is an *isomorphism*, or that it determines an isomorphism between domains $\mathfrak{A}$ and $\mathfrak{B}$ (in symbols: $\mathfrak{A}\,\text{ism}_h\,\mathfrak{B}$). When two domains are isomorphic, then we say, too, that the functions (relations) F, G, S are isomorphic with F', G', S', respectively.

Conditions (80) can be formulated in words as follows: with the mapping h of the domain $\mathfrak{A}$ into the domain $\mathfrak{B}$ the operations F, G, S on elements of the set Z of the domain $\mathfrak{A}$ have analogues in the operations F', G', S' on those elements of the set Z' of the domain $\mathfrak{B}$ which are associated with the elements of Z by the function h. The study of the homorphism of relations which are not functions does not yield interesting results. That is why in practice homomorphism is defined only by conditions (i) and (ii), which apply to functions. On the other hand, isomorphisms of relations that are not functions are being studied as often as are isomorphisms of functions. But if a function h is a one-to-one mapping of the set Z onto the set Z', that is, if h is an isomorphism, then condition (iii) is equivalent to something much simpler:

$$\text{(iii')} \qquad xSy \equiv h(x)\,S'h(y) \quad \text{for any } x, y \in Z,$$

which states that the relation S holds between elements of Z if and only if the relation S' holds between their images in Z'. The relation S is thus mapped to the relation S'.

EXAMPLES

1. Every domain is isomorphic with itself. The function which establishes such an isomorphism is, for instance, the identity function: $h(x) = x$.

2. The domain $\langle \mathcal{N}^*, +, \cdot \rangle$ is isomorphic with the domain $\langle \mathcal{N}, +, \cdot \rangle$. The function which establishes such an isomorphism is the mapping $\leftrightarrow$ from p. 46. Compare formulae (54), (55), and (56) with formula (80).

3. The domain of integers $\langle \mathscr{C}, +, \cdot \rangle$ is isomorphic with the domain $\langle \mathscr{C}^*, \mp, \bar{\times} \rangle$, cf. formula (79). This isomorphism is established by the relation $h(p) = [\langle p, 1 \rangle]_{\text{sts}}$.

4. The function $h(x) = [x]_{\text{sts}}$ establishes a homomorphism of the domain $\langle \mathcal{U}, \oplus, \otimes \rangle$ onto the domain $\langle \mathcal{W}, \mp, \bar{\times} \rangle$.

This is a special case of the following general theorem:

THEOREM 1 (on homomorphism). *If R is a congruence relation on a domain $\mathfrak{A}$, then the function $h(x) = [x]_R$ establishes a homomorphism of the domain $\mathfrak{A}$ onto the quotient domain $\mathfrak{A}/R$.*

The proof follows instantly from the definitions. It suffices to compare formulae (75) with formulae (80). For relations we obtain condition (iii′), which is stronger than (iii).

A theorem which is in a sense converse to Theorem 1 is the following:

THEOREM 2 (on homomorphism). *If a function h establishes a homomorphism of a domain $\mathfrak{A}$ onto a domain $\mathfrak{B}$, and if the relations S and S', which are not functions, satisfy condition (iii′), then the relation*

(81) $\qquad x R y \equiv h(x) = h(y)$

is a congruence relation in the domain $\mathfrak{A}$ and the domain $\mathfrak{B}$ is isomorphic with the quotient domain $\mathfrak{A}/R$.

PROOF. It follows directly from definition (81) and from the fact that the identity relation is an equivalence relation, that R is an equivalence. Suppose that $\mathfrak{A} = \langle Z, F, G, S \rangle$ and $\mathfrak{B} = \langle Z', F', G', S' \rangle$. To prove that R is a congruence relation in $\mathfrak{A}$, it suffices to demonstrate that (73) and (74) hold. The relation R in fact preserves the operation F: if $x R y$, then, by (81), $h(x) = h(y)$, and hence $F'(h(x)) = F'(h(y))$. The first part of formula (80) yields that $h(F(x)) = h(F(y))$, which in view of (81) gives $F(x) R F(y)$. We have thus proved that if $x R y$, then $F(x) R F(y)$, which is the first part of formula (73). The second part of (73) is obtained in a similar way: if $x R y$ and $z R v$, then $h(x) = h(y)$ and $h(z) = h(v)$ by (81), and hence $G'(h(x), h(z)) = G'(h(y), h(v))$. It follows from (80) that $h(G(x, z)) = h(G(y, v))$, i.e., $G(x, z) R G(y, v)$. Formula (74) is obtained in an analogous way: if $x R y$ and $z R v$, then $h(x) = h(y)$ and

$h(z) = h(v)$. Suppose now that $x S z$. Then, by (80), $h(x) S' h(z)$, and hence $h(y) S' h(v)$, since $h(x) = h(y)$ and $h(z) = h(v)$. By (iii'), $y S v$. Conversely, if $y S v$, then by (80), $h(y) S' h(v)$. Thus $h(x) S' h(z)$ holds, and again, by (iii'), $x S z$. Thus the condition $(x S z \equiv y S v)$ holds if and only if $x R y$ and $z R v$.

The relation R is therefore a congruence relation. We shall form the quotient domain $\langle \mathrm{Abs}_R, \bar{F}^R, \bar{G}^R, \bar{S}^R \rangle = \mathfrak{A}/R$, and we shall demonstrate that an isomorphism between $\mathfrak{A}$ and $\mathfrak{A}/R$ is established by the function

(82) $\qquad \chi(x) = O_h^{-1}(\{x\})$,

defined for $x \in Z'$ and associating with every $x \in Z'$ its inverse image with respect to the function h. By definition (17) of inverse image, formula (82) is equivalent to the following formula:

(83) $\qquad y \in \chi(x) \equiv h(y) = x$

for any $y \in Z$ and $x \in Z'$.

We shall derive from (83) an accessory formula:

(84) $\qquad$ *if* $y_0 \in Z$, *then* $\chi(h(y_0)) = [y_0]_R$.

In fact, on substituting $h(y_0)$ for x in (83), we obtain

$$y \in \chi(h(y_0)) \equiv h(y) = h(y_0).$$

It follows from this equivalence and from (81) that

$$y \in \chi(h(y_0)) \equiv y R y_0.$$

This equivalence, together with (67), yields

$$y \in \chi(h(y_0)) \equiv y \in [y_0]_R.$$

Since this equivalence holds for any $y, y_0 \in Z$, it follows from the axiom of extensionality that $\chi(h(y_0)) = [y_0]_R$.

It follows at once from (84) that the function χ maps the set Z' onto the entire set Abs_R. Namely, if $x \in Z'$, then there is an $y_0 \in Z$ such that $x = h(y_0)$, since the function h by definition maps the set Z onto the entire set Z'. If $x = h(y_0)$, then by (84), $\chi(x) = \chi(h(y_0)) = [y_0]_R$. Hence $\chi(x) \in \mathrm{Abs}_R$, since $[y_0]_R \in \mathrm{Abs}_R$ in view of (68).

Conversely, if $a \in \mathrm{Abs}_R$, i.e., for some $y_0 \in Z$, $a = [y_0]_R$, then, by (84), $a = \chi(h(y_0))$, and hence a is the value of the function χ for the element $h(y_0) \in Z'$.

Next, it can easily be noticed that the function χ is one-to-one. Namely, if $\chi(x) = \chi(z)$ and $x \in Z'$, then by assumption there is an y_0 such that $x = h(y_0)$, since h maps Z onto Z'. Hence, by (83), we have $y_0 \in \chi(x)$. If $\chi(x) = \chi(z)$ then $y_0 \in \chi(z)$, i.e., by (83), $h(y_0) = z$. We have already proved that $x = h(y_0)$, and so the last two formulae yield $x = z$. We have thus proved that if $\chi(x) = \chi(z)$, then $x = z$, which, by (14), means that the function χ is one-to-one.

It remains to prove that the function χ satisfies conditions (80), reformulated for our special case as follows:

$$\chi(F'(x)) = \bar{F}^R(\chi(x)),$$
(85) $$\quad \chi(G'(x, y)) = \bar{G}^R(\chi(x), \chi(y)),$$
$$\chi(x)\,\bar{S}^R\,\chi(y) \equiv x\,S'\,y.$$

In fact, if $y \in \chi(F'(x))$ and $x \in Z'$, then $x = h(y_0)$ for some $y_0 \in Z$, and hence $y \in \chi(F'(h(y_0)))$, which by (83), means that $h(y) = F'(h(y_0))$. But since h is a homomorphism, by (80), $F'(h(y_0)) = h(F(y_0))$. By the identity law we have $h(y) = h(F(y_0))$. By (81), $y\,R\,F(y_0)$, and by (67) $y \in [F(y_0)]_R$. Hence $y \in F^R([y_0]_R)$, by (75). Again by (81), $y \in \bar{F}^R(\chi(h(y_0)))$. But $h(y_0) = x$, and hence $y \in \bar{F}^R(\chi(x))$.

Conversely, if $y \in \bar{F}^R(\chi(x))$ and $x \in Z'$, then, for some $y_0 \in Z$, $x = h(y_0)$ and $y \in \bar{F}^R(\chi(h(y_0)))$ hence, by (84), $y \in \bar{F}^R([y_0]_R)$, so that by (75), $y \in [F(y_0)]_R$. This means that $y\,R\,F(y_0)$, hence, $h(y) = h(F(y_0))$. But, by (80), we have $h(F(y_0)) = F'(h(y_0))$, hence $h(y) = F'(h(y_0))$, so that by (83), $y \in \chi(F'(h(y_0)))$. But $h(y_0) = x$, hence $y \in \chi(F'(x))$.

We have just proved that if $y \in \chi(F'(x))$, then $y \in \bar{F}^R(\chi(x))$, and conversely, if $y \in \bar{F}^R(\chi(x))$, then $y \in \chi(F'(x))$. In view of the laws of equivalence already known, this means that $y \in \chi(F'(x)) \equiv y \in \bar{F}^R(\chi(x))$ for any y and $x \in Z'$. By the axiom of extensionality $\chi(F'(x)) = \bar{F}^R(\chi(x))$ for any $x \in Z'$. In this way we have proved the first of the formulae (85). The second and the third formula can be proved in an analogous way by reference to the corresponding original formulae.

Theorems 1 and 2 on homomorphism show that there is an intimate relation between a homomorphism of two domains and a quotient domain. We shall list without proof some other simple properties of homomorphisms and isomorphisms, using the abbreviations introduced above:

If $\mathfrak{A} \text{hom}_h \mathfrak{B}$ and $\mathfrak{B} \text{hom}_g \mathfrak{C}$, then $\mathfrak{A} \text{hom}_{gh} \mathfrak{C}$.

If $\mathfrak{A} \text{ism}_h \mathfrak{B}$, then $\mathfrak{B} \text{ism}_{h^{-1}} \mathfrak{A}$.

If $\mathfrak{A} \text{ism}_h \mathfrak{B}$ and $\mathfrak{B} \text{ism}_g \mathfrak{C}$, then $\mathfrak{A} \text{ism}_{gh} \mathfrak{C}$.

An isomorphism is an equivalence relation. A much more important property of isomorphisms, to be proved later, is that if a domain $\mathfrak{A} = \langle Z, F, G, S \rangle$ is isomorphic with a domain $\mathfrak{B} = \langle Z', F', G', S' \rangle$, and if a property which can be formulated in terms of Z, F, G, and S is an attribute of the domain $\mathfrak{A}$, then an analogous property formulated in terms of Z', F', G', and S' is an attribute of $\mathfrak{B}$. For instance if F is a one-to-one function, then so is F'. If S is an equivalence relation, then so is S'. If S is an ordering in Z, then S' is an ordering in Z'.

An isomorphism of two domains is thus a faithful similarity which associates with the operations F and G and the relation S the corresponding operations F' and G' and the relation S'. It is obvious that the definitions and the properties of isomorphism and homomorphism can be expanded so as to cover domains that contain many more functions and relations than three.

The properties of such general concepts as isomorphism, homomorphism, subdomain, and the product of domains fall under that part of set theory which has in fact become a new mathematical discipline called universal algebra. The basic monographs for that discipline are the books of Birkhoff [5] and van der Waerden [102].

EXERCISES

1. Prove in detail that the domains $\langle \mathscr{C}, +, \cdot \rangle$ and $\langle \mathscr{C}^*, \mp, \bar{\times} \rangle$ are isomorphic.

2. Complete the proof of the second theorem on homomorphism by proving the second and third formulae of (85).

3. Prove that the relation of isomorphism is an equivalence relation.

4. Prove that if $\langle X, f, R \rangle \text{ism}_h \langle X', f', R' \rangle$, then:

    a. If R is a single-valued relation in X, then R' is single-valued in X';

    b. If R is an ordering in X, then R' is an ordering in X';

    c. If R is an equivalence relation, then so is R';

    d. If f is a one-to-one function, then so is f'.

11. PRODUCTS. REAL NUMBERS

The set of all ordered pairs $\langle x, y \rangle$ where x, y are elements of a set X, was denoted by X^2. By refering to the concept of ordered pair it is easy to define an ordered triple. An ordered triple can be constructed for instance, in the following way:

$$\langle x, y, z \rangle = \langle x, \langle y, z \rangle \rangle$$

where the second member of the ordered pair is itself an ordered pair. Similarly, we can define ordered quadruples, quintuples, and arbitrary ordered n-tuples:

$$\langle x_1, x_2, ..., x_n \rangle = \langle x_1, \langle x_2, \langle ..., x_n \rangle ... \rangle \rangle.$$

The set of all triples whose elements are in a set X is denoted by X^3. In general, the set of all ordered n-tuples with elements in a set X is denoted by X^n. With every ordered pair $\langle x, y \rangle$ we can associate a certain unique function f, defined on a two-element set, for instance $\{1, 2\}$, such that $f(1) = x$ and $f(2) = y$. This correspondence is one-to-one. The set X^2 may therefore be considered identical with the set $X^{\{1, 2\}}$. Likewise, the set X^3 may be identified with the set $X^{\{1, 2, 3\}}$ in full agreement with formula (41), since X^Y has been chosen to denote the set of all functions defined on Y and with values in X. The sets $X^2, X^3, X^4, ...$ are called *finite products* of the set X. These sets consist of finite sequences of elements of a set X. The sets which consist of infinite sequences of elements of a set X are called *infinite products* of that set X. There are many kinds of infinite products. The simplest infinite product of a set X is the set of all countable sequences $\{a_n\}$ whose terms belong to the set X. Countable sequences, $\langle a_0, a_1, a_2, a_3, ... \rangle$, can be considered functions defined on the set $\mathcal{N}$ of natural numbers such that $f(0) = a_0$, $f(1) = a_1, ..., f(n) = a_n, ..., n \in \mathcal{N}$. By Definition (41), the set of all countable sequences whose terms belong to the set $\mathcal{N}$ is denoted by $X^{\mathcal{N}}$ which is in agreement with the notation adopted for finite products. In the most general form, the *product* of a set X with indices ranging over a set Y is the set of all functions defined on Y and with values in X. According to the notation adopted above that product is denoted by X^Y.

The formation of products is often used in constructing new mathematical domains. If $\mathcal{R}$ denotes the set of real numbers, then $\mathcal{R}^2$ may stand for the plane, since with each pair $\langle x, y \rangle$ of real numbers we can associate a point on the plane with the co-ordinates x and y. Likewise, $\mathscr{R}^3$ may be considered the three dimensional space, and $\mathscr{R}^n$ may be considered the n-dimensional Euclidean space. The formation of products thus serves as a basic tool in constructing new geometric domains.

The real numbers, $\mathscr{R}$, themselves can be obtained from rational numbers by the formation of products and abstraction classes, but no finite products suffice for that purpose. It is only by resorting to the infinite product $\mathscr{W}^{\mathscr{N}}$, that we can define real numbers. This can be done in many ways. One of them, to be sketched below, is based on the observation that every real number a is a limit of a sequence $\{a_n\}$ of rational numbers. We might therefore identify real numbers with sequences of rational numbers converging to them. To make this idea concrete, we begin with the domain $\langle \mathscr{W}, \mp, \bar{\times} \rangle$ of rational numbers, and we define the domain $\langle \mathscr{W}^{\mathscr{N}}, +', \times' \rangle$, which consists of sequences of rational numbers, with operations defined as follows:

$$\{a_n\} +' \{b_n\} = \{a_n \mp b_n\},$$
$$\{a_n\} \times' \{b_n\} = \{a_n \bar{\times} b_n\}.$$

The operations on sequences are therefore defined by operations on terms of these sequences. Next, we distinguish in the domain $\langle \mathscr{W}^{\mathscr{N}}, +', \times' \rangle$ the *subdomain of convergent sequences* $\langle \text{Conv}, +', \times' \rangle$. The set Conv of convergent sequences can be defined thus:

$$\{a_n\} \in \text{Conv} \equiv \bigwedge k \in \mathscr{N} \left[\bigvee m \in \mathscr{N} \bigwedge n \{ n > m \rightarrow \right.$$
$$\left. \left(|a_n - a_m| < 1/(k+1) \right) \} \right].$$

The operations in this domain are the same as in the former domain. It would, however, be difficult to consider every convergent sequence to be a real number, because for any real number there is an infinity of various rational sequences converging to that number. It is, however, possible to combine all the sequences that are convergent to the same real number, into a single set and to consider that set as standing for that real number. We accordingly define the relation of *co-convergence* (convergence to the same limit) between sequences

$$\{\dot{a}_n\} \text{ coconv } \{b_n\} \equiv \bigwedge k \in \mathscr{N} \left[\bigvee m \in \mathscr{N} \bigwedge n \{ n > m \rightarrow \right.$$
$$\left. \left(|a_n - b_n| < 1/(k+1) \right) \} \right].$$

By this definition, two sequences of rational numbers are *co-convergent* if they converge to the same limit, that is, intuitively, if the same real number is their limit.

It can be proved that the relation coconv is a congruence relation in the domain $\langle \text{Conv}, +', \times' \rangle$. It seems, therefore, natural to consider the abstraction classes of the relation coconv to stand for real numbers. In this way the field of real numbers may be defined as the quotient domain

$$\langle \text{Conv}, +', \times' \rangle / \text{coconv}.$$

Other kinds of numbers, such as complex numbers, or quaternions, can be defined as easily by analogous constructions.

Thus, from the premises of set theory, by assuming that there exists at least one infinite set, it is possible to define exact analogues of all concepts of classical mathematics. This fact was discovered at the end of the 19th century, and it was rigorously proved in the first quarter of the 20th century.

This fact seems to be interesting and in agreement with the intuitions of philosophers who had been claiming since antiquity that quantitative relationships in the universe can be deduced from a general vision of reality. Such a completely general theory of reality, which agrees with formal philosophical ontologies of the past, is provided precisely by logic and set theory.

EXERCISES

1. Define integers in the following way:

a. In the product $\mathcal{N}^2$ we assume that the pair $\langle m, n \rangle$ of natural numbers stands for the integers $m - n$, so that it stands for a positive number, if $m > n$, and for a negative number, if $n < m$. Of course, one and the same integer, for instance -1, can be represented by infinitely many pairs $\langle 0, 1 \rangle$, $\langle 1, 2 \rangle$, $\langle 2, 3 \rangle$, ...

b. Define the operations $\mp$ and $\bar{\times}$ on pairs so that they correspond to the operations on integers.

c. Define a congruence relation R as follows: $\langle p, q \rangle R \langle r, s \rangle \equiv$ the pair $\langle p, q \rangle$ stands for the same integer as the pair $\langle r, s \rangle$.

d. Define the quotient domain and prove that it is isomorphic with the domain $\langle \mathcal{C}, +, \cdot \rangle$, defined in Section 8.

2. Prove rigorously that the relation coconv is a congruence relation in the domain $\langle \mathcal{W}, +', \times' \rangle$.

3. In the domain of real numbers as defined in this Section, distinguish a subdomain isomorphic with the domain of rational numbers, and a subdomain isomorphic with the domain of integers.

Chapter I

THE CLASSICAL LOGICAL CALCULUS

Logic selects inference methods used in science and considered to be correct, and then constructs logical systems on them. These systems are sets of various laws and rules which, when followed, yield all the inferences we can immediately accept. There is perfect agreement in mathematics as to which methods of inference are acceptable and which are not. It does not happen that some people consider some rules of inference to be correct, while others consider them to be entirely wrong. There are only slight differences of opinion about the meaning of some rules and about the extent of their applicability. The set of almost all the methods of inference that are used in mathematics is called *classical logic*.

The present chapter will be concerned with an exposition of that system, its applications, and some syntactic research on it. Classical logic can best be described by first dividing it into two logical calculi: the classical sentential calculus and the classical predicate calculus.

1. THE CLASSICAL CHARACTERISTICS OF THE SENTENTIAL CONNECTIVES

The *sentential calculus*, the first part of classical logic, is to a great extent the theory of sentential connectives. The *sentential connectives* are those expressions which combine two sentences into one compound sentence. For instance, each of the expressions:

> ... *and* ...,
>
> ... *or* ...,
>
> *if* ... *then* ...,
>
> ... *if and only if* ...,

is a sentential connective, since if each string of dots is replaced by a sentence, each expression will yield a compound sentence. These expressions thus define certain two-argument operations on sentences, which yield new sentences. A similar function is performed by the expression:

it is not true that ...,

which defines a one-argument function on sentences, and which has new sentences as its values. This is why in logic this expression too is called a sentential connective.

The compound sentence: it is not true that p (in symbols: $\sim p$) is called the *negation* of the sentence p. The compound sentence: p and q (in symbols: $p \wedge q$) is called the *conjunction* of the sentence p and the sentence q. The sentence: p or q (in symbols: $p \vee q$) is called the *disjunction* of the sentences p and q. The sentence: if p, then q (in symbols: $p \rightarrow q$) is called the *implication* with the antecedent p and the consequent q. The sentence: p if and only if q (in symbols: $p \equiv q$) is called the *equivalence* between the elements p and q. The role of these sentential connectives in the formulation of mathematical theorems has been partially evident in the Introduction.

The classical sentential calculus can best be approached if we try to characterize the sentential connectives by reference to *truth* and *falsehood*. The concepts of truth and falsehood are commonly understood. The classical definition of truth is that a sentence is true if it describes a state of affairs which indeed takes place. A sentence is false if it express something in disagreement with reality. The sentences: *Warsaw lies on the Vistula* and $2+2 = 4$, are true since they describe real situations. On the other hand, the sentences: *Warsaw lies on the river Odra* and $2+2 = 7$, are false, since the actual state of things is different from what they state.

This concept of truth and falsehood has the characteristic property that we hold every sentence which states a uniquely determined thought to be either true or false, regardless of whether it concerns the past, the present, or the future, and whether the subject matter is known or unknown to us. This belief is formulated as the *principle of two-valued propositions*. Every proposition, every sentence that is unambiguous, can, according to this principle, assume only one of the two logical values: either the value truth, or the value falsehood, and no sentence can at the same time assume both values.

From this viewpoint the role of the connective *it is not true that* ...

can be characterized as follows. Consider a sentence P whose content is unambiguous. Suppose this sentence is: *Warsaw lies on the Vistula*, or any other sentence. If the sentence P is true, then its negation, the sentence: *It is not true that P* (*It is not true that Warsaw lies on the Vistula*), is false. Conversely, if P were a false sentence, e.g. *Warsaw lies on the Odra*, then its negation: It is not true that P (*It is not true that Warsaw lies on the Odra*), would be a true sentence. Thus the negation phrase *it is not true that* ... turns a true sentence into a false one, and a false sentence into a true one.

That relation can be expressed symbolically:

(1)
    1. *If p is true, then (not-p) is false,*
    2. *If p is false, then (not-p) is true.*

For simplicity's sake we adopt once and for all the following convention:

    Instead of *p is true* we shall use $p = 1$,

    instead of *p is false* we shall use $p = 0$.

Under this convention the sentences 1 and 2 of (1) can be written briefly:

(2)
    $1'$. *If p = 1, then (not-p) = 0,*
    $2'$. *If p = 0, then (not-p) = 1.*

There is still another way to put it. We can use a table as below.

p	$not\text{-}p$
1	0
0	1

The column headed by p has the symbols which stand for the values which the sentence p may assume. Since p is an arbitrary sentence, it may be true, but it may also be false, and we must therefore write in the first column both the symbol for truth and the symbol for falsehood. In the column headed by *not-p*, we write the symbols of the values assumed by the negation of the sentence p, the values assumed by *not-p*. These values correspond to the values assumed by the sentence p in the same row. In this way the table shows very clearly what has been expressed by (1) and (2).

All the remaining logical connectives can be characterized in a similar manner. This will be done mainly by analysing examples drawn from common usage. In ordinary speech the logical connectives have the same meaning as in mathematics. There is no significant difference between the mathematical language and the ordinary one, except for the fact that mathematics makes use of more complicated concepts. It is advisable to bear in mind that there is a close relationship between mathematical logic on the one hand and the logic of other sciences and of all other honest extra-scientific thinking on the other hand.

Let us begin with conjunction. Consider under what conditions we should accept the following sentence to be a true sentence:

It will be freezing tomorrow and it will be snowing tomorrow.

We can certainly accept this sentence as true if what it states really does happen, that is, if it is freezing tomorrow and it is snowing tomorrow. In other words, we can consider this sentence true if both the first part *it will be freezing tomorrow* and the second part *it will be snowing tomorrow* of this conjunction prove to be true sentences. But if it is snowing, and the snow is melting because it is not freezing, that is if the first part of the conjunction proves to be false, then the whole conjunction is false, even though its second part is true. Likewise, the conjunction is considered false, if it is freezing tomorrow but it is not snowing. In that case the second part of the conjunction is not satisfied. If neither the first nor the second part is satisfied, then the whole conjunction obviously does not describe the facts and we consider it false. The above example shows that

(3)   *A conjunction is considered true only if both its parts are considered true.*

An analysis of any other example will give the same result.

It is interesting to ask about the converse. Is it always the case that when both parts of a conjunction are true, then the conjunction itself is considered true?

If a person accepts that *Warsaw lies on the Vistula* is true, and he accepts that *Paris lies on the Seine* is true, then he will accept that *War-*

saw lies on the Vistula and Paris lies on the Seine is true. If he accepts that $2 \times 2 = 4$, too, then he will accept as true the compound sentence *Warsaw lies on the Vistula and* $2 \times 2 = 4$. It seems that every one could agree that a conjunction of two sentences that are true is true, even if it appears that there is no relation whatever between these sentences. If we accept each of these sentences separately, we accept them when taken together. One may add that the sentence *Warsaw lies on the Vistula and* $2 \times 2 = 4$, true as it is, does not lead to any interesting conclusion, but that is another question. It may be a conjunction that serves no purpose, still it is true. The sentence *Warsaw lies on the Vistula and Paris lies on the Seine* may lead to an interesting conclusion, even if it is only in the listing of capitals that lie on rivers. That sentence may have some application in geography, whereas the sentence *Warsaw lies on the Vistula and* $2 \times 2 = 4$ can find application neither in geography nor in mathematics. Thus, conjunctions of arbitrary true sentences are accepted as true, even though many of them are useless both in science and in everyday usage. It may, therefore, be said that the converse implication to (3) is true:

(4)   *If both parts of a conjunction are considered true sentences, then the whole conjunction is considered a true sentence.*

To summarize what has just been said about conjunction, it is enough to combine sentences (3) and (4) into one sentence: *a conjunction is considered true if and only if both its parts are true, and it is considered false if and only if at least one of its parts is false.* A sentence *p and q* is true if and only if both sentences *p* and *q* are true. In each of the three remaining cases the sentence *p and q* is false.

Using the notation adopted, we can express the above in the following way:

(5)
1. *If* $p = 1$ *and* $q = 1$, *then* $(p \text{ and } q) = 1$,
2. *If* $p = 1$ *and* $q = 0$, *then* $(p \text{ and } q) = 0$,
3. *If* $p = 0$ *and* $q = 1$, *then* $(p \text{ and } q) = 0$,
4. *If* $p = 0$ *and* $q = 0$, *then* $(p \text{ and } q) = 0$.

Sentences 1–4 characterize conjunctions much in the same way as the sentences of (2) characterize negation. The sentences of (5) can be expressed by the following table:

p	q	p and q
1	1	1
1	0	0
0	1	0
0	0	0

The column headed by "p" and the column headed by "q" state the values which these sentences may respectively assume. The column headed by "p and q" states the values which the conjunction assumes for the values of p and q stated in a given row.

To characterize disjunction, we shall proceed in a similar fashion. Let us imagene when the sentence

Columbus was in India or in Egypt

can be considered true. It seems that everyone will accept this sentence as true if it turns out that Columbus was in India but was not in Egypt, and also if it turns out that he was not in India, but he was in Egypt, and also, if it turns out that he was both in India and in Egypt. On the other hand, no one will accept this sentence as true if it turns out that Columbus was neither in India nor in Egypt. Likewise, the sentence:

(6)   *In the summer we will go to the mountains or we will go to the shore*

will be accepted as true only if we either go to the mountains and do not go to the shore, or if we do not go to the mountains but do go to the shore, or if we go to both. In each of these three cases the disjunction is considered true. It is considered false only in the case of the last remaining possibility, that is, if we go neither to the mountains nor to the shore.

We can now lay down the following general rule:

(7)   *A disjunction is considered true only if at least one of its parts is true.*

Again, we inquire about the converse. Is it always the case that a disjunction is considered true provided that at least one of its parts is true?

For instance, are the following disjunctions true:

(8)   *Warsaw lies on the Vistula or $2 \times 2 = 5$.*
       Cats are usually sly or $2 \times 2 = 4$.

The sentences of (8) and other similar sentences make sense, and we would like to say whether they are true or false. Are we to consider them true, or false? Everyone will agree that there is no reason to reject these sentences as false. They look harmless. If someone hesitates to accept such sentences as true, it is mainly because there is no need in everyday practice and in science to make use of disjunctions whose parts are unrelated as to their meaning. In fact, sentences like these just quoted are never used to any serious purpose.

Moreover, in ordinary language we use a disjunction only if we do not know which of its parts is true. For instance, if we are sure that we will go to the seaside in the summer, and we want to tell someone what our plans are, then we do not use disjunction (6), but we just use that sentence which is true: *we go to the seaside in summer*. The mere act of stating something plays a social role. In certain milieus, there is a custom of telling the whole known truth. A disjunction always suggests then that it is not known which part of that alternation is true. In the case of the sentences of (8) everyone knows which part of the disjunction is true. When someone uses such disjunctions, his intention is not clear for most people. Assuming that they make sense, the problem is what to do with such sentences? In logic we consider them true. We decide that no matter whether the parts of a disjunction have something or nothing in common as to their meaning, and no matter for what purpose and under what conditions a disjunction is used,

(9)   *If at least one part of a disjunction is true, then the whole disjunction is true.*

The sentences of (8) are therefore considered true, since each of them has one part which is true.

The result of our discussion on disjunction can be stated in one sentence which has the form of an equivalence:

(10)   *A disjunction is true if and only if at least one of its parts is true.*

This equivalence is composed of implications (7) and (9). Using abbreviations that have already been adopted, we can express (10) also as follows:

(11)
1. *If p = 1 and q = 1, then (p or q) = 1,*
2. *If p = 1 and q = 0, then (p or q) = 1,*
3. *If p = 0 and q = 1, then (p or q) = 1,*
4. *If p = 0 and q = 0, then (p or q) = 0,*

or else, we can use a table:

p	q	$p \ or \ q$
1	1	1
1	0	1
0	1	1
0	0	0

We shall ask two analogous questions about implication.

1. What logical values are assumed by the antecedent and the consequent of an implication if this implication is considered true?

2. If the antecedent and the consequent have those logical values, is it always so that the whole implication is considered true?

To be able to answer the first of these questions correctly, we shall consider two examples, taken from everyday life and from mathematics. Let us consider first a promise subject to a certain condition:

(12)   *If you improve your output, you will get a bonus.*

What may happen? Suppose you improve your output and indeed get a bonus. In such a case you undoubtedly think that your boss has kept his promise and that the implication he expressed has turned out to be true. If you improve your output, and get no bonus, you may certainly claim that your boss has not kept his word. His promise has turned out to be false. If you show no improvement and get no bonus, you can claim nothing. You did not satisfy the condition your boss set. He gave you no bonus, but he kept his word, nevertheless. He kept his word by just doing nothing, because in that case he was not obliged to do any-thing. However, it seems that even if you do not improve your output, your boss can give you a bonus if he feels like it. He did not say that he would give you a bonus only if you improved your output. He only said that if you did improve your output, you would get it for sure, but perhaps it is possible that you might get it in some other circumstances, too. Your boss can give you a bonus without contradicting himself, even if you did not improve your output. The statement expressing the decision of your boss is true even if you do not improve your output, but still get a bonus. We therefore conclude that the implication your boss used is false only if you improved your output and did not get the bonus, that is, if its antecedent *improve your output* is true and the consequent *you will get a bonus* is false.

Consider now a mathematical example. The implication:

(13)   *If a number x is divisible by 6, then that number x is divisible by 2,*

is true. Substitute now for *x* arbitrary numbers, for instance 4, 12, and 17. This is legitimate, since if we accept an implication in its general formulation, then we accept any of its special cases, too. We thus obtain the following true sentences:

(14)   *If 4 is divisible by 6, then 4 is divisible by 2.*

(15)   *If 12 is divisible by 6, then 12 is divisible by 2.*

(16)   *If 17 is divisible by 6, then 17 is divisible by 2.*

In (14) the antecedent, that is, the sentence *4 is divisble by 6*, is a false sentence, and the consequent, that is, the sentence *4 is divisible by 2*, is a true sentence. An implication with a false antecedent and a true consequent thus proves to be true. In (15) the antecedent, *12 is divisible by 6*, is a true sentence. The consequent, *12 is divisible by 2*, it true, too. Thus an implication with a true antecedent and a true consequent is true. In (16) the antecedent, *17 is divisible by 6*, is false, and the conse-quent, *17 is divisible by 2*, is false, too. Thus an implication with a false antecedent and a false consequent proves to be true, also. When would the above implication be false? If there were a number *x* such that *x* were divisible by 6 and were not be divisible by 2. If we should find such a number *x*, we would reject the general statement at once. Thus, we would consider an implication to be false only if it turned out that its antecedent is true and its consequent is false. Since, however, there is no number x such for which the sentence x *is divisible by 6* would be true, and the sentence x *is divisible by 2* would be false, the implication (13) is always true. This mathematical example also shows that an implication is false if its antecedent is true and its consequent is false. This yields the answer to the first question:

(17)   *An implication is accepted as true only if either both parts of the implication are true, or if both are false, or if the antecedent is false and the consequent is true. If the antecedent is true and the consequent is false, then that implication is considered to be false.*

This still leaves the other question: Is an implication always accepted as true if its two parts are both true or both false, or if the antecedent is false and the consequent is true? The situation here resembles that which we encountered when analysing sentential connectives above. In everyday life and in science, inferences make use only of such implications in which the antecedent and the consequent are related in meaning, though that relationship is difficult to define more precisely. Those implications in which there is no such relationship are usually disregarded in inferences. This is why we do not use them and we find it difficult to have a clear attitude toward them. Hence, we may choose any attitude we wish. It turns out that logical considerations do not contradict assuming that, if the antecedent and the consequent of an implication are arbitrary meaningful sentences, and if they are either both true, or both false, or if the antecedent is false and the consequent is true, then the implication is true regardless of the meaning of its component parts and regardless of the situation in which it is formulated. In this case, too, as in the case of the sentential connectives discussed above, logical practice shows that such a convention does not result in any absurd consequences, and even simplifies the properties of the sentential connective under consideration and makes the connective *if ... then ...* a good instrument of logical research, especially in mathematics. Thus, implication may be characterized as follows:

(18)   *An implication is false if and only if its antecedent is true and its consequent is false.*

74

Accordingly, the following implications are true:

> *If* $2 \times 2 = 4$, *then Warsaw lies on the Vistula,*
> *If Warsaw lies on the Seine, then* $2 \times 2 = 4$,
> *If Warsaw lies on the Seine, then* $2 \times 2 = 5$,

even though they are of no practical importance whatever.

On the other hand, the following implications are false:

> *If* $2 \times 2 = 4$, *then Warsaw lies on the Seine,*
> *If Warsaw lies on the Vistula, then* $2 \times 2 = 5$.

The above properties of implication can also be recorded in the form of a table (in which, for brevity's sake, the expression *if ...*, *then ...* will be replaced by $p \rightarrow q$):

(19)
1. *If* $p = 1$ *and* $q = 1$, *then* $(p \rightarrow q) = 1$,
2. *If* $p = 1$ *and* $q = 0$, *then* $(p \rightarrow q) = 0$,
3. *If* $p = 0$ *and* $q = 1$, *then* $(p \rightarrow q) = 1$,
4. *If* $p = 0$ *and* $q = 0$, *then* $(p \rightarrow q) = 1$.

p	q	$p \rightarrow q$
1	1	1
1	0	0
0	1	1
0	0	1

Like other disciplines, logic draws its concepts from everyday language and then imparts them with strictly defined meanings. This imparting with meaning is always conventional to a large extent, since in everyday language concepts do not have strictly defined meanings; they are always ambiguous and vague, whereas in science we need concepts that are unambiguous and much less vague.

An similar analysis of equivalence leads to the conclusion that:

(20)   *An equivalence is true if and only if its two parts are either both true or both false.*

On replacing the expression *p if and only if q* by the formula $p \equiv q$ we can characterize equivalence in a semi-symbolic notation as follows:

If $p = 1$ and $q = 1$, then $(p \equiv q) = 1$,
If $p = 1$ and $q = 0$, then $(p \equiv q) = 0$,
If $p = 0$ and $q = 1$, then $(p \equiv q) = 0$,
If $p = 0$ and $q = 0$, then $(p \equiv q) = 1$.

The precise formulation of the meanings of the sentential connectives, given in this section, is the foundation of the classical logical calculus. We must mention that there are also other interpretations of sentential connectives, which yield other, nonclassical logical calculi. Yet the classical interpretation is the most frequent and has won almost universal acceptance in practice.

EXERCISES

1. Analyse sentences of the type *either p or q*. Make a table for the sentential connective "either...or...".

2. Using the tables for conjunction, negation, and disjunction make tables for compound expressions with the following schemata:

$$\sim (p \lor q), \quad \sim (p \land q), \quad (p \lor q) \land \sim (p \land q),$$
$$\sim p \land \sim q, \quad p \lor \sim p, \quad \sim p \to q, \quad \sim p \lor q,$$
$$\sim (\sim p \land \sim q), \quad \sim (\sim p \lor \sim q), \quad p \to (q \to p).$$

Note which schemata have the same tables.

3. Since we use only two logical values, there are 16 different possible tables for sentential connectives with two sentential arguments. Using the symbols of negation and disjunction construct schemata corresponding to those 16 tables.

4. Construct 16 analogous schemata using only the symbols of negation and implication, and then using only the symbols of negation and conjunction.

5. Note that the expression *neither p nor q*, defined as follows:

$$neither\ p\ nor\ q \equiv (\sim p \land \sim q),$$

alone suffices to define all 16 sentential connectives.

6. There is another expression which has the same property as the expression *neither p nor q*. Formulate it and prove that there are no more sentential connectives with that property.

H i n t. Consider how negation could be defined.

2. TAUTOLOGIES IN THE CLASSICAL SENTENTIAL CALCULUS AND THEIR APPLICATIONS TO CERTAIN MATHEMATICAL CONSIDERATIONS

Sentential connectives can be used to describe the various schemata of compound sentences. For instance, the formula:

(21) $p \to (q \lor r)$

is a schema of the arithmetical theorem:

If $(a \cdot b = 0)$, then $((a = 0)$ or $(b = 0))$.

The same formula (21) also is a schema of an arithmetically false sentence, for instance,

If $(a \cdot b = 10)$, then $((a = 7)$ or $(b = 13))$.

The formula

(22) $(p \lor q) \to (r \land s)$

is a schema of the sentence

If [(it rains tomorrow) or (it is cold tomorrow)], then [(tomorrow I will stay at home) and (tomorrow I will read books)

which may prove to be sometimes true, and sometimes false.

An important role in correct inference is played by schemata of compound sentences which are such that every sentence falling under them is always a true compound sentence. Such schemata are called *tautologies* in the classical sentential calculus, or just *theorems* (*laws*) of the c.s.c. (c.s.c. = the classical sentential calculus).

DEFINITION 1. *A schema* 𝔄 *is a tautology in the c.s.c. if and only if every sentence falling under the schema* 𝔄 *is true.*

The schema

(23) $p \lor \sim p$,

called the *law of excluded middle*, may serve as an example of a tautology in the c.s.c. The following sentences are examples of sentences falling under that schema:

[*There is somebody in the next room*] or [*it is not true that (there is somebody in the next room*)],

[*Columbus was in India*] or [*it is not true that* (*Columbus was in India*)],

x is a rational number or *x is not a rational number*,

$a = 0$ or $a \neq 0$.

These sentences seem to be true. Likewise, every other sentence that falls under the schema (23) seems to be true in an obvious manner. Using the tables of disjunction and negation from the preceding section and referring to the principle of two logical values one can even prove in a general way that every sentence satisfying schema (23) is a true compound sentence. Suppose that p is a true sentence: $p = 1$; then, even though $\sim p$ is false: $\sim p = 0$, disjunction (23) is nevertheless true, in accordance with the truth table for disjunction, since one of its parts is true. When p is false: $p = 0$, then $\sim p$ is the true part of disjunction (23). Since there are only these two possibilities: $p = 0$ or $p = 1$, in either case one part of disjunction (23) is true. Thus, the entire disjunction (23) is always true.

The law of excluded middle is frequently used when a number of mutually exclusive possibilities are to be examined. For instance, when we prove that real numbers have a certain property, we make separate proofs for those x's which are rational, and separate proof for those x's which are irrational, and next, referring to the law of excluded middle we infer that every real number has the property in question.

The schema

(24) $(p \rightarrow q) \rightarrow (\sim q \rightarrow \sim p)$,

called the *law of transposition*, also is a tautology. It is also called the *law of contraposition* and is an analogue of the traditional *modus tollendo tollens*. This schema covers such sentences as:

If [*if* (*x is divisible by* 6), *then* (*x is divisible by* 3)], *then* [*if* (*x is not divisible by* 3), *then* (*x is not divisible by* 6)].

If it is true that [*if* (*the reviewer suggests that the book be published*), *then* (*the book will be published*)], *then* [*if* (*the book is not published*), *then this means that* (*the reviewer did not suggest that the book be published*)].

We can, thus, see that the compound sentences falling under schema (24) intuitively seem to be true. Using the tables for implication and negation one can prove in a general way, as was the case with schema (23). that every compound sentence falling under schema (24) is true. Since the schema includes two variables, p and q, each of which stands for any true or false sentence, we have to examine four possible cases:

1. $p = 1$ and $q = 1$,
2. $p = 1$ and $q = 0$,
3. $p = 0$ and $q = 1$,
4. $p = 0$ and $q = 0$.

In each of these cases we can, using the tables for implication and negation, compute the value of the entire sentence by substituting values for letters in (24):

1. $(1 \rightarrow 1) \rightarrow (\sim 1 \rightarrow \sim 1) =$
   $(1 \rightarrow 1) \rightarrow (0 \rightarrow 0) = 1 \rightarrow 1 = 1$,
2. $(1 \rightarrow 0) \rightarrow (\sim 0 \rightarrow \sim 1) =$
   $(1 \rightarrow 0) \rightarrow (1 \rightarrow 0) = 0 \rightarrow 0 = 1$,
3. $(0 \rightarrow 1) \rightarrow (\sim 1 \rightarrow \sim 0) =$
   $(0 \rightarrow 1) \rightarrow (0 \rightarrow 1) = 1 \rightarrow 1 = 1$,
4. $(0 \rightarrow 0) \rightarrow (\sim 0 \rightarrow \sim 0) =$
   $(0 \rightarrow 0) \rightarrow (1 \rightarrow 1) = 1 \rightarrow 1 = 1$.

We can, thus, see that in each of the four possible cases the entire compound sentence falling under schema (24) proves to be true. Hence, by Definition 1, schema (24) is a tautology in the sentential calculus.

Schema (24) is used in reasoning such as the following: Having proved that

Every rational number is an algebraic number,

we draw the conclusion that

Every non-algebraic number is an irrational number.

As we can see, tautologies, that is, laws of sentential logic, play an important role in the processes of inference. This is why it is important to know how to tell those schemata which are tautologies from those which are not. It turns out that the method we have used to prove that schemata (23) and (24) are tautologies, is general enough and makes it possible to distinguish tautologies from all other schemata. The following theorem can be proved concerning the tautologies:

THEOREM 1 (on the decidability of the c.s.c.). *There exists an effective method (an algorithm) which makes it possible to verify, in a finite number of steps, whether an arbitrary schema α consisting of sentential connectives, variables, and parentheses, is, or is not, a law of (tautology in) the c.s.c.*

The general meanings of the concept of decidability and of effective method will be made precise in Chapter III.

PROOF. The method mentioned above will be described in general and the proof of the theorem will be outlined first, and then practical hints concerning a simplified application of the method will be given.

The method in question is universally called the *method of zero-one verification*. The way it is used is very simple. In any schema α, every variable stands for a true or a false sentence. Hence in schema α that variable ought to be replaced by the symbol 1 or the symbol 0, everywhere the same. By proceeding in this way with every variable we obtain a number of substitutions of numbers for variables in the schema. If the schema in question has n variables, we obtain 2^n cases. For instance, if schema α is schema (24), then schema α has 2 variables. This yields the four cases specified on p. 79. Next, using the tables for the sentential connectives we compute the values of all substitutions. If every substitution yields the value 1, this means that every compound sentence covered by the schema α is true, so that, by Definition 1, α is a law of the sentential calculus. But if at least one of the substitutions yields the value 0, then this means that there are false compound sentences falling under the schema α, so that α is not a tautology. Since both the number of the cases to be examined is finite, and the computation of the value of each substitution can be performed in a finite number of steps, the verification of every schema can be performed in a finite number of steps. Such methods are called *effective (algorithmic)*. The proof of Theorem 1 is thus completed.

It is worthwhile examining some other examples of the method of the verification of tautologies in the c.s.c. For instance, to verify the formula

$$p \rightarrow \sim p,$$

we have to consider only two cases:

1. $p = 1$: $1 \rightarrow \sim 1 = 1 \rightarrow 0 = 0,$
2. $p = 0$: $0 \rightarrow \sim 0 = 0 \rightarrow 1 = 1;$

the first of these yields the value 0. Hence the formula in question is not a tautology, since those sentences which fall under that schema and in which the sentence p is true, are false.

When verifying the formula

(25) $p \rightarrow \sim \sim p$

we also need only consider two cases:

$$1 \rightarrow \sim \sim 1 = 1 \rightarrow \sim 0 = 1 \rightarrow 1 = 1,$$
$$0 \rightarrow \sim \sim 0 = 0 \rightarrow \sim 1 = 0 \rightarrow 0 = 1;$$

in both of them the formula turns out to be true, and hence it is a tautology. Formula (25) is called the *law of double negation*, as is also the formula which is converse to it:

(26) $\sim \sim p \rightarrow p.$

These two schemata are often used in various reasonings, and their meaning is usually summarized in the statement that two negations yield an affirmation.

In the case of long formulae, which contain a large number of variables, verification can be simplified if we resort to certain short cuts in reasoning.

Such reasoning reduces the number of verifications: we do not always have to verify all the substitutions of 0's and 1's, but often only one of them. This reasoning consists in looking for a substitution for which the formula in question takes on the value 0. This will be shown in an example. We have to verify the following formula:

(27) $((\sim p) \rightarrow q) \rightarrow ((\sim q) \rightarrow p).$

We reason as follows:

1. For a formula not to be a logical theorem, it must for some substitution turn out to be a false sentence, that is, its value must turn out to be 0.

2. Since the formula in question is an implication, it can prove to be a false sentence only if for some substitution its antecedent turns out to be true, while its consequent turns out to be false; this would happen if $(\sim p) \rightarrow q$ equalled 1 while at the same time $(\sim q) \rightarrow p$ equalled 0.

3. For the consequent $(\sim q) \rightarrow p$ to equal 0, its antecedent must be

true, and its consequent must be false, that is, $(\sim q) = 1$ and $p = 0$; it is only then that $((\sim q) \to p) = 0$.

4. Hence the formula can prove false only when $p = 0$ and $q = 0$; it suffices to verify whether it is true in that case.

5. Now we have to verify only whether the antecedent of the entire formula, that is, only the sentence $(\sim p) \to q$, is true or false for that substitution, since we know that the consequent of the formula, that is, the sentence $(\sim q) \to p$, is false for such a substitution, because we just looked for such a substitution. But for $p = 0$ and $q = 0$ we have $(\sim p) = 1$, and hence $((\sim p) \to q) = 0$. This means that on the only substitution for which the consequent of the formula equals 0, the antecedent also equals 0, and hence the entire formula is true. Since the formula proves to be true on the only substitution for which it could be false, this means that it is always true, and hence that it is a logical theorem.

All this reasoning can be automated and carried out by a machine. Let us verify also the formula

(28) $(p \to q) \to ((q \to r) \to (p \to r))$,
    $q = 0, \quad r = 0, \quad p = 1$.

For that formula to equal 0 it must be that $(p \to q) = 1$ and $((q \to r) \to (p \to r)) = 0$. But for $((q \to r) \to (p \to r))$ to equal 0 it must be that $(q \to r) = 1$ and $(p \to r) = 0$. But in turn for $(p \to r)$ to equal 0 it must be that $p = 1$ and $r = 0$. Hence it is only for this substitution that the formula may prove false $(= 0)$. We still have to find the condition for the sentence q. Since $q \to r$ must equal 1, and $r = 0$, q must equal 0, since otherwise $q \to r$ would not equal 1. Hence the only case in which the formula may equal 0 is that when $p = 1$, $r = 0$, and $q = 0$. Yet, if $p = 1$ and $q = 0$, then $(p \to q) = 0$. Thus, even though this is the only case in which $((q \to r) \to (p \to r)) = 0$, in this case $(p \to q) = 0$ also. Accordingly, the entire formula is true in the only case in which it might be false. Thus the formula under consideration is a logical theorem. Note that should we use the verification method without the short cut resorted to above, we would have to compute the value of the formula for eight cases.

Let us attempt to verify the formula

$(p \to (q \lor r)) \to ((q \land r) \to p)$,
    $q = 1, \quad r = 1, \quad p = 0$.

2. TAUTOLOGIES

If it is to equal 0, it is necessary that $(p \rightarrow (q \lor r)) = 1$ and $((q \land r) \rightarrow p) = 0$. For $((q \land r) \rightarrow p)$ to equal 0 it must be that $(q \land r) = 1$ and $p = 0$. For $(q \land r)$ to equal 1 it must be that $q = 1$ and $r = 1$. Hence, if the formula is to equal 0, we must have: $p = 0$, $q = 1$, and $r = 1$. We examine the antecedent: $(q \lor r) = 1$, $p = 0$, hence $(p \rightarrow (q \lor r)) = 1$. The antecedent is true and the consequent is false, as we wanted it to be, and the formula proves to be false in that case for which we suspected it might be false. Hence, the formula under consideration is not a logical theorem. Should we use the verification method without short cuts, it might happen that we would unnecessarily examine as many as seven other substitutions before coming to that substitution for which the formula is false. By resorting to short cuts in our procedure we found that substitution immediately.

In describing the abbreviated verification method we have become familiar with law (27), which is also classified as one of the *laws of transposition* (corresponding to the traditional *modus tollendo ponens*), and law (28), called the *hypothetical syllogism of the Stoics*. The following laws are also classified as laws of transposition:

$$(29) \qquad (p \rightarrow \sim q) \rightarrow (q \rightarrow \sim p)$$

(corresponding to the traditional *modus ponendo tollens*),

$$(30) \qquad (\sim p \rightarrow \sim q) \rightarrow (q \rightarrow p).$$

All four laws of transposition: (24), (27), (29), and (30), have the property in common that they each consist of two implications combined by a main implication symbol. Moreover, in the second implication the variables occur in the reverse order to that in which they occur in the first, hence, the origin of the name of these laws.

Law (28), already known in antiquity to the Stoics, formulates the transitivity of the relation of implication. This law will be used very often. For instance, suppose we want to prove the implication $(a > b) \rightarrow (-b > -a)$, using the laws of addition of equals to unequals, associativity and commutativity of addition, and cancellation of those terms which yield zero. By making use of the law of addition of equals to unequals we obtain the implication

$$(31) \qquad (a > b) \rightarrow (a-b-a > b-b-a).$$
$$ p \qquad \rightarrow \qquad q$$

Using the laws of associativity and commutativity of addition, and of those terms which yield zero we obtain the implication

(32) $(a-b-a > b-b-a) \rightarrow (-b > -a)$.
         q $\rightarrow$ r

The antecedent of this implication is equiform with the consequent of the preceding implication. These two implications yield a conclusion which has the form of an implication whose antecedent is the same as that of the first implication, and whose consequent is the same as that of the second implication:

(33) $(a > b) \rightarrow (-b > -a)$.
         p $\rightarrow$ r

The schema $(p \rightarrow q) \rightarrow ((q \rightarrow r) \rightarrow (p \rightarrow r))$ summarizes, in symbolic notation, the transition from the first two implications, (31) and (32), to the third, (33).

If we want to describe rigorously the use of law (28) in the reasoning carried out above, we have to make use of a certain logical rule of acceptance of theorems, namely the *rule of detachment* (or in Latin *modus ponens*), which states:

RULE OF DETACHMENT. *If a sentence which has the form of an implication*

$$\alpha \rightarrow \beta$$

is accepted as true, and if the antecedent α of that implication is accepted as true, then the consequent β of the implication must be accepted as true also.

The rule of detachment is used very often in various arguments in mathematics and in everyday life. For instance, suppose that it was announced that:

If it rains on Friday, the match will not be played.

This announcement is accepted as a true implication. Now Friday comes and we see that it is raining. Accordingly, we accept the sentence

It rains on Friday,

and, in accordance with the rule of detachment, we conclude from the two sentences that

The match will not be played.

2. TAUTOLOGIES

The rule of detachment may be summarized in the sentence: *If we accept the antecedent of an accepted implication, then we also must accept its consequent.*

Now the arithmetic reasoning quoted above can be given the following form. Since schema (28), as has been verified, is a tautology, the following sentence falling under that schema is true:

$$(34) \qquad ((a > b) \to (a-b-a > b-b-a)) \to \{((a-b-a >$$
$$\qquad\qquad\quad\ p \quad \to \qquad\quad q \qquad\qquad\qquad q$$
$$b-b-a) \to (-b > -a)) \to ((a > b) \to (-b > -a))\}.$$
$$\qquad \to \qquad r \qquad\qquad p \quad \to \quad r$$

Since we accept sentence (34) as a special case of law (28) and since we accept sentence (31), which is the antecedent of (34), then, on the strength of the rule of detachment, we have to accept the consequent of (34), that is, the sentence

$$(35) \qquad ((a-b-a > b-b-a) \to (-b > -a)) \to ((a > b) \to$$
$$(-b > -a)).$$

Since we accept sentence (35) and sentence (32), which is the antecedent of (35), then, on the strength of the rule of detachment, we have to accept sentence (33), which is the consequent of (35). We thus deduce sentence (33) from sentences (31) and (32) by referring to the logical law (28) and by applying the rule of detachment twice.

We shall now analyse some other laws of the sentential calculus and some simple applications.

The formulae of the form

$$(36) \qquad (p \to \sim p) \to \sim p,$$
$$(\sim p \to p) \to p,$$

are called *Clavius' laws*. They are sometimes formulated as the statement:

If a statement yields its own negation, then that statement is false.

Arguments of this type are often encountered in proofs. For instance, if we want to prove the theorem

$$(37) \qquad \sim (x < x),$$

assuming that the relation $<$ is an ordering in the set of real numbers, we reason as follows. The relation $<$, like any order relation, is anti-symmetric, i.e., it satisfies the condition

$$x < y \rightarrow \, \sim (y < x)$$

for all real numbers. In particular, this condition is satisfied for $y = x$, so that

(38) $x < x \rightarrow \, \sim (x < x)$.

Clavius' first law allows us to accept the sentence

(39) $\left(x < x \rightarrow \, \sim (x < x) \right) \rightarrow \, \sim (x < x)$

as a special case, where the sentence p is replaced by the sentence $x < x$. The antecedent of formula (39) is the accepted sentence (38), and hence, on the strength of the rule of detachment, we accept the consequent, that is, sentence (37). Thus, Clavius' law yields the following directive concerning proofs of theorems:

To prove A deduce A from the assumption not-A.

The following law has an application similar to that of Clavius' law:

(40) $\left(p \rightarrow (q \wedge \, \sim q) \right) \rightarrow \, \sim p$.

The conjunction of a sentence q with its negation $\sim q$ is called a *contradictory sentence*. This is why law (40) is sometimes formulated as: *an assumption which yields a contradiction is false.* Law (40) yields the following directive:

To prove not-A deduce a contradiction from A.

Arguments of this type can be encountered in proofs utilizing *reductio ad absurdum*. Every contradictory sentence is self-evidently false. This idea is expressed by the following law:

(41) $\sim (p \wedge \, \sim p)$,

which is called the *law of contradiction*. In classical logic, contradiction has the property that a contradictory sentence implies any other sentence. That property is formulated as the law:

(42) $(p \wedge \, \sim p) \rightarrow q$,

called *Duns Scotus' law*. This law is a consequence of the accepted truth table for implication. We have agreed to consider every implication with a false antecedent to be true. We have also agreed to consider every implication with a true consequent to be true. This yields the law

(43) $p \rightarrow (q \rightarrow p)$,

which is called the *law of simplification*. If any true sentence is substituted for p, and any sentence for q, we obtain, by applying the rule of detachment, an implication with a true consequent and with an arbitrary antecedent. Laws (42) and (43), which express all the paradox of the concept of formal implication, do not play any important role in proofs.

In the Introduction we made use of the following laws concerning implication and equivalence:

$$(p \equiv q) \rightarrow (p \rightarrow q),$$
$$(p \equiv q) \rightarrow (q \equiv p),$$
$$(p \rightarrow q) \rightarrow ((q \rightarrow p) \rightarrow (p \equiv q)),$$
$$(p \equiv q) \rightarrow ((q \equiv r) \rightarrow (p \equiv r)).$$

Many other laws will be introduced later, when there is a need to use them.

Consider now a number of such laws of the sentential calculus which find applications in proofs. The verification that they are tautologies is left to the reader as an exercise.

The laws of importation and exportation:

$$(p \rightarrow (q \rightarrow r)) \rightarrow ((p \wedge q) \rightarrow r),$$
$$((p \wedge q) \rightarrow r) \rightarrow (p \rightarrow (q \rightarrow r)).$$

They allow us to combine by the conjunction symbols a number of antecedents into a single antecedent, and conversely, to treat the various premisses of the antecedent as separate antecedents conditioning .the conclusion. These laws are frequently used in proving those implications in which the consequent is itself an implication. The following law (called *the law of commutativity*):

$$(p \rightarrow (q \rightarrow r)) \rightarrow (q \rightarrow (p \rightarrow r)),$$

finds a similar application: it allows us to change the order of the antecedents which jointly condition the final conclusion.

The law of combining antecedents into a disjunction:

$$((p \rightarrow r) \wedge (q \rightarrow r)) \rightarrow ((p \vee q) \rightarrow r).$$

This law finds many applications to inferences in everyday life. If, for instance, a person promises a student that if he passes his examination in geometry, then he will be given a bicycle, and at the same time another person promises to give him a bicycle if he passes his examination in physics, then the student is entitled to conclude that he will get a bicycle if he passes his examination in geometry or if he passes his examination in physics.

The law of combining the consequents into a conjunction:

$$((p \rightarrow q) \wedge (p \rightarrow r)) \rightarrow (p \rightarrow (q \wedge r)),$$

resembles the former. It yields the directive that if a conclusion which is a conjunction $q \wedge r$ is to be deduced from an assumption p, then it suffices to deduce separately q from p and r from p.

The defining of disjunction by implication:

$$(\sim p \rightarrow q) \rightarrow (p \vee q).$$

To prove the disjunction $p \vee q$ it suffices to deduce one of its terms from the negation of other. It was in this way that, e.g., formula (70) was proved in the Introduction. To prove that $a \cap b = \Lambda$, or $a = b$, it was proved that if $a \cap b \neq \Lambda$, then $a = b$. Likewise, if two straight lines in a plane do not intersect, then they are parallel, and hence any two straight lines in a plane either intersect or are parallel.

The converse implication:

$$(p \vee q) \rightarrow (\sim p \rightarrow q),$$

is also true. For instance, if a locality is inhabited only by people who are fishermen or fishmongers, then an inhabitant who is not a fisherman is a fishmonger.

These two implications may be combined into the equivalence:

$$(p \vee q) \equiv (\sim p \rightarrow q),$$

which defines disjunction in terms of negation and implication. It can easily be verified that negation and disjunction can be used to define implication:

$$(p \rightarrow q) \equiv (\sim p \vee q).$$

2. TAUTOLOGIES

This definition does not, however, satisfy our ordinary intuitions, but rather brings out the paradoxical nature of the classical interpretation of implication.

The following laws:

$$\sim (p \wedge q) \equiv (\sim p \vee \sim q),$$
$$\sim (p \vee q) \equiv (\sim p \wedge \sim q),$$

are called *De Morgan's laws*. They state that the negation of a conjunction is equivalent to the disjunction of the negations of the terms of that conjunction, and conversely, the negation of a disjunction is equivalent to the conjunction of the negations of the terms of the disjunction. For instance, if a person is not both (a painter and a sculptor), then either he (is not a painter) or he (is not a sculptor). And conversely, if a person (is not a painter) or he (is not a sculptor), then he is not (both a painter and a sculptor). Likewise, if a person is not (a fisherman or a fishmonger), then he (is not a fisherman) and also he (is not a fishmonger).

These laws, together with the laws of double negation, yield a definition of conjunction in terms of negation and disjunction, and a definition of disjunction in terms of negation and conjunction:

$$(p \wedge q) \equiv \sim (\sim p \vee \sim q),$$
$$(p \vee q) \equiv \sim (\sim p \wedge \sim q).$$

If we consider the equivalences described earlier, we can note easily that all the sentential connectives under consideration can be defined by means of implication and negation:

$$(p \vee q) \equiv (\sim p \rightarrow q),$$
$$(p \wedge q) \equiv \sim (p \rightarrow \sim q).$$

The tautologies of the sentential calculus suffice to carry out certain proofs, in which *bound variables* do not occur. Tautologies will be used below in a proof. But to be able to do that we have to formulate one rule more, namely the rule of substitution:

In a very general and inexact formulation the *rule of substitution* states:

If a general schema (a general law) is accepted as true, then every special case of that schema (every definite application of that law) is to be accepted as true, too.

In this sense the rule of substitution, like the rule of detachment, is used everyday when we prove theorems. For instance, if we accept the general law of the commutativity of addition

(44) $a+b = b+a,$

then we have to accept every case to which that law applies, both such in which reference is made definite numbers, e.g.,

(45) $2+5 = 5+2,$

and such which themselves are a kind of a schema, e.g.,

$$a+7 = 7+a,$$
(46)
$$a+(b+3c) = (b+3c)+a,$$

but nevertheless falls under the original schema. Sentence (45) is said to be obtained from sentence (44) by the substitution of the constant 2 for the variable a, and of the constant 5 for the variable b. The first sentence of (46) is obtained by the substitution of the constant 7 for the variable b, and the second, by the substitution of the compound formula $b+3c$ for the variable b.

The long formulation

> the sentence A is obtained by the substitution of the formula
> a for the variable x

will hereafter be replaced by the   abbreviated one:

> the sentence A is obtained by the substitution x/a.

In order to formulate the rule of substitution rigorously and in a sufficiently general way, we assume that the formulae under consideration contain variables of the form:

$$x_1, x_2, x_3, ...,$$

each of them formed out of the letter x with a numerical subscript (for practical purposes we write: x for x_1, y for x_2, and z for x_3); the formulae in question include, besides the variables, only the sentential connectives: $\wedge, \vee, \rightarrow, \sim \equiv$, the parentheses: $(,)$, the comma, and the mathematical symbols: $+, \cdot, = , <, -, 0.$

DEFINITION OF SUBSTITUTION. *Formula A is obtained from formula B by the substitution x_i/a if formula A differs from formula B only in* *that in every place where the variable x_i occurs in formula B the expression a occurs in formula A, and the remaining parts of formula B remain unchanged.*

In this sense, (45) is in fact the substitution: $a/2$, $b/5$ of formula (44). Likewise, the sentence $x_3 + x_4 = x_2$ is the substitution $x_1/x_3 + x_4$ of the sentence $x_1 = x_2$.

RULE OF SUBSTITUTION (in a very narrow formulation). *If we accept sentence A, consisting of symbols specified above and containing the variable x_i, then we also accept any sentence B obtained from A by the substitution $x_i/0$ or x_i/x_j (one variable for another variable), and also any sentence B obtained from sentence A by the substitution $x_i/x_j + x_k$ and also by the substitution $x_i/x_j \cdot x_k$, where i, j, k are any natural numbers.*

Some fairly long algebraic arguments can be carried out rigorously when use is made of the rules of detachment and substitution and of sentential calculus tautologies.

For instance, the cancellation law:

$$x+y = x+z \rightarrow y = z,$$

will be deduced from the following assumptions which state that addition is a group operation:

1. $x = x$,
2. $x = y \rightarrow y = x$,
3. $(x = y \wedge y = z) \rightarrow x = z$,
4. $x = y \rightarrow x+z = y+z$,
5. $x = y \rightarrow z+x = z+y$,
6. $x+(y+z) = (x+y)+z$,
7. $x+0 = x$,
8. $0+x = x$,
9. $x+(-x) = 0$,
10. $-x+x = 0$.

The proof will consist of a sequence of mathematical sentences, each of which will either be a substitution in a tautology of the sentential calculus, or will be obtained from such a substitution and axioms 1–10 by the successive application of the rules of detachment and substitution:

11. $(x+y = x+z) \rightarrow \left(-x+(x+y) = -x+(x+z)\right)$
     (from 5 by subst. $x/(x+y), y/(x+z), z/-x$),

12. $-x+(x+y) = (-x+x)+y$
     (from 6 by subst. $x/-x, y/x, z/y$),

13. $-x+x = 0 \rightarrow (-x+x)+y = 0+y$
     (from 4 by subst. $x/(-x+x), y/0, z/y$),

14. $(-x+x)+y = 0+y$
     (from 10 and 13 by detachment),

15. $0+y = y$
     (from 8 by subst. x/y),

16. $\left((-x+x)+y = 0+y\right) \rightarrow$
     $\left((0+y = y) \rightarrow \left(((-x+x)+y) = (0+y) \wedge (0+y = y)\right)\right)$
     (substitution in the tautology: $p \rightarrow (q \rightarrow (p \wedge q))$),

17. $0+y = y \rightarrow \left(((-x+x)+y = 0+y) \wedge (0+y = y)\right)$
     (from 14 and 16 by detachment),

18. $\left((-x+x)+y = 0+y\right) \wedge (0+y = y)$
     (from 15 and 17 by detachment),

19. $\left(((-x+x)+y = 0+y) \wedge (0+y = y)\right) \rightarrow (-x+x)+y = y$
     (from 3 by subst. $x/(-x+x)+y, y/0+y, z/y$),

20. $(-x+x)+y = y$
     (from 18 and 19 by detachment),

21. $-x+(x+y) = (-x+y)+y \rightarrow \left((-x+x)+y = y \rightarrow\right.$
     $\left.(-x+(x+y) = (-x+x)+y \wedge (-x+x)+y = y)\right)$
     (tautology: $p \rightarrow (q \rightarrow (p \wedge q))$),

22. $(-x+x)+y = y \rightarrow \left(-x+(x+y)+y \wedge (-x+x)+y = y\right)$
     (from 12 and 21 by detachment),

23. $-x+(x+y) = (-x+x)+y \wedge (-x+x)+y = y$
     (from 21 and 22 by detachment),

24. $\left(-x+(x+y) = (-x+x)+y \wedge (-x+x)+y = y\right) \rightarrow$
     $-x+(x+y) = y$
     (from 3 by subst. $x/\left(-x+(x+y)\right), y/\left((-x+x)+y\right), z/y$),

25. $-x+(x+y) = y$
    (from 23 and 24 by detachment),

26. $-x+(x+z) = z$
    (from 25 by subst. y/z),

27. $((x = y \wedge y = z) \rightarrow (x = z)) \rightarrow (x = y \rightarrow (y = z \rightarrow x = z))$
    (tautology: $((p \wedge q) \rightarrow r) \rightarrow (p \rightarrow (q \rightarrow r)))$,

28. $x = y \rightarrow (y = z \rightarrow x = z)$
    (from 27 and 3 by detachment),

29. $-x+(x+y) = y \rightarrow y = x-x+(x+y)$
    (from 2 by subst. $x/(-x+(x+y)))$,

30. $y = -x+(x+y)$
    (from 25 and 29 by detachment),

31. $y = -x+(x+y) \rightarrow (-x+(x+y) = -x+(x+z) \rightarrow y = -x+(x+z))$
    (from 28 by subst. x/y, $y/(-x+(x+y))$, $z/(-x+(x+z)))$,

32. $-x+(x+y) = -x+(x+z) \rightarrow y = -x+(x+z)$
    (from 30 and 31 by detachment),

33. $(x = y \rightarrow (y = z \rightarrow x = z)) \rightarrow (y = z \rightarrow (x = y \rightarrow x = z))$
    (tautology: $(p \rightarrow (q \rightarrow r)) \rightarrow (q \rightarrow (p \rightarrow r)))$,

34. $y = z \rightarrow (x = y \rightarrow x = z)$
    (from 28 and 33 by detachment),

35. $-x+(x+z) = z \rightarrow (y = -x+(x+z) \rightarrow y = z)$
    (from 34 by subst. $y/(-x+(x+z))$, $x/y)$,

36. $y = -x+(x+z) \rightarrow (y = z)$
    (from 35 and 26 by detachment),

37. $(-x+(x+y) = -x+(x+z) \rightarrow y = -x+(x+z)) \rightarrow ((y = -x+(x+z) \rightarrow y = z) \rightarrow (-x+(x+y) = -x+(x+z) \rightarrow y = z))$
    (tautology: $(p \rightarrow q) \rightarrow ((q \rightarrow r) \rightarrow (p \rightarrow r)))$,

38. $(y = -x+(x+z) \rightarrow (y = z)) \rightarrow (-x+(x+y) = -x+(x+z) \rightarrow y = z)$
    (from 37 and 32 by detachment),

39. $-x+(x+y) = -x+(x+z) \rightarrow y = z$
    (from 36 and 38 by detachment),

40. $\big(x+y = x+z \rightarrow -x+(x+y) = -x+(x+z)\big) \rightarrow$
    $\big((-x+(x+y) = -x+(x+z) \rightarrow y = z) \rightarrow$
    $(x+y = x+z \rightarrow y = z)\big)$
    (tautology: $(p \rightarrow q) \rightarrow ((q \rightarrow r) \rightarrow (p \rightarrow r)))$,

41. $\big(-x+(x+y) = -x+(x+z) \rightarrow y = z\big) \rightarrow$
    $(x+y = x+z \rightarrow y = z)$
    (from 11 and 40 by detachment),

42. $x+y = x+z \rightarrow y = z$
    (from 39 and 41 by detachment).

As the above example shows, when we make a mathematical proof in the domain of the sentential calculus and when we have certain premisses, for instance, 32 and 37, we select such a schema from the sentential calculus which on substitution makes it possible to deduce conclusion 39 by two applications of the operation of detachment. The formula $(p \rightarrow q) \rightarrow ((q \rightarrow r) \rightarrow (p \rightarrow r))$ proves to be such a schema; we verify it by the truth-table method and when it turns out to be a tautology we may use it. This is how tautologies in the sentential calculus are used: we select a schema which is needed, we verify it, and if it proves to be a tautology, we apply it. Of course, mathematicians in their books and papers do not construct such formal proofs, since that would result in total stagnation in mathematics. They just refer to their well-trained mathematical intuition. The task of logic is not to provide mathematicians with convenient tools for their everyday work. The task of logic is rather to reconstruct and to systematize those modes of reasoning which are constantly used in mathematics, and in other disciplines as well. The realization of this goal makes it, of course, possible to check mathematical reasoning which might raise objections.

Not all mathematical arguments can be formalized in terms of the sentential calculus alone. Those arguments which refer to quantifiers (see below) cannot be interpreted in terms of the sentential calculus and the rules of substitution and detachment alone. They also require laws or rules of the use of quantifiers.

2. TAUTOLOGIES

Exercises

1. Deduce the theorem: $-(-x) = x$ from assumptions 1–10, making use only of substitution, detachment and tautologies.

H i n t. Refer to the cancellation law, as proved above, and to Axioms 9 and 10.

2. Deduce the analogical law of right-side cancellation:

$$x+z = y+z \rightarrow x = y.$$

N o t e. Commutativity: $x+y = y+x$ is not assumed in the axioms and does not follow from them.

3. Deduce the laws:

$$x+y = 0 \rightarrow x = -y,$$
$$x = y \rightarrow -x = -y.$$

4. Add to Axioms 1–10 the following ones:

$$x = y \rightarrow (x < z \rightarrow y < z),$$
$$x = y \rightarrow (z < x \rightarrow z < y),$$
$$x < y \rightarrow x+z < y+z,$$
$$x < y \rightarrow z+x < z+y,$$

Deduce the theorem: $x < y \rightarrow (-y < -x)$.

5. Verify whether the following formulae are tautologies:

$$((p \rightarrow q) \wedge p) \rightarrow q,$$
$$((p \rightarrow q) \wedge \sim q) \rightarrow \sim p,$$
$$(p \vee q) \rightarrow (\sim q \rightarrow p),$$
$$(p \rightarrow (p \rightarrow q)) \rightarrow q,$$
$$((q \rightarrow p) \rightarrow p) \rightarrow p,$$
$$((p \rightarrow q) \rightarrow p) \rightarrow p.$$

6. Verify the following formulae, which are algebraic in nature:

$$\left. \begin{array}{l} (p \vee q) \equiv (q \vee p) \\ (p \wedge q) \equiv (q \wedge p) \end{array} \right\} \text{ laws of commutativity,}$$

$$\left. \begin{array}{l} p \vee (q \vee r) \equiv (p \vee q) \vee r \\ p \wedge (q \wedge r) \equiv (p \wedge q) \wedge r \end{array} \right\} \text{ laws of associativity,}$$

$$p \vee (q \wedge r) \equiv (p \vee q) \wedge (p \vee r) \text{ distributivity of disjunction with respect to conjunction,}$$

$$p \wedge (q \vee r) \equiv (p \wedge q) \vee (p \wedge r) \text{ distributivity of conjunction with respect to disjunction,}$$

$$(p \vee p) \equiv p,$$
$$(p \wedge p) \equiv p.$$

I. CLASSICAL LOGICAL CALCULUS

Note that disjunction is sometimes called *logical addition*, while conjunction is called *logical multiplication*. Contrary to arithmetic, the role of disjunction and that of conjunction in logic are fully symmetric. For instance, distributivity of addition with respect to multiplication and distributivity of multiplication with respect to addition both hold.

The laws of associativity make it possible to disregard and/or omit parentheses in the case of iterated disjunctions and conjunctions, since it follows from these laws that insertion of parentheses in formulae of the type:

$$p_1 \lor p_2 \lor p_3 \lor p_4 \lor \ldots \lor p_n,$$

$$p_1 \land p_2 \land p_3 \land p_4 \land \ldots \land p_n$$

does not change the truth value of the formulae concerned.

7. Prove in a general manner that for any n formulae of the form:

$$(p_1 \to p_2) \to \left((p_2 \to p_3) \to \left((p_3 \to p_4) \to \right. \right.$$

$$\left. \left. \left(\ldots((p_{n-1} \to p_n) \to (p_1 \to p_n)) \ldots \right) \right) \right),$$

$$\left(p_1 \to \left(p_2 \to \left(p_3 \to (\ldots(p_n \to q)\ldots) \right) \right) \right) \equiv ((p_1 \land p_2 \land p_3 \land \ldots \land p_n) \to q),$$

$$\sim (p_1 \land p_2 \land \ldots \land p_n) \equiv \sim p_1 \lor \sim p_2 \lor \ldots \lor \sim p_n,$$

$$\sim (p_1 \lor p_2 \lor \ldots \lor p_n) \equiv \sim p_1 \land \sim p_2 \land \ldots \land \sim p_n,$$

are theorems in the c.s.c.

8. If a system of the sentential calculus includes only the following sentential connectives: conjunction, disjunction, equivalence, and negation, then a formula α is called *dual* to a formula β if and only if α differs from β only in that in those places where α has symbols of disjunction, β has symbols of conjunction, and conversely, that in those places where α has symbols of conjunction, β has symbols of disjunction. What are the dual formulae to tautologies?

3. AN AXIOMATIC APPROACH TO THE SENTENTIAL CALCULUS

The set of laws (tautologies) of the classical sentential calculus was defined in the foregoing section in a synthetic manner: the theorem on the decidability of the c.s.c. gave an easy method of constructing an arbitrary number of laws of that calculus. Usually, however, mathematical theories develop in a somewhat different way. They develop in an axiomatic manner. The founders of a given calculus or theory come to realize the fundamental properties of the objects considered in that theory, usually, those properties which are the most obvious and the least questionable. The sentences stating those properties are often called the *assumptions* or the *axioms* of a given theory. The axioms are the initial theorems of a theory. All other theorems of that theory are deduced from axioms by logical inference. Logical systems can also be given such an axiomatic structure. We can single out certain fundamental logical laws, which appeal to our intuition, and adopt them as axioms. All other laws are deduced from those axioms in a predefined way which makes the laws thus deduced as true as the axioms. This way of deducing new logical laws from those laws which have already been accepted will be described soon. But we shall first discuss the axioms of the classical sentential calculus, classifying them from the outset into certain groups. The system of axioms, as selected in the present book, consists of a fairly large number of statements; this has the purpose of bringing out the role of certain logical laws and avoiding the tedious and rather unnecessary process of deducing new theorems. There are very many different but equivalent systems of axioms of the sentential calculus. We could even formulate an axiom system of the sentential calculus consisting of only one axiom. Such an axiom, however, does not lend itself to a clear interpretation. If the number of axioms is larger, the role of each axiom is easier to grasp.

A. *Positive Axioms for Implication*:

1. $p \to (q \to p)$,
2. $(p \to (q \to r)) \to ((p \to q) \to (p \to r))$.

B. *Axioms Describing Equivalence in Terms of Implication*:

3. $(p \equiv q) \to (p \to q)$,
4. $(p \equiv q) \to (q \to p)$,
5. $(p \to q) \to ((q \to p) \to (p \equiv q))$.

C. *Axioms Describing Conjunction and Disjunction*:

6. $(p \lor q) \to (q \lor p)$ ⎫
7. $(p \land q) \to (q \land p)$ ⎬ commutativity,

8. $p \to (p \lor q)$ ⎫
9. $(p \land q) \to p$ ⎬ absorption,

10. $p \to (q \to (p \land q))$ linking by the conjunction symbol,
11. $((p \to r) \land (q \to r)) \to ((p \lor q) \to r)$ linking by the disjunction symbol.

D. *Laws Describing Negation*:

12. $(p \rightarrow (q \wedge \sim q)) \rightarrow \sim p$,

13. $(p \wedge \sim p) \rightarrow q$.

E. *Law of Excluded Middle*:

14. $p \vee \sim p$.

Axioms 1–11 form the axiom system of what is called *positive logic*. Axioms 1–13 form the axiom system of what is called *intuitionistic logic*. Positive logic does not include theorems concerned with negation. Intuitionistic logic does not include the law of excluded middle. (These systems, narrower than the c.s.c., will be mentioned again below.)

The meaning of those axioms which we have not encountered so far will be discussed briefly. Axiom 2 can be expressed with the statement that if p entails the implication $(q \rightarrow r)$ and entails the antecedent q of that implication, then it also entails its consequent r. The axioms on equivalence describe equivalence as holding between p and q if and only if the implications $(p \rightarrow q)$ and $(q \rightarrow p)$ both hold. The commutativity of conjunction and disjunction is obvious, since it does not matter in what order the various possibilities or sentences accepted as true are specified. Axiom 10 allows us to accept jointly any two sentences which are both accepted as true separately.

When we have accepted a number of logical laws as obvious, we deduce other logical laws from those accepted previously; we do so by making use of the two rules:

1. The rule of detachment,

2. The rule of substitution.

The rule of detachment was formulated in a general way in the preceding section. Its application to the laws of logic is only a special case. As applied to the laws of logic, the rule of detachment may be formulated in the following statement:

If we accept a schema which has the form of an implication

$$\alpha \rightarrow \beta$$

to be a law of logic, and if we accept as a law of logic the schema α, which is the antecedent of the first schema, then we have to accept the schema β, which is the consequent of the first schema, as a law of logic, too.

As applied to the laws of logic, the rule of substitution states that all those schemata which are particular cases of the schemata accepted as laws of logic are to be accepted as laws of logic. Thus, for instance, having accepted the law of excluded middle: $p \lor \sim p$, we have to accept as a law of logic the schema

$$(p \to q) \lor \sim (p \to q),$$

which is obtained from the former by the substitution $p/(p \to q)$, and also the schema

$$(p \to (q \land r)) \lor \sim (p \to (q \land r)),$$

which is obtained from the preceding schema by the substitution $q/(q \land r)$, or directly from the law of excluded middle by the substitution $p/(p \to (q \land r))$.

We shall now give several examples of the alternate application of the rules of detachment and substitution. For the time being, we shall confine ourselves to deducing a number of conclusions from the axioms included in the first group, in particular, the positive axioms for implication:

1. $p \to (q \to p)$,
2. $(p \to (q \to r)) \to ((p \to q) \to (p \to r))$.

The theorems deduced from Axioms 1–14 will be marked with numbers equal to or greater than 20. We shall first prove the theorem $(p \to p)$, deducing first for that purpose three auxiliary theorems:

20. $(p \to (q \to p)) \to ((p \to q) \to (p \to p))$
     (from 2 by subst. r/p).

21. $(p \to q) \to (p \to p)$
     (from 1 and 20 by detachment).

22. $(p \to (q \to p)) \to (p \to p)$
     (from 21 by subst. $q/(q \to p)$).

23. $p \to p$
     (from 1 and 22 by detachment).

This theorem is sometimes called the *identity law for the sentential calculus*. Now that we have 23 we shall prove 29, which is concerned with the transitivity of implication.

24. $2 \rightarrow \big((q \rightarrow r) \rightarrow 2\big)$
   (from 1 by subst. $p/2$, $q/(q \rightarrow r)$).

25. $(q \rightarrow r) \rightarrow 2$, i.e., $(q \rightarrow r) \rightarrow \big((p \rightarrow (q \rightarrow r)) \rightarrow ((p \rightarrow q) \rightarrow (p \rightarrow r))\big)$
   (from 24 and 2 by detachment).

26. $25 \rightarrow \big(\big((q \rightarrow r) \rightarrow (p \rightarrow (q \rightarrow r))\big) \rightarrow \big((q \rightarrow r) \rightarrow ((p \rightarrow q) \rightarrow (p \rightarrow r))\big)\big)$
   (from 2 by subst. $p/(q \rightarrow r)$, $q/(p \rightarrow (q \rightarrow r))$,
   $r/((p \rightarrow q) \rightarrow (p \rightarrow r))$).

27. $\big((q \rightarrow r) \rightarrow (p \rightarrow (q \rightarrow r))\big) \rightarrow \big((q \rightarrow r) \rightarrow ((p \rightarrow q) \rightarrow (p \rightarrow r))\big)$
   (from 24 and 25 by detachment).

28. $(q \rightarrow r) \rightarrow \big(p \rightarrow (q \rightarrow r)\big)$
   (from 1 by subst. $p/(q \rightarrow r)$, q/p).

29. $(q \rightarrow r) \rightarrow \big((p \rightarrow q) \rightarrow (p \rightarrow r)\big)$
   (from 27 and 28 by detachment).

Law 29 is a form of the theorem on the *transitivity of implication* formulated in terms of implication itself.

To shorten the proofs of other theorems we shall prove in general that, on the strength of the theorems for positive implication obtained so far,

(47)   *From a theorem of the form* $\alpha \rightarrow (\beta \rightarrow \gamma)$ *we can obtain a theorem of the form* $\beta \rightarrow (\alpha \rightarrow \gamma)$.

To prove (47), assume that we have a theorem of the form

(48) $\alpha \rightarrow (\beta \rightarrow \gamma)$.

We deduce the following theorems:

(49) $(\alpha \rightarrow (\beta \rightarrow \gamma)) \rightarrow ((\alpha \rightarrow \beta) \rightarrow (\alpha \rightarrow \gamma))$
   (from 2 by subst. p/α, q/β, r/γ).

(50) $(\alpha \rightarrow \beta) \rightarrow (\alpha \rightarrow \gamma)$
   (from (49) and (48) by detachment).

(51) $((\alpha \rightarrow \beta) \rightarrow (\alpha \rightarrow \gamma)) \rightarrow (\beta \rightarrow (\alpha \rightarrow \beta)) \rightarrow (\beta \rightarrow (\alpha \rightarrow \gamma))$
   (from 29 by subst. $q/(\alpha \rightarrow \beta)$, $r/(\alpha \rightarrow \gamma)$, p/β).

(52) $(\beta \to (\alpha_{\cdot}^{\cdot} \to \beta)) \to (\beta \to (\alpha \to \gamma))$
        (from (51) and (50) by detachment).

(53) $\beta \to (\alpha \to \beta)$
        (from 1 by subst. p/β, q/α).

(54) $\beta \to (\alpha \to \gamma)$
        (from (52) and (53) by detachment).

We have thus proved (47). This rule allows us to change the order of antecedents $\alpha, \beta, \delta, \ldots$ in a statement of the type $\alpha \to \big(\beta \to (\delta \to (\ldots \gamma))\big)$. It may be applied, for instance to Law 29. On substituting $\alpha = (q \to r)$, $\beta = (p \to q)$, $\gamma = (p \to r)$, we obtain the conclusion that since 29 can be deduced from Axioms 1 and 2, then the following sentence also is deducible from them:

30. $(p \to q) \to \big((q \to r) \to (p \to r)\big)$
        (from 29 by the application of rule (47)).

Law 30 is the most natural form of the theorem on the trasitivity of implication, formulated in terms of implication itself.

31. $(p \to q) \to \big(p \to (q \to r)\big) \to (p \to r)$
        (from 2 by the application of (47), by substituting $\alpha = \big(p \to (q \to r)\big)$, $\beta = (p \to q)$, $\gamma = (p \to r)$).

32. $\Big((p \to q) \to \big((p \to (q \to r)) \to (p \to r)\big)\Big) \to \Big((q \to (p \to q)) \to$
     $\big(q \to ((p \to (q \to r)) \to (p \to r))\big)\Big)$
        (from 29 by subst. p/q, $q/(p \to q)$, $r/\big((p \to (q \to r)) \to (p \to r)\big)$).

33. $(q \to (p \to q)) \to \big(q \to (p \to (q \to r)) \to (p \to r)\big)$
        (from 32 and 31 by detachment).

34. $q \to (p \to q)$
        (from 1 by subst. p/q, q/p).

35. $\big(q \to (p \to (q \to r)) \to (p \to r)\big)$
        (from 33 and 34 by detachment).

36. $(p \to (q \to r)) \to (q \to (p \to r))$
        (from 35 by applying (47) and by substituting $\alpha = q$, $\beta = (p \to (q \to r))$, $\gamma = (p \to r)$).

Law 36 formulates, as a law of the sentential calculus, the same idea which we earlier stated with formula (47): different antecedents which condition the sentence r may be interchanged, since their order is inessential. This is a form of the commutation law, introduced earlier.

37. $(p \to p) \to \big((p \to (p \to q)) \to (p \to q)\big)$
    (from 31 by subst. q/p and r/q).

38. $\big(p \to (p \to q)\big) \to (p \to q)$
    (from 37 and 23 by detachment).

Law 38 formulates the idea that a repetition of an antecedent is superfluous.

39. $\big(p \to (q \to p)\big) \to \big(((q \to p) \to r) \to (p \to r)\big)$
    (from 30 by subst. $q/(q \to p)$).

40. $\big((q \to p) \to r\big) \to (p \to r)$
    (from 39 and 1 by detachment).

(Sentences 1 and 2 above are also traditionally called positive axioms of implication. They include, however, Axiom 1, which, convenient as it is and often used in the proofs above, does not play any important role in applications. An almost equally convenient system of implication can be obtained by adopting as axioms sentences 23, 30, 36, 38 and the sentence: $(p \to q) \to (p \to (p \to q))$. The law of simplification (Axiom 1) does not follow from them, nor do other, equally paradoxical, consequences of that axiom, e.g., 40.)

These examples of proofs show that the verification of the correctness of proofs is easy. A closer examination of a theorem just deduced reveals whether it really is obtained by the operations of substitution and/or detachment, as described above. On the other hand, it is not an easy matter to come upon the idea of such a proof. There is no general method which would indicate which substitutions and/or detachments are to be performed if the desired theorem is to be obtained after a number of steps. Carrying out such a proof often requires ingenuity (and luck).

Most proofs constructed by mathematicians consist in deducing new theorems from those proved before. But in average mathematical proofs reasoning is very much abbreviated, and usually many uninteresting intermediate theorems, which lead to the theorem to be proved, are just disregarded. If every formula had to be proved rigorously, short mathematical papers would swell to the size of thick volumes. It usually suffices if a proof is such that an average specialist can grasp it so that, if necessary, he is able to formulate every part of the proof with desired precision. The reduction of a theory T to a form such that every theorem of that theory T, obtained so far, either is an axiom or has been obtained from axioms of T by a number of applications of the rules of inference, is called a *formalization of the theory T*. In accordance with what has been said above, not every fragment of a proof is to be formalized. Yet every proof ought to be clear enough to allow any specialist to formalize any fragment of the proof without difficulty. Regardless of the forms which various proofs have in practice, from a theoretical point of view it is advantageous to consider every mathematical theory as fully formalized and stated in a complete form, that is, as including all those theorems which are deducible from the axioms by substitution and detachment, even though the majority of them will never be so deduced. This kind of idealization makes us independent of the random character of the history of science and the mentality of scientists. Other disciplines, too, include theories which resemble those encountered in mathematics. Hence the text below will refer to theories in general, and the terms: theory, formalized theory, deductive theory, deductive system, and formalized system, will be used alternately.

To formulate now a few simple observations concerning the classical sentential calculus we shall give more precision to the concept of theory as outlined above. That concept is usually defined by reference to the concept of consequence, or conclusion. The term *consequence* of (*conclusion* from) a set X of sentences stands for any sentence that can be proved on the strength of the set X, that is, can be deduced from a number of sentences of the set X, using the rules of inference. The term *theory* stands for any set X of sentences, which has the property that the consequences of the set X are still in that set. Thus, a theory is a set which is closed under consequence. The definition of consequence in turn refers to the concept of proof. This is why a rigorous definition will now begin with a definition of a proof. In the present section, the term *proof* will mean only a proof in the sentential calculus, that is such which can be carried out by the rules of inference in the sentential calculus: detachment and substitution. Another concept of proof and a related concept of consequence, namely such which is applicable not to the system of the sentential calculus, but to mathematical theories, will be introduced in Section 6.

DEFINITION 2. *D is a proof of sentence B on the strength of a set X of formulae adopted as assumptions if and only if D is a finite sequence of formulae*

$$(55) \qquad D = \{D_1, D_2, ..., D_n\}$$

such that the last formula in that sequence is identical with the sentence B: $D_n = B$, and every formula D_k in the sequence D $(1 \leqslant k \leqslant n)$ 1) either is contained in the set X, or 2) is obtained by correct substitution from a formula D_j earlier than D_k in that sequence $(j < k)$, or 3) is obtained from two formulae contained in D, namely D_j and D_i, both earlier than D_k $(j < k, i < k)$, by detachment: $D_j = (D_i \rightarrow D_k)$.

For instance, the finite sequence of theorems: 1, 2, 20, 21, 22, 23 is a proof of the formula 23 on the strength of Axioms 1 and 2. Likewise, the sequence 1, 2, 24, 25, 26, 27, 28, 29 is a proof of the transitivity law 29 on the strength of Axioms 1 and 2.

DEFINITION 3 (Tarski [90], [93]). *A formula A is a consequence of a set X* (in symbolic notation: $A \in \text{Cnq}(X)$) *if and only if there exists a finite sequence D of formulae such that D is a proof of A on the strength of the set X of formulae.*

Hence all the formulae 20–38 are consequences of the set $X = \{1, 2\}$ of the positive axioms for implication. Among the most elementary properties of the concept of consequence the one which is the easiest to see is that every set of formulae is included in the set of its consequences. If a sentence A is contained in a set X, then A, as a one-term sequence, may be considered a proof of A on the strength of the set X. Thus $A \in \text{Cnq}(X)$ if $A \in X$. Using the inclusion symbol we can write this conclusion in symbolic notation:

$$(56) \qquad X \subset \text{Cnq}(X).$$

Thus Axioms 1 and 2 also are consequences of the set $\{1, 2\}$.

DEFINITION 4. *A set X of formulae is a theory if and only if the consequences of the set X are members of the set X, that is, if the formula*

$$(57) \qquad \text{Cnq}(X) \subset X$$

is satisfied.

Since the inclusion (56), converse to (57), is true for any set of sentences, as has been demonstrated above, the definition of a theory may also be formulated as follows:

A set X is a theory if and only if it is identical with the set of its consequences, that is, satisfies the formula:

(58) $X = \mathrm{Cnq}(X)$.

THEOREM 2. *The set of the tautologies of the c.s.c. is a theory.*

PROOF. We shall first prove two properties of the c.s.c.:

1. *Every schema which is a substitution for a tautology is a tautology.*

2. *A schema obtained from tautologies by detachment is also a tautology.*

Property 1 follows directly from the definition of tautology. If a schema B is obtained from a schema A by substitution, in the way which, for instance, the schema

$$(p \rightarrow q) \vee \sim (p \rightarrow q)$$

is obtained by substitution from the schema $p \vee \sim p$, then every sentence which falls under the schema B thereby falls under the schema A, since B is a particular case of A. Hence, if there existed a false sentence falling under the schema B, it would also be a false sentence falling under the schema A. Now if the schema A is a tautology, which means, by Definition 1, that every sentence falling under A is true, which means in turns that no false sentence falls under the schema A, then there can be no false sentence falling under the schema B. Hence, by Definition 1, the sentences falling under B also are tautologies in the c.s.c. Thus the set of the tautologies in the c.s.c. is closed under the operation of substitution.

To prove property 2 suppose that schemata A and $A \rightarrow B$ are tautologies in the c.s.c. Should the schema B not be a tautology, this would mean that there exists a substitution for which the schema B becomes a false sentence. Let that substitution be performed in the entire schema $A \rightarrow B$. Since A is a tautology, then for that substitution A becomes a true sentence. This would yield $A = 1$ and $B = 0$, so that the implication $A \rightarrow B$ would, for that substitution, be false, and hence the schema $A \rightarrow B$ could not be a tautology. Thus, the set of tautologies is closed under the operation of detachment, too.

Now that the two properties are proved, the proof of the theorem under consideration is as follows. It is to be demonstrated that the consequences of tautologies are themselves tautologies, that is, in accordance with the definitions given above, it is to be proved that if a sequence D of formulae

$$D_0, D_1, D_2, ..., D_n$$

is a proof of a formula $A = D_n$ on the strength of a tautology, then the formula A itself is a tautology. By the definition of proof, every formula D_j in the proof D either is a tautology or is obtained from earlier terms D_i, D_h $(i, h < j)$ of that sequence by substitution or detachment. Hence, all the formulae D_i in the sequence D either are tautologies or are obtained from tautologies by an iterated application of the logical rules of substitution and detachment. Since, as has been demonstrated above, the logical rules lead from a tautology in the c.s.c. to a tautology in the c.s.c., every formula D_i of the proof D is a tautology in the c.s.c. In particular, the terminal formula $A = D_n$ also is a tautology in the c.s.c. Thus, the consequences of tautologies in the c.s.c. are themselves tautologies in the c.s.c., which, in agreement with the definition of a theory, means that the tautologies of the c.s.c. form a theory.

By interpreting the last part of the proof in a more abstract manner we have a proof of the following theorem:

THEOREM 3. *If a set X is closed under the logical rules of substitution and detachment, then the set X is a theory.*

DEFINITION 5. *A set X is an axiom system of a set Y if and only if Y is the set of all the consequences of X, that is,*

$$Y = \text{Cnq}(X).$$

For instance the set of sentences 1–14 is an axiom system of the set of all the tautologies of the classical sentential calculus. This theorem will not be proved here. Its proof is left to the reader as an exercise which will be easy to perform on the strength of the theorems to be proved below.

It will be demonstrated now that if X is an axiom system of a set Y, then the set Y is a theory.

THEOREM 4. *For any set X, the set $\text{Cnq}(X)$ is a theory.*

PROOF. It will be demonstrated first that the set Cnq(X) is closed under the two logical rules.

1. If $A \in$ Cnq(X), then, by Definition 3, there exists a proof $D_0, ..., D_n = A$ of the formula A, based on the set X. Thus, if a formula C is obtained from the formula A by substitution, then the set $D_0, ..., D_n, D_{n+1} = C$ is a proof of the formula C, based on the set X. The set Cnq(X) is thus closed under the operation of substitution.

2. If $A \in$ Cnq(X) and $(A \to B) \in$ Cnq(X), then, by Definition 3, there exist the proofs $D_0, ..., D_n = A$ and $D'_0, ..., D'_m = (A \to B)$ of the formulae A and $(A \to B)$, based on the set X. The sequence

$$D_0, ..., D_n, D'_0, ..., D'_m, B,$$

formed of these two sequences, is then a proof of the formula B, based on the set X. As can easily be seen, the conditions laid down in Definition 3, are satisfied. The formula B is thus a consequence of the set X. The set Cnq(X) is thus closed under the operation of detachment.

If the set Cnq(X) is closed under the logical rules of substitution and detachment, then, by Theorem 3, it is a theory and hence, on the substitution of Cnq(X) for X, it satisfies formula (58)

(59) $\text{Cnq}(\text{Cnq}(X)) = \text{Cnq}(X)$.

Thus every theory forms its own axiom system. This statement, however, is not interesting. In the case of every theory we usually want to know which sets of sentences, preferably the least numerous ones, are axiom systems of that theory. The problem whether a finite set of axioms exists is of particular interest. A theory for which there exists a finite axiom system is called *finitely axiomatizable*.

DEFINITION 6. *A set Y is finitely axiomatizable if and only if there exists a finite set X such that $Y = $ Cnq(X).*

THEOREM 5. *The set of the tautologies of the c.s.c. is finitely axiomatizable.*

PROOF. It will be demonstrated that every formula which is a sentential schema is reducible to a special equivalent form which is called a normal form. Such a reduction can be made by reference to a finite number of theorems, which will be stated explicitly. Next we note that if a formula which has a normal form is a tautology, then it follows

107

easily from a number of explicitly stated tautologies. The proof will then be complete.

DEFINITION 7. *A formula α has a conjunctive-disjunctive normal form if and only if α is a conjunction of a number of disjunctions $\alpha_1, \ldots, \alpha_n$, the terms of these disjunctions being sentential variables or negations of sentential variables*:

$$(60) \qquad \alpha = \underbrace{(p \vee q \vee \ldots \vee \sim p \vee \sim r)}_{\alpha_1} \wedge \underbrace{(s \vee \sim t \vee \ldots)}_{\alpha_2} \wedge \ldots \wedge$$

$$\underbrace{(p \vee \sim s \vee \ldots)}_{\alpha_n}.$$

In particular, α may be a single disjunction, and every disjunction may in a particular case be reduced to a single term.

Concerning the formulae which have the normal form as described above it can easily be noted that

LEMMA 1. *If α has a c.-d. normal (conjunctive-disjunctive) form and if α is a conjunction of disjunctions $\alpha_1, \ldots, \alpha_n$, then α is a tautology if and only if in each of the disjunctions $\alpha_1, \ldots, \alpha_n$ some variable occurs once without the negation symbol and once with the negation symbol*:

$$(61) \qquad \alpha = \underbrace{(p \vee q \vee \sim p \vee \ldots)}_{\alpha_1} \wedge \underbrace{(q \vee r \vee \sim r \vee \ldots)}_{\alpha_2} \wedge \ldots \wedge$$

$$\underbrace{(s \vee \sim s \vee r \vee \ldots)}_{\alpha_n}.$$

In fact, the conjunction α is a tautology if and only if all its terms $\alpha_1, \ldots, \alpha_n$ are tautologies. Hence it suffices to examine any disjunction

$$(62) \qquad \alpha_i = (p \vee \sim p \vee q \vee \ldots \vee r).$$

If a variable, for instance p, occurs in α_i on one occasion without the negation symbol, and on another occasion with the negation symbol, then if we assume $p = 0$, then $\sim p = 1$; now the zero-one verification method for any substitution yields either $p = 1$ or $\sim p = 1$, so that at least one term of the disjunction is always true. The disjunction α_i is thus true for any substitution, and hence it is a tautology. If the disjunction α_i has no variable which occurs in it in both contexts: with the negation symbol and without it, then one can easily make substitutions which turn it into a false sentence. It suffices to substitute 0 for all those variables p_t which occur without the negation symbol ($p_t = 0$), and 1 for those variables p_u which occur with the negation symbol ($p_u = 1$). Then $\sim p_u = 0$, all the terms of the disjunction α_i are false, the entire disjunction takes on the value 0, and thus cannot be a tautology.

LEMMA 2. *If α has a c.-d normal form and if α is a tautology, then α is a consequence of tautologies 1–14 and of*:

41. $(p \vee \sim p) \vee q$,

42. $p \vee q \equiv q \vee p$,

43. $p \vee (q \vee r) \equiv (p \vee q) \vee r$,

44. $(p \equiv q) \rightarrow ((p \vee r) \equiv (q \vee r))$,

45. $(p \equiv q) \rightarrow ((q \equiv r) \equiv (p \equiv r))$.

In fact, if α is a tautology, then, by Lemma 1, every disjunction α has the form (62) and, accordingly, has a variable which on one occasion occurs with the negation symbol, and on another occasion occurs without the negation symbol. If that disjunction has exactly the form (62), that is, if it falls under the schema

$$p \vee \sim p \vee \gamma,$$

then it can be obtained from 41 by substitution q/γ. If the variables and their negations are grouped in a way different from that in which they occur in formula (62), then they can be regrouped in an arbitrary way on the strength of the laws of commutativity 42 and associativity 43 for disjunction, and theorems 44, 45, and 1–5. Having thus obtained all the disjunctions $\alpha_1, \ldots, \alpha_n$ we may combine them into a conjunction in any order we wish on the strength of Axiom 10.

LEMMA 3. *For every schema α there exists a schema α' of the c.-d. normal form, such that the equivalence $\alpha \equiv \alpha'$ is a consequence of tautologies 1–14 and 42–45 and*:

46. $(p \wedge q) \equiv (q \wedge p)$,

47. $(p \wedge (q \wedge r)) \equiv ((p \wedge q) \wedge r)$,

48. $p \wedge (q \vee r) \equiv (p \wedge q) \vee (p \wedge r)$,

49. $p \vee (q \wedge r) \equiv (p \vee q) \wedge (p \vee r)$,

50. $(p \rightarrow q) \equiv (\sim p \vee q)$,

51. $(p \equiv q) \equiv ((p \to q) \wedge (q \to p))$,

52. $\sim (p \wedge q) \equiv (\sim p \vee \sim q)$,

53. $\sim (p \vee q) \equiv (\sim p \wedge \sim q)$,

54. $\sim \sim p \equiv p$,

55. $(p \equiv q) \to ((p \to r) \equiv (q \to r))$,

56. $(p \equiv q) \to ((r \to p) \equiv (r \to q))$,

57. $(p \equiv q) \to ((p \wedge r) \equiv (q \wedge r))$,

58. $(p \equiv q) \to (\sim p \equiv \sim q)$.

Theorems 44, 45, 55–58 allow us to interchange equivalent parts in any formula: those occurring in the antecedent of implication 55, in the consequent of 56, in disjunction 44, etc. On the strength of these theorems it can be proved in general that if $\alpha(p)$ is any formula which contains the variable p and does not contain the variable q, and if $\alpha(q)$ is obtained from $\alpha(p)$ by subst. p/q, then the tautologies given above yield the theorem of the form

(63) $(p \equiv q) \to (\alpha(p) \equiv \alpha(q))$,

which states that a formula may be replaced by an equivalent in any place. In carrying out such equivalence replacements we will refer to theorems 50–54 as well.

Theorems 50 and 51 allow us to remove the symbols of equivalence and implication from a formula α. The applications of such equivalence transformations will be illustrated below by the reduction of the following formula α:

(64) $(p \to q) \to ((p \wedge r) \to (q \wedge r))$

to a normal form. Using 50 we transform (64) into

(65) $\sim (\sim p \vee q) \vee (\sim (p \wedge r) \vee (q \wedge r))$.

The formulation: *we transform* (64) *into* (65) means here the same as: *on the strength of* 50 *and* (63) *we can deduce the equivalence* (64) $\equiv$ (65). This equivalence is a consequence of the theorems referred to above. This leaves in α only the symbols of disjunction, conjunction, and negation. On the removal of implications and equivalences the symbols of negation may occur in various places. They may precede conjunctions and disjunctions, as is the case in (65). Theorems 52 and 53 allow us to replace a negation of conjunctions by a disjunction of negations, and a negation of disjunctions by a conjunction of negations. Thus, iterated applications of 52 and 53 can remove those symbols of negation which precede disjunctions and conjunctions. For instance, by applying 52 and 53 to formula (65) we obtain

(66) $\quad (\sim \sim p \wedge \sim q) \vee ((\sim p \vee \sim r) \vee (q \wedge r))$.

Then a symbol of negation will precede only a variable or another occurrence of the symbol of negation. But the negation of a negation yields an affirmation by 54, and, hence, ultimately all negation symbols can be removed from all places except those in which a negation symbol directly precedes a variable. For instance, by applying 54 to formula (66) we obtain

(67) $\quad (p \wedge \sim q) \vee ((\sim p \vee \sim r) \vee (q \wedge r))$.

Thus, we now know how to reduce a formula α to a form in which only variables, their negations, and various conjunctions and disjunctions occur. But now we still have to arrange it so that every term of a disjunction is either a single variable or the negation of a single variable. Such a transformation may be achieved on the strength of Theorem 49, which states that disjunction is distributive with respect to conjunction. Hence, if a term of a disjunction is a conjunction, we transform that disjunction, by Theorem 49, into a conjunction of disjunctions. Iterated application of that procedure yields the formula α', which has a c.-d. normal form. By applying 49 to (67) we are thus in a position to gradually remove the conjunction symbols from the various terms of a given disjunction. For instance, the last disjunction in (67) has the conjunction $(q \wedge r)$ as its second term. By applying 49 we transform (67) into

(68) $\quad (p \wedge \sim q) \vee ((\sim p \vee \sim r \vee q) \wedge (\sim p \vee \sim r \vee r))$.

The right-hand term of the main disjunction is now a conjunction. By applying Theorem 49 to it we obtain the formula

(69) $\quad ((p \wedge \sim q) \vee \sim p \vee \sim r \vee q) \wedge ((p \wedge \sim q) \vee \sim p \vee \sim r \vee r)$.

This formula still includes two conjunctions of the form $(p \wedge \sim q)$, which occur as terms of disjunctions, so we apply 49 again and ultimately obtain

$$\alpha' = (p \vee \sim p \vee \sim r \vee q) \wedge (\sim q \vee \sim p \vee \sim r \vee q) \wedge$$
$$(p \vee \sim p \vee \sim r \vee r) \wedge (\sim q \vee \sim p \vee \sim r \vee r).$$

Obviously, in making these transformations we repeatedly avail ourselves of the laws of commutativity and associativity of disjunction and conjunction.

Since all the transformations performed above were based on equivalences specified in the lemma in question, the resulting equivalence

$$\alpha \equiv \alpha'$$

also is a consequence of tautologies specified in the lemma.

Using Lemmas 2 and 3 together we immediately obtain a proof of the theorem in question. Formula α' has a normal form, and hence, if it is a tautology, then it follows from theorems specified in Lemma 2. The equivalence $\alpha \equiv \alpha'$ follows from theorems specified in Lemma 3, and α follows from α' and $\alpha \equiv \alpha'$ on the strength of 1–5. Hence, if α' is a tautology, then α is a consequence of tautologies specified in the two lemmas. The equivalence $\alpha \equiv \alpha'$ is a tautology, because it follows from the tautologies specified above, and by Theorem 2, the set of tautologies is a theory. Thus, if α is a tautology, then α' must be a tautology, too. But since α' is a tautology, then α is a consequence of the theorems specified above. Hence, if α is a tautology, then α is a consequence of theorems specified in the two lemmas. The number of those theorems is finite, and hence the set of tautologies is finitely axiomatizable.

We shall now reduce a number of formulae to c.-d. normal form. For instance, $(p \rightarrow q) \wedge (q \rightarrow p)$ can immediately be transformed into $(\sim p \vee q) \wedge (\sim q \vee p)$. The formula $(p \rightarrow q) \rightarrow q$ changes in turn into: $\sim (\sim p \vee q) \vee q$, $(\sim \sim p \vee \sim q) \vee q$, $(p \wedge \sim q) \vee q$, $(p \vee q) \wedge (\sim q \vee q)$. Reduction to a normal form might thus serve as a method of verifying the truth of the formula of the c.s.c.

As mentioned above, the axiom system of the c.s.c. can be made much simpler. The axiom system of the c.s.c. becomes particularly simple if we reduce the number of primitive terms and assume the possibility of making definitions. Every theory includes a number of concepts (terms) which are referred to at the very outset, in the axioms of that theory, and then includes a large number of other terms, which, if necessary, may be introduced in the theory by way of definitions. The same procedure can be used in the sentential calculus. For instance, the c.s.c. may be based on two primitive terms: implication and negation, when the following axioms (given by the Polish logician, J. Łuka-siewicz):

$$(p \to q) \to ((q \to r) \to (p \to r)),$$
$$p \to (\sim p \to q),$$
$$(\sim p \to p) \to p,$$

the following definitions:

$$p \lor q \overset{\text{df}}{=} \sim p \to q,$$
$$p \land q \overset{\text{df}}{=} \sim (p \to \sim q),$$
$$p \equiv_i^r q \overset{\text{df}}{=} (p \to q) \land (q \to p),$$

and a special rule on the use of such definitions are adopted. The rule in question states:

In any theorem of the c.s.c., any part of that theorem, which is equiform with one side of a definition, may be replaced by a formula which is equiform with the other side of that definition.

This rule is called the *rule of definitional replacement*. When applied, for instance, to the second axiom of those listed above, it yields the theorem $p \to (p \lor q)$.

The set of the consequences of the three axioms listed above, when the rule of definitional replacement is added to the rules of substitution and detachment, is identical with the classical sentential calculus.

Theorems on the consistency and completeness of the c.s.c. are important theorems and will be discussed later, when the problems of consistency and completeness are considered in full.

This completes our exposition of the sentential calculus. It is, however, worth mentioning that research on sentential calculi is not confined to what has been written here. That research covers very many issues, both those which are formal in nature and those which are concerned with interpretations. There are many systems of the sentential calculus, especially those which make it a point to describe implication in a weaker way than that used in the classical sentential calculus, and also weaker than that used in the intuitionistic calculus. Such systems are called *systems of strict implication*. Some authors describe strict implication by reference to the concept of possibility by stating that

p implies q means that *it is not possible that* $(p \land \sim q)$.

113

I. CLASSICAL LOGICAL CALCULUS

The concept of possibility is described axiomatically in systems which are called *modal logics*. While there are mathematicians who in practice do not go beyond the apparatus of intuitionistic logic, there are no mathematicians who would in practice use modal logic. Nevertheless, modal logics offer many problems which are still open and which are interesting in themselves.

As well as logics with formulae of finite length, logics with formulae of infinite length are also studied. Such formulae do not, of course, occur in practice, but their theory is a mathematical theory of some interest.

EXERCISES

1. Prove the following general property of the concept of consequence:

$$X \subset Y \to \mathrm{Cnq}(X) \subset \mathrm{Cnq}(Y).$$

2. The fragmentary systems of logic: those of positive implication, of positive logic, and of intuitionistic logic, are defined as the sets of consequences of some of Axioms 1–14, namely:

the system of positive implication	= Cnq (1, 2),
positive logic	= Cnq (1–11),
intuitionistic logic	= Cnq (1–13).

What relations of inclusion between these systems follow from the theorem proved in exercise 1?

3. Deduce, in positive logic, the following theorems:

59. $(p \wedge q) \to q$
    (from 7, 9, 30).

60. $(p \to (q \to s)) \to ((p \wedge q) \to s)$
    (from 9, 59, and from a theorem in the system of positive implication, which must be deduced separately, namely $(t \to p) \to \big((t \to q) \to$
    $((p \to (q \to s)) \to (t \to s))\big)$ (from 2, 30, 29, 36)).

Likewise, in any other proof in positive logic we have first to select appropriate theorems in the system of positive implication, and next apply them to theorems in which conjunctions and disjunctions occur. In this way we deduce the following theorems:

61. $((p \wedge q) \to s) \to (p \to (q \to s))$
    (from 10).

62. $(t \to p) \to \big((t \to q) \to (t \to (p \wedge q))\big)$
    (from 10).

63. $q \to (p \wedge q)$
    (from 6, 8).

114

64. $(p \to r) \to \big((q \to r) \to ((p \lor q) \to r)\big)$
    (from 61, 11).

4. Partly on the strength of the foregoing exercises, prove that positive logic is identical with the set Cnq(1–5, 8, 9, 59, 62–64).

5. Prove, in the field of positive logic, those laws which are concerned with conjunction and disjunction, specified in Exercise 6, p. 95, which are algebraic in nature (associativity and commutativity of conjunction and disjunction, and distributivity of conjunction with respect to disjunction and disjunction with respect to conjunction).

6. On the strength of the theorems of positivie logic thus obtained deduce in intutionistic logic the formulae listed below:

65. $(p \to q) \to ((p \to \sim q) \to \sim p)$
    (from 62, 12).

66. $p \to (\sim p \to (p \land \sim p))$
    (from 10).

67. $p \to \sim \sim p$
    (from 66, 12, $p/\sim p$, q/p, 30).

68. $\sim (p \land \sim p)$
    (from 65, $p/(p \land \sim p)$, q/p, 9, 59).

69. $(p \to \sim q) \to (q \to \sim p)$
    (from 1, 65).

70. $(p \to q) \to (\sim q \to \sim p)$
    (similarly).

71. $\sim (p \lor q) \to \sim p$
    (from 70, 8).

72. $\sim (p \lor q) \to (\sim p \land \sim q)$
    (from 71 and analogically p/q, q/p).

73. $\sim (p \lor q) \equiv (\sim p \land \sim q)$
    (from 72 and conversely).

74. $(\sim p \lor \sim q) \to \sim (p \land q)$
    (from 9, 59, 70, 64).

75. $(p \to \sim p) \to \sim p$
    (from 65, 23).

76. $\sim \sim \sim p \to \sim p$
    (from 67, 70).

77. $(p \lor q) \to (\sim p \to q)$
    (from 1, 13, 59).

78. $(p \lor \sim p) \to (\sim \sim p \to p)$
    (from 77, $p/\sim p$, q/p).

79. $((p \lor \sim p) \to \sim q) \to \sim q$
    (from 69, $p/p \lor \sim p$, 80, 12).

N o t e. The implications converse to 67, 70, 74, 77, 78, which are theorems in the c.s.c., are not theorems in the intuitionistic calculus. But the proof of *non sequitur* is much less elementary. Likewise, no equivalence that could serve as a definition of any of the three terms: $\to$, $\land$, $\lor$, by means of the remaining terms and negation is a theorem in the intuitionistic calculus.

7. Prove that for every law α of classical logic the formula $\sim\sim\alpha$ is a theorem in intuitionistic logic. Base the proof on Theorem 67 and on the following two theorems:

80. $\sim\sim(p \lor \sim p)$
    (from 72, $q/\sim p$, 68, $p/\sim p$, 12).
81. $\sim\sim(p \to q) \to (\sim\sim p \to \sim\sim q)$
    (from 70 repeatedly and 36)

and also on the fact that the set 1–14 is an axiom system of the c.s.c.

8. Prove that the c.s.c. is identical with the set of consequences of 1–14.

9. A formula α has a disjunctive-conjunctive normal form, dual to a normal conjunctive-disjunctive form, if and only if it is a disjunction of a number of conjunctions such that in each conjunction the terms are variables and/or their negations only, e.g.:

$$(p \land \sim q \land \ldots) \lor (r \land \sim s \land \ldots) \lor \ldots \lor (q \land \sim p \land \ldots).$$

Prove that every formula of sentential logic can be transformed into an equivalent formula of a disjunctive-conjunctive normal form. What is the characteristic property of those formulae of such a normal form which are equivalent to the laws of the c.s.c.?

4. THE CLASSICAL CONCEPT OF QUANTIFIER

The beginnings of the sentential calculus date back to antiquity. However, the concept of quantifier developed only in the second half of the 19th century, and the predicate calculus in its present form was formulated only in the early 20th century.[1] In order to introduce that concept in a systematic way it is necessary to reflect for a while on the language of those theories in which we encounter quantifiers. Those theories

---

[1] Cf. the historical outline in the Supplement. More comprehensive historical data are to be found in books by Church [8], Bocheński [6], and Scholz [82], and also in the article by Łukasiewicz [49]. Church's book [8] provides historical data concerning all major observations made in logic.

refer to various mathematical domains. Thus, the variables occurring in those theories stand for any individuals in the given domain. Those variables are as it were names of those objects to which a given theory refers, for instance, numbers, points, sets, etc.; those objects are referred to by special phrases. For instance, when referring to numbers we commonly use such phrases as: $x = y$, $x = (y+z)$, $x = (y \cdot z)$, $x > y$. Using such phrases, sentential connectives, and quantifiers, which will be discussed below, we can define many other mathematical phrases, such as: $x = 2y(x+z)$, x is even, $x > 0$, etc. Those phrases describe certain relations between numbers or certain properties of numbers. Such formulae, which contain variables and belong to mathematical theories, are called sentential functions or predicates. Thus, a *predicate* or a *sentential function* is a formula which contains some variables and describes a property or a relation. The predicate $x > 0$ describes the property of being a positive number. The predicate $x = y$ describes the identity relation. The sentential function $x = 2y(x+z)$ describes a relation which holds between three numbers and is defined by the operations of addition and multiplication. In any mathematical theory, we can always single out from its predicates the atomic predicates, that is, such that no part of them is itself a predicate (original or basic formulae), and compound predicates, which are more or less complex combinations of basic formulae. In arithmetic, the predicate $x = y+z$ is usually treated as basic, whereas the phrase x *is a prime number*, which can be stated in terms of addition and multiplication, usually is not a basic formula.

In this section we shall not be interested in predicates of any definite mathematical theory. We shall assume in general that mathematical theories always include a number of predicates which describe certain properties of, and relations among, elements of a given mathematical domain. For the time being we shall not be interested in the definite meaning of those properties and relations. This is why, both in this section and in the following one, predicates will be symbolized by formulae lacking any fixed mathematical meaning, such as:

$$P(x), R(x, y), S(x, y, z), T(x, y, z, v), \text{ etc.}$$

Such formulae will stand for any mathematical predicates with as many variables as letters in the parentheses to the right of the capital letter.

117

Thus, $P(x)$ may stand both for the property $x > 0$ and for the property "x is even", and $R(x, y)$ may stand both for the relation $x > y$ and for the relation $x = y$, etc.

Suppose for a moment that we are interested in natural numbers only. Those numbers have names, namely the inscriptions: 0, 1, 2, 3, 4, ... If a property described by the predicate $P(x)$ is an attribute of the number 1, then this fact is described by the formula $P(1)$. The sentences: $P(1)$, $P(2)$, $P(3)$, $P(4)$ state that the numbers 1, 2, 3, 4 have the property described by the predicate $P(x)$. That property may be, for instance, $x > 0$; then the sentences given above are true and have the form: $1 > 0, 2 > 0, 3 > 0, 4 > 0$. If we want to state in a single sentence that all the four numbers 1, 2, 3, 4 have the property P, then we combine these four sentences into a conjunction:

(70) $P(1) \wedge P(2) \wedge P(3) \wedge P(4).$

If we would use this procedure to state that every natural number has the property P, we would have to write down an infinite conjunction

(71) $P(0) \wedge P(1) \wedge P(2) \wedge P(3) \wedge P(4) \wedge P(5) \wedge \ldots$

Since it is not possible to write down an infinite conjunction, we have to resort to a finite formulation. The simplest way out is to use the sentence

(72)   *For every x, $P(x)$.*

The phrase *for every x...* is called the *universal quantifier* binding the variable x; it is often abbreviated with the symbol $\bigwedge x$. Thus, in a symbolic notation, sentence (72) is written as follows:

(73) $\bigwedge x P(x).$

The infinite sentence (71) may thus be considered equivalent to the sentence (73) if natural numbers are the only objects under consideration:

(74) $\bigwedge x P(x) \equiv P(0) \wedge P(1) \wedge P(2) \wedge P(3) \wedge \ldots$

This equivalence may be considered as informal definition of the universal quantifier. The universal quantifier is thus, as it were, an infinite conjunction.

Similar reflections lead us to the conclusion that the existential quantifier, that is, the phrase *for some x...*, is, as it were, an infinite disjunction. If we want to say that among those numbers which are less than 5 there exists a number which has the property P, we may formulate that idea as the disjunction

$$P(0) \lor P(1) \lor P(2) \lor P(3) \lor P(4).$$

It is so because only one of the numbers 0, 1, 2, 3, 4 can be that number less than five. Should we state, in a similar way, that there exists in general a natural number which has the property P, we would have to resort to an infinite disjunction

(75) $\qquad P(0) \lor P(1) \lor P(2) \lor P(3) \lor P(4) \lor P(5) \lor \ldots$

The phrase *there exists an x such that...* is considered to be equivalent to the phrase *for some x....* In a symbolic notation these phrases are written in one and the same way as $\bigvee x \ldots$, which is called the *particular* or *existential quantifier*. If natural numbers are the only objects under consideration, then the infinite disjunction (75) may be considered equivalent to the formula $\bigvee x P(x)$:

(76) $\qquad \bigvee x P(x) \equiv P(0) \lor P(1) \lor P(2) \lor P(3) \lor \ldots$

This equivalence may be considered as informal definition of the existential quantifier as applied to natural numbers.

The phrases which we have called quantifiers play an important role in the construction of mathematical theories. That branch of logic which we are going to discuss now and which is called the predicate or quantificational calculus, gives the correct uses of quantifiers. Before, however, we proceed to describe the predicate calculus we must first discuss in greater detail the concept of predicate as a phrase defining a property or a relation; for certain properties of predicates are indispensable elements in the description of the predicate calculus.

The quantifiers are the most important instrument with which new predicates are constructed. The predicate $x \leqslant z$ describes a relation. But the formula

(77) $\qquad \bigwedge z(x \leqslant z)$

does not define any relation, but defines a certain property of the number x, namely that every number z is greater than or equal to the number x. Thus, formula (77) defines the property which might be stated in words as follows: x is the least of all the numbers under consideration. If the numbers under consideration (that is those which belong to the domain in question) are natural numbers only, then formula (77) defines a property which is an attribute of the number 0, since zero is the least natural number. Thus, by applying a quantifier to a formula that describes a relation we obtain a formula that defines a property. It is true that in formula (77), which defines the property of being the least number, there are two letters which stand for variables, but their role is not the same. Only the variable x is a genuine variable, or, as it is usually said, a *free variable*; the letter z in formula (77) does not stand for a free variable. The letter z serves only to describe that property of the number x to which formula (77) refers. The letter z is, as is usually said, *bound* by the quantifier $\bigwedge z$ ... The distinction between free and bound variables plays an important role in the predicate calculus. We shall discuss this in greater detail. Consider one more example. The formula

(78) $x+y = z$

has three free variables, and accordingly it describes a three-place relation. If in (78) we bind the variable y by the existential quantifier, we obtain the formula

(79) $\bigvee y(x+y = z)$,

which now describes only a two-place relation between the numbers x and z. If we consider only the domain of natural numbers, then the relation $x \leqslant z$ would be what is described in (79). If to formula (79) we apply the universal quantifier binding the variable z: $\bigwedge z...$, then we obtain the formula

(80) $\bigwedge z \bigvee y(x+y = z)$,

which describes only a certain property of the single number x. If we confine ourselves to the domain of natural numbers, this is the property of being the least natural number which was also described by formula (77). The variables z and y are bound in (80), and the variable x is the only free variable in (80). When we apply to (80) the existential quantifier $\bigvee x...$, which binds the only remaining free variable, we obtain the formula

(81) $\bigvee x \bigwedge z \bigvee y(x+y = z)$,

which no longer contains any free variables. It does not describe any property of the numbers x, y, z, and is not a predicate. Formula (81) is a mathematical sentence. Should natural numbers be the only objects under consideration, sentence (81) would express the same mathematical idea as the sentence

(82) $\quad \bigvee x \bigwedge z(x \leqslant z).$

In everyday language that idea can be expressed as

(83)   *There exists a least natural number.*

If all the individuals are interpreted as natural numbers, sentences (81) and (82) state a mathematical truth and are thus true sentences.

This kind of analysis of mathematical formulae leads to the conclusion that a mathematical formula defines a relation holding among as many variables (a relation of as many arguments) as there are free variables in the formula in question. Free variables are those variables which are not bound by any quantifier. Those formulae which do not contain any free variables are not predicates and hence do not define any relations or properties, but are sentences: they state certain definite mathematical ideas, true or false.

What has so far been said about variables, predicates and sentences does not include precise definitions of those concepts. More precise definitions will be useful in the text below, and so they will be introduced at once. Note first that these concepts have a visual sense: they may be interpreted as referring to the graphic form of the mathematical formulae. The question whether a formula X in a mathematical theory is a predicate (that is, defines a property or a relation) or a sentence (that is, states a complete idea) can be answered without any reference to the meaning of the formula X, but just on the strength of its graphic structure. The definitions of a free and a bound variable are purely structural in nature. Many other concepts concerned with mathematical theories are of the same nature. In general, a concept concerned with mathematical theories is called *syntactic* if the definition of that concept refers to the forms of mathematical formulae, only, and not to their meanings. On the other hand, if a definition of a concept concerned with certain formulae cannot dispense with referring to the meanings of those formulae, then such a concept is called *semantic*. The concepts of free and bound variable, sentence, and predicate, as applied to mathematical theories, are syntactic in nature.

(The differentiation of concepts concerned with mathematical theories into the syntactical and the semantic ones dates from the 1930's. The studies by A. Tarski contributed much to that distinction. These concepts, however, are not quite precise, which will be discussed later. It seems that the distinction between the syntactic and the semantic concepts is rather a preliminary one on the way toward a logical classification of concepts, to be outlined in Chapter II.)

To bring out the syntactic character of the above concepts we shall reformulate their definitions to make them more precise. Yet to define these concepts we shall make use of only very simple graphic relations holding between expressions. These relations are:

(84)
an expression X is equiform with an expression Y,

an expression X is contained in an expression Y (X is a part of Y),

an expression X follows directly an expression Y (Y directly precedes X).

The meaning of these relations can easily be grasped from the empirical point of view. An expression *X is equiform with an expression Y* if it would be said in everyday parlance that *X* and *Y* are two instances of the same expression.

For instance, *X* has the form of the letter "*A*" and *Y* has the form of letter "*A*", but the two formulae occur in different places. For instance, in the title of the present book, in the next to last word the first letter is equiform with the sixth letter. Likewise, in formula (82) the second and the sixth symbol are equiform, since they are two instances of the variable "*x*". Formula (82) contains a part which is equiform with formula (77). Formula (81) has a part which is equiform with formula (80), and also a part equiform with formula (79). In formula (81), that part of it which has the form "$\bigvee x$" precedes that part which is equiform with formula (80), etc. The above examples explain the relations of equiformity, part and immediate precedence as holding between formulae. The use of the quotation marks is also to be explained in this connection. By a formula inside quotation marks we mean the set of all those formulae which are equiform with that occurring inside the quotation marks. Thus "$\bigvee$" is the set of all formulae equiform with the first symbol in formula (81). Hence the first symbol in formula (81) and the first symbol in formula (82) belong to the set "$\bigvee$".

Now that we have explained these concepts we can state the following syntactic definitions. These definitions are stated in an informal language, to make them appeal better to the intuition, but they can easily be formulated with pedantic precision.

DEFINITION 8.

A. *Every symbol belonging to the set "$\bigwedge$" is called a universal quantifier.*

B. *Every symbol belonging to the set "$\bigvee$" is called an existential quantifier.*

C. *Any letter which directly follows a quantifier is called a variable under a quantifier.*

D. *A quantifier is then said to bind that variable.*

DEFINITION 9. *The whole formula in parentheses directly following the variable under a quantifier is called the scope of the quantifier binding that variable.*

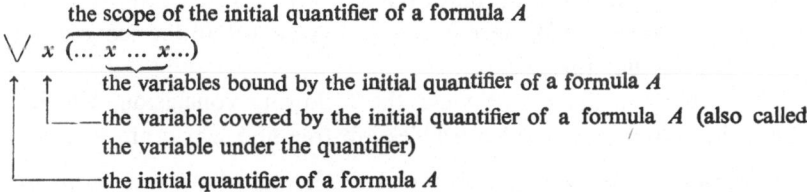

Fig. 3. Schema of a formula A which begins with a quantifier and has no quantifiers in the scope of the initial quantifier

If there is no risk of misunderstanding as to the scope of a given quantifier, its scope will not be parenthesized in order not to overload formulae with parentheses. For instance, in formulae (80), (81), (82) the scope of each quantifier obviously covers the entire formula from the variable under the given quantifier to the end of the given formula.

DEFINITION 10.

A. *If no other quantifier occurs within the scope of a quantifier that binds a variable, then all the variable letters equiform with the variable under that quantifier and occurring within its scope are variables bound by that quantifier.*

B. *If quantifiers occur within the scope of the initial quantifier of a formula A, then all those variables equiform with the variable under the initial quantifer and occurring within its scope which are not bound by any quantifier occurring within the scope of the initial quantifier of formula A are variables bound by the initial quantifier of formula A.*

C. *All those variables which occur in a formula A and are bound by quantifiers occurring in formula A are called bound variables in formula A.*

DEFINITION 11. *A letter is a free variable in a formula A if it is not a bound variable in A.*

It is worth noting in this connection that, in accordance with the definitions given above, out of two letters equiform with one another and occurring in the same formula A one may be free while the other is bound. For instance, in the formula

(85) $\qquad \bigwedge x (x = x) \lor x = 0$

the fourth and the sixth symbol are bound variables in (85), since they occur within the scope of the quantifier that binds a variable which is equiform with them, but the ninth symbol, which is equiform with them, is a free variable in (85), since it does not occur within the scope of any quantifier in that formula.

The foregoing definitions yield the following conclusion, which is useful in finding out which variables are free and which are bound:

A variable is bound by a quantifier Y if that variable is free in the formula that forms the scope of the quantifier Y, and if it is bound in the formula consisting of the quantifier Y, the variable under that quantifier, and the scope of that quantifier.

Consider the following example:

(86) $\qquad \bigwedge x (\bigwedge x (x = x) \lor x = 0);$

in formula (86) the variable which occurs as the twelfth symbol is bound by the first quantifier of formula (86), since it is free in the formula which forms the scope of the quantifier, as we have stated when analysing formula (85), which is equiform with the scope of the initial quantifier of formula (86). On the other hand, the seventh and the ninth symbol in formula (86) are bound by the second quantifier in (86). The twelfth and the seventh symbol, even though they are equiform, are bound by different quantifiers. This is why they play the role of variables of different forms. The same idea which is stated by (86) may be formulated with the form of one of these variables modified and the shape of the other left unchanged, e.g., in the sentence

$$(87) \qquad \bigwedge z (\bigwedge x(x = x) \lor z = 0).$$

Such situations as occur in the case of formulae (85) and (86) could be avoided once and for all if they were replaced by formulae of the type of (87), but this would make inference rules somewhat more complicated.

DEFINITION 12. *A formula X is a predicate of n arguments (n variables) if it is constructed correctly and contains n free variables, no two of which are equiform with one another, and every other free variable in X is equiform with one of those n distinct variables.*

DEFINITION 13. *A formula X is a sentence (a closed sentence) if X is constructed correctly and does not contain free variables.*

For instance, formula (85) is a predicate of one argument, whereas formula (86) is a sentence.

Definitions of the same concepts can be formulated without any reference to the concept of equiformity. For instance, when describing formula (86) one can say that the variable x occurs in it in the 12th place bound by the first quantifier, and in the seventh and the ninth places bound by the second quantifier. But more precise definitions of phrases used in this way require reference to the concept of equiformity.

The concept of predicate was initially described above in terms of semantics (by reference to the meanings of formulae). It was said that formula that defines a relation or a property, for instance, $x < y$, $x < 0$, etc., is a predicate. Now, however, we have defined predicate in terms of syntax only, as a formula that contains free variables. Likewise, sentence was first defined as a formula that states a complete idea. This is why every sentence is true or false. Now, however, with reference to mathematical theories we have defined a sentence as a formula without free variables, that is, in purely syntactic terms. Non-occurrence of free variables accounts for the fact that, in a mathematical theory, a formula states a complete and fully defined idea. Thus, when it comes to formulae in mathematical theories, certain concepts which commonly are defined in terms of semantics can be defined in terms of syntax. We shall witness this phenomenon again.

The definitions formulated above did not state with precision what is meant by a *correctly constructed* (or *meaningful*) *formula*. We shall return to that topic soon, but for the time being we shall confine ourselves to the intuitive explanation that every formula in which all mathematical symbols are in their proper place is constructed correctly. Thus, the formulae: $x < y$, $\bigwedge z((x < z) \to (z < u))$, $x+y = z \lor x < z$, as well as formulae (77)–(87) are constructed correctly. On the contrary, the formulae: $x < +$, $(x+z) \to (x < z)$, $\bigwedge ((z < z) \to (z < u))$, $(x \to u) + 0 = x$, $\to < x$, are not constructed correctly.

Now that we have formulated the concept of predicate with more precision we return to how to interpret the function of quantifiers. As applied to natural numbers 0, 1, 2, ..., the quantifiers are best interpreted as infinite conjunctions and disjunctions. As applied to real numbers, the quantifiers have a very clear geometrical interpretation as the *operations of projection* and *of inner projection*. This interpretation will be outlined below.

Suppose that real numbers are the mathematical domain under consideration. Let $R(x, y)$ be a predicate of two arguments, defined for all pairs $\langle x, y \rangle$ of real numbers. This predicate thus defines a certain relation R holding between real numbers. A relation of two arguments, being a set of pairs, can easily be mapped onto a plane with designated co-ordinates. Since there is a one-to-one correspondence between every point of that plane and a pair $\langle x, y \rangle$ of real numbers, there is a correspondence between the relation R, defined by the predicate $R(x, y)$, and the set of those points z of that plane such that the relation R holds between their co-ordinates $\langle x_z, y_z \rangle$, that is, which satisfy the formula: $R(x_z, y_z)$. Let that set be denoted by R and called the *graph of the predicate* $R(x, y)$. Consider now the interpretation of the predicate of one argument:

$$(88) \qquad \bigvee y(R(x, y)),$$

obtained by the application of the existential quantifier to the predicate $R(x, y)$. Since the predicate (88) has one argument, its graph is not a set of pairs, but just a set of real numbers x for which there exist

126

numbers y such that the point $\langle x, y \rangle$ belongs to the plane set R. The vertical projection of the set R on the x axis (cf. Figure 4) is such a unidimensional set. In fact, it follows from the definition of a projection that for every point x of a projection there is a coordinate y such that the pair $\langle x, y \rangle$ belongs to the set R. An interpretation of the predicate of one argument

(89) $\bigwedge y(R(x, y))$

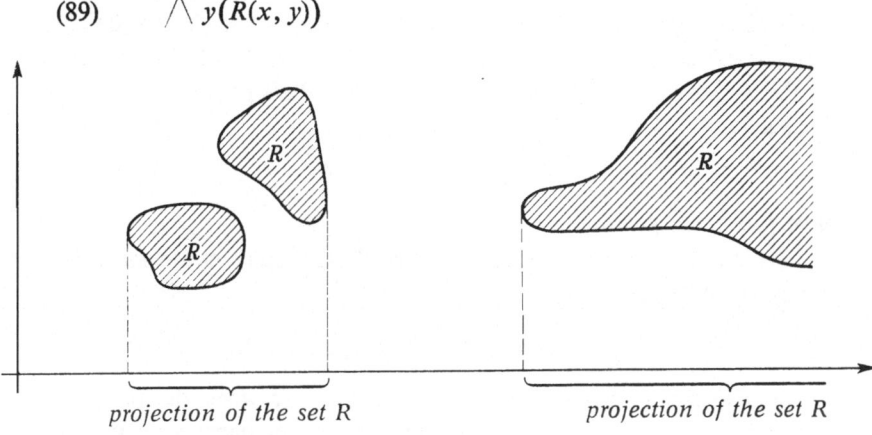

projection of the set R   projection of the set R

Fig. 4

is given by a set situated on the x axis and consisting of such numbers that for every y the point $\langle x, y \rangle$ belongs to the set R. It is thus the set of those x's for which the entire vertical straight line passing through

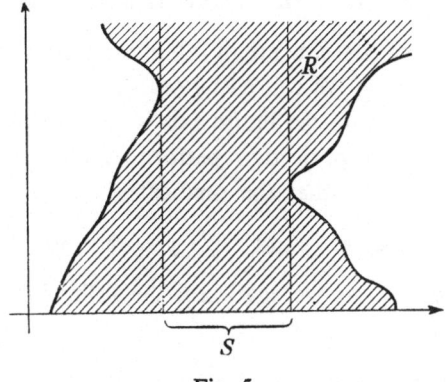

S

Fig. 5

127

x lies within the graph of the predicate $R(x, y)$ (the set S in Figure 5). This set will be called the result of the operation of inner projection on the set R, since it is obtained as the projection of all those vertical lines which pass through the set R without being interrupted by points belonging to the complement of the set R.

There is a relationship between the operation of projection and that of inner projection. Figure 5 shows clearly that the complement of the result of the operation of inner projection on the set R, that is the complement of the set S, on the entire x-axis, is identical with the projection on the x-axis of the complement of the set R. This proves that the predicates

$$\sim \bigwedge yR(x, y) \quad \text{and} \quad \bigvee y \sim R(x, y)$$

have the same graph.

As we shall see, the sentence stating that these predicates are equivalent, called *De Morgan's law*,

$$\bigwedge x \left(\bigvee y \sim R(x, y) \equiv \sim \bigwedge yR(x, y) \right),$$

is a logical theorem. Consider the simplified case when the parameter x is not essential (that is, the relation R does not depend on it); then the above equivalence yields

$$\bigvee y \sim R(y) \equiv \sim \bigwedge yR(y).$$

If, for instance, $R(y)$ stands for *y is a rational number*, we obtain a sentence which is true in the domain of real numbers:

There exist irrational numbers $\equiv$ not all numbers are rational.

All other examples of that law are in full agreement with our intuitions. If we consider Figure 4, we can easily see that the complement of the projection of the set R on the entire axis is identical with the result of the operation of inner projection performed on the complement of the set R. This indicates that the predicates

$$\sim \bigvee yR(x, y) \quad \text{and} \quad \bigwedge y \sim R(x, y)$$

have the same graph. As we shall see, the following sentence, which states that they are equivalent:

$$\bigwedge x \left(\sim \bigvee yR(x, y) \equiv \bigwedge y \sim R(x, y) \right),$$

is a logical theorem. For instance, for $R(x, y) = $ "$x > y$" this sentence states:

> *There do not exist numbers less than $x \equiv$ all numbers are not less than x.*

This law, too, is called *De Morgan's law*. For a fixed x, e.g., $x = 0$, and the domain of natural numbers, we obtain an example which is in full agreement with our intuition:

> *There do not exist numbers less than $0 \equiv$ all numbers are not less than 0.*

We shall soon prove that law in its form without the parameter x:

$$\sim \bigvee y R(y) \equiv \bigwedge y \sim R(y).$$

If $P(x_1, x_2, ..., x_n)$ is a predicate of n arguments, defined in the domain of real numbers, then the graph of the predicate P is a set P in an n-dimensional space. The graph of the predicate $\bigvee x_i P(x_1, x_2, ..., x_n)$ $(1 \leqslant i \leqslant n)$ is an $(n-1)$-dimensional set in that space, a set which is a projection parallel to the i-th axis of the set P on the hyperplane defined by the formula $x_i = 0$. The graph of the predicate $\bigwedge x_i P(x_1, x_2, ..., x_n)$ is a set in the same hyperplane, a set which is a result of the operation of inner projection on the set P by straight lines parallel to the i-th axis. Since every domain may be raised to its Cartesian powers, this interpre-

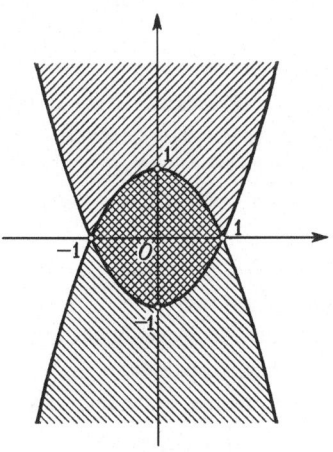

Fig. 6

tation of the quantifiers applies to all those mathematical domains to which the variables and the predicates are taken to refer.

A geometrical interpretation of predicates is very convenient for many reasons. All logical operations on predicates find intuitive analogues. This applies to the operations used in the sentential calculus as well. The graph of the negation of the predicate $P(x \ldots)$ is the complement of the graph of the predicate $P(x \ldots)$. The graph of the disjunction $P(x \ldots) \vee Q(x \ldots)$ is the union of the graphs of the predicates $P(x \ldots)$ and $Q(x \ldots)$. The graph of the conjunction $P(x \ldots) \wedge Q(x \ldots)$ is the intersection of the graphs of the predicates $P(x \ldots)$ and $Q(x \ldots)$. For instance, the graph of the predicate $y > x^2 - 1$ is the area inside the parabola $y = x^2 - 1$, while the graph of the predicate $y < 1 - x^2$ is the area inside the parabola $y = 1 - x^2$. Hence the graph of the predicate of two arguments

$$y > x^2 - 1 \vee y < 1 - x^2$$

is the union of these two areas, that is the whole shaded area in Fig. 6.

The graph of the predicate $y > x^2 - 1 \wedge y < 1 - x^2$ is the part common to both shaded areas. The graph of the predicate of one argument

$$\bigwedge y(y > x^2 - 1 \vee y < 1 - x^2)$$

is accordingly the set of x's contained in the open interval $(-1, 1)$. The same set is also the graph of the predicate

$$\bigvee y(y > x^2 - 1 \wedge y < 1 - x^2).$$

EXERCISES

1. Plot graphs for the following predicates of two arguments: $x = y$, $x < y$, $x > y$ $y \geq |x|, y \geq |x| \wedge y \leq 1 - |x|, \sim (y \geq |x|), \sim (y \geq |x|) \wedge \sim (y \leq 1 - |x|)$.

2. Plot graphs for the following predicates of one argument: $\bigvee x(y < x)$, $\bigwedge y(y < x), \bigwedge y(y \geq |x| \vee y \leq 1 - |x|), \bigvee y(\sim (y \geq |x|) \vee \sim (y \leq 1 - |x|)), \bigwedge y(x^2 + y^2 > 1), \sim \bigvee y(x^2 + y^2 \leq 1)$.

3. Note for which of the predicates listed above the graphs complement one another to form a complete straight line.

4. For any definite predicate $P(x, y)$, verify whether the graphs of the predicates: $\sim \bigvee y(P(x, y))$ and $\bigwedge y \sim P(x, y)$ coincide.

5. Indicate free and bound variables in the following formulae:

$$\bigwedge x(\bigvee y(x + y = 2) \to x = 2 - y),$$
$$\bigwedge y(\bigwedge x(x + y = x) \to x + y = y).$$

130

6. Adopt certain conventional abbreviations for relations (84), for instance: $x \text{ef} y$, $x \text{prt} y$, $x \text{prec} y$. Using these abbreviations and formulae of the type: $x \in$ "$\bigvee$", $x \in$ "$\bigwedge$", $x \in$ "(", $x \in$ ")", x is a *variable*, formulate, in a symbolic notation, the syntactic definitions occurring in this section.

5. THE PREDICATE CALCULUS IN THE TRADITIONAL PRESENTATION

As the sentential calculus describes the correct ways of using the sentential connectives, so the predicate calculus describes the correct ways of using the two phrases which we characterized in the preceding section and which we called quantifiers. As in the case of the sentential calculus, the correct way of handling quantifiers will be those types of inference which lead from true premisses to true conclusions when we reason on any subject, that is, when we consider any domain of objects, or when we interpret the variables as names of objects from any domain, on which we focus our attention. As in the case of the sentential calculus, the laws of the predicate calculus will be schemata of sentences that are always true. But *always true* now means somewhat more, since the predicate calculus contains variables interpreted as names of certain objects, so that a sentence is called always true if it is true when the variables are interpreted as names of objects from any domain. A sentence schema in the predicate calculus is always true if it is true in every domain.

The natural, historical development of logic has in fact resulted in the formulation of such a system of logic which can be proved to cover all schemata of logically correct reasoning on any subject. This property of the system is called *semantic completeness*. The semantic completeness of the classical logical calculus will be discussed in the next chapter. For the time being we shall formulate an interpretation of the predicate calculus which is the most commonly accepted and is the most convenient in reasoning, and then we shall analyse it from a certain point of view.

The predicate calculus is a much more important branch of logic than is the sentential calculus. Moreover, it must include the sentential calculus and may not be treated in separation from the latter. Hence the predicate calculus forms the whole of logic, and this is why the system of the predicate calculus shall just be called the *logical calculus*.

I. CLASSICAL LOGICAL CALCULUS

Since the classical predicate calculus, to be presented below, is based on the classical sentential calculus, it shall be called the *classical logical calculus*, and the term shall be abbreviated as c.l.c.

For the classical logical calculus there is no mechanical method of verifying the truth of theorems as there is in the case of the classical sentential calculus. The predicate calculus, as will be demonstrated in the last chapter, is an undecidable system, as opposed to the sentential calculus, for which we have proved the decidability theorem. Hence, in order to systematically present all the theorems and rules of the predicate calculus we have to rely alone on the ordinary method of an axiomatic presentation of a theory. Many authors present the classical logical calculus by defining it through an interpretation adopted in a meta-system. One then has to formulate precisely what is meant by a sentence true in a domain, and then the set of tautologies in the predicate calculus is defined as the set of sentences true in every domain. Such a procedure, however, assumes that the whole classical predicate calculus exists in the metasystem. In order to describe the metasystem with adequate precision one would then have to describe the predicate calculus as used in the metasystem. This results in a vicious circle. Or else the logical calculus is assumed in the metasystem by way of a convention, that is, axiomatically, without analysis. It may be objected that any presentation of logic involves a vicious circle, since nothing can be presented with precision without reference to logic. Yet the presentation of an axiomatic approach to the c.l.c. requires much poorer logical constructions than does the presentation of a semantic approach. This is why a description of the axiomatic approach involves a much less vicious circle.

One of the best known axiomatic approaches to the c.l.c. will be shown first. In the next section a certain modification of that approach will be analysed.

In mathematical circles, the most common approach of the classical logical calculus is by adoption of directives, that is, rules of inference. The usefulness of a rule becomes clear when that rule is being used. Thus the approach by adoption of rules is associated with the pragmatic approach, which is common in mathematics. The rules show how accepted formulae may be transformed into other formulae. Hence, the rules have the form of hypothetical assertions:

If sentences α and β are accepted, then a certain new sentence $\Phi(\alpha, \beta)$ is to be accepted, too.

Mathematicians have developed the convenient device of writing the rules as vertical schemata: the premisses α and β are written above a horizontal line, and the conclusion $\Phi(\alpha, \beta)$ is written below the line. A rule accordingly has the following form:

(R) $$\frac{\alpha, \beta}{\Phi(\alpha, \beta).}$$

The rule of detachment, when written in this way, is a schema of the following form:

$$\frac{\alpha, \alpha \rightarrow \beta}{\beta}.$$

The premisses have the forms: α and $\alpha \rightarrow \beta$, and the conclusion has the form β.

Contrary to the assumption made in the preceding section and to the approach to be taken in the next section, in the case of the traditional approach we assume that mathematical sentences may include free variables. The formula $x+y = y+x$ is, accordingly, accepted as a sentence. The rules of inference allow us to deduce from it the formula

$$\bigwedge x \bigwedge y(x+y = y+x),$$

and conversely. Thus the difference is not essential and is purely technical in nature. In the present approach an open formula may thus be accepted as a theorem.

In the axiomatic approach to the sentential calculus, as shown in Section 3, we were in a position to adopt many other rules in addition to the rules of substitution and detachment. The rules, being (R)-type relations between accepted sentences, are certain properties of a system, and many rules, which are called *derived rules* of a given system in order to distinguish them from those assumed at the outset, hold in every system based on the sentential calculus and the rule of detachment. For instance, the following rule of acceptance of conjunction holds:

If sentences α and β are accepted, then the sentence α ∧ β is to be accepted, too.

It can easily be proved that this relation is, in fact, valid, for on the strength of the sentential calculus we accept the theorem:

$$\alpha \rightarrow \left(\beta \rightarrow (\alpha \wedge \beta) \right).$$

As the sentences α and β are accepted, we may detach them, first one, then the other, until the conjunction $\alpha \wedge \beta$ is left as accepted. Thus, the above rule is a derived rule in every system based on the c.s.c. and the rule of detachment. It would be easy to construct such an axiomatic system of the sentential calculus in which the above rule would be a basic rule and in which the axioms would not include the theorem $p \rightarrow (q \rightarrow (p \wedge q))$.

Many calculi can be described either by rules or by axioms. In every system based on the sentential calculus and the rule of detachment the following rule also holds:

$$\frac{\alpha \rightarrow \beta,\ \beta \rightarrow \gamma}{\alpha \rightarrow \gamma}.$$

This rule makes use of the transitivity of implication. It is always worthwhile knowing which general rules are valid in a given system: this knowledge allows us to abbreviate proofs, since we do not have to repeat the same proof procedures several times.

Thus, the derived rules are those which are not assumed in the description of a system, but which, when used in reasoning, do not lead outside that system.

Let the traditional approach to the c.l.c. by denoted by L^+. The letter L will be reserved for the approach taken in the next section.

AXIOMS OF L^+. All substitutions of any meaningful formulae in the schemata of the classical sentential calculus are axioms of L^+. (The concept of a meaningful, that is, well-formed, formula will be used in an intuitive way for the time being.)

In addition, the axioms of L^+ include all those well-formed formulae which fall under the following schemata:

L1 $\qquad \alpha(x_i) \rightarrow \bigvee x_j \alpha(x_j),$

L2 $\qquad \bigwedge x_j \alpha(x_j) \rightarrow \alpha(x_i).$

The applications of these schemata will soon be illustrated by examples.

5. THE PREDICATE CALCULUS

RULES OF L^+. These include two rules explicitly concerned with the handling of quantifiers and the two universally accepted rules already discussed, namely those of detachment and substitution, which makes four rules in all. The quantifier rules are as follows:

$$\text{R1} \qquad \frac{\alpha(x_i) \to \Psi}{\bigvee x_i \alpha(x_i) \to \Psi},$$

$$\text{R2} \qquad \frac{\Psi \to \alpha(x_i)}{\Psi \to \bigwedge x_i \alpha(x_i)};$$

each rule may be applied only if Ψ does not contain x_i as a free variable.

R1 is the rule of preceding the antecedent of an implication with the existential quantifier $\bigvee x_i$ if the consequent of that implication does not contain x_i as a free variable. R2 is the rule of preceding the consequent of an implication with the universal quantifier $\bigwedge x_i$ if the antecedent of that implication does not contain x_i as a free variable. All the variables are considered to consist of the letter x with numerical subscripts: $x_0, x_1, x_2, \ldots$ The letters x, y, z, v are used as if they were identical with x_0, x_1, x_2, x_3.

The following rules of detachment and substitution are also adopted:

$$\text{Det:} \qquad \frac{\alpha, \alpha \to \beta}{\beta},$$

$$\text{Subst:} \qquad \frac{\alpha(x_i)}{\alpha(a)}.$$

In a general formulation of the rule of substitution, a may stand for any variable other than a bound variable occurring in $\alpha(x_i)$, and also for a constant or an expression symbolizing the value of a function. In the description of the logical calculus L^+, a will stand for variables only, but in the applications to the construction of definite mathematical theories it often stands for constants or names of values of functions.

In the construction of the predicate calculus the rule of substitution is subject to a restriction which was not applicable when the rule of substitution was analysed in its applications to quantifier-free formulae. The rule of substitution for variables is now given the following full formulation:

135

RULE OF SUBSTITUTION. *Any free variable x_j other than x_i may be substituted in a formula $\alpha(x_i)$ for the free variable x_i if the variable x_i, free in $\alpha(x_i)$, does not occur in any place in the formula $\alpha(x_i)$ within the scope of a quantifier that binds the variable x_j and is contained in the formula $\alpha(x_i)$.*

In other words, any variable x_j may be substituted for a variable x_i, free in a formula $\alpha(x_i)$, if following that substitution the variable x_j does not become a bound variable in any place in which it is substituted for the variable x_i. Thus, for instance, in the sentence

(i) $\qquad \bigwedge x_j (x_i \leqslant x_j) \to x_i = 0$

the variable x_j may not be substituted for the variable x_i, since such a substitution leads to conclusions that are false in the domain of natural numbers. The sentence (i) is true in the domain of natural numbers, and the forbidden substitution x_i/x_j yields the sentence

(ii) $\qquad \bigwedge x_j (x_j \leqslant x_j) \to x_j = 0,$

the antecedent of which is true in the domain of natural numbers; on detachment this would yield the theorem

(iii) $\qquad x_j = 0,$

which in turn, on the substitution $x_j/1$, would yield the contradiction

$\qquad 1 = 0.$

Such an application of the rule of substitution must be rejected as in certain domains it leads from true premises to false conclusions. The task of logic is to provide rules of inference that are correct in any domain, that is, with reference to any subject matter. Obviously, the safest procedure is to substitute only variables which differ from all bound variables.

Applications of quantifier laws and rules will be illustrated by examples from the field of arithmetic. Note first that the application of the rule of substitution to L1 yields a substitution in which x_i is replaced by a numerical expression a, mentioned above in connection with the rule of substitution:

L1 $\qquad \Phi(a) \to \bigvee x_j \Phi(x_j).$

As applied to numbers, this law may be formulated as follows:

If a formula Φ is satisfied by a number denoted by a variable, a constant, or by any numerical expression a, then there exists a number x such that x satisfies the formula Φ.

This law serves as the foundation of all *constructive proofs of existence.* It allows us to assert that there exists a number with a certain property if we have a numerical formula that defines a number with that property. In a more general formulation, it allows us to assert that there exists an object with a certain property if we can describe that object (at least one of those with that property) in our mathematical language, or, as it is often put, if we can construct it. The meaning of this law can be formulated in still another way: if we can indicate an object with a certain property, this means that objects with that property exist. Thus, this law has an obvious intuitive meaning. It is evidently a general schema that covers many implications in which Φ is any well-formed formula.

EXAMPLES

We have

$$y = y \rightarrow \bigvee x(y = x).$$

Since $y = y$, we obtain the theorem $\bigvee x(y = x)$, which states that for every object y there exist an object x which is identical with the former. This is obvious, since y itself is an object x that has that property.

The law

$$0 = 0+0 \rightarrow \bigvee x(x = x+x)$$

yields the arithmetical theorem stating that there exists a number x which has the property that $x+x = x$.

The theorem

$$0 = 0+0 \rightarrow \bigvee x(0 = x+x)$$

is an equally good application of L1.

It might be objected that the introduction of the existential quantifier only for the purpose of saying that an x has a property Φ while we know very well which x has it is of no significance. Yet in many cases the construction of an object with certain properties ceases to be of

137

interest to us in further analysis and we rest satisfied with the knowledge that an object with the given properties does exist.

Likewise, by substituting the numerical formula a for x_i in L2 we arrive at the law:

L2 $\qquad \bigwedge x_j \Phi(x_j) \to \Phi(a).$

The meaning of L2 can now be formulated as follows:

If every number x has a property Φ, then the number a also has the property Φ.

By a property is meant a property described by a formula, hence L2 may also be formulated as follows:

If every number x satisfies a formula Φ, then the number a also satisfies the formula Φ.

$$\bigwedge x(x+0 = x) \to 2+0 = 2,$$
$$\bigwedge x(x \cdot y = y \cdot x) \to (z+u) \cdot y = y \cdot (z+u),$$
$$\bigwedge x(x+y) = x \cdot y \to 0+y = 0 \cdot y.$$

These examples show that in L2 the consequent is just a substitution of the antecedent and so it follows, as it were, from the antecedent under the rule of substitution. Hence, the implication may be accepted as true.

Examples of applications of rules R1 and R2 in arithmetical analyses will be given below.

The meaning of rule R1 may be given the following verbal formulation:

If it is true that an object x having a property α entails Ψ, then it is also true that the existence of an object with the property α entails Ψ.

$$\frac{x+y = x \to y = 0}{\bigvee x(x+y = x) \to y = 0},$$
$$\frac{x+y = 0 \to y = 0}{\bigvee x(x+y = 0) \to y = 0}.$$

Law L1 and rule R1 will be applied to prove that addition is monotonic with respect to the relation *less than*, that is, the law

$$x \leqslant y \to x+z \leqslant y+z.$$

To do so we shall make use of the definition

0. $x \leqslant y \equiv \bigvee v(y = x+v)$

and we shall accept as proved the following arithmetical laws:

1. $a = b \rightarrow a+z = b+z$,

2a. $(x+u)+z = x+(u+z)$,

2b. $u+z = z+u$.

Law 1, which allows addition of equals to equals, is in fact an application of logical properties of identity. Law 2 expresses the associativity and commutativity of addition. These laws, by substitution and the sentential calculus, yield the theorem

3. $y = x+u \rightarrow y+z = (x+z)+u$.

The following is a case of L1:

4. $y+z = (x+z)+u \rightarrow \bigvee v (y+z = (x+z)+v)$.

By Definition 0 of the relation *less than*, the following equivalence holds:

5. $x+z \leqslant y+z \equiv \bigvee v(y+z = (x+z)+v)$.

Theorems 4 and 5 and the sentential calculus yield

6. $y+z = (x+z)+u \rightarrow x+z \leqslant y+z$.

Theorems 3 and 6 and the sentential calculus yield

7. $y = x+u \rightarrow x+z \leqslant y+z$.

Under the rule of substitution, the variable u in 7 may be replaced by the variable v, which yields

8. $y = x+v \rightarrow x+z \leqslant y+z$.

Under the quantifier rule R1, 8 yields

9. $\bigvee v(y = x+v) \rightarrow x+z \leqslant y+z$.

Definition 0 of the relation *less than* may be applied to 9, which yields the desired sentence:

10. $x \leqslant y \rightarrow x+z \leqslant y+z$.

To give an example of the application of rule R2 we shall prove the law

$$x < y \rightarrow \bigwedge z(z < x \rightarrow z < y),$$

which may be expressed as the statement that if $x < y$, then whatever is less than x it is also less than y. As a starting point we may take Theorem 0 on the transitivity of the relation *less than*:

0. $(x < y \wedge y < u) \rightarrow x < u$.

First, under the rule of substitution, we obtain from 0 the theorem

1. $(z < x \wedge x < y) \rightarrow z < y$,

by the substitutions: x/z, y/x, u/y. Next, laws of the sentential calculus are used to transform Theorem 1 into

2. $x < y \rightarrow (z < x \rightarrow z < y)$.

Rule R2 may be applied to Theorem 2, which yields the theorem

3. $x < y \rightarrow \bigwedge z(z < x \rightarrow z < y)$,

which we wanted to obtain.

We shall now give some applications of rules R1 and R2 and laws L1 and L2 to proofs of theorems of the predicate calculus. We shall first deduce one derived rule, namely

R3 $\qquad \dfrac{\Phi(x_i)}{\bigwedge x_i \Phi(x_i)}$.

This rule allows us to precede any theorem by the universal quantifier binding a variable occurring free in that theorem. The rule can be understood on the strength of the very meaning of the theorems. A theorem in which a variable x_i occurs free is considered to refer to any object from the domain in question, and hence, to state the same thing as the theorem preceded by the universal quantifier: *for every x_i.*

P r o o f. It is assumed that $\Phi(x)$ is accepted:

1. $\Phi(x_i)$.

2. $\Phi(x_i) \rightarrow \left[\left(\bigwedge x_0 P(x_0) \rightarrow \bigwedge x_0 P(x_0) \right) \rightarrow \Phi(x_i) \right]$.

This premiss is obtained from the sentential calculus as a substitution instance of the sentential calculus schema $p \rightarrow ((q \rightarrow q) \rightarrow p)$.

3. $\left(\bigwedge x_0 P(x_0) \rightarrow \bigwedge x_0 P(x_0) \right) \rightarrow \Phi(x_i)$
    (from 1 and 2 by detachment).

4. $\left(\bigwedge x_0 P(x_0) \rightarrow \bigwedge x_0 P(x_0) \right) \rightarrow \bigwedge x_i \Phi(x_i)$
    (from 3 and R2).

5. $\bigwedge x_0 P(x_0) \rightarrow \bigwedge x_0 P(x_0)$

  (from the sentential calculus schema $p \rightarrow p$).

6. $\bigwedge x_i \Phi(x_i)$

  (from 4 and 5 by detachment).

Thus, if we accept $\Phi(x_i)$ as a theorem we may also, under laws and rules adopted above, accept $\bigwedge x_i \Phi(x_i)$.

We shall now obtain two laws on the replacement of bound variables of a certain shape by variables of a different shape:

L3 $\bigwedge x_i \Phi(x_i) \rightarrow \bigwedge x_j \Phi(x_j)$ $\Big\{$ if x_j is not free in $\Phi(x_i)$ and

L4 $\bigvee x_i \Phi(x_i) \rightarrow \bigvee x_j \Phi(x_j)$ $\Big\{$ if x_i is not free in $\Phi(x_j)$.

PROOFS. By L2

$$\bigwedge x_i \Phi(x_i) \rightarrow \Phi(x_j)$$

is accepted. If x_j is not free in $\Phi(x_i)$, then we may apply rule R2 and accept L3. Likewise, by L1,

$$\Phi(x_i) \rightarrow \bigvee x_j \Phi(x_j)$$

is accepted. If x_i not free in $\Phi(x_j)$, then we may apply rule R1 and accept L4.

We shall now prove De Morgan's laws, discussed in the preceding section.

As a proof of

L5 $\bigvee x_j \sim P(x_j) \rightarrow \sim \bigwedge x_i P(x_i)$.

we have

1. $\bigwedge x_i P(x_i) \rightarrow P(x_j)$
  (L2).

2. $\left(\bigwedge x_i P(x_i) \rightarrow P(x_j)\right) \rightarrow \left(\sim P(x_j) \rightarrow \sim \bigwedge x_i P(x_i)\right)$
  (c.s.c. $(p \rightarrow q) \rightarrow (\sim q \rightarrow \sim p)$).

3. $\sim P(x_j) \rightarrow \sim \bigwedge x_i P(x_i)$
  (2, 1, Det.).

4. $\bigvee x_j \sim P(x_j) \rightarrow \sim \bigwedge x_i P(x_i)$
  (3, Det.).

141

Since in L1 and L2 the variables x_i and x_j may be one and the same variable, in the proofs of the rest of De Morgan's laws they will be treated as identical and, to simplify matters, will be replaced by the variable x without any subscript.

As proofs of

L6 $\qquad \sim \bigvee x \sim P(x) \to \bigwedge x P(x),$

L7 $\qquad \sim \bigwedge x P(x) \to \bigvee x \sim P(x).$

we have

1. $\sim P(x) \to \bigvee x \sim P(x)$
   (L1).

2. $(\sim P(x) \to \bigvee x \sim P(x)) \to (\sim \bigvee x \sim P(x) \to P(x))$
   (c.s.c. $(\sim p \to q) \to (\sim q \to p)$).

3. $\sim \bigvee x \sim P(x) \to P(x)$
   (1, 2, Det.).

4. $\sim \bigvee x \sim P(x) \to \bigwedge x P(x)$
   (3, R2).

5. $(\sim \bigvee x \sim P(x) \to \bigwedge x P(x)) \to (\sim \bigwedge x P(x) \to \bigvee x \sim P(x))$
   (c.s.c. $(\sim p \to q) \to (\sim q \to p)$).

6. $\sim \bigwedge x P(x) \to \bigvee x \sim P(x)$
   (4, 5, Det.).

Laws L5 and L7 may be combined into the following equivalence:

L8 $\qquad \bigvee x \sim P(x) \equiv \sim \bigwedge x P(x).$

This law was discussed in the preceding section and illustrated by examples and a graph (cf. Fig. 5).

7. $(\bigvee x \sim P(x) \to \sim \bigwedge x P(x)) \to (\bigwedge x P(x) \to \sim \bigwedge x \sim P(x))$
   (c.s.c. $(p \to \sim q) \to (q \to \sim p)$).

8. $\bigwedge x P(x) \to \sim \bigvee x \sim P(x)$
   (7, L5, Det.).

L9 $\qquad \bigwedge x P(x) \equiv \sim \bigvee x \sim P(x)$ $\qquad$ (8, L6, c.s.c.).

This law may be interpreted as a definition of the universal quantifier by the existential quantifier. It is also quite intuitive. Consider an example from the domain of human beings: for $P(x) = x$ *is mortal* we obtain

All men are mortal $\equiv$ *there does not exist an immortal man.*

We shall now prove L11

$$\sim \bigvee xP(x) \equiv \bigwedge x \sim P(x).$$

We have

1. $\bigwedge x \sim P(x) \to \sim P(x)$
   (L2).

2. $\left(\bigwedge x \sim P(x) \to \sim P(x)\right) \to \left(P(x) \to \sim \bigwedge x \sim P(x)\right)$
   $((p \to \sim q) \to (q \to \sim p))$.

3. $P(x) \to \sim \bigwedge x \sim P(x)$
   (1, 2, Det.).

4. $\bigvee xP(x) \to \sim \bigwedge x \sim P(x)$
   (3, R1).

5. $\left(\bigvee xP(x) \to \sim \bigwedge x \sim P(x)\right) \to \left(\bigwedge x \sim P(x) \to \sim \bigvee xP(x)\right)$
   $((p \to \sim q) \to (q \to \sim p))$.

6. $\bigwedge x \sim P(x) \to \sim \bigvee xP(x)$
   (5, 4, Det.).

7. $P(x) \to \bigvee xP(x)$
   (L1).

8. $\left(P(x) \to \bigvee xP(x)\right) \to \left(\sim \bigvee xP(x) \to \sim P(x)\right)$
   $((p \to q) \to (\sim q \to \sim p))$.

9. $\sim \bigvee xP(x) \to \sim P(x)$
   (7, 8, Det.).

10. $\sim \bigvee xP(x) \to \bigwedge x \sim P(x)$
    (9, R2).

11. $\left(\sim \bigvee xP(x) \to \bigwedge x \sim P(x)\right) \to \left(\sim \bigwedge x \sim P(x) \to \bigvee xP(x)\right)$
    $((\sim p \to q) \to (\sim q \to p))$.

L10 $\sim \bigwedge x \sim P(x) \rightarrow \bigvee xP(x)$ (10, 11, Det.).

L11 $\sim \bigvee xP(x) \equiv \bigwedge x \sim P(x)$ (10, 6, c.s.c.).

L12 $\bigvee xP(x) \equiv \sim \bigwedge x \sim P(x)$ (4, L10, c.s.c.).

Law L11 was discussed in the preceding section and illustrated by intuitive examples. Law L12 may be interpreted as a definition of the existential quantifier in terms of the universal quantifier. Its meaning is also quite intuitive. For instance, for $P(x) = x$ *is even* it becomes the following sentence:

There exist even numbers $\equiv$ not all numbers are odd.

Both law L10 and law L1, written as $P(a) \rightarrow \bigvee xP(x)$, have existence statements in the consequent. Yet, by L1, we may state that there exists an object with a property P if we have proved, for an object a having a definite name, that it has the property P. All constructive proofs of existence are based on this schema. In such proofs, an object with a given property is constructed first, and it is only then that it is stated that objects with that property do exist.

Theorem L10 indicates a quite different method of proving existence. By L10, in order to prove the existence of objects with a property P it suffices to prove that it is not true that all objects do not have the property P. Thus, it suffices to demonstrate that the assumption that all objects do not have the property P results in a contradiction. There is thus no need to indicate (construct) an object that has the property P; it suffices to demonstrate that the assumption to the contrary results in a contradiction.

Theorem L10 is thus a principle of non-constructive proofs of existence. (proofs not based on the construction of an object). Many mathematicians (especially the intuitionists) give priority to constructive proofs of existence. It can easily be seen that the proofs of Theorems L7 and L10 have made use of the sentential calculus theorem: $(\sim p \rightarrow q) \rightarrow (\sim q \rightarrow p)$, which does not belong to intuitionistic sentential logic

The non-constructive principles of existence L7 and L10 cannot be deduced in intuitionistic logic.

Every proof of existence, if based on intuitionistic logic, must be constructive, obviously, if additional existential assumptions are not introduced.

We shall now deduce the laws of distributivity for quantifiers. Consider first distributivity with respect to implication. It is formulated in the following two laws:

L13 $\quad \bigwedge x_i(\alpha \rightarrow \beta) \rightarrow (\bigwedge x_i\alpha \rightarrow \bigwedge x_i\beta),$

L14 $\quad \bigwedge x_i(\alpha \rightarrow \beta) \rightarrow (\bigvee x_i\alpha \rightarrow \bigvee x_i\beta).$

These laws may be read as follows. If for every x_i a property α implies a property β, then if all objects have the property α, then they all have the property β as well, and, under the same assumption, if there exist objects with the property α, then there also exist objects with the property β.

In the above formulation and in many proofs as well we shall not specify the variable that follows α and β (that is, we shall not write $\alpha(x_i)$, $\beta(x_i)$), because the formulations of the laws on the use of quantifiers are so general that the variable which is bound by a quantifier may even not occur within the scope of that quantifier. Such a quantifier is, obviously, superfluous, but logic allows the existence of superfluous quantifiers. Arguments concerning those formulae in which variables do occur are, however, much more intuitive, and this is why we shall use them frequently, too.

1. $\bigwedge x_i(\alpha(x_i) \rightarrow \beta(x_i)) \rightarrow (\alpha(x_j) \rightarrow \beta(x_j))$
    (L2),

2. $\alpha(x_j) \rightarrow \left(\bigwedge x_i(\alpha(x_i) \rightarrow \beta(x_i)) \rightarrow \beta(x_j) \right)$
    (from 1 and c.s.c.),

3. $\bigwedge x_i\alpha(x_i) \rightarrow \alpha(x_j)$
    (L2),

4. $\bigwedge x_i\alpha(x_i) \rightarrow \left(\bigwedge x_i(\alpha(x_i) \rightarrow \beta(x_i)) \rightarrow \beta(x_j) \right)$
    (from 2 and 3, and c.s.c., transitivity of implication),

5. $\left(\bigwedge x_i(\alpha(x_i) \rightarrow \beta(x_i)) \wedge \bigwedge x_i\alpha(x_i) \right) \rightarrow \beta(x_i)$
    (from 4 and c.s.c., law of importation),

6. $\left(\bigwedge x_i(\alpha(x_i) \rightarrow \beta(x_i)) \wedge \bigwedge x_i\alpha(x_i) \right) \rightarrow \bigwedge x_j\beta(x_j)$
    (from 5 and R2),

7. $\bigwedge x_j\beta(x_j) \rightarrow \bigwedge x_i\beta(x_i)$
    (L3),

8. $\left(\bigwedge x_i(\alpha(x_i) \to \beta(x_i)) \wedge \bigwedge x_i\alpha(x_i)\right) \to \bigwedge x_i\beta(x_i)$
   (from 6 and 7, and c.s.c. transitivity of implication),

9. $\bigwedge x_i(\alpha \to \beta) \to \left(\bigwedge x_i\alpha \to \bigwedge x_i\beta\right)$
   (from 8 and c.s.c. law of exportation),

10. $\left(\alpha(x_j) \wedge \bigwedge x_i(\alpha(x_i) \to \beta(x_i))\right) \to \beta(x_j)$
    (from 1 and c.s.c.)

11. $\beta(x_j) \to \bigvee x_i\beta(x_i)$
    (L1),

12. $\left(\alpha(x_j) \wedge \bigwedge x_i(\alpha(x_i) \to \beta(x_i))\right) \to \bigvee x_i\beta(x_i)$
    (from 10, 11, and c.s.c. transitivity of implication),

13. $\alpha(x_j) \to \left(\bigwedge x_i(\alpha(x_i) \to \beta(x_i)) \to \bigvee x_i\beta(x_i)\right)$
    (from 12 and c.s.c. law of exportation),

14. $\bigvee x_j\alpha(x_j) \to \left(\bigwedge x_i(\alpha(x_i) \to \beta(x_i)) \to \bigvee x_i\beta(x_i)\right)$
    (from 13 and R1),

15. $\bigvee x_i\alpha(x_i) \to \bigvee x_j\alpha(x_j)$
    (L4),

16. $\bigvee x_i\alpha(x_i) \to \left(\bigwedge x_i(\alpha(x_i) \to \beta(x_i)) \to \bigvee x_i\beta(x_i)\right)$
    (from 15, 14, and c.s.c. transitivity of implication),

17. $\bigwedge x_i(\alpha \to \beta) \to \left(\bigvee x_i\alpha \to \bigvee x_i\beta\right)$
    (from 16 and c.s.c. law of exchange of antecedents).

Next we shall prove the laws of distributivity of quantifiers with respect to conjunction and disjunction. These laws are as follows:

L15 $\bigwedge x_i(\alpha \wedge \beta) \equiv \left(\bigwedge x_i\alpha \wedge \bigwedge x_i\beta\right),$

L16 $\bigvee x_i(\alpha \vee \beta) \equiv \left(\bigvee x_i\alpha \vee \bigvee x_i\beta\right).$

Their meaning is quite intuitive. Every object in a given domain has properties α and β if and only if every such object has the property α and every such object has the property β. Likewise, there exists an object with a property α or a property β if and only if there exists an object with the property α or there exists an object with the property β.

1. $(\alpha \wedge \beta) \to \alpha$
   (from c.s.c.),

2. $(\alpha \wedge \beta) \to \beta$
   (from c.s.c),

3. $\bigwedge x_i((\alpha \wedge \beta) \to \alpha)$
   (from 1 and R3),

4. $\bigwedge x_i((\alpha \wedge \beta) \to \beta)$
   (from 2 and R3),

5. $\bigwedge x_i((\alpha \wedge \beta) \to \alpha) \to (\bigwedge x_i(\alpha \wedge \beta) \to \bigwedge x_i \alpha)$
   (from L13),

6. $\bigwedge x_i((\alpha \wedge \beta) \to \beta) \to (\bigwedge x_i(\alpha \wedge \beta) \to \bigwedge x_i \beta)$
   (from L13),

7. $\bigwedge x_i(\alpha \wedge \beta) \to \bigwedge x_i \alpha$
   (from 3, 5, Det.),

8. $\bigwedge x_i(\alpha \wedge \beta) \to \bigwedge x_i \beta$
   (from 4, 6, Det.),

9. $(\bigwedge x_i(\alpha \wedge \beta) \to \bigwedge x_i \alpha) \to ((\bigwedge x_i(\alpha \wedge \beta) \to \bigwedge x_i \beta) \to$
   $(\bigwedge x_i(\alpha \wedge \beta) \to (\bigwedge x_i \alpha \wedge \bigwedge x_i \beta)))$
   (c.s.c. $(p \to q) \to ((p \to r) \to (p \to (q \wedge r)))$),

10. $\bigwedge x_i(\alpha \wedge \beta) \to (\bigwedge x_i \alpha \wedge \bigwedge x_i \beta)$
    (from 9, 7, 8, Det., Det.),

11. $\bigwedge x_i \alpha \to \alpha$
    (from L2),

12. $\bigwedge x_i \beta \to \beta$
    (from L2),

13. $(\bigwedge x_i \alpha \wedge \bigwedge x_i \beta) \to (\alpha \wedge \beta)$
    (from 11, 12, and c.s.c. $(p \to q) \to ((r \to s) \to (p \wedge r \to q \wedge s)))$,

14. $(\bigwedge x_i \alpha \wedge \bigwedge x_i \beta) \to \bigwedge x_i(\alpha \wedge \beta)$
    (from 13, R2).

Law L15 results from 10, 14, and c.s.c.

1. $\alpha \to \bigvee x_i \alpha$
   (from L1),

2. $\beta \to \bigvee x_i \beta$
   (from L1),

3. $(\alpha \lor \beta) \to (\bigvee x_i \alpha \lor \bigvee x_i \beta)$
   (from 1, 2, and c.s.c. $(p \to q) \to ((r \to s) \to ((p \lor r) \to (q \lor s)))$),

4. $\bigvee x_i(\alpha \lor \beta) \to (\bigvee x_i \alpha \lor \bigvee x_i \beta)$
   (from 3, R1),

5. $\alpha \to (\alpha \lor \beta)$
   (from c.s.c.),

6. $\beta \to (\alpha \lor \beta)$
   (from c.s.c.),

7. $\bigwedge x_i(\alpha \to (\alpha \lor \beta))$
   (from 5, R3),

8. $\bigwedge x_i(\beta \to (\alpha \lor \beta))$
   (from 6, R3),

9. $\bigwedge x_i(\alpha \to (\alpha \lor \beta)) \to (\bigvee x_i \alpha \to \bigvee x_i(\alpha \lor \beta))$
   (from L14),

10. $\bigwedge x_i(\beta \to (\alpha \lor \beta)) \to (\bigvee x_i \beta \to \bigvee x_i(\alpha \lor \beta))$
    (from L14),

11. $\bigvee x_i \alpha \to \bigvee x_i(\alpha \lor \beta)$
    (from 7, 9, Det.),

12. $\bigvee x_i \beta \to \bigvee x_i(\alpha \lor \beta)$
    (from 8, 10, Det.),

13. $(\bigvee x_i \alpha \lor \bigvee x_i \beta) \to \bigvee x_i(\alpha \lor \beta)$
    (from 11, 12, c.s.c. $(p \to q) \to ((r \to q) \to ((p \lor r) \to q))$).

Law L16 results from 4, 13, and c.s.c.
Some other obvious theorems of the predicate calculus will be proved.

The laws of interchanging similar quantifiers:

L17 $\bigwedge x_i \bigwedge x_j \alpha \to \bigwedge x_j \bigwedge x_i \alpha,$

L18 $\bigvee x_i \bigvee x_j \alpha \to \bigvee x_j \bigvee x_i \alpha.$

PROOFS. We have

1. $\bigwedge x_j \alpha \to \alpha$
   (from L2),

2. $\bigwedge x_i (\bigwedge x_j \alpha \to \alpha)$
   (from 1, R3),

3. $\bigwedge x_i (\bigwedge x_j \alpha \to \alpha) \to (\bigwedge x_i \bigwedge x_j \alpha \to \bigwedge x_i \alpha)$
   (from L13),

4. $\bigwedge x_i \bigwedge x_j \alpha \to \bigwedge x_i \alpha$
   (from 2, 3, Det.),

5. $\bigwedge x_i \bigwedge x_j \alpha \to \bigwedge x_j \bigwedge x_i \alpha$
   (from 4, R2),

6. $\alpha \to \bigvee x_i \alpha$
   (from L1),

7. $\bigwedge x_j (\alpha \to \bigvee x_i \alpha)$
   (from 6, R3),

8. $\bigwedge x_j (\alpha \to \bigvee x_i \alpha) \to (\bigvee x_j \alpha \to \bigvee x_j \bigvee x_i \alpha)$
   (from L14),

9. $\bigvee x_j \alpha \to \bigvee x_j \bigvee x_i \alpha$
   (from 7, 8, Det.),

10. $\bigvee x_i \bigvee x_j \alpha \to \bigvee x_j \bigvee x_i \alpha$
    (from 9, R1).

On the contrary, the universal quantifier may not be interchanged with the existential quantifier in an arbitrary manner. The formulae

$$\bigvee x \bigwedge y \Phi(x, y) \quad \text{and} \quad \bigwedge y \bigvee x \Phi(x, y)$$

are not equivalent: the latter follows from the former, but the former does not follow from the latter. The implication:

L19 $\bigvee x \bigwedge y \Phi(x, y) \rightarrow \bigwedge y \bigvee x \Phi(x, y)$

is a law of logic.

PROOF. We have

1. $\Phi(x, y) \rightarrow \bigvee x \Phi(x, y)$
      (from L1),

2. $\bigwedge y(\Phi(x, y) \rightarrow \bigvee x \Phi(x, y))$
      (from 1, R3),

3. $\bigwedge y(\Phi(x, y) \rightarrow \bigvee x \Phi(x, y)) \rightarrow (\bigwedge y \Phi(x, y) \rightarrow \bigwedge y \bigvee x \Phi(x, y))$
      (from L13),

4. $\bigwedge y \Phi(x, y) \rightarrow \bigwedge y \bigvee x \Phi(x, y)$
      (from 2, 3, Det.),

5. $\bigvee x \bigwedge y \Phi(x, y) \rightarrow \bigwedge y \bigvee x \Phi(x, y)$
      (from 4, R1).

On the contrary, the implication converse to L19 is not a law of logic. We also have to prove the following laws the intuitive meaning of which will be explained in the next section:

L20 $\bigwedge x \bigwedge y \alpha(x, y) \rightarrow \bigwedge x \alpha(x, x)$,

L21 $\bigvee x \alpha(x, x) \rightarrow \bigvee x \bigvee y \alpha(x, y)$.

PROOFS. We have

1. $\bigwedge y \alpha(x, y) \rightarrow \alpha(x, y)$
      (from L2),

2. $\bigwedge y \alpha(x, y) \rightarrow \alpha(x, x)$

(Theorem 2 results from Theorem 1 by a substitution which is in accordance with the rule of substitution; the variable y in the consequent of 1 is free and x is substituted for it, under the assumption that in no place does it become bound by any quantifier occurring in the formula $\alpha(x,x)$)

3. $\bigwedge x(\bigwedge y \alpha(x, y) \rightarrow \alpha(x, x))$
      (from 2, R3),

4. $\bigwedge x(\bigwedge y \alpha(x, y) \rightarrow \alpha(x, x)) \rightarrow (\bigwedge x \bigwedge y \alpha(x, y) \rightarrow \bigwedge x \alpha(x, x))$
      (from L13).

150

L20 is obtained from 3 and 4 by detachment.

5. $\alpha(x,y) \rightarrow \bigvee y\alpha(x,y)$
   (from L1),

6. $\alpha(x, x) \rightarrow \bigvee y\alpha(x, y)$
   (from 5 by substitution, similarly to how 2 was obtained from 1),

7. $\bigwedge x(\alpha(x, x) \rightarrow \bigvee y\alpha(x, y))$
   (from 6, R3),

8. $\bigwedge x(\alpha(x, x) \rightarrow \bigvee y\alpha(x, y)) \rightarrow (\bigvee x\alpha(x, x) \rightarrow \bigvee x\bigvee y\alpha(x, y))$
   (from L14).

L21 is obtained from 7 and 8 by detachment.

Note also that, under the assumption that the formula α does not contain the free variable x, the following theorems hold:

L22 $\alpha \rightarrow \bigwedge x\alpha,$

L23 $\bigvee x\alpha \rightarrow \alpha.$

They are obtained from the sentential calculus theorem $\alpha \rightarrow \alpha$ by the application of rule R2 in the case of L22, and rule R1 in the case of L23.

The theorem that states the existence of an object that has the tautological property:

L24 $\bigvee x(P(x) \equiv P(x))$

also holds. It follows from L1:

$$(P(x) \equiv P(x)) \rightarrow \bigvee x(P(x) \equiv P(x))$$

by the detachment of the antecedent, which is a tautology in the sentential calculus.

We shall also prove theorems which allow us to shift quantifiers from one term of a conjunction to the front of the whole conjunction if the other term does not contain any free variable of the same shape as the variable bound by the quantifier in question. These theorems are equivalences of the following form:

L25 $(\Psi \wedge \bigwedge x\Phi(x)) \equiv \bigwedge x(\Psi \wedge \Phi(x)),$

L26 $(\Psi \wedge \bigvee x\Phi(x)) \equiv \bigvee x(\Psi \wedge \Phi(x)),$

under the assumption that Ψ does not contain a free variable x.

PROOFS. We have

1. $\bigwedge x\Phi(x) \to \Phi(x)$
   (from L2),

2. $(\Psi \wedge \bigwedge x\Phi(x)) \to (\Psi \wedge \Phi(x))$
   (from 1 and c.s.c. theorem: $(p \to q) \to ((r \wedge p) \to (r \wedge q))$),

3. $(\Psi \wedge \bigwedge x\Phi(x)) \to \bigwedge x(\Psi \wedge \Phi(x))$
   (from 2, R2),

4. $\bigwedge x(\Psi \wedge \Phi(x)) \to (\Psi \wedge \Phi(x))$
   (from L2),

5. $\bigwedge x(\Psi \wedge \Phi(x)) \to \Psi$
   (from 4, c.s.c.),

6. $\bigwedge x(\Psi \wedge \Phi(x)) \to \Phi(x)$
   (from 4, c.s.c.),

7. $\bigwedge x(\Psi \wedge \Phi(x)) \to \bigwedge x\Phi(x)$
   (from 6, R2),

8. $\bigwedge x(\Psi \wedge \Phi(x)) \to (\Psi \wedge \bigwedge x\Phi(x))$
   (from 5, 7, c.s.c.).

Theorem 4 yields 5 and 6 under the c.s.c. theorems $(p \to (q \wedge r)) \to (p \to q)$ and $(p \to (q \wedge r)) \to (p \to r)$. Then the implications 5 and 7 are again combined into one under the converse theorem: $(p \to q) \to ((p \to r) \to (p \to (q \wedge r)))$. Theorems 3 and 8 combined yield L25.

1. $(\Psi \wedge \Phi(x)) \to \bigvee x(\Psi \wedge \Phi(x))$
   (from L1),

2. $\Phi(x) \to (\Psi \to \bigvee x(\Psi \wedge \Phi(x)))$
   (from 1, c.s.c.),

3. $\bigvee x\Phi(x) \to (\Psi \to \bigvee x(\Psi \wedge \Phi(x)))$
   (from 2, R1),

4. $(\Psi \wedge \bigvee x\Phi(x)) \to \bigvee x(\Psi \wedge \Phi(x))$
   (from 3, c.s.c.),

5. $\Phi(x) \rightarrow \bigvee x \Phi(x)$
   (from L1),

6. $(\Psi \wedge \Phi(x)) \rightarrow (\Psi \wedge \bigvee x \Phi(x))$
   (from 5, c.s.c.),

7. $\bigvee x(\Psi \wedge \Phi(x)) \rightarrow (\Psi \wedge \bigvee x \Phi(x))$
   (from 6, R2).

Theorems 4 and 7 combined yield L26.

Theorems L25 and L26 can serve as a basis for a demonstration that in every formula the quantifiers may be shifted to the beginning of that formula. From the terms of a conjunction they may be shifted before the conjunction, and when they follow negation symbols they may, in each case, be shifted so as to stand before the negation symbol, provided, however, that each quantifiers is changed into the opposite one. This is how De Morgan's laws may be interpreted:

$$\sim \bigvee x \Phi \equiv \bigwedge x \sim \Phi,$$
$$\sim \bigwedge x \Phi \equiv \bigvee x \sim \Phi.$$

Since all other sentential connectives may be defined by means of negation and conjunction, hence in any formula the quantifiers may gradually be shifted—provided that each quantifier is changed into the opposite one whenever it is shifted from a position following a negation symbol to a position preceding such a symbol—until all of them are brought to stand at the beginning of the formula in question. This statement is called the *theorem on the reduction to a normal form*, to be discussed in greater detail later in the text.

EXERCISES

1. Prove the following laws of the predicate calculus:

$$\bigwedge x_i(\alpha \rightarrow \beta(x_i)) \rightarrow (\alpha \rightarrow \bigwedge x_i \beta(x_i)),$$
$$\bigwedge x_i(\beta(x_i) \rightarrow \alpha) \rightarrow (\bigvee x_i \beta(x_i) \rightarrow \alpha),$$

under the assumption that the free variable x does not occur in α.

2. Demonstrate that if the schemata from exercise 1 are accepted as c.l.c. axioms and if, in addition the derived rule R3 is accepted, instead of rules R1 and R2, then a system of logic that is equivalent to L^+ is obtained.

3. Prove the following derived rule:

$$\frac{\bigwedge x_i \alpha(x_i)}{\alpha(x_i)}.$$

4. Prove the following laws of the predicate calculus:

$$\bigvee x(\alpha \to \beta) \to (\bigwedge x\alpha \to \beta),$$

$$\bigvee x(\beta \to \alpha) \to (\beta \to \bigvee x\alpha).$$

Demonstrate that the intuitionistic sentential calculus suffices to prove these laws whereas the theorems of the intuitionistic sentential calculus do not suffice to prove the converse implications, which, too, are theorems in the c.l.c. (What assumptions regarding the variables are required?)

5. Give an example, drawn, e.g., from the domain of natural numbers, showing that the implications converse to L13 and L14 are not theorems in the c.l.c.

6. REDUCTION OF QUANTIFIER RULES TO AXIOMS. C.L.C. TAUTOLOGIES TRUE IN THE EMPTY DOMAIN

A somewhat different axiomatic approach to the c.l.c. will now be given in this section. This approach is distinguished by two important characteristic. First of all, in this approach—in addition to axioms—use is made of only one rule of inference, namely that of detachment. Both the use of quantifiers and substitution are regulated by axioms.

The reduction of all steps in the proofs to successive operations of detachment, which is realizable in this approach, simplifies, as we shall see, the concept of proof and makes it possible to present more elegantly certain ideas on logic as a mathematical theory.

The other important characteristic is that in this approach the axioms are probably more convincing from the intuitive point of view than the rules which occur in other approaches based on rules and axioms.

The approach to be discussed now and to be symbolized by L will be described independently of that given in the preceding section, and only then will the equivalence of the two be proved. Let us now forget about the method of introducing the predicate calculus as described in the preceding section, and let us consider quite independently the following introduction of the logical calculus.

We shall first be concerned with a more precise description of the language of logic and the concept of logical tautology. The concept of tautology in the full system of logic is based on the same intuitions which were the starting point for the formulation of the concept of tautology in the sentential calculus. Tautologies are, as before, schemata of sentences that are always true. The only difference is that in the case of the sentential calculus the analysis was confined to only those schemata which can be written down by means of sentential connectives and sentential variables alone. The schemata to be discussed now are much more complex, since in addition to the sentential connectives and parentheses: $\vee$, $\wedge$, $\sim$, $\rightarrow$, $\equiv$, $(,)$, they also include quantifiers: $\bigvee$, $\bigwedge$, the symbols: P, Q, R, S, T, ..., P_1, Q_1, R_1, ..., P_2, Q_2, R_2, ..., which stand for arbitrarily chosen predicates, and the variables: x, y, z, ..., ..., x_1, y_1, z_1, ..., which stand for individual objects from an arbitrarily chosen mathematical domain. We, thus, are dealing with a much richer language of the schemata under consideration. This is why the concept of tautology, when formulated with precision, must become somewhat more complicated. In order to better describe the concept of tautology in the comprehensive logical calculus we must give more precision to the language of those logical formulae with which the tautologies will be recorded. The basic graphic elements of the language of logical schemata (formulae) have been listed above. Obviously, not every sequence of symbols taken from the above list is a logical schema of a mathematical sentence. A sequence of such symbols is a logical schema if it yields a meaningful mathematical formula when the capital letters standing for predicates: P, Q, ..., P_1, Q_1, ..., are replaced by definite mathematical predicates. For instance, the formula

(90) $\bigvee x \bigwedge y R(x, y)$

is a logical schema, for when a definite mathematical predicate, e.g., $x \geqslant y$, is substituted for the predicate $R(x, y)$, we obtain a definite mathematical sentence, namely

$$\bigvee x \bigwedge y (x \geqslant y),$$

which refers to the existence of a maximum element. Thus, the sentence in question has a clearly defined meaning. The sentence is false in the domain of real numbers, and in the domain of natural numbers as well. But should we decide that the variables x and y range only over the set of negative integers, then the sentence would be true, since there exists

155

a greatest negative integer. Thus schema (90) is not a logical tautology, since it may not be said that it is always true: there is a mathematical domain (that of real numbers) and a relation $\geqslant$ in that domain such that, if the predicate $R(x, y)$ is interpreted as describing that relation, the schema in question becomes a sentence that is false from the mathematical point of view.

The concept of logical schema, that is, correct logical formula (well-formed formula) has been defined above with not too great precision, and semantically rather than formally, since we to some extent referred to understanding the formulae. But a well-formed formula can be described in purely syntactic terms. The following description may be treated as an outline of such a purely syntactic definition:

DEFINITION 14.

1. *Every formula consisting of a symbol that stands for a predicate:* $P, Q, R, ..., P_1, Q_1, ...,$ *and is followed directly by parentheses which contain a number of individual variables separated by commas, is a well-formed formula; such formulae will be called atomic.*[1]

2. *Every linking of two well-formed formulae by one of the sentential connectives of two arguments:* $\vee$, $\wedge$, $\rightarrow$, $\equiv$, *is a well-formed formula if in that formula equiform symbols standing for predicates are followed by parentheses containing the same number of variables.*

3. *Every formula consisting of a well-formed formula preceded by the symbol of negation* $\sim$ *is well-formed.*

4. *Every formula consisting of a quantifier covering a variable followed directly by parentheses containing a well-formed formula is well-formed.*

5. *Every well-formed formula is obtained from atomic formulae by operations* 2, 3 *and* 4, *applied a finite number of times.*

This definition is not quite precise. Its form comes close to what is called an inductive definition. Nevertheless, it provides sufficient information about the meaning of the concept defined. We shall see below how such definitions ought to be changed into precise ones.

In the above definition of a well-formed formula the parentheses serve the purpose of uniquely indicating the scope of logical operations

---

[1] We will also use the term *atomic predicate* to denote an atomic formula in which any two different variables are different in form. Hence, $R(x, y)$ is an atomic predicate, whereas $R(x, x)$ is not, although it is an atomic formula.

(the scope of quantifiers and sentential connectives). Accordingly, if the scope of a given logical operation does not raise doubts the parentheses will often be dropped. Likewise, the simple term: *formula* will often be used instead of: *well-formed formula*.

By the definition above, formulae (89) and (90) are well-formed. On the contrary, the following expressions are not well-formed:

$$\bigwedge x(x \equiv \bigwedge),$$

$$\bigvee (P(x)),$$

$$\bigvee x\big(P(\bigwedge)\big).$$

The following

$$(91) \qquad \sim \bigvee y(P(y)) \equiv \bigwedge y(\sim P(y))$$

is an example of a well-formed formula. It is even a logical tautology. To make sure about that note first that for any meaning of the predicate P sentence (91) seems to be intuitively true. For instance, let the variables range over the set of men, and let the predicate $P(y)$ mean: y is horned. Then formula (91) becomes the following sentence:

(There is no man y who is horned) if and only if (every man is not horned).

This sentence is intuitively true. Schema (91) covers the following sentence, too:

There is no quadruple of numbers: x, y, z, n such that n > 2 and x^n + y^n = z^n, if and only if for every quadruple of the numbers x, y, z, n it is not true that n > 2 and x^n + y^n = z^n.

The above sentence is intuitively true in the domain of natural numbers. Likewise, any other sentence falling under schema (91) seems to be intuitively true. Schema (91) is called *De Morgan's law for quantifiers*; the same term applies to

$$(92) \qquad \sim \bigwedge y(P(y)) \equiv \bigvee y \sim P(y).$$

Both laws are well founded in everyday intuition. The sentence:

It is not true that all real numbers are rational if and only if there exist irrational real numbers,

is an example of the application of the second of these laws to the domain of real numbers.

Of course, De Morgan's laws apply to formulae of any number of variables, e.g.:

$$(93) \quad \begin{aligned} &\sim \bigvee y R(x, y) \equiv \bigwedge y \sim R(x, y), \\ &\sim \bigwedge y R(x, y) \equiv \bigvee y \sim R(x, y). \end{aligned}$$

Laws (91) and (92) are generalizations of De Morgan's laws in the sentential calculus for infinite conjunctions and disjunctions. The following schemata are tautologies in the sentential calculus:

$$\sim (p_0 \vee p_1 \vee \ldots \vee p_n) \equiv \sim p_0 \wedge \sim p_1 \wedge \ldots \wedge \sim p_n,$$
$$\sim (p_0 \wedge p_1 \wedge \ldots \wedge p_n) \equiv \sim p_0 \vee \sim p_1 \vee \ldots \vee \sim p_n;$$

they are known as *De Morgan's laws*. By substituting sentences of the form $P(0)$, $P(1)$, ..., $P(n)$ for the sentential variables $p_0, p_1, \ldots, p_n$ we obtain from these tautologies the following equivalences:

$$\sim \big(P(0) \vee P(1) \vee \ldots \vee P(n)\big) \equiv$$
$$\sim P(0) \wedge \sim P(1) \wedge \ldots \wedge \sim P(n),$$
$$\sim \big(P(0) \wedge P(1) \wedge \ldots \wedge P(n)\big) \equiv$$
$$\sim P(0) \vee \sim P(1) \vee \ldots \vee \sim P(n).$$

In accordance with the intuitive definition of quantifiers as infinite conjunctions and disjunctions De Morgan's laws for natural numbers are analogies of the above theorems in the sentential calculus.

Not all the schemata, however, obtained by a generalization from a finite number of cases to an infinite number are universally valid. The above analogy can not serve as a foundation for the definition of a general concept of tautology.

Nevertheless, in view of the above examples we are now in a position to define the concept of logical tautology with more precision. As has been said above, tautologies are schemata of those sentences which are always true. Since schemata in the predicate calculus include variables that stand for objects from a given domain, and letters P, Q, ..., that stand for any predicates denoting properties and relations in that domain, the statement that a schema is always true must now mean that it is *true in every domain*.

We have to define first when it may be said that a schema A is true in a domain $\mathfrak{M}$. A precise definition of this relation will be postponed, and for the time being we will give only certain intuitive explanations.

6. REDUCTION OF QUANTIFIER RULES

First, to simplify matters, we shall confine ourselves to an analysis of logical formulae that include only three predicate symbols: $P(x, y)$, $R(x, y)$, $S(x, y, z)$ and which are sentences (that is, include no free variables). In this connection it suffices to consider domains $\mathfrak{M} \langle X, \pi, \varrho, \sigma \rangle$, each consisting of a set X, two relations π, ϱ of two arguments, and one relation σ of three arguments.

DEFINITION 15*. *It is said that a sentence A is true in a domain $\mathfrak{M} \langle X, \pi, \varrho, \sigma \rangle$, with the interpretation of the predicate symbols as names of relations π, ϱ, σ, respectively, if and only if the relations π, ϱ, σ hold between elements of the set X in the way stated in the sentence A.*

Thus, for instance, the sentence

$$(94) \qquad \bigwedge x \left(\bigvee y \left(P(x, y) \wedge \sim (R(x, y)) \right) \right)$$

is true in the domain $\mathfrak{N}_0 \langle \mathcal{N}, \leqslant, \mathrm{Suc}, + \rangle$ (where $\mathcal{N}$ = the set of natural numbers: $\leqslant$ = the relation *not greater than*: $\leqslant (x, y) \equiv x \leqslant y$; Suc = the successor relation: Suc $(x, y) \equiv x + 1 = y$; + = the relation of addition: $+(x, y, z) \equiv x + y = z$), with the interpretation of the predicate symbols P and R as names of the relations $\leqslant$ and Suc, respectively, because in the domain of natural numbers it is true that

$$(95) \qquad \bigwedge x \bigvee y \left(\leqslant (x, y) \wedge \sim (\mathrm{Suc}(x, y)) \right).$$

It is so because this sentence means that for every number x there exists a natural number y which is greater than x and does not equal $x + 1$. In fact, there are very many numbers that are greater than x and do not equal $x + 1$; they are: $x + 2$, $x + 3$, ... On the other hand, the same sentence (94) is not true in the same domain if the predicate symbol P is interpreted as a name of the successor relation, and the predicate symbol R as a name of the relation $\leqslant$. For such an interpretation of the symbols involved, that sentence would state that for every natural number x there exists a number y which is the successor of x, that is, equals $x + 1$, and is less than x. Such a case obviously does not occur, since $x + 1$ is always greater than x.

Likewise, the sentence

$$\bigwedge x \bigvee y \bigvee z (P(x, y) \wedge S(x, y, z))$$

is true in the domain $\mathfrak{N}_0 \langle \mathcal{N}, \leqslant, \mathrm{Suc}, + \rangle$ if the predicate symbols P and S are interpreted, respectively, as names of the relations $\leqslant$ and +

since in fact for every natural number x there exist numbers y and z such that y is greater than x and z is the sum $x+y$. For instance, $y = x+1$ and $z = x+x+1$. It is, however, easy to find a domain $\mathfrak{M}$ and an interpretation of the predicate symbols P, R, and S, for which that sentence is false.

Definition 15* is thus easy to comprehend, although we use in it formulations which still lack precise meanings. To emphasize the insufficient precision of that definition we have provided its number with an asterisk, and shall do so in the case of all later definitions which will refer to it.

DEFINITION 16*. *A sentence A, containing the predicate symbols P,R, S, is a tautology in the c.l.c. if and only if the sentence A is true in every non-empty domain $\mathfrak{M}$ for any interpretation of the symbols P, R, and S as names of certain relations in that domain.*

Henceforth, only certain closed sentences, that is, formulae without free variables, shall be called tautologies in the c.l.c., or theorems in an axiomatic system of the c.l.c. But in order to avoid writing too many quantifiers before every formula we adopt the following convention concerning the writing down of theorems:

If a formula A is called a theorem in an axiomatic system of the c.l.c. (or a tautology in the c.l.c.) and is written down so that free variables $x, y, ..., z$ do occur in it:

$$A(x, y, ..., z),$$

this means that the formula in question is merely considered to be an abbreviated form of the formula

$$\bigwedge x\left(\bigwedge y\left(... \bigwedge z(A(x, y, ..., z))\right)\right),$$

obtained from the formula A by preceding it with universal quantifiers binding all the free variables in A.

Thus, for instance, formulae (93), called De Morgan's laws for quantifiers, are considered abbreviated forms of the formulae:

$$(96) \quad \begin{aligned} &\bigwedge x(\sim \bigvee y R(x, y) \equiv \bigwedge y \sim R(x, y)), \\ &\bigwedge x(\sim \bigwedge y R(x, y) \equiv \bigvee y \sim R(x, y)), \end{aligned}$$

which, as compared with formulae (93), are preceded by the universal quantifier $\bigwedge x$ which binds the variable x, free in formulae (93).

6. REDUCTION OF QUANTIFIER RULES

In some cases, in order to indicate that we take note of the quantifiers at the beginning of a formula, we shall use the symbol: $\bigwedge$ ..., which is supposed to stand for the quantifiers binding all the free variables occurring in the formula which follows that symbol. A sequence of universal quantifiers will be combined into a single quantifier in order to simplify the notation. The same will apply to existential quantifiers.

To simplify the notation of schemata which cover many c.l.c. formulae we shall use formulae of the form.:

(97) $\quad \alpha, \beta, \gamma, \alpha(x), \beta(x), \gamma(x, y)$, etc.,

consisting of Greek letters, sometimes followed by parentheses with variables. Formulae of the form of (97) without parentheses with variables will stand, in logical schemata, for any meaningful formulae containing any number of any variables or not containing any variables at all. Formulae of the form (97) with variables in parentheses, such as $\alpha(x)$, will stand for any well-formed formulae which, however, contain as free variables at least those variables which occur in the parentheses. Hence, for instance, the formula

$$\sim \bigvee y \alpha(y) \equiv \bigwedge y \sim \alpha(y)$$

is a schema which covers law (91) and also the laws

$$\sim \bigvee y (R(y) \to S(y)) \equiv \bigwedge y \sim (R(y) \to S(y)),$$
$$\sim \bigvee y (\sim P(y)) \equiv \bigwedge y \sim \sim P(y);$$

on the other hand, it does not cover the formula

$$\sim \bigvee y P(x) \equiv \bigwedge y \sim P(x),$$

which is well-formed and falls under the schema

$$\sim \bigvee y \alpha \equiv \bigwedge y \sim \alpha.$$

DEFINITION 17. *All those well-formed formulae of the c.l.c. which fall under the following schemata are axioms in the c.l.c. system L:*

I. *Any schemata of the classical sentential calculus in which:*
   A. *well-formed formulae of the c.l.c. are substituted for the sentential variables, and*
   B. *the entire formula is preceded by universal quantifiers which bind all the free variables occurring in the formula.*

161

II. *The schemata of handling universal quantifiers:*

   1. $\bigwedge \ldots \left(\bigwedge x(\alpha \rightarrow \beta) \rightarrow (\bigwedge x\alpha \rightarrow \bigwedge x\beta) \right)$

      (the law of distributivity with respect to implication).

   2. $\bigwedge \ldots \left(\bigwedge x\alpha(x) \rightarrow \alpha(x) \right)$

      (the law of transition from a general to a particular case).

   3. $\bigwedge \ldots \left(\bigwedge x \bigwedge y\alpha \rightarrow \bigwedge y \bigwedge x\alpha \right)$

      (the law of commutativity of quantifiers).

   4. $\bigwedge \ldots \left(\bigwedge x \bigwedge y\alpha \rightarrow \bigwedge x\alpha(y/x) \right)$

(if the formula α does not contain any quantifier binding the variable x in the scope of which a variable y, free in α, occurs, and if $\alpha(y/x)$ differs from α only in having all the variables y free in α replaced by the variables x)

      (the law of combining universal quantifiers).

   5. $\bigwedge \ldots \left(\alpha \rightarrow \bigwedge x\alpha \right)$

(if x does not occur in α as a free variable)

      (the law of joining a superfluous quantifier).

III. *The schemata of handling existential quantifiers:*

   6. $\bigwedge \ldots \left(\bigwedge x(\alpha \rightarrow \beta) \rightarrow (\bigvee x\alpha \rightarrow \bigvee x\beta) \right)$

      (the law of distributivity with respect to implication).

   7. $\bigwedge \ldots \left(\alpha(x) \rightarrow \bigvee x\alpha(x) \right)$

      (the law of abstraction from concreteness).

   8. $\bigwedge \ldots \left(\bigvee x \bigvee y\alpha \rightarrow \bigvee y \bigvee x\alpha \right)$

      (the law of commutativity of quantifiers).

   9. $\bigwedge \ldots \left(\bigvee x\alpha(y/x) \rightarrow \bigvee x \bigvee y\alpha \right)$

(with the same proviso as in the case of Axiom II.4)

      (the law of splitting existential quantifiers).

   10. $\bigwedge \ldots \left(\bigvee x\alpha \rightarrow \alpha \right)$

(if x does not occur in α as a free variable)

      (the law of omitting a superfluous quantifier).

6. REDUCTION OF QUANTIFIER RULES

Likewise, any other schemata obtained from Schemata 1–10 by the replacement of the variables x and y by any other pair of variables.

IV. *The assumption of non-emptiness*:

0. $\bigvee x(P(x) \to P(x))$.

The axiom system described above is based on research carried out by the following authors: A. Mostowski [54], T. Hailperin [20], W. V. Quine [66].

If we wish to deduce a c.l.c. theorem from c.l.c. axioms, we use only the rule of detachment. In this approach to the c.l.c., any proof reduces to a repetition of detachment operations. But to achieve that we would have to assume an infinite number of axioms: all substitutions are provided for by the axioms.

We shall now briefly discuss the meanings of the various groups of c.l.c. axioms and we shall demonstrate, in an intuitive manner, that they are obviously true in any non-empty domain.

I. Schemata in the classical sentential calculus

Schemata in the sentential calculus were interpreted as schemata of sentences that are always true, regardless of the meaning of the simple sentences that occur in the compound sentences which fall under these schemata. Thus sentences referring to any domain of objects may fall under one and the same c.s.c. schema. Every c.s.c. axiom in the first groups is thus true in any mathematical domain. For instance, the formulae:

$$\bigwedge x, y\Big(R(x, y) \to \big(S(x, y) \to (R(x, y) \wedge S(x, y))\big)\Big),$$

$$\bigwedge x, y\Big((P(x) \to Q(x)) \to \big((Q(x) \to R(x, y)) \to (P(x) \to R(x, y))\big)\Big),$$

$$\bigwedge y, z, u\Big((Q(y) \to S(z, u)) \to (\sim S(z, u) \to \sim Q(y))\Big),$$

are thus axioms of the first group. As can easily be seen, they are obtainable from the following c.s.c. tautologies:

$$p \to (q \to (p \wedge q)),$$
$$(p \to q) \to ((q \to r) \to (p \to r)),$$
$$(p \to q) \to (\sim q \to \sim p),$$

163

by substitution of equiform predicates for equiform sentential variables in a given c.s.c. law and by preceding every formula obtained in this way with universal quantifiers binding all the individual variables free in the formula thus obtained.

In sentential calculus formulae, complex formulae, which contain quantifiers, may be substituted for sentential variables. For instance, the formula

$$\bigwedge x\big((\bigwedge yR(x, y)) \to (\bigwedge yR(x, y))\big),$$

obtained from the law $p \to p$ by the substitution $p/\bigwedge yR(x, y)$ and by the binding of the variable x with an additional universal quantifier placed at the beginning of the formula, is an axiom of the first group. Likewise,

$$\bigwedge y \bigwedge z(\bigvee xS(x, y, z) \lor \sim \bigvee xS(x, y, z))$$

is obtained from the law: $p \lor \sim p$ by the substitution $p/\bigvee xS(x, y, z)$ and by preceding the result with the quantifiers $\bigwedge y$, $\bigwedge z$.

II. Schemata for handling universal quantifiers

II. 1. *The schema of distributing the universal quantifier over an implication* is a law which is quite obvious in every domain. Consider, for instance, the domain of animals. The implication

(98) $\bigwedge x$ (*if x moves, then x can search for food*),

is true in that domain, since every animal that moves can search for food. Assume now that we accept the sentence stating that every animal moves, so that the sentence

(99) $\bigwedge x(x \text{ moves})$

is true in the domain under consideration. It would then be obvious that every animal can search for food. Thus the sentence

(100) $\bigwedge x(x \text{ can search for food})$

would be true in the domain of animals. Thus, by reasoning in a quite spontaneous way, we have accepted the conclusion which also follows from the following law of distributivity with respect to implication:

(101) $\bigwedge x(P(x) \to Q(x)) \to (\bigwedge xP(x) \to \bigwedge xQ(x))$

by the substitutions $P(x)/x$ *moves*, $Q(x)/x$ *can search for food*, and by two operations of detachment of the accepted sentences (98) and (99). Law (101) is a special case of schema II.1.

Consider now the domain of sets of real numbers. In general terms, it can be proved in set theory that

(102) $\bigwedge x$ (*if x is a well ordered set, then ordinal numbers may be associated with the elements of the set x*).

Thus, this sentence is in particular true for the sets of real numbers. Assume now that we have accepted the axiom of choice, which states that every set may be well ordered. Hence the sentence

(103) $\bigwedge x$ (*x is a well-ordered set*)

is true in the domain under consideration; sentences (102) and (103) yield, under law (101), the conclusion:

$\bigwedge x$ (*ordinal numbers may be associated with the elements of the set x*).

This conclusion is in full agreement with intuition, since intuition tells us that if an implication $P \rightarrow Q$ is always true in a given domain, and if the antecedent P of that implication is always true, then the consequent Q of that implication is always true, too. This is exactly the meaning of schema II.1.

II.2. *The schema of substitution* is also satisfied in every domain in an obvious manner. Consider, for instance, the domain of natural numbers. Those conclusions from II.2 which are obtained, with the rule of substitution, by the substitution of mathematical constants for individual variables, seem particularly clear. That rule states that if there is a general mathematical theorem, that is, a theorem with the universal quantifier at the beginning, it is legitimate to substitute any definite mathematical constant for the variables bound by that universal quantifier occurring at the beginning of the formula, of course, with the simultaneous elimination of that quantifier. For instance, the general law: $\bigwedge x \bigwedge y(x+y = y+x)$ yields on substitution the theorems $\bigwedge y(2+y = y+2)$, $\bigwedge y(7+y = y+7)$, etc. Assuming that the formula $\alpha(x)$ has the form: $x+3 = 3+x$, we obtain, by II.2, as a special case, the implication

$$\bigwedge x(\bigwedge x(x+3 = 3+x) \rightarrow x+3 = 3+x),$$

165

to which the rule of substitution may be applied; we may accordingly substitute the symbol 2 for the variables x, bound by the initial quantifier. We then obtain

$$\bigwedge x(x+3 = 3+x) \to 2+3 = 3+2.$$

This sentence states that if the relation $x+3 = 3+x$ holds for every x, then the same relation also holds for $x = 2$. The meaning of the entire law II.2 is similar.

Law II.2 might be read as follows: If a relation α holds for every x, then it must also hold for a given x. The variables bound by the universal quantifier occurring at the beginning of a formula may be interpreted as "given", since we may substitute for them any names of definite objects from the domain under consideration. Law II.2, thus, allows us to pass from general theorems to definite cases. Hence the names: *law of substitution*; *law of specification*; *law of transition from the general to the particular*—all seem to be proper.

II.3. *Law of alteration of universal quantifiers.* This law states that the order in which universal quantifiers occur in a formula is irrelevant. It can easily be verified that for every property of any domain the relevant special cases of law II.3 are obvious. For instance, for the real numbers it is true that

$$\bigwedge x\bigwedge y(x \geqslant y \vee y \geqslant x),$$

but the same fact may be stated in a sentence with the quantifiers interchanged:

$$\bigwedge y\bigwedge x(x \geqslant y \vee y \geqslant x).$$

II.4. *Law of identification of variables* (or: *law of combining quantifiers*). This law states that if we have a theorem of the form

$$\bigwedge x\bigwedge yP(x, y),$$

then we may accept the theorem

$$\bigwedge xP(x, x),$$

which differs from the former in having the free variable x substituted for the variable y, free in P. In fact, the former sentence states that a relation P holds between any x and y, while the latter states that it holds between x and y which are identical with one another. Thus, the latter sentence is a special case of the former.

For instance, if we accept the law of commutativity of addition

$$\bigwedge x \bigwedge y(x+y = y+x),$$

then we may deduce from it the law

$$\bigwedge x(x+x = x+x),$$

which is, as it were, a special case of the former. As can be seen, it also is a special case of the law: $\bigwedge x(x = x)$. The identification of variables thus usually yields theorems which are less interesting and are like special cases of the theorem in its form before the identification of variables. The mathematical theorem:

$$\bigwedge x, y, z(x+(y+z) = (x+y)+z) \rightarrow$$
$$\bigwedge x, z(x+(x+z) = (x+x)+z),$$

and the theorem drawn from everyday life:

If (for all men x and y, if x strikes y then y is hurt), then (for any man x it also is true that (if x strikes x (that is, x strikes himself), then x is hurt)), also are applications of II.4.

The law of combining universal quantifiers had the restriction that the formula α may not contain any quantifier binding the variable x, in the scope of which the variable y, free in α, occurs. It is easy to give an example in which that restriction is not satisfied and the resulting theorem is in fact intuitively false. Let α be the following formula with two free variables, x and y:

$$\alpha = x \wedge \bigvee x(x < y).$$

We shall apply II.4 to it, disregarding the restriction in question. We then obtain the sentence

$$\bigwedge x \bigwedge y(x = x \wedge \bigvee x(x < y)) \rightarrow \bigwedge x(x = x \wedge \bigvee x(x < x)).$$

This sentence is not true in every domain. For instance, it is not true in the domain of real numbers, if the symbols are interpreted in the usual way. The antecedent is true in the domain of real numbers, since for any real numbers x and y obviously $x = x$ and there exists for y a number x which is less than y, for instance $y-1$. On the other hand, the consequent is not true in the domain of real numbers, since there is no number x which is less than itself, as claimed by one of the terms of the conjunction which forms the consequent, namely the term which has the form $\bigvee x(x < x)$.

II.5. *Law of joining a superfluous universal quantifier.* This law resembles the simplification law: $p \rightarrow (q \rightarrow p)$. The simplification law allows us to join any superfluous antecedent to a previously accepted sentence p. Similarly, law II.5 allows us to join a superfluous phrase of the type: for every x. Thus, for instance:

If $2+2 = 4$, then it is also true that, for every x, $2+2 = 4$.

A sentence containing a superfluous quantifier, that is, a quantifier that does not bind any variable, is treated as meaningful. The rules of handling quantifiers adopted allow that a superfluous universal quantifier may always be adjoined to a sentence, but it is not always allowed to omit it. On the other hand, a superfluous existential quantifier may always be omitted, but it is not always allowed to add it.

III. Schemata for handling existential quantifiers

III.6. *Law of distributivity with respect to implication.* It is an analogue of law II.1. Its meaning also agrees very well with our intuition. A free application of this law is illustrated, e.g., by the following sentence:

If (all spiders are arthropods), then (if there is a spider in this room, then there is an arthropod in this room).

The sentence: *all spiders are arthropods* means the same as: *for every x, if x is a spider, then x is an arthropod.* The above statement about the arthropods is just a special case of a more general set-theoretical theorem, also obtainable from law III.3, namely the theorem:

If (a set A is contained in a set B), then (if there exist objects which are elements of the set A, then there also exist objects which are elements of the set B).

In symbolic notation, this theorem may be written as follows:

$$\bigwedge x(x \in A \rightarrow x \in B) \rightarrow \left(\bigvee x(x \in A) \rightarrow \bigvee x(x \in B) \right).$$

The notation shows that this theorem results from an application of law III.6, if the predicate $\alpha(x, A)$ is interpreted as $x \in A$, and if the predicate β is interpreted in a similar way.

168

6. REDUCTION OF QUANTIFIER RULES

Law III.6, thus, seems intuitively to be satisfied in every domain.

III.7. *Law of abstraction from concreteness.* As in the case of law II.2, those applications of law III.7 seem particularly convincing, in which a constant is substituted for the variable x at the beginning of the theorem. For instance, the sentence:

If the Moon is a satellite of the Earth, then there exists an x such that x is a satellite of the Earth,

is of the type:

$$(104) \qquad P(\textit{the Moon}) \to \bigvee x P(x),$$

where $P(x)$ is interpreted as: *x is a satellite of the Earth.* Sentence (104) is obtained directly from the axiom

$$(105) \qquad \bigwedge x\big(P(x) \to \bigvee x P(x)\big),$$

which falls under schema III.7, by the substitution of the definite name *the Moon* for the variable x, bound by the quantifier placed at the beginning of the formula. The sentence:

If π is irrational, then irrational numbers exist,

results from the application of the same axiom, since it is obtained from axiom (105) by the substitution of π for x, the predicate $P(x)$ being interpreted as: *x is an irrational number.* In all effective proofs of existence of mathematical objects we constantly have to do with applications of axioms of III.7, since those axioms, used in combination with the rule of substitution, make it possible to assert that there exists an object with certain properties if we are able to indicate a definite object which has those properties. The meaning of Law III.7 is, thus, intuitively true in an obvious manner.

III.8. *Law of alteration of existential quantifiers.* This law allows us to arrange existential quantifiers in an arbitrary order. The order in which the objects whose existence we assert is specified is irrelevant. Thus, the sentence:

If (there exists a fish x and an aquatic plant y such that the plant y is poisonous for the fish x), then (there exists an aquatic plant y and a fish x such that the plant y is poisonous for the fish x)

is an application of Law III.8.

III.9. *Law of splitting existential quantifiers.* An example of this law is the following argument. The number zero has the property: $0+0 = 0$. Hence, by III.7, we may obtain the theorem

$$\bigvee x(x+x = 0).$$

which, by III.9, yields

$$\bigvee x \bigvee y(x+y = 0).$$

This reasoning is intuitively convincing, since if there exists a number x such that, when added to itself, it is zero, then it may be said that there exist two numbers such that when added together they yield zero. The latter sentence states less than the former. The former asserts that the two numbers whose sum is zero are one and the same number (that is, they are identical with one another). The latter does not assert that, but is confined to the statement that the sum of these two numbers is zero.

The theorem:

If $(\bigvee x(x$ *is an irrational number* $\wedge$ x *is an algebraic number)), then* $(\bigvee x \bigvee y(x$ *is an irrational number* $\wedge$ y *is an algebraic number)),*

also is an application of Law III.9.

Of course, if there exists a number which is both irrational and algebraic, then there exists an irrational number and an algebraic number.

It would also be easy to give an example, as in the case of Law II.4, showing that disregard of the restriction about bound variables would result in assertion of sentences that are false in certain domains.

III.10. *Law of omitting superfluous existential quantifiers.* This law allows us to omit an existential quantifier which occurs at the beginning of a theorem and which does not bind any variable. Thus, it allows us to obtain the theorem: $2+2 = 4$ from the theorem: $\bigvee x(2+2 = 4)$. The latter provides more information than the former since in addition to the information that $2+2 = 4$ it also states the existence of a number x. Law III.10, thus, corresponds to the law $(p \wedge q) \rightarrow p$ in the sentential calculus.

As can easily be seen, the schemata of III for handling existential quantifiers are analogues of the respective schemata of II for handling universal quantifiers.

170

IV. The assumption of non-emptiness

The axiom of non-emptiness, which occurs last on the list above, simply states that at least one object exists. Properly speaking, that axiom should not belong to logic, since it would seem natural to free the system of logic from the assumption of the existence of any object. Usage, however, has us use the expression *the classical system of logic* for that logical calculus which includes Axiom 0. In view of the alien character of Axiom 0 the system of axioms given above is formulated so that Axiom 0 does not follow from the others, and that those remaining axioms form an axiom system of all those and only those logical laws from which Axiom 0 does not follow. Axiom 0 is not true in the empty domain as it refers to the existence of a certain object. On the contrary, all the remaining axioms are true in the empty domain. Hence the consequences of the axioms other than Axiom 0 form a logical calculus true in every domain, while the consequences of all the axioms, Axiom 0 included, form a logical calculus true in all non-empty domains. In practical research, the non-empty domains are obviously more interesting than the empty domain is. Moreover, there are fairly important logical laws (which make it possible to reduce formulae to what is called the normal form) which are true in non-empty domains only. This is why it would be difficult to confine the study of the classical logical calculus to a calculus that would be true in all domains, the empty domain included. In this section, however, we shall be concerned exclusively with the consequences of the c.l.c. axioms other than Axiom 0.

The idea of restricting logic to predicate calculus laws true in all domains, including the empty domain, is due to A. Mostowski, who justifies it as follows:

"Making a distinction between those formulae which are true in all domains, the empty domain included, and those which are only true in non-empty domains may, at the first glance, seem to be unnecessary pedantry. This impression vanishes if we realize that in the applications of logic to mathematics we usually have to do with formulae in which quantifiers are restricted to certain sets (domains). In mathematics, we hardly ever say "for every x", but we almost always say "for every point x", "for every number x", etc. No less frequent are the cases in which the domain to which the quantifier is restricted depends on a parameter or parameters; for instance, in the case of the formulation

171

"for every real root of the equation..." the coefficients of that equation are such parameters. In such a case the domain may become empty for new values of the parameter, and this is why only those logical theorems which are true in every domain, the empty domain included, are applicable to it without any reservation whatever."

DEFINITION 18. *Every formula obtained from axioms of the c.l.c. system L by the application of the rule of detachment is a theorem in the axiomatic c.l.c. system L.*

As stated in Section 2, the rule of detachment is intuitively true regardless of the meanings of the sentences to which it is applied. Hence, it leads from sentences that are always true to sentences that are always true. We extend this reasoning so as to cover our present analysis and accordingly we note that the rule of detachment must lead from sentences true in a certain domain to sentences true in that domain. Hence, as the c.l.c. axioms are intuitively true in every domain, obviously:

THEOREM 6. *Every theorem in the c.l.c. axiomatic system L is true in every non-empty domain, and hence it is a c.l.c. tautology.*

The theorem converse to 6, stating that every c.l.c. tautology is a theorem in the c.l.c. axiomatic system will be proved in the next chapter, since it is more difficult and requires a more precise proof. It is namely the theorem on semantic completeness, mentioned above.

We shall now prove a number of c.l.c. theorems, including those already proved in the system L^+. Those theorems will be marked with numbers greater than 10, and the more important theorems will be marked with an asterisk.

11. $\left(\bigwedge x(A \to B(x)) \to (\bigwedge xA \to \bigwedge xB(x))\right)$

> (a special case of Schema 1; it is assumed that A is a sentence, and that $B(x)$ has only one free variable x).

12. $(A \to \bigwedge xA)$

> (a case of Schema 5).

13. $\left(\left(\left(\bigwedge x(A \to B(x)) \to (\bigwedge xA \to \bigwedge xB(x))\right) \to \right.\right.$
$\left.\left.\left((A \to \bigwedge xA) \to \left(\bigwedge x(A \to B(x)) \to (A \to \bigwedge xB(x))\right)\right)\right)\right)$

> (a special case of the sentential calculus schema
> $\left(p \to (q \to r)\right) \to \left((s \to q) \to (p \to (s \to r))\right)$).

14. $(A \to \bigwedge xA) \to \left(\bigwedge x(A \to B(x)) \to (A \to \bigwedge xB(x)) \right)$
(from 11 and 13 by detachment).

15. $\left(\bigwedge x(A \to B(x)) \to (A \to \bigwedge xB(x)) \right)$
(from 14 and 12 by detachment).

It can easily be shown in a general way that a proof similar to the proof of Theorem 15 can be used to deduce from c.l.c. axioms any theorem that falls under the general schema

15*. $\bigwedge \dots \left(\bigwedge x(\alpha \to \beta) \to (\alpha \to \bigwedge x\beta) \right)$

(if α does not contain x as a free variable).

Schema 15* is an important schema in the predicate calculus. It allows us to shift a universal quantifier from before an implication $\alpha \to \beta$ to the consequent of that implication if that quantifier does not bind any variable occurring in the antecedent α of that implication.

All the sentences falling under the schema converse to 15*, that is,

22*. $\bigwedge \dots \left((\alpha \to \bigwedge x\beta(x)) \to \bigwedge x(\alpha \to \beta(x)) \right)$

also are theorems in the c.l.c. under the assumption that α does not contain x as a free variable.

Here is an outline of a proof of the simple case converse to 15*. We start from schema II.2:

16. $\bigwedge x(\bigwedge xB(x) \to B(x))$.

Next we obtain the theorem

17. $\bigwedge x \left((\bigwedge xB(x) \to B(x)) \to \left((A \to \bigwedge xB(x)) \to (A \to B(x)) \right) \right)$,

which is a special case of the sentential schema $(p \to q) \to ((s \to p) \to (s \to q))$. In order to split the universal quantifier in Theorem 17 into a universal quantifier in the antecedent and a universal quantifier in the consequent we refer to a special case of schema II.1:

18. $\bigwedge x \left((\bigwedge xB(x) \to B(x)) \to \left((A \to \bigwedge xB(x)) \to (A \to B(x)) \right) \right) \to$
$\left(\bigwedge x(\bigwedge xB(x) \to B(x)) \to \bigwedge x \left((A \to \bigwedge xB(x)) \to (A \to B(x)) \right) \right)$.

Next, Theorem 17 is detached:

19. $\bigwedge x(\bigwedge xB(x) \to B(x)) \to \bigwedge x \left((A \to \bigwedge xB(x)) \to (A \to B(x)) \right)$
(from 17 and 18 by detachment).

Theorem 19 differs from 17 only in that the universal quantifier $\bigwedge x$, which stands at the beginning of 17, is split in 19 into one before the antecedent and one before the consequent. By referring to schema 1 we always may pass in this way from a theorem of the form $\bigwedge x(\alpha \to \beta)$ to a theorem of the form $\bigwedge x\alpha \to \bigwedge x\beta$. To shorten the writing of proofs we will not write out such transitions in full, but will just refer to schema 1.

20. $\bigwedge x\big((A \to \bigwedge xB(x)) \to (A \to B(x))\big)$
(from 19 and 16 by detachment).

21. $\bigwedge x\big((A \to \bigwedge xB(x)) \to (A \to B(x))\big) \to \big((A \to \bigwedge xB(x)) \to \bigwedge x(A \to B(x))\big)$.

(A special case of schema 15*, which we consider proved. In order to apply it we must, of course, assume that x does not occur in A as a free variable. Then x does not occur in the entire formula $(A \to \bigwedge xB(x))$ as a free variable.)

22. $(A \to \bigwedge xB(x)) \to \bigwedge x(A \to B(x))$
(from 20 and 21 by detachment),

if x does not occur in A as a free variable.

It can be proved in an analogous way that every sentence falling under the following general schema:

22*. $\bigwedge \ldots \big((\alpha \to \bigwedge x\beta(x)) \to \bigwedge x(\alpha \to \beta(x))\big)$

is a law of logic if x does not occur in α as a free variable.

Schemata 15* and 22* allow us to shift a universal quantifier $\bigwedge x$ from the consequent to the position preceding the entire implication, and from the position preceding the entire implication to the consequent, provided that x does not occur in the antecedent of that implication as a free variable.

23. $\bigwedge x(\bigwedge xQ(x) \to Q(x))$
(from 2).

24. $\bigwedge x\big((\bigwedge xB(x) \wedge \bigwedge xQ(x)) \to (B(x) \wedge Q(x))\big)$
(from 16, 23, the sentential theorem $(p \to q) \to ((r \to s) \to (p \wedge r \to q \wedge s))$, schema 1 and detachment).

174

25*. $(\bigwedge xB(x) \wedge \bigwedge xQ(x)) \rightarrow \bigwedge x(B(x) \wedge Q(x))$
  (from 24 and schema 15, and detachment).

26. $\bigwedge x(P(x) \wedge Q(x) \rightarrow P(x))$
  (from $p \wedge q \rightarrow p$).

27. $\bigwedge x(P(x) \wedge Q(x)) \rightarrow \bigwedge xP(x)$
  (from 26, schema 1, detachment).

28. $\bigwedge x(P(x) \wedge Q(x)) \rightarrow \bigwedge xQ(x)$
  (similar to 27).

29*. $\bigwedge x(P(x) \wedge Q(x)) \rightarrow (\bigwedge xP(x) \wedge \bigwedge xQ(x))$
  (from 27, 28, and the theorem $(p \rightarrow q) \rightarrow ((p \rightarrow r) \rightarrow$
  $(p \rightarrow (q \wedge r)))$).

30*. $\bigwedge \ldots (\bigwedge x(\alpha(x) \wedge \beta(x)) \equiv (\bigwedge x\alpha(x) \wedge \bigwedge x\beta(x)))$.

(Any theorem falling under schema 30* can be proved in a way analo
gous to proofs of Theorems 25* and 29*. In 30* two implications of
the type of 25* and 29* are combined into an equivalence.)

31. $\bigwedge x(P(x) \rightarrow (P(x) \vee Q(x)))$
  (from $p \rightarrow (p \vee q)$).

32. $\bigwedge x(Q(x) \rightarrow (P(x) \vee Q(x)))$.

33. $\bigwedge xP(x) \rightarrow \bigwedge x(P(x) \vee Q(x))$
  (from 31, schema 1, detachment).

34. $\bigwedge xQ(x) \rightarrow \bigwedge x(P(x) \vee Q(x))$
  (from 32, 1, detachment).

35. $(\bigwedge xP(x) \vee \bigwedge xQ(x)) \rightarrow \bigwedge x(P(x) \vee Q(x))$
  (from 33, 34, and the sentential calculus theorem $(p \rightarrow q) \rightarrow$
  $((r \rightarrow q) \rightarrow (p \vee r \rightarrow q))$).

Any theorem falling under the schema:

35*. $\bigwedge \ldots ((\bigwedge x\alpha \vee \bigwedge x\beta) \rightarrow \bigwedge x(\alpha \vee \beta))$

can be proved in a way analogous to 35. On the other hand, the theorems
converse to 35* are not tautologies in the c.l.c.

Hereafter, the universal quantifiers occurring at the beginning of theorems will usually be omitted, except for those cases in which our intention is to emphasize their role.

36. $P(x) \rightarrow \bigvee xP(x)$
(from 7).

37. $(\bigvee xP(x) \rightarrow Q) \rightarrow (P(x) \rightarrow Q)$
(from 36 and the theorem $(p \rightarrow r) \rightarrow ((r \rightarrow q) \rightarrow (p \rightarrow q)))$.

38. $(\bigvee xP(x) \rightarrow Q) \rightarrow \bigwedge x(P(x) \rightarrow Q)$
(from 37 and 15*)

if x does not occur in Q as a free variable.

Likewise, any other theorem falling under the following schema is a law in the c.l.c.

38*. $(\bigvee x\alpha(x) \rightarrow \beta) \rightarrow \bigwedge x(\alpha(x) \rightarrow \beta)$,

if x does not occur in β as a free variable.

This schema allows us to replace an existential quantifier preceding the antecedent of an implication by a universal quantifier preceding the entire implication if the consequent of the implication does not contain any free variable which would become bound by that quantifier. The schema converse to 38* is valid, too:

39. $\bigwedge x(\alpha(x) \rightarrow \beta) \rightarrow (\bigvee x\alpha(x) \rightarrow \bigvee x\beta)$
(from 6).

40*. $\bigwedge x(\alpha(x) \rightarrow \beta) \rightarrow (\bigvee x\alpha(x) \rightarrow \beta)$
(from 39 and 10 (the law of omitting a superfluous existential quantifier) and an appropriate sentential calculus theorem),

if x does not occur in β as a free variable.

41. $P(x) \rightarrow \bigvee xP(x)$
(from 7).

42. $Q(x) \rightarrow \bigvee xQ(x)$
(from 7).

43. $\bigwedge x\big((P(x) \vee Q(x)) \rightarrow (\bigvee xP(x) \vee \bigvee xQ(x))\big)$
(from 41, 42 and the sentential calculus theorem $(p \rightarrow r) \rightarrow$
$\big((q \rightarrow s) \rightarrow ((p \vee q) \rightarrow (r \vee s))\big))$.

44. $\bigvee x(P(x) \vee Q(x)) \rightarrow (\bigvee xP(x) \vee \bigvee xQ(x))$
    (from 43 and schema 40*).

45. $\bigwedge x(P(x) \rightarrow (P(x) \vee Q(x)))$
    (from $p \rightarrow (p \vee q)$).

46. $\bigvee xP(x) \rightarrow \bigvee x(P(x) \vee Q(x))$
    (from 45 and 6).

47. $\bigvee xQ(x) \rightarrow \bigvee x(P(x) \vee Q(x))$
    (similar to 46).

48. $(\bigvee xP(x) \vee \bigvee xQ(x)) \rightarrow \bigvee x(P(x) \vee Q(x))$
    (from 46, 47, and the sentential calculus theorem $(p \rightarrow r) \rightarrow$
    $((q \rightarrow r) \rightarrow ((p \vee q) \rightarrow r)))$.

49*. $(\bigvee xP(x) \vee \bigvee xQ(x)) \equiv \bigvee x(P(x) \vee Q(x))$
    (from 44 and 48).

Obviously, any sentence falling under the schema

50*. $\bigvee x(\alpha(x) \vee \beta(x)) \equiv (\bigvee x\alpha(x) \vee \bigvee x\beta(x))$

can be proved in a way similar to the proof of 49*.

Schema 50* is analogous to schema 30*. The former states that the existential quantifier may be distributed with respect to disjunction, whereas the latter states that the universal quantifier may be distributed with respect to conjunction. The algebraic meaning of laws 30* and 50* is very clear. The universal quantifier is an infinite logical product, and the existential quantifier is an infinite logical sum. Thus laws 30* and 50* state that infinite logical operations may be distributed with respect to the corresponding finite logical operations.

51. $\bigwedge x((P(x) \wedge Q(x)) \rightarrow P(x))$
    (from $p \wedge q \rightarrow p$).

52. $\bigvee x(P(x) \wedge Q(x)) \rightarrow \bigvee xP(x)$
    (from 51 and 6).

53. $\bigvee x(P(x) \wedge Q(x)) \rightarrow \bigvee xQ(x)$
    (similar to 52).

177

54*. $\bigvee x(P(x) \wedge Q(x)) \rightarrow (\bigvee xP(x) \wedge \bigvee xQ(x))$

      (from 52, 53, and the sentential calculus theorem $(r \rightarrow p) \rightarrow$
$((r \rightarrow q) \rightarrow (r \rightarrow p \vee q)))$.

On the contrary, the implication converse to 54* is not a law of logic. This can easily be illustrated by a simple example drawn from every day life. Let $P(x)$ stand for: *x is an aviator*, and let $Q(x)$ stand for: *x is a baker*. For this interpretation of the predicates P and Q sentence 54* becoms the following theorem:

If (there exists a man who is an aviator and a baker), then (there exists a man who is an aviator and there exists a man who is a baker).

But the implication converse to the above is not necessarily true, since even if aviators exist and bakers exist, there may nevertheless be no man who is both an aviator and a baker. A mathematical counter-example can be drawn from the domain of natural numbers, with $P(x)$ standing for: *x is even*, and $Q(x)$ standing for: *x is odd*. There are even numbers, and there are odd numbers, but there is no number that is both even and odd. Hence the implication

(106) $(\bigvee xP(x) \wedge \bigvee xQ(x)) \rightarrow \bigvee x(P(x) \wedge Q(x))$

is false in the domain of natural numbers, since its antecedent is true while its consequent is false. Since, as has been demonstrated above, all c.l.c. theorems are true in every domain (Theorem 6), formula (106), being false in a certain domain, is not a theorem in the c.l.c. As will be seen below, if a logical formula is not a theorem in the c.l.c., a counter-example which refutes it can always be found in the domain of natural numbers.

55. $\bigwedge xP(x) \rightarrow P(x)$

      (from 2).

56. $(P(x) \rightarrow Q) \rightarrow (\bigwedge xP(x) \rightarrow Q)$

      (from 55 and the sentential calculus).

57. $\bigvee x(P(x) \rightarrow Q) \rightarrow \bigvee x(\bigwedge xP(x) \rightarrow Q)$

      (from 56 and 6).

58. $\bigvee x(\bigwedge xP(x) \rightarrow Q) \rightarrow (\bigwedge xP(x) \rightarrow Q)$

      (from 10),

if x does not occur in Q as a free variable.

178

59*. $\bigvee x(P(x) \to Q) \to (\bigwedge xP(x) \to Q)$
  (from 57 and 58),

if x does not occur in Q as a free variable.

Any other theorem that falls under the schema which is a generalization of 59* can be proved in a way similar to the proof of 59*. Such generalizations will not hereafter be written out, since they can be obtained in an obvious manner.

60. $\bigwedge x(\sim P(x) \to \sim \bigwedge xP(x))$
  (from 55 and the theorem $(p \to q) \to (\sim q \to \sim p)$).

61*. $\bigvee x \sim P(x) \to \sim \bigwedge xP(x)$
  (from 60 and 40*).

62*. $\bigwedge xP(x) \to \sim \bigvee x \sim P(x)$
  (from 61* and $(p \to \sim q) \to (q \to \sim p)$).

63. $\bigwedge x(\sim \bigvee xP(x) \to \sim P(x))$
  (from 41).

64*. $\sim \bigvee xP(x) \to \bigwedge x \sim P(x)$
  (from 63 and 15*).

65*. $\sim \bigwedge x \sim P(x) \to \bigvee xP(x)$
  (from 64* and $(\sim p \to q) \to (\sim q \to p)$).

66. $\bigwedge x(\sim P(x) \to \bigvee x \sim P(x))$
  (from 7).

67. $\bigwedge x(\sim \bigvee x \sim P(x) \to P(x))$
  (from 66 and transposition).

68*. $\sim \bigvee x \sim P(x) \to \bigwedge xP(x)$
  (from 67 and 15*).

69. $\bigwedge x(\bigwedge x \sim P(x) \to \sim P(x))$
  (from 2).

70. $\bigwedge x(P(x) \to \sim \bigwedge x \sim P(x))$
  (from 69 and transposition).

71* $\bigvee xP(x) \to \sim \bigwedge x \sim P(x)$
  (from 70 and 40*).

72*. $\bigvee xP(x) \equiv \sim \bigwedge x \sim P(x)$
(from 71* and 65).

73*. $\bigwedge xP(x) \equiv \sim \bigvee x \sim P(x)$
(from 62* and 68*).

The laws marked by the numbers 61*, 62*, 64*, 65*, 68*, 71*, 72*, 73* are sometimes called *De Morgan's laws*. The intuitive meaning of some of them in the domain of real numbers has been indicated at the beginning of this section. Now these laws have been deduced from axioms which are intuitively true in every domain. Hence we are sure that these laws, as theorems in the c.l.c., are intuitively true in every domain.

Of course, every sentence falling under one of the two schemata given below can be deduced in a way similar to that in which 72* and 73* have been deduced above:

74*. $\bigwedge \ldots (\bigvee x\alpha(x) \equiv \sim \bigwedge x \sim \alpha(x))$.

75*. $\bigwedge \ldots (\bigwedge x\alpha(x) \equiv \sim \bigvee x \sim \alpha(x))$.

These schemata might serve as kinds of definition of the existential quantifier in terms of the universal quantifier, and vice versa. It would suffice to adopt only schemata II (1–5) and schema 74*, and all the axioms of III (6–10) could be deduced from them.

Note also the following law:

76*. $\bigwedge \ldots (\bigwedge y\alpha(y) \equiv \bigwedge x\alpha(x))$,

which is true under the condition that $\alpha(x)$ differs from $\alpha(y)$ only in this that the variable y occurs in $\alpha(y)$ in those places in which the variable x occurs in $\alpha(x)$, while y does not occur in $\alpha(x)$ as a free variable, and x does not occur in $\alpha(y)$ as a free variable, provided that neither $\alpha(y)$ contains a quantifier binding x, nor $\alpha(x)$ contains a quantifier binding y. This law allows us to replace a bound variable by any other variable which does not occur in the formula in question. The law results from axioms of II.4 and II.5. Axioms of II.5 allow us to join a superfluous quantifier:

77. $\bigwedge \ldots (\bigwedge y\alpha(y) \rightarrow \bigwedge x\bigwedge y\alpha(y))$.

Under the assumption made, the implication

78. $\bigwedge \ldots (\bigwedge x\bigwedge y\alpha(y) \rightarrow \bigwedge x\alpha(x))$

180

is a special case of Axiom II.4. By the transitivity of implication and, obviously, II.1, these two implications yield the theorem

79. $\bigwedge \ldots (\bigwedge y \alpha(y) \rightarrow \bigwedge x \alpha(x))$.

The converse implication is obtained in a similar way, and these two implications yield Theorem 76*.

The theorems proved so far have been deduced from axioms which are true in the empty domain.

The following sections will be concerned with consequences of the entire axiom system, including Axiom 0, with practical applications of the c.l.c. theorems proved in this section, and with certain general properties of the entire logical calculus.

For the time being, the axiom on non-emptiness will be referred to only to prove the equivalence of the calculi L and L^+. Since theorems in L^+ need not be closed sentences, i.e., they may contain free variables, the theorem on equivalence will be formulated as follows:

THEOREM ON THE EQUIVALENCE OF L AND L^+. *If a formula α contains free variables $x_{t_1}, \ldots, x_{t_k}$, then*

$$\bigwedge x_{t_1}, \ldots, x_{t_k} \alpha \in L \equiv \alpha \in L^+.$$

If $x_{t_1}, \ldots, x_{t_k}$ are all the free variables in α, then the formula $\bigwedge x_{t_1}, \ldots, x_{t_k} \alpha$ is called the *closure of the formula* α and will be denoted by $\bar{\alpha}$. Hence, the above theorem may be stated as the following equivalence:

$$\bar{\alpha} \in L \equiv \alpha \in L^+.$$

PROOF. To prove the implication: $\bar{\alpha} \in L \rightarrow \alpha \in L^+$, it suffices to recall that in Section 5 we have proved in L^+ sentences whose closures are the axioms of L. Hence, Ax $L \subset L^+$. Since the rule of detachment is adopted in L^+, hence $L \subset L^+$.

Conversely, since the closures of axioms in L^+ (substitutions for sentential calculus tautologies, and L1 and L2) are in L, in order to prove the implication: $\alpha \in L^+ \rightarrow \bar{\alpha} \in L$, we have to demonstrate that if $\bar{\alpha}$ and $\bar{\beta}$ are closures of the formulae α and β, and γ is obtained from α and β under any of the rules R1, R2, Det., Subst., and $\bar{\alpha}, \bar{\beta} \in L$, then $\bar{\gamma} \in L$.

We examine these rules one by one.

R1. If $\overline{\alpha(x_i) \rightarrow \Psi} \in L$, then, by Theorem 40 and axiom schema 1, $\overline{\bigvee x_i \alpha(x_i) \rightarrow \Psi} \in L$.

R2. If $\overline{\Psi \rightarrow \alpha(x_i)} \in L$, then, by Theorem 15 and axiom schema 1, $\overline{\Psi \rightarrow \bigwedge x_i \alpha(x_i)} \in L$.

As stated in Section 5, if the rule of substitution is used only to deduce theorems in L^+, then its application reduces to the replacement of a variable x_i by another variable x_j. Two cases are possible: x_j does, or does not, occur as a free variable in α. These two applications of the rule of substitution fall under their respective schemata:

$$\frac{\alpha(x_i, x_j)}{\alpha(x_j, x_j)},$$

$$\frac{\alpha(x_i)}{\alpha(x_j)}.$$

In the former case, if $\overline{\alpha(x_i, x_j)} \in L$, then by axiom schema 4 and axiom schema 1, $\overline{\alpha(x_j, x_j)} \in L$. In the latter case, by Theorem 78, if $\overline{\alpha(x_i)} \in L$, then $\overline{\alpha(x_j)} \in L$. The rule of substitution is covered by the same restriction which applies to the corresponding theorems in the system L.

R u l e   o f   d e t a c h m e n t. If $\overline{\alpha} \in L$ and $\overline{\alpha \rightarrow \beta} \in L$, then let the free variables occurring in the formula $\alpha \rightarrow \beta$ be denoted by $x_1, ..., x_n$. By axiom schema 1, the theorem

(107) $\bigwedge x_1, ..., x_n \alpha \rightarrow \bigwedge x_1, ..., x_n \beta$

also belongs to L. The formula α can have as many free variables as the formula $\alpha \rightarrow \beta$ has, or fewer. Hence, the formula $\overline{\alpha}$ can have at its beginning fewer universal quantifiers than there are in the antecedent of Theorem (107). The remaining quantifiers in the antecedent of Theorem (107) do not bind any free variables in the formula α, nad hence, under axiom schema 5, they may be joined to that formula, which yields

$$\overline{\alpha} \rightarrow \bigwedge x_1, ..., x_n \alpha$$

as a theorem in L. Hence, by (107) and by the assumption that $\overline{\alpha} \in L$, we have that $\bigwedge x_1, ..., x_n \beta \in L$. The formula β also may have fewer free variables than the formula $\alpha \rightarrow \beta$ has. Accordingly, the theorem

6. REDUCTION OF QUANTIFIER RULES

$\bigwedge x_1, \ldots, x_n\beta$ may have more universal quantifiers than $\bar{\beta}$ has. To obtain $\bar{\beta}$ we must disregard certain quantifiers which do not bind any variables in β. For that purpose we must make use of the axiom on non-emptiness. Suppose that we have a theorem of the form $\bigwedge y\beta$, where β does not have y as a free variable. The theorem β is now deduced as follows:

$$\bigwedge y\big(\beta \to \big((P(y) \equiv P(y)) \to \beta\big)\big)$$

is, under the sentential calculus, an axiom in L. This, by axiom schema 1, yields

$$\bigwedge y\beta \to \bigwedge y\big((P(y) \equiv P(y)) \to \beta\big).$$

On detaching the theorem $\bigwedge y\beta$ we obtain

$$\bigwedge y\big((P(y) \equiv P(y)) \to \beta\big).$$

Since β by assumption does not have y as a free variable, hence, by Theorem 40*, we obtain

$$\bigvee y(P(y) \equiv P(y)) \to \beta,$$

which, on the detachment of axiom 0, yields the desired theorem β.

We have thus proved the equivalence of the calculi L and L^+. All the logical rules have thus been reduced to a single rule of detachment and to the appropriate axioms, which have a fairly obvious intuitive foundation. This foundation, thus, includes the entire calculus L^+.

EXERCISES

1. Prove the c.l.c. theorem: $\bigwedge x, y(\bigwedge x, yR(x, y) \to R(x, y))$. This theorem is not a special case of Axiom 2, but results from it by the splitting of the quantifier $\bigwedge y$ that stands at the beginning of axiom schema 2 and by the shifting of the quantifier $\bigwedge y$ from the consequent to the position preceding the entire implication. Write down the detailed proof.

2. Check the fact that all c.l.c. theorems up to 64 were obtained on the strength of intuitionistic theorems of the sentential calculus. Among De Morgan's laws the intuitionistic ones are: 61*, 62*, 64*, 71* and

80. $\bigwedge x \sim P(x) \to \sim \bigvee xP(x).$

Demonstrate that 71* and 80 can be proved by reference to only intuitionistic schemata of the sentential calculus.

3. Demonstrate that the implication converse to 35* is not a tautology. Try to apply schemata that serve to disprove formula (106).

4. Prove the following c.l.c. law:

81. $\bigvee x(P \to Q(x)) \to (P \to \bigvee xQ(x))$.

Generalizations of that law allow us to shift an existential quantifier from the beginning of an implication to its consequent, provided that the variable involved does not occur free in the antecedent.

H i n t. Start from the theorem $P \to \big((P \to Q(x)) \to Q(x)\big)$, apply schema 1 to the entire theorem, and then apply schema 6 to its consequent. This yields the theorem $\bigwedge xP \to \big(\bigvee x(P \to Q(x)) \to \bigvee xQ(x)\big)$, which, on the elimination, on the strength of schema 5, of the quantifier $\bigwedge x$ standing before P, in turn yields 81 by means of easy sentential calculus transformations.

5. Prove the theorem

82. $\bigvee x(Q \wedge P(x)) \equiv Q \wedge \bigvee xP(x)$,

on the assumption that x does not occur in Q as a free variable.

H i n t. The implication $\to$ follows directly from Theorem 54 and axiom schema 10. The converse implication is to be deduced from the following c.l.c. theorem:

83. $\bigwedge x\big(Q \to \big(P(x) \to (Q \wedge P(x))\big)\big)$.

We first split the quantifier:

84. $\bigwedge xQ \to \bigwedge x\big(P(x) \to (Q \wedge P(x))\big)$,

and then apply axiom schema 5; this yields:

85. $Q \to \bigwedge x\big(P(x) \to (Q \wedge P(x))\big)$.

Next the application of Axiom 6 yields

86. $Q \to \big(\bigvee xP(x) \to \bigvee x(Q \wedge P(x))\big)$,

which by reference to the appropriate c.l.c. theorem yields the implication $\leftarrow$.
Write down the above reasoning in detail.

6. Prove the theorem

87. $\bigwedge x(Q \vee P(x)) \equiv (Q \vee \bigwedge xP(x))$,

on the assumption that x does not occur in Q as a free variable.

H i n t. Express disjunction by means of implication, which will reduce 87 to 15* and 22 (the laws on shifting the universal quantifier from the position before the implication to the consequent of that implication, and conversely).

7. Note that every sentence beginning with a universal quantifier is true in the empty domain, whereas every sentence beginning with an existential quantifier is false in the empty domain.

Refer to this fact and demonstrate that Theorems 82 and 87 are true in the empty domain, whereas the following Theorems 88 and 89, similar to the former ones, are not true in the empty domain:

88. $\bigwedge x(\beta(x) \wedge \alpha) \equiv (\bigwedge x\beta(x) \wedge \alpha)$,

89. $\bigvee x(\beta(x) \vee \alpha) \equiv (\bigvee x\beta(x) \vee \alpha)$.

They are c.l.c. theorems on the assumption that x does not occur in α as a free variable.

8. Remaining in the sphere of laws true in the empty domain, prove the theorem

90. $\bigvee x(P(x) \equiv P(x)) \to (\bigwedge x\alpha \to \alpha)$,

on the assumption that x does not occur in α as a free variable.

H i n t. First prove the theorem of the form

91. $\bigwedge x\alpha \to (\bigvee x(P(x) \equiv P(x)) \to \alpha)$,

which is obtained from the sentential formula $\alpha \to ((P(x) \equiv P(x)) \to \alpha)$ by the application, first, of Schema II. 1, and, next, of schema 40, on the assumption that x does not occur in α as a free variable. Next, formula 91 is transformed into formula 90 by referring to sentential calculus theorems.

Next prove the theorems:

92. $\bigvee x(P(x) \equiv P(x)) \to (\bigwedge x\alpha \equiv \alpha)$,

93. $\bigvee x(P(x) \equiv P(x)) \to (\bigvee x\alpha \equiv \alpha)$,

94. $\bigvee x(P(x) \equiv P(x)) \to (\bigwedge x(\beta(x) \wedge \alpha) \equiv (\bigwedge x\beta(x) \wedge \alpha))$,

95. $\bigvee x(P(x) \equiv P(x)) \to (\bigvee x(\beta(x) \vee \alpha) \equiv (\bigvee x\beta(x) \vee \alpha))$,

96. $\bigvee x(P(x) \equiv P(x)) \to (\bigwedge x\beta(x) \to \bigvee x\beta(x))$,

97. $\bigvee x(P(x) \equiv P(x)) \to ((\bigwedge x\beta(x) \to \alpha) \to \bigvee x(\beta(x) \to \alpha))$,

98. $\bigvee x(P(x) \equiv P(x)) \to ((\alpha \to \bigvee x\beta(x)) \to \bigvee x(\alpha \to \beta(x)))$.

In all these schemata it is assumed that x does not occur in α as a free variable.

H i n t s. 92 results from 90 and 2; 93 can be obtained from 92 by De Morgan's laws; 94 and 95 result from 92 and 93 and laws 30* and 50*; 97 can best be proved by first proving a theorem in which the last quantifier $\bigvee x$ is replaced by $\sim \bigwedge x$, and by applying then the appropriate De Morgan's law. The first part of the proof refers to Theorem 94. Theorem 98 can best be obtained in the same way.

7. THE CONCEPTS OF CONSEQUENCE AND THEORY. APPLICATIONS OF THE LOGICAL CALCULUS TO THE FORMALIZATION OF MATHEMATICAL THEORIES

The classical logical calculus is the foundation of all exact deductions in classical mathematics. In practice, arguments are not formalized, since that would make mathematical papers intolerably lengthy, nevertheless, it is worth while to have practical certainty that every argument which is intuitively correct can fully be formalized in the logical calculus. Below is an example of a partial formalization, in the system L of the classical logical calculus, of an argument which is intuitively self-evident, but in the case of a strict formalization requires several dozen intermediate theorems. The example is a very weak mathematical theory, namely the theory of the relation *less than*. The example, for all its simplicity, is sufficiently characteristic. The theory in question is an axiomatic theory. Every branch of mathematics, if worked out with sufficient precision, is always expounded as an axiomatic theory. Every mathematical theory T has certain primitive predicates, which in the case of the theory under consideration are: *equality* and the *less than* relation:

$$I(x, y), \text{ or } x = y \quad (\textit{the number x equals the number y}),$$

$$< (x, y), \text{ or } x < y \quad (\textit{the number x is less than the number y}).$$

Certain assumptions, that is, axioms, are made about those predicates; these assumptions are the original theorems of the theory. The concept of equality, that is, identity, is a primitive concept in many mathematical theories, and hence it is sometimes treated as a logical concept, so that the theory of identity is often included in the classical logical calculus, as we shall see below. It is always assumed about equality that it is an *equivalence relation*:

101. $I(x, x)$,

102. $I(x, y) \rightarrow I(y, x)$,

103. $\big(I(x, y) \wedge I(y, z)\big) \rightarrow I(x, z)$,

and that it does not affect other predicates, that is, in our case, that it preserves the *less than* relation:

104. $\big(I(x, y) \wedge\ < (x, z)\big) \rightarrow\ < (y, z)$,

105. $\big(I(x, y) \wedge\ < (z, x)\big) \rightarrow\ < (z, y)$.

It is assumed about the *less than* relation that it is an *ordering*:

106. $< (x, y) \rightarrow\ \sim (< (y, x))$,

107. $\big(< (x, y) \wedge\ < (y, z)\big) \rightarrow\ < (x, z)$,

108. $I(y, x) \vee\ < (x, y) \vee\ < (y, x)$,

186

and that it is a *dense ordering*, which means that between any two elements x, y such that $x < y$ there is a third element z such that $x < z$ and $z < y$:

109. $< (x, y) \rightarrow \bigvee z(< (x, z) \wedge < (z, y))$,

and that neither a least element nor a greatest element exists:

110. $\sim \left(\bigvee x \bigwedge y(I(y, x) \vee < (x, y)) \right)$,

111. $\sim \left(\bigvee x \bigwedge y(I(y, x) \vee < (y, x)) \right)$.

Sentences 101–111 form the complete axiom system of the *elementary theory of the less than relation*. Obviously, sentences 101–109 have at the beginning implied universal quantifiers which bind all free variables.

The theory under consideration has only 11 special axioms of its own, but the majority of interesting mathematical theories all have an infinite number of axioms. They always fall, however, under a finite number of simple schemata.

The axioms of the theory of the *less than* relation are satisfied, for instance, by the *less than* relation between real numbers and also between rational numbers.

Further theorems are deduced from these axioms on the strength of the classical logical calculus. In the case of the calculus L that mechanism can be described as follows:

1) the specific predicates of a given theory T (hence, in the case under consideration, the predicates I, $<$) are treated as included in the infinite list of c.l.c. predicates (it is assumed in general that the infinite sequence of predicate symbols P, Q, R, ..., of the c.l.c. includes all the predicate symbols, whether primitive or defined, of all those mathematical theories which we may consider);

2) the c.l.c. axioms are adopted for any formulae α, β, ..., which are well formed and constructed out of specific predicates of the theory T;

3) the special axioms of the theory T are adopted;

4) theorems are deduced from the axioms of 2) and 3) by detachment.

The term *predicate* was adopted (Definition 12) to denote any formula with free variables. The symbols P, Q, R, ..., I, $<$, etc., are in principle called *predicate symbols*. A predicate which consists of a predicate

187

symbol followed by parentheses with variables of different shapes is called an *atomic predicate* (cf. Definition 14). But predicate symbols are often called just predicates without any risk of misunderstanding.

Thus, in the case of this theory of the *less than* relation its theorems are all those theorems which we can deduce from the axioms of the classical logical calculus and Axioms 101–111, specific to the theory under consideration, using the rule of detachment. Several theorems will be deduced below by way of example. For instance, from Axiom 110, stating that the least element does not exist, we shall deduce the conclusion, which is obvious in the light of De Morgan's laws, stating that for any element x there exists an element which is less than x:

$$\bigwedge x \bigvee y (y < x).$$

In principle we observe the convention that the predicate symbol precedes the variables, but to make the reasonings more intuitive we shall sometimes use the ordinary symbolism: $x = y$ and $x < y$, in which the relation symbol stands between the variables. When referring to c.l.c. theorems from the preceding section we shall preserve the numbers under which they were listed there.

112. $\bigwedge x (\bigwedge y (y = x \vee x < y) \rightarrow \bigvee x \bigwedge y (y = x \vee x < y))$

(a special case of axiom schema 7).

113. $\bigwedge x (\sim \bigvee x \bigwedge y (y = x \vee x < y) \rightarrow \sim \bigwedge y (y = x \vee x < y))$

(from the schema $\bigwedge x ((\alpha \rightarrow \beta) \rightarrow (\sim \beta \rightarrow \sim \alpha))$ by splitting the universal quantifier into a universal quantifier in the antecedent and a universal quantifier in the consequent (axiom schema 1) and by detaching 112).

114. $\sim \bigvee x \bigwedge y (y = x \vee x < y) \rightarrow \bigwedge x \sim \bigwedge y (y = x \vee x < y)$

(from schema 15 and Theorem 113 (shifting of the universal quantifier to the consequent)).

115. $\bigwedge x \sim \bigwedge y (y = x \vee x < y)$

(from 114 and 110 by detachment).

116. $(y = x \vee x < y \vee y < x) \rightarrow (\sim (y < x) \rightarrow (y = x \vee x < y))$

(from the sentential calculus).

117. $\bigwedge x \bigwedge y(\sim (y < x) \rightarrow (y = x \vee x < y))$

   (splitting of the universal quantifier in 116 into a universal quantifier in the antecedent and a universal quantifier in the consequent, and detachment of 108).

118. $\bigwedge x \big(\bigwedge y(\sim (y < x) \rightarrow (y = x \vee x < y)) \rightarrow$
   $\bigwedge y \sim (y < x) \rightarrow \bigwedge y(y = x \vee x < y) \big)$

   (axiom schema 1).

119. $\bigwedge x \bigwedge y(\sim (y < x) \rightarrow (y = x \vee x < y)) \rightarrow$
   $\bigwedge x(\bigwedge y \sim (y < x) \rightarrow \bigwedge y(y = x \vee x < y))$

   (from 118 by splitting the universal quantifier $\bigwedge x$).

120. $\bigwedge x(\bigwedge y \sim (y < x) \rightarrow \bigwedge y(y = x \vee x < y))$

   (from 117, 119, detachment).

121. $\bigwedge x \big((\bigwedge y \sim (y < x) \rightarrow \bigwedge y(y = x \vee x < y)) \rightarrow$
   $(\sim \bigwedge y(y = x \vee x < y) \rightarrow \sim \bigwedge y \sim (y < x)) \big)$

   (sentential calculus theorem $(\alpha \rightarrow \beta) \rightarrow (\sim \beta \rightarrow \sim \alpha)$).

122. $\bigwedge x(\bigwedge y \sim (y < x) \rightarrow \bigwedge y(y = x \vee x < y)) \rightarrow$
   $\bigwedge x(\sim \bigwedge y(y = x \vee x < y) \rightarrow \sim \bigwedge y \sim (y < x))$

   (from 121, splitting of the quantifier $\bigwedge x$).

123. $\bigwedge x(\sim \bigwedge y(y = x \vee x < y) \rightarrow \sim \bigwedge y \sim (y < x))$

   (from 120, 122).

124. $\bigwedge x(\sim \bigwedge y(y = x \vee x < y) \rightarrow \sim \bigwedge y \sim (y < x)) \rightarrow$
   $(\bigwedge x \sim \bigwedge y(y = x \vee x < y) \rightarrow \bigwedge x \sim \bigwedge y \sim (y < x))$

   (axiom schema 1).

125. $\bigwedge x \sim \bigwedge y(y = x \vee x < y) \rightarrow \bigwedge x \sim \bigwedge y \sim (y < x)$

   (from 124, 123).

126. $\bigwedge x \sim \bigwedge y \sim (y < x)$

   (from 125, 115).

127. $\bigwedge x(\sim \bigwedge y \sim (y < x) \rightarrow \bigvee y(y < x))$

   (a special case of one of the implications contained in equivalence 74; De Morgan's law).

128. $\bigwedge x(\sim \bigwedge y \sim (y < x) \to \bigvee y(y < x)) \to$
$(\bigwedge x \sim \bigwedge y \sim (y < x) \to \bigwedge x \bigvee y(y < x))$
(axiom schema 1).

129. $\bigwedge x \sim \bigwedge y \sim (y < x) \to \bigwedge x \bigvee y(y < x)$
(from 127, 128).

130. $\bigwedge x \bigvee y(y < x)$
(from 126, 129).

Finally, we have the desired theorem. Some stages of the proof were not written down in full because of the length of the formulae, but we gave hints as to how to obtain more complete proofs based exclusively on the rule of detachment. If we wish to write down all the auxiliary c.l.c. theorems used in the proof, then the proof by detachment of Theorem 130, based on c.l.c. axioms and the axioms of the theory of the *less than* relation, would have some 40 theorems. From the intuitive point of view Theorems 110 and 130 are self-evidently equivalent. The conclusions which we draw immediately if we have a minimum intellectual training, are in a logical procedure broken into elementary steps so that all correct reasoning is reduced to the application of c.l.c. axioms and the rule of detachment. Analysis of many mathematical proofs shows that these means fully suffice to reproduce any mathematical proof that is intuitively correct.

Concept of consequence in the c.l.c.

Every proof in a theory based on the c.l.c., like any proof within the c.l.c. itself, can thus be interpreted as a finite sequence of operations of detachment. All substitutions are implicitly contained in the schemata of the c.l.c. axioms. The concepts of proof and consequence can thus now be simplified as compared with the concepts introduced on the occasion of the analysis of sentential calculi (cf. Definitions 2 and 3).

DEFINITION 19. *D is a proof by detachment of a sentence B from a set X of formulae adopted as assumptions if and only if D is a finite sequence of formulae*

$$D = \{D_1, D_2, ..., D_n\}$$

such that the last formula in that sequence is identical with the sentence $B: D_n = B$, *and every formula* D_k *in the sequence D* $(1 \leqslant k \leqslant n)$ *either (1) is in the set X; or* (2) *is obtained from two formulae* D_j, D_i *earlier than* D_k ($j < k$, $i < k$) *by correct use of detachment:* $D_j = (D_i \rightarrow D_k)$.

For instance, the sequence of formulae $\{126$–$130\}$ is a proof of 130 from the set $\{126, 127, 128\}$. Likewise, the sequence of formulae $\{110, 114, 115, 117, 119, 120, 122$–$130\}$ is a proof by detachment of 130 from the set $\{110, 114, 117, 119, 122, 124, 127, 128\}$.

DEFINITION 20 (Tarski [90], [93]). *A formula A is a consequence by detachment of a set X of formulae* (in symbols: $A \in Cn_0(X)$) *if and only if there exists a finite sequence D of formulae such that D is a proof by detachment of the formula A from the set X of formulae adopted as assumptions.*

Thus formula 130 is a consequence by detachment of the set $\{126, 127, 128\}$, and also of the set $\{110, 114, 117, 119, 122, 124, 127, 128\}$. Formula 130 also is a consequence by detachment of the set $(Ax L \cup Ax M)$, where $Ax L$ stands for the set of axioms of the c.l.c., and $Ax M$, for the set of axioms of the theory of the *less than* relation (101–111).

THEOREM 7. *The concept of consequence* Cn_0 *satisfies the following formulae, similar to those satisfied by the concept of consequence based on substitution and detachment,* Cnq:

1. $X \subset Cn_0(X)$,
2. $X \subset Y \rightarrow Cn_0(X) \subset Cn_0(Y)$,
3. $Cn_0(Cn_0(X)) \subset Cn_0(X)$,
4. $A \in Cn_0(X) \rightarrow \bigvee B_1, ..., B_k(B_1, ..., B_k \in X \wedge A \in Cn_0(\{B_1, ..., B_k\}))$,
5. $A \in Cn_0(X) \wedge (A \rightarrow B) \in Cn_0(X) \rightarrow B \in Cn_0(X)$.

PROOF. 1. If the formula $B \in X$, then, by Definition 19, B may be considered a proof of the formula B from X, so that, by Definition 20, $B \in Cn_0(X)$.

2. If $X \subset Y$, then by Definition 19 every proof based on the formulae of the set X is a proof based on the formulae of the set Y. Hence, by Definition 20, $Cn_0(X) \subset Cn_0(Y)$.

3. If $A \in Cn_0(Cn_0(X))$, then, by Definition 20, A has a proof D based on the set $Cn_0(X)$:

$$D = \{D_1, ..., D_k\}.$$

191

By the definition of a proof, every formula D_i $(1 \leqslant i \leqslant k)$ either can be derived from earlier ones by detachment or is in the set $\mathrm{Cn}_0(X)$. If D_i is in $\mathrm{Cn}_0(X)$, then it has its proof E_i. The proof of the formula A from the set X is constructed so that the proof D is preceded by all the proofs $E_{i_1}, \ldots, E_{i_l}$ of all the formulae $D_{i_1}, \ldots, D_{i_l}$, which by assumption are in $\mathrm{Cn}_0(X)$. Thus the proof of the formula A from the set X has the form

$$\{E_{i_1}, \ldots, E_{i_l}, D\}.$$

4. Formula 4 states that every consequence of a set X is a consequence of a finite subset $B_1, \ldots, B_k$ of the set X. In fact, if A is in $\mathrm{Cn}_0(X)$, then A has a proof D from the set X. That proof is a finite sequence, and hence it can include only a finite subset $B_1, \ldots, B_k$ of formulae belonging to the set X, and obviously $A \in \mathrm{Cn}_0(\{B_1, \ldots, B_k\})$, by definition.

5. If A and $(A \to B)$ are in $\mathrm{Cn}_0(X)$, then A and $A \to B$ have proofs from the set X. Let $\{D_1, \ldots, D_n\}$ be a proof of A from X, and let $\{E_1, \ldots, \ldots, E_k\}$ be a proof of $(A \to B)$ from X. It can easily be noted that the sequence

$$\{D_1, \ldots, D_n, E_1, \ldots, E_k, B\}$$

is a proof of B from X. All the sentences $D_1, \ldots, D_n$ and $E_1, \ldots, E_k$ satisfy by assumption the conditions laid down by the definition of a proof by detachment. These conditions are also satisfied by the sentence B, since the latter is obtained by detachment from earlier sentences: $D_n = A$ and $E_k = (A \to B)$.

Let $\mathrm{Ax}\,LI$ stand for the set of all axioms of group I of the system of the c.l.c. (substitutions for sentential calculus tautologies). Let $\mathrm{Ax}\,L$ stand for the set of all c.l.c. axioms, both those of group I and those of group II: 1–5, group III: 6–10, and group IV of the system of the c.l.c., which were explained in the preceding section. From the point of view of the concept of consequence that set has the following, rather important property:

THEOREM 8 (Deduction Theorem); (Tarski [90] and Herbrand [23]). *If A and B are closed sentences, then*

$$B \in \mathrm{Cn}_0(\mathrm{Ax}\,LI \cup X \cup \{A\}) \to \ulcorner(A \to B)\urcorner \in \mathrm{Cn}_0(\mathrm{Ax}\,LI \cup X).$$

The inscription $\ulcorner(A \to B)\urcorner$, which occurs in this formula, means an implication with the antecedent A and the consequent B, as the inscription to which we refer. This symbolism will be explained in greater detail in Chapter II.

PROOF. If $B \in \text{Cn}_0(\text{Ax}LI \cup X \cup \{A\})$, then, by Definition 19, there is a proof $D = \{D_1, ..., D_n\}$, $(D_n = B)$, from the set $\text{Ax}LI \cup X \cup \{A\}$.

It will now be proved by induction that for every i $(1 \leqslant i \leqslant n)$ $(A \to D_i) \in \text{Cn}_0(\text{Ax}LI \cup X)$. First, for $i = 1$, $(A \to D_i) \in \text{Cn}_0(\text{Ax}LI \cup X)$, for, since $D_1, ..., D_n$ is a proof under the set $(\text{Ax}LI \cup X \cup \{A\})$, then either $D_1 = A$, or $D_1 \in \text{Ax}LI \cup X$. This is so because D_1 cannot be obtained from earlier sentences by detachment, since it is the earliest sentence of the proof. If $D_1 = A$, then $(A \to A) \in \text{Ax}LI$, and hence, by properties 1 and 2 of Theorem 7, $(A \to A) \in \text{Cn}_0(\text{Ax}LI)$ and $(A \to A) \in \text{Cn}_0(\text{Ax}LI \cup X)$. Likewise, if $D_1 \in (\text{Ax}LI \cup X)$, then $D_1 \in \text{Cn}_0(\text{Ax}LI \cup X)$, and also $(D_1 \to (A \to D_1)) \in \text{Ax}LI$; consequently, $(D_1 \to (A \to D_1)) \in \text{Cn}_0(\text{Ax}LI \cup X)$ and hence, by property 5 of Theorem 7, also $(A \to D_1) \in \text{Cn}_0(\text{Ax}LI \cup X)$.

Hence, in both cases $(A \to D_1) \in \text{Cn}_0(\text{Ax}LI \cup X)$.

Suppose now that we have already proved, for all $i < k$, that $(A \to D_i) \in \text{Cn}_0(\text{Ax}LI \cup X)$; we shall prove now also that $(A \to D_k) \in \text{Cn}_0(\text{Ax}LI \cup X)$. The sentence D_k, as it occurs in a proof under the set $\text{Ax}LI \cup X \cup \{A\}$, satisfies one of the following conditions:

(a) $D_k = A$,

(b) $D_k \in \text{Ax}LI \cup X$,

(c) there exist i, j such that $i < k, j < k$, and

(108) $D_i = (D_j \to D_k)$.

In cases (a) and (b), $(A \to D_k) \in \text{Cn}_0(\text{Ax}LI \cup X)$ for the same reasons for which $(A \to D_1) \in \text{Cn}_0(\text{Ax}LI \cup X)$. In case (c), $(A \to D_i) \in \text{Cn}_0(\text{Ax}LI \cup X)$ by assumption, and hence, by (108),

(109) $(A \to (D_j \to D_k)) \in \text{Cn}_0(\text{Ax}LI \cup X)$

and

(110) $(A \to D_j) \in \text{Cn}_0(\text{Ax}LI \cup X)$.

Further, the definition of $\text{Ax}LI$ informs us that the sentence

(111) $(A \to (D_j \to D_k)) \to ((A \to D_j) \to (A \to D_k))$

is in $\text{Ax}LI$. (111) and formulae (109) and (110) on double detachment yield the formula $A \rightarrow D_k$; hence, in view of the properties 1, 2 and 5 of the concept of consequence in Theorem 7, $(A \rightarrow D_k) \in \mathrm{Cn}_0(\mathrm{Ax}LI \cup X)$.

It has thus been proved (1) that for $i = 1$, $(A \rightarrow D_1) \in \mathrm{Cn}_0(\mathrm{Ax}LI \cup X)$, and (2) that if for every i less than a certain k the property under consideration holds: $(A \rightarrow D_i) \in \mathrm{Cn}_0(\mathrm{Ax}LI \cup X)$, then that property also holds for D_k: $(A \rightarrow D_k) \in \mathrm{Cn}_0(\mathrm{Ax}LI \cup X)$.

To make this idea clearer, let us denote the property under consideration by W:

$$W(D_k) \equiv (A \rightarrow D_k) \in \mathrm{Cn}_0(\mathrm{Ax}LI \cup X).$$

Premiss (1) states that $W(D_1)$, and premiss (2) states that if $W(D_i)$ for all $i < k$, then $W(D_k)$. It is inferred from premisses (1) and (2) that every element D_k of the proof under consideration has the property W; this refers in particular to the last element $D_n = B$. Hence $W(B)$ must hold, so that $(A \rightarrow B) \in \mathrm{Cn}_0(\mathrm{Ax}LI \cup X)$.

Note that use has been made here only of those elements of $\mathrm{Ax}LI$ which are theorems in the system of positive implication.

Below are a few simple consequences of the Deduction Theorem.

The substitution in Theorem 8 of $\mathrm{Ax}L$ for X yields, in view of the inclusion $\mathrm{Ax}LI \subset \mathrm{Ax}L$, the equivalence

(112) $B \in \mathrm{Cn}_0(\mathrm{Ax}L \cup \{A\}) \equiv (A \rightarrow B) \in \mathrm{Cn}_0(\mathrm{Ax}L)$.

Since the axioms of logic allow us to combine any finite number of accepted sentences into a conjunction and to deduce any single sentence from the conjunction, the following equivalence is true:

(113) $B \in \mathrm{Cn}_0(\mathrm{Ax}L \cup \{A_1, ..., A_n\}) \equiv$
$\big((A_1 \wedge ... \wedge A_n) \rightarrow B\big) \in \mathrm{Cn}_0(\mathrm{Ax}L)$

derivable from the preceding one and from the fact that sentences under the schemata

$$\alpha \rightarrow \big(\beta \rightarrow (\alpha \wedge \beta)\big),$$
$$(\alpha \wedge \beta) \rightarrow \alpha,$$
$$(\alpha \wedge \beta) \rightarrow \beta$$

are in $\mathrm{Ax}LI$.

Since every consequence of a set is always a consequence of a finite number of sentences of that set (Theorem 7, property 4), this ultimately yields the strongest equivalence that may also be considered a formulation of the theorem on deduction:

(114) $B \in Cn_0(Ax L \cup X) \equiv$ *there exists a finite set of sentences*
$\{A_1, ..., A_n\} \subset X$ *such that*
$((A_1 \wedge ... \wedge A_n) \rightarrow B) \in Cn_0(Ax L)$.

Each of the equivalences given above will be considered a form of the Deduction Theorem for consequence by detachment.

The Deduction Theorem shows that if a mathematical theory T is extended, on the basis of the c.l.c., in the way described earlier in the section, the same conclusions are obtained as would be obtained if logical theorems of the form $(A_1 \wedge A_2 \wedge ... \wedge A_n \rightarrow B)$, where A_1, $A_2, ..., A_n$ is a finite sequence of axioms of the theory T, were derived first, and then the axioms $A_1, ..., A_n$ of the theory T were used only once to form their conjunction to use once in the operation of detachment.

The concept of consequence by detachment Cn_0 is rather narrow. In logic, we most often use the concept of logical consequence Cn, for which the whole c.l.c. is assumed.

DEFINITION 21. *A formula A is a logical consequence of a set X ($A \in$ $Cn(X)$) if and only if*

$A \in Cn_0(X \cup Ax L)$.

Definition 21 and the preceding ones yield immediately the statement that the concept Cn satisfies the same formulae 1–5 of Theorem 7 which are satisfied by the concept Cn_0.

The Deduction Theorem for Cn is true for every statement (112)–(114) if in those statements Cn_0 is replaced by Cn and $Ax L$ by L, where L stands for the set of theorems of the classical logical calculus. This theorem may also be formulated as

(115)
$B \in Cn(X \cup \{A\}) \equiv (A \rightarrow B) \in Cn(X)$,
$B \in Cn(\{A\}) \equiv (A \rightarrow B) \in L$.

The following equations are also obvious:

C.l.c. theorems $= L = Cn_0(Ax L) = Cn(\Lambda)$,

where Λ stands for the empty set of formulae. Obviously, $130 \in Cn(Ax M)$ and similarly all the sentences with numbers 101–130 are in $Cn(Ax M)$, where $\text{Ax}\,M$ stands for the set of axioms 101–111 of the theory of the *less than* relation.

DEFINITION 22. *A set X is a classical theory with terms $P_1, ..., P_n$ if $P_1, ..., P_n$ are the only predicate symbols occurring in the sentences of the set X, and if all the consequences of the set X containing the predicates $P_1, ..., P_n$ are in X. The symbols $P_1, ..., P_n$ are then said to be the terms of the theory X.*

If $\text{Cn}_{P_1, ..., P_n}(X)$ denotes those logical consequences of the set X which contain the terms $P_1, ..., P_n$, then X is a theory with the terms $P_1, ..., P_n$ if and only if

$$X = \text{Cn}_{P_1, ..., P_n}(X).$$

Every classical branch of the mathematics is a classical theory in the sense of Definition 22. In view of the truth of the formula $\text{Cn}(\text{Cn}X) = \text{Cn}(X)$ the set $\text{Cn}_{P_1, ..., P_n}(X)$ of the logical consequences of any set X, consequences containing the terms $P_1, ..., P_n$, is a classical theory. The set of the logical consequences of any axiom system is, thus, a theory. Hence, the set $\text{Cn}_{I, <}(\text{Ax}\,M)$ is a theory. •

Two other mathematical theories, the theory of the relation of one point lying between two other points, and set theory, will be described below by way of examples.

An elementary theory of the between relation

This theory describes the ordering of points on a straight line. The primitive concept is the relation: the point y lies between the points x and z, which will be expressed concisely by the formula $M(x, y, z)$. In the theory of the *between* relation, the following axioms are adopted:

I. Axioms of *identity* 101, 102, 103 and and the axioms of identity stating that the identity relation does not modify the position of a point on a straight line:

204. $I(x, y) \wedge M(x, v, w) \rightarrow M(y, v, w)$,
205. $I(x, y) \wedge M(v, x, w) \rightarrow M(v, y, w)$,
206. $I(x, y) \wedge M(v, w, x) \rightarrow M(v, w, y)$.

II. Axioms specific to the theory of the *between* relation:

207. $M(x, y, z) \rightarrow \sim \big(I(x, y) \vee I(x, z)\big)$,
208. $M(x, y, z) \rightarrow M(z, y, x)$,

209. $\sim \big(I(x, z) \big) \to \bigvee y M(x, y, z),$

210. $\sim \big(I(x, y) \big) \to \bigvee z M(x, y, z),$

211. $\sim \big(I(x, y) \vee I(y, z) \vee I(x, z) \big) \to \big(M(x, y, z) \vee M(y, z, x) \vee M(z, x, y) \big),$

212. $M(x, y, z) \to \sim \big(M(y, x, z) \vee M(x, z, y) \big),$

213. $\big(M(x, y, t) \wedge M(x, z, t) \big) \to \big(M(x, z, y) \vee z = y \vee M(y, z, t) \big),$

214. $\big(M(x, y, t) \wedge \big(M(x, z, y) \vee M(y, z, t) \big) \big) \to M(x, z, t),$

215. $\big(M(x, y, z) \wedge M(x, y, t) \big) \to \big(M(y, t, z) \vee t = z \vee M(y, z, t) \big).$

The *theory of the between relation* is the set of sentences which have M and I as the only predicate symbols and are logical consequences of all the axioms 101–103, 204–215, listed above. Many theorems on the ordering of points on a straight line in the ordinary Euclidean space can be proved in this theory. Since a straight line has no designated direction, no two-place relation ordering points on a straight line can be defined in the theory of the *between* relation. Yet, every model of this theory, that is, every domain in which the axioms of the theory of the *between* relation are satisfied, is simultaneously a model of the theory of the *less than* relation.

Axiomatic set theory

Axiomatic set theory is a much stronger mathematical theory than the two so far analysed in this section. In a sense it contains the whole of classical mathematics. Set theory has been presented in part in the Introduction. Its primitive terms are:

$$Z(x) = x \text{ is a set,}$$

$$x \in y = x \text{ is an element of the set } y$$

(for clarity's sake we shall henceforth write the symbols $\in$ and $=$ between the variables).

I. Axioms of identity 101–103 and

304. $\big(x = y \wedge Z(x) \big) \to Z(y),$

305. $(x = y \wedge x \in z) \to y \in z,$

306. $(x = y \wedge z \in x) \to z \in y.$

II. Specifically set-theoretical axioms:

307. $\left(Z(x) \wedge Z(y) \wedge \bigwedge z(z \in x \equiv z \in y)\right) \to x = y$
(*axiom of extensionality*),

308. $\left(Z(x) \wedge \bigwedge y(y \in x \to Z(y))\right) \to \bigvee w\left(Z(w) \wedge \bigwedge z(z \in w \equiv\right.$
$\left.\bigvee y(z \in y \wedge y \in x))\right)$
(*axiom of the union*),

309. $Z(x) \to \bigvee w\left(Z(w) \wedge \bigwedge y\left(y \in w \equiv (Z(y) \wedge \bigwedge u(u \in y \to u \in x))\right)\right)$
(*axiom of the power set*),

310. $\bigvee x\left(Z(x) \wedge \bigvee y(y \in x) \wedge \bigwedge z(z \in x \to Z(z)) \wedge \bigwedge y\left(y \in x \to\right.\right.$
$\left.\left.\bigvee v(v \in x \wedge \sim (v = y) \wedge \bigwedge z(z \in y \to z \in v))\right)\right)$
(*axiom of infinity*).

311. *If* $\alpha(x, y)$ *is a formula with at least two free variables and the*
variable w does not occur in $\alpha(x, y)$, *then the following formula*
is an axiom:
$\left(Z(u) \wedge \bigwedge x, y, z\left((\alpha(x, y) \wedge \alpha(x, z)) \to y = z\right)\right) \to$
$\bigvee w\left(Z(w) \wedge \bigwedge y\left(y \in w \equiv \bigwedge x(x \in u \wedge \alpha(x, y))\right)\right)$
where the remaining free variables which occur in α are treated as bound
by a universal quantifier that stands at the beginning of the axiom
(*schema of axioms of substitution*),

312. $\left(Z(x) \wedge \bigwedge y(y \in x \to Z(y))\right) \wedge$
$\bigwedge y, z, v\left(y \in x \wedge z \in x \wedge \sim (y = z) \wedge v \in y) \to \sim (v \in z)\right) \to$
$\bigwedge w\left(Z(w) \wedge \bigwedge y\left((y \in x \wedge \bigvee z(z \in y)) \to\right.\right.$
$\left.\left.\bigvee z(z \in y \wedge z \in w \wedge \bigwedge v((v \in y \wedge v \in w) \to v = z))\right)\right)$
(*axiom of choice*),

313. *If* $\alpha(x)$ *is a formula with at least one free variable, then the follow-*
ing formula is an axiom:
$\left(\bigvee z\alpha(z) \wedge \bigwedge x, y\left((x \in y \wedge \alpha(y)) \to \alpha(x)\right)\right) \to$
$\bigvee z(\alpha(z) \wedge \bigwedge y \sim (y \in z))$
(*schema of Fundierungsaxiom*).

Axioms 307–310 are already familiar from the Introduction. Axiom 311 states that if a predicate $\alpha(x, y)$ is single-valued, then it maps a set onto a set, namely a set u onto a set w, which is the image of the set u under the mapping α. In other words: An image of a set is a set if the mapping is determined by a single-valued predicate. Axiom 311 is stronger than the axiom of set construction, presented in the Introduction. The axiom of set construction follows immediately from 311. If in a set u we want to designate a subset consisting of those elements which satisfy the formula $\alpha(y)$, then we construct a formula $\alpha'(x, y)$ of the form

(116) $\qquad \big(x = y \wedge \alpha(y)\big).$

The predicate $\alpha'(x, y)$ satisfies the condition of uniqueness, that is, single-valuedness. Thus, 312 yields the sentence

$$Z(u) \rightarrow \bigvee w\Big(Z(w) \wedge \bigwedge y\big(y \in w \equiv \bigvee x(x \in u \wedge \alpha'(x, y))\big)\Big).$$

But it can easily be seen that the sentence

$$\bigvee x\big(x \in u \wedge \alpha'(x, y)\big) \equiv \big(y \in u \wedge \alpha(y)\big)$$

is a theorem in the theory under consideration. The two formulae given above yield the axiom of set construction

(117) $\qquad Z(u) \rightarrow \bigvee w\Big(Z(w) \wedge \bigwedge y\big(y \in w \equiv (y \in u \wedge \alpha(y))\big)\Big).$

Many interesting theorems, which cannot be obtained from the axiom of set construction, follow from Axiom 311.

The axiom of choice 312 states that for every set x, whose elements are disjoint sets, there exists a set w such that it has only one common element with each non-empty set of the family x. Thus the set w is a set of "representatives", as it contains one representative of each set of the family x. The axiom of choice, thus, allows us to form a set of representatives for every family of disjoint sets. This axiom is much disputed as it is less self-evident than the remaining ones.

The set of sentences which contain only the predicates $=$ and $\in$, and which are logical consequences of axioms 101–103, 304–313 is called *Zermelo–Fraenkel's axiomatic set theory*. Zermelo's original system can be restored if the axioms of substitution are replaced by axioms of set construction, if the Fundierungsaxioms are omitted, and the axiom on the existence of pairs added (cf. Section 5 in the Introduction):

314. $\qquad \bigwedge x \bigwedge y \bigvee w\Big(Z(w) \wedge \bigwedge z(z \in w \equiv (z = x \vee z = z))\Big).$

I. CLASSICAL LOGICAL CALCULUS

When working in the field of axiomatic set theory we have to distinguish clearly when we are dealing with a set of elements, and when we are dealing merely with a predicate describing a certain property, and, similarly, when we are dealing with a relation as a set of pairs, and when we are dealing merely with a predicate of two arguments. We have to do so because not every predicate defines a set. For instance, we can easily define the predicate of equinumerosity of two sets:

$$(118) \quad x \operatorname{eqnum} y \equiv \bigvee z \Big(Z(z) \wedge \bigwedge v \big(v \in z \rightarrow \bigvee u,w(v = \langle u,w \rangle \wedge$$
$$u \in x \wedge w \in y)\big) \wedge$$
$$\bigwedge u \big(u \in x \rightarrow \bigvee w(\langle u,w \rangle \in z)\big) \wedge$$
$$\bigwedge w \big(w \in y \rightarrow \bigvee u(\langle u,w \rangle \in z)\big) \wedge$$
$$\bigwedge u,w,r \big((\langle u,w \rangle \in z \wedge \langle u,r \rangle \in z) \rightarrow w = r \big) \wedge$$
$$\bigwedge u,w,r \big((\langle u,w \rangle \in z \wedge \langle r,w \rangle \in z) \rightarrow u = r \big) \Big).$$

However, it can be proved in axiomatic set theory that there does not exist any set s such which would satisfy the formula $\bigwedge x,y(\langle x,y \rangle \in s \equiv x \operatorname{eqnum} y)$. Predicates may be defined by any formulae, but the existence of a set can be proved only on the strength of axioms. The axioms make it possible to prove that there exists a relation of equinumerosity (interpreted as a set of pairs), but that it holds only between sets contained in a family of sets which itself is a set. On the contrary, equinumerosity between all arbitrary sets is not a relation as a set of pairs, but a predicate. Likewise, the very concept of set $Z(x)$ is a predicate, whereas the set of all sets does not exist.

To illustrate the applications of logic to proofs of theorems in set theory we shall now prove the theorem stating that the set of all sets does not exist:

$$\sim \bigvee x \Big(Z(x) \wedge \bigwedge z(z \in x \equiv Z(z)) \Big).$$

The idea of the proof is that if a set of all sets should exist we could designate in it a subset y, consisting of those sets z which satisfy the sentential function $\sim (z \in z)$.

The set y would then have the property

$$z \in y \equiv \sim (z \in z).$$

If we consider y itself as z we obtain the false equivalence $y \in y \equiv\; \sim (y \in y)$. Hence, such a set y cannot exist, and accordingly the set x of all sets cannot exist either. This conclusion can be arrived at with fair precision, when attention is paid to the applications of laws of logic:

320. $\Big(\big(z \in y \equiv (z \in x \;\wedge\; \sim (z \in z))\big) \rightarrow \big(z \in x \rightarrow (z \in y \equiv\; \sim (z \in z))\big)\Big)$
   (from the sentential calculus).

321. $\bigwedge y,x \Big(\bigwedge z \big(z \in y \equiv (z \in x \;\wedge\; \sim (z \in z))\big) \rightarrow$
   $\bigwedge z \big(z \in x \rightarrow (z \in y \equiv\; \sim (z \in z))\big)\Big)$
   (from 320, distributing of the quantifier).

322. $\bigwedge y \bigwedge z \bigwedge x \Big(\bigwedge z \big(z \in y \equiv (z \in x \;\wedge\; \sim (z \in z))\big) \rightarrow$
   $\bigwedge \big(z \in x \rightarrow (z \in y \equiv\; \sim (z \in z))\big)\Big)$

   (from 321, shifting of the quantifier $\bigwedge z$ from the consequent to the position preceding the entire implication and
   change in the order of quantifiers).

323. $\bigwedge y \bigwedge x \Big(\bigwedge z \big(z \in y \equiv (z \in x \;\wedge\; \sim (z \in z))\big) \rightarrow$
   $\big(y \in x \rightarrow (y \in y \equiv\; \sim (y \in y))\big)\Big)$

   (from 322 and the combining of universal quantifiers
   (axiom schema 4)).

324. $\sim (y \in y \equiv\; \sim (y \in y))$ (from the sentential calculus).

325. $\bigwedge z \big(z \in y \equiv (z \in x \;\wedge\; \sim (z \in z))\big) \rightarrow\; \sim (y \in x)$
   (from 323, 324, and the sentential caculus theorem
   $\big((p \rightarrow (q \rightarrow r)) \rightarrow (\sim r \rightarrow (p \rightarrow\; \sim q))\big)$).

326. $\bigwedge x,z \big((z \in x \equiv Z(z)) \rightarrow (Z(z) \rightarrow z \in x)\big)$
   (from the sentential calculus).

327. $\bigwedge x \Big(\bigwedge z (z \in x \rightarrow Z(z)) \rightarrow \bigwedge z (Z(z) \rightarrow z \in x)\Big)$
   (from 326 and the distributing of the universal quantifier).

201

328. $\bigwedge x \bigwedge z \left(\left(\bigwedge z (z \in x \equiv Z(z)) \right) \rightarrow (Z(z) \rightarrow z \in x) \right)$

(from 327 and the shifting of the quantifier $\bigwedge z$ to the position preceding the entire implication).

329. $\bigwedge z \bigwedge x \left(\left(\bigwedge z (z \in x \equiv Z(z)) \right) \rightarrow (Z(z) \rightarrow z \in x) \right)$

(from 328, interchange of the order of universal quantifiers).

330. $\bigwedge y \bigwedge x \left(\left(\bigwedge z (z \in x \equiv Z(z)) \right) \rightarrow (Z(y) \rightarrow y \in x) \right)$

(from 329, L3 (change of the variable bound by the initial quantifier)).

331. $\bigwedge z (z \in x \equiv Z(z)) \rightarrow \sim \left(Z(y) \wedge \bigwedge z (z \in y \equiv (z \in x \wedge \sim (z \in z))) \right)$

(from 330, 325, and the sentential calculus theorem $(p \rightarrow (q \rightarrow t)) \rightarrow \left((s \rightarrow \sim r) \rightarrow (p \rightarrow \sim (q \wedge s)) \right)$; on the appropriate substitutions we first detach 330, and then 325).

332. $\bigwedge z (z \in x \equiv Z(z)) \rightarrow \bigwedge y \sim \left(Z(y) \wedge \bigwedge z (z \in y \equiv (z \in x \wedge \sim (z \in z))) \right)$

(from 331 and the shifting of the quantifier $\bigwedge y$ to the consequent).

333. $\bigwedge z (z \in x \equiv Z(z)) \rightarrow \sim \bigvee y \left(Z(y) \wedge \bigwedge z (z \in y \equiv (z \in x \wedge \sim (z \in z))) \right)$

(from 332 and De Morgan's law).

334. $Z(x) \rightarrow \bigvee y \left(Z(y) \wedge \bigwedge z (z \in y \equiv (z \in x \wedge \sim (z \in z))) \right)$

(a special case of the axiom on set construction: in the set x we designate the subset consisting of those elements which are not elements of themselves).

335. $\sim \left(Z(x) \wedge \bigwedge z (z \in x \equiv Z(z)) \right)$

(from 333, 334, and the sentential calculus theorem $(p \rightarrow q) \rightarrow ((r \rightarrow \sim q) \rightarrow \sim (p \wedge r)))$.

336. $\sim \bigvee x \left(Z(x) \wedge \bigwedge z (z \in x \equiv Z(z)) \right)$

(from 335 and De Morgan's law).

Set theory begun and grew in the second half of the 19th century. At first it was developed without any precise formulation of its axioms, and under the tacit assumption that every predicate defines the set of those elements which satisfy that predicate. At the turn of the 19th century it was noticed that this assumption results in contradictions. One of the contradictions was that recorded above, which is based on B. Russell's idea dating from about 1900; his reasoning has since been known as *Russell's antonomy*. As a result of the contradictions discovered, the trend developed toward formulating set theory more precisely. One of the formalizations is due to B. Russell himself [80], [103]. He formalized, however, only a part of set theory, and he put it in the form of what is called *type theory*. Another, much more comprehensive, formalization of set theory was given in 1908 by E. Zermelo [105] who introduced the axioms on set construction, given in the Introduction. His axiom system was later completed by A. Fraenkel, who added the axiom of substitution and the Fundierungsaxiom.

EXERCISES

1. Prove the existence of the empty set by locating it in the infinite set, the existence of which is assumed by Axiom 310, and then by proving that the empty set does not depend on the set in which it was located.

2. Deduce Axiom 315 on the existence of pairs from Axioms 304–313.

H i n t. First locate the empty set Λ in the existing infinite set. Next obtain the sets $\{\Lambda\} = 2^\Lambda$ and $\{\Lambda, \{\Lambda\}\} = 2^{2^\Lambda}$. To the last of these sets we apply the correspondence

$$\alpha(x, y) \equiv ((x = \Lambda \wedge y = a) \wedge (x = \{\Lambda\} \wedge y = b)).$$

The axiom on substitution yields any pair $\{a, b\}$.

3. Assume, in the theory of the *less than* relation, the following definition of the *between* relation:

$$M(x, y, z) \equiv (x < y \vee y < z),$$

and then deduce all the axioms of the theory of the *between* relation as theorems in the theory of the *less than* relation.

8. THE LOGICAL FUNCTIONAL CALCULUS L^* AND ITS APPLICATIONS TO THE FORMALIZATION OF THEORIES WITH FUNCTIONS

In the present section we shall extend the logical calculus under consideration in an inessential way. It is mainly the language which is being extended; the other changes will just be consequences of the changes in the language.

I. CLASSICAL LOGICAL CALCULUS

So far we have considered only theories which had certain predicate symbols as their primitive terms, such as $<$ and $\in$. In many mathematical theories symbols denoting functions, e.g., $x+y, y \cdot x, S(y)$, etc., are adopted as primitive terms. Moreover, various functions are defined and denoted by function symbols, too: $\sqrt{x}$, $\text{sgn}(x)$, $\tan x$, $x!$, etc. Even our everyday language includes names of functions, such as: the father of x, the mother of x, and many others. Hence, it appears that logic ought to cover the general ways of using functional formulae. Functional formulae may be simple, such as $x+y$, $x \cdot y$, or compound, such as $x \cdot (y+z)$, $((x+y) \cdot (y+u))^2$, the latter being derived by superposition of various simple formulae. They always preserve the property of denoting a certain function, more or less complicated. Names of individuals, such as 0, 1, 2, 3, ..., π, e, which may be treated as functions of zero arguments, are the simplest functional formulae. Functional formulae are also called *term formulae*, since they do not express any thought, but simply lay out the names of objects referred to in the theory. In extending the system of logic, we consider arbitrary function symbols, and, thus, now add to the system infinitely many function symbols of the form $F(x), G(x, y), H(x, y, z), ..., F_1(x), G_1(x, y), ..., F_2(x), G_2(x, y), ...$, and also individual constants $a, b, c, ...$, to which we do not ascribe any definite meaning. It is assumed that they include all those function symbols and individual constants which we intend to consider. In particular, they include such symbols as $+$ and $\cdot$. Obviously, in the logical system under consideration we do not assume about these symbols anything more than we do about the other symbols.

Function symbols may be superimposed one upon another and thus form formulae of increasing degree of complication, which also are either functional formulae or constants, that is, names of objects. The formulae of the logical calculus thus form two sets of meaningful, or well-formed formulae:

1. *term formulae*, derived from variables and individual constants by superposition of various function symbols, and always denoting certain individuals or functions defined on elements of a given domain, with values in that domain;

2. *sentential formulae*, which always state an idea, not necessarily a closed one, derived from term formulae and predicates.

Rigorous inductive definitions of the formulae of both kinds are given below. Assume first that all function symbols consist only of letters F, G, H with indices.

DEFINITION 23.

1. *Every variable is a term formula.*

2. *Every individual constant a, b, c, ... is a term formula.*

3. *A formula consisting of a function symbol followed by parentheses with a number of variables, fixed for that function symbol, is a term formula.*

4. *A formula derived by substitution of a term formula for a variable in a term formula is a term formula.*

5. *Every term formula is derived from the formulae of* 1, 2 *and* 3 *by the application, a finite number of times, of the operation of* 4.

Now that we have defined the concept of term formula, we can define a sentential formula in a way similar to the previous definition (Definition 14).

DEFINITION 24 (of sentential formula in the new sense).

1. *The atomic predicates* $x = y$, $P(x)$, $Q(x, y)$, *... are sentential formulae.*

2. *Substitution of a term formula for any free variable, in a sentential formula, yields a sentential formula.*

3. *If* α *and* β *are sentential formulae, then the formulae:* $\alpha \rightarrow \beta$, $\sim \alpha$, $\alpha \equiv \beta$, $\alpha \vee \beta$, $\alpha \wedge \beta$, $\bigwedge x_i \alpha$, $\bigvee x_i \alpha$, *where* x_i *is any variable, are sentential formulae.*

When we formulate the logical calculus for such an extended language it suffices, in the traditional approach, if we admit, in the case of the rule of substitution, the possibility of substituting any meaningful term formulae for term variables, with a restriction similar to that made in the case of the rule of substitution as defined in Section 5:

EXTENDED RULE OF SUBSTITUTION: *In a formula* $\alpha(x_i)$ *any meaningful term formula* $\varphi(x_{j_1}, ..., x_{j_K})$ *may be substituted for the free variable* x_i *provided that the variable* x_i, *free in* $\alpha(x_i)$, *does not occur in any place within the scope of a quantifier binding any of the variables* $x_{j_1}, ..., x_{j_K}$.

If the logical calculus is formalized using the rule of detachment and Axioms 0–10 only, then only Axiom schema II.2 and Axiom schema III.7 require a modification: they must be given the form stated below:

AXIOM SCHEMA II.2:

$$\bigwedge \ldots (\bigwedge x_i \alpha(x_i) \rightarrow \alpha(\varphi)).$$

AXIOM SCHEMA III. 7:

$$\bigwedge \ldots (\alpha(\varphi) \rightarrow \bigvee x_i \alpha(x_i)).$$

In these axioms, $\alpha(x_i)$ differs from $\alpha(\varphi)$ only in that in those places in which the free variable x_i occurs in the formula $\alpha(x_i)$ the term formula φ occurs in the formula $\alpha(\varphi)$, and none of the free variables, if any, occurring in the term formula φ is bound by a quantifier contained in $\alpha(\varphi)$.

As can easily be seen, when introducing term constants in logic we are not in a position to confine ourselves to those laws of logic which would be true in the empty domain as well. At present, every sentence of the form

$$\bigwedge x \alpha(x) \rightarrow \alpha(a),$$

where a is a term constant, falls under Axiom schema II.2, while every sentence of the form

$$\alpha(a) \rightarrow \bigvee x \alpha(x)$$

falls under Axiom schema III.7. In view of the transitivity of implication we may combine these two formulae into the following one:

$$\bigwedge x \alpha(x) \rightarrow \bigvee x \alpha(x),$$

which is not true in the empty domain. On substituting $P(x) \rightarrow P(x)$ for $\alpha(x)$, we obtain from the above sentence the non-emptiness Axiom 0.

This fact is quite clear. The introduction of constants indicates that we consider domains in which those constants can be interpreted as names of objects of some kind, that is, that we consider only non-empty domains.

We shall hereafter also need a symbol for the set of those logical theorems which are consequences of the logical axioms in the new sense. That set will be denoted by L^*:

$$L^* = \mathrm{Cn}_0(\mathrm{Ax}\,L^*).$$

Logical consequences in the new sense (Cn*) will be the consequences based on the set of axioms of the classical logical calculus with functions:

$$\mathrm{Cn}^*(X) = \mathrm{Cn}_0(\mathrm{Ax}\,L^* \cup X),$$

where $\mathrm{Ax}\,L^*$ stands for the set of sentences which are well-formed formulae in the new sense, without free variables, and falling either (I) under the schema of any c.l.c. tautologies, or (II) under any of the axiom schema 0–10, where axiom schemata 2 and 7 are interpreted in the way indicated above.

The calculus L^* will be called *the functional logical calculus,* and L *the predicate calculus.* The calculus L^* is much more convenient to use in formalizations of many theories which have functions as primitive terms. Group theory will be described below as an example of such theories.

Group theory

Group theory has a function of two arguments as its primitive term. Let that function be denoted by + because of its similarity to ordinary addition. To make the meaning of the axioms even clearer we shall write the symbol + between arguments.

The other primitive term, which, however, is not specifically group-theoretical, is that of identity, which occurs in nearly every mathematical theory and, hence, is often treated as a general logical concept. We shall first list the axioms concerning the concept of identity:

101. $\bigwedge xx = x,$

102. $\bigwedge x, y(x = y \rightarrow y = x),$

103. $\bigwedge x, y, z\big((x = y \wedge y = z) \rightarrow x = z\big).$

The above axioms on identity are assumed in every theory; moreover, in every theory we assume axioms of extensionality, which are different in each case but fall under the same schema. In the case of group theory they have the form:

104. $\bigwedge z, x, y(x = y \rightarrow x{+}z = y{+}z),$

105. $\bigwedge x, y, z(x = y \rightarrow z{+}x = z{+}y).$

The specific axioms of group theory are:

404. $\bigwedge x, y, z\big(x{+}(y{+}z) = (x{+}y){+}z\big)$
      (*associativity of the* + *operation*),

405. $\bigwedge x, y \bigvee z(x+z = y)$
    (*solvability with respect to the second argument*),

406. $\bigwedge x, y \bigvee v(v+x = y)$
    (*solvability with respect to the first argument*).

Group theory is the set: $Cn^*(101–105, 404–406)$.

Ring theory

Likewise, ring theory is the set of consequences Cn^* of the axioms referred to above plus the three following axioms which describe the third concept: $\cdot$, which, in addition to those listed above, is a primitive concept in ring theory:

$$\bigwedge x, y, z(x = y \rightarrow x \cdot z = y \cdot z),$$

$$\bigwedge x, y, z(x = y \rightarrow z \cdot x = z \cdot y),$$

504. $\bigwedge x, y, z(x \cdot (y \cdot z) = (x \cdot y) \cdot z)$
    (*associativity of* $\cdot$),

505. $x \cdot (y+z) = x \cdot y + x \cdot z$
    (*distributivity*),

506. $(y+z) \cdot x = y \cdot x + z \cdot x$
    (*distributivity*).

Before proceeding to analyse other examples of mathematical theories let us consider once more, in a general manner, the theory of identity. As indicated by the examples above and in Section 7 on identity, Axioms 101, 102, and 103, stating that identity is an equivalence relation, are always assumed; further, for every primitive predicate and for every primitive function in a given theory the following axioms on extensionality are assumed, too:

Ex1 $(x = y \wedge P(x_1, ..., x, ..., x_k)) \rightarrow P(x_1, ..., y, ..., x_k)$,

Ex2 $x = y \rightarrow F(x_1, ..., x, ..., x_k) = F(x_1, ..., y, ..., x_k)$.

Sentences 101, 102, 103, and schemata Ex1 and Ex2 shall be called the *axioms of identitiy theory* and briefly denoted by AxId. If these are added to the logical system L^*, we obtain a logical system L_1, which shall be called *a system of logic with identity and functions* or the *functional calculus with identity*:

$$L_1 = Cn^*(AxId) = Cn_0(AxL^* \cup AxId).$$

208

The consequences based on the set of axioms of the c.l.c. with identity and functions:

$$\mathrm{Cn}_1(X) = \mathrm{Cn}_0(\mathrm{Ax}L^* \cup \mathrm{Ax}\,\mathrm{Id} \cup X) = \mathrm{Cn}^*(\mathrm{Ax}\,\mathrm{Id} \cup X)$$

shall be called *consequences under the c.l.c. with identity and functions*. As can be seen, both concepts of consequence, Cn^* and Cn_1, satisfy Theorem 7 and the theorem on deduction in the same form in which it holds for the concept of consequence Cn. The equations $L_1 = \mathrm{Cn}_1(\varLambda)$ and $L^* = \mathrm{Cn}^*(\varLambda)$ are trivial, too.

The concept of theory is expanded in a similar way. Every set closed under consequence Cn_1, which contains only the identity predicate and the predicates $P_1, \ldots, P_n$ out of all predicate terms, and only the terms $F_1, \ldots, F_k$ out of all function terms, is called a *theory with identity and functions*, with terms $P_1, \ldots, P_n, F_1, \ldots, F_k$, some of which are the predicates $P_1, \ldots, P_n$, while the other are the functions $F_1, \ldots, F_k$.

Consider now the following theories, which shall be treated as theories with identity and functions.

Elementary lattice theory

This theory may be treated as a theory of two functions of two arguments each, the functions being usually denoted by $\cup$ and $\cap$. The following equations are axioms:

601. $x \cap x = x,$
602. $x \cup x = x,$
603. $x \cap y = y \cap x,$
604. $x \cup y = y \cup x,$
605. $x \cap (y \cap z) = (x \cap y) \cap z,$
606. $x \cup (y \cup z) = (x \cup y) \cup z,$
607. $x \cap (x \cup y) = x,$
608. $x \cup (x \cap y) = x.$

An elementary theory of Boolean algebra

This theory may be treated as a theory with the same primitive concepts as the elementary lattice theory, but it is convenient to add three other primitive terms: 0, 1, and $-$, since then all the axioms can be written down as equations. These axioms are: Axioms 601–608 of lattice theory, and

609. $x \cap (y \cup z) = (x \cap y) \cup (x \cap z)$;

610. $x \cup (y \cap z) = (x \cup y) \cap (x \cup z)$,

611. a. $0 \cup x = x$,　　b. $0 \cap x = 0$,

612. a. $1 \cup x = 1$,　　b. $1 \cap x = x$,

613. a. $x \cup -x = 1$,　　b. $x \cap -x = 0$.

The algebra of sets is the most important interpretation of Boolean algebra: $x \cup y$ may be treated as the union of the sets x and y; $x \cap y$ may be treated as the intersection of the sets x and y; 0 is taken to be the empty set; 1, the universe; and $-x$, the complement of the set x in the universe. All the calculable properties of union, intersection, complement, 0 and 1 can be deduced in this theory.

An elementary theory of real numbers[1]

The primitive terms of this theory are: addition $+$, multiplication $\cdot$, the *less than* relation $<$, zero 0, and one 1. The axioms are: Axioms 106–111 of the theory of the *less than* relation, Axioms 404–406 of group theory, Axioms 504–506 of ring theory, and the following ones:

701. a. $x+y = y+x$,　　b. $x \cdot y = y \cdot x$,

702. $x < y \to x+z < y+z$,

703. $(x < y \wedge 0 < z) \to x \cdot z < y \cdot z$,

704. $\sim (x = 0) \to \bigvee z(x \cdot z = y)$,

705. $0 < 1$,

706. a. $0+x = x$,　　b. $1 \cdot x = x$,

707. $\left(\bigvee x,y(\alpha(x) \wedge \beta(y)) \wedge \bigwedge x,y\big((\alpha(x) \wedge \beta(y)) \to x < y\big) \right) \to$
$\bigvee z \bigwedge x,y\big((\alpha(x) \wedge \beta(y)) \to (x \leqslant z \wedge z \leqslant y)\big)$.

Axiom 707 is called the *axiom of continuity*. It is a schema of an infinite number of axioms for any sentential formulae $\alpha(x)$ and $\beta(y)$.

In an elementary theory of real numbers we can express the theory of n linear equations of n unknowns for any definite $n = 1, 2, 3, \ldots$,

---

[1] Cf. Tarski [94] and [98].

but we cannot construct in it a theory of equations of higher degress. No distinction can be made either between natural, rational, and irrational numbers; nor can a theory of exponential and logarithmic functions be formulated. Thus, the theory does not provide an apparatus sufficient for a formalization of mathematical analysis.

The examples given above were those of comparatively weak mathematical theories, which do not make use of the concepts of set theory. Advanced mathematical research concerned with groups, rings, Boolean algebras and real numbers cannot in fact be formalized in terms of the elementary theories described above. Advanced mathematical research always refers to set-theoretical concepts, namely those of set and elementhood. In group theory we are interested in the concept of coset, in Boolean algebra, in that of ideal, and in the theory of real numbers we want to be in a position to discuss the various subsets of the set of real numbers: rational numbers, algebraic numbers, etc. Those theories in which the concept of elementhood does not occur and is not definable by its terms are called *elementary*. Accordingly, we cannot discuss arbitrary sets of elements considered in such a theory.

Comments on non-elementary theories

If in a theory we want to discuss sets of elements (sets of points, sets of numbers, etc.), then we usually have to add to that theory new primitive terms, most importantly a term which denotes elementhood. In doing so we may differentiate the degrees to which we introduce set theory. For instance, we may introduce the whole of set theory, or only the sets of this first order, that is, in the case of real numbers, sets of real numbers alone, without any possibility of constructing sets of such sets, etc. We may introduce sets of real numbers and sets of such sets, and stop at that, that is, introduce the sets of the first and the second order. A theory which would cover the whole of the arithmetic of real numbers and the whole of set theory can be described as follows. The primitive terms of that theory are the symbols: Z, $\in$, $+$, $\cdot$, 0, 1, $<$, $\mathscr{R}$. The terms Z and $\in$ stand for the concepts of set and elementhood, respectively. All the set-theoretical axioms 307–313 concerning these concepts are adopted. The symbols $+$, $\cdot$, 0, 1, $<$ all have the same respective meanings as in the elementary theory of real numbers. $\mathscr{R}$ stands for the set of real numbers. An axiom concerning $\mathscr{R}$ is adopted, which states that $\mathscr{R}$ is a set which contains 0 and 1 and is closed under addition and multiplication:

$$Z(\mathscr{R}), \quad 0 \in \mathscr{R}, \quad 1 \in \mathscr{R},$$
$$y, x \in \mathscr{R} \rightarrow ((x+y) \in \mathscr{R} \wedge (x \cdot y) \in \mathscr{R}).$$

Analogues of all axioms of the elementary theory of real numbers are also adopted. Those analogues are constructed in the following manner: if A is a closed sentence which is an axiom of the elementary theory of real numbers, then its analogue A' is constructed by means of the following modifications: if an existential quantifier together with its scope

$$\bigvee x\beta,$$

occurs in A, then the entire formula is replaced by a formula which has the form

$$\bigvee x(x \in \mathscr{R} \wedge \beta),$$

and every formula of the form:

$$\bigwedge x\beta,$$

that is, one consisting of a universal quantifier and its scope, is replaced by the formula

$$\bigwedge x(x \in \mathscr{R} \rightarrow \beta).$$

In this way, Axiom 704, which in a rigorous notation has the form

$$\bigwedge x, y(\sim (x = 0) \rightarrow \bigvee z(x \cdot z = y)),$$

is replaced by the axiom

$$\bigwedge x, y\big((x \in \mathscr{R} \wedge y \in \mathscr{R}) \rightarrow (\sim (x = 0) \rightarrow$$
$$\bigvee z(z \in \mathscr{R} \wedge x \cdot z = y))\big)$$

This transformation is called a *restriction of the quantifiers to the predicate* $x \in \mathscr{R}$. Thus, all the axioms of the elementary theory of real numbers are adopted with restricted quantifiers, except for schema 707, which can now be replaced by a single axiom:

707*. $\bigwedge x, y, w, v\big((x, y \in \mathscr{R} \wedge Z(w) \wedge Z(v) \wedge x \in w \wedge y \in v \wedge$
$\bigwedge x, y((x, y \in \mathscr{R} \wedge x \in w \wedge y \in v) \rightarrow x < y)) \rightarrow$
$\bigvee z(z \in \mathscr{R} \wedge \bigwedge x, y((x, y \in \mathscr{R} \wedge x \in w \wedge y \in v) \rightarrow$
$(x \leqslant z \wedge z \leqslant y))\big))$.

This axiom refers to the sets w and v. The sentential formulae $\alpha(x)$ and $\beta(y)$, which occur in Axiom 707, are replaced in 707* by the formulae $x \in w$ and $y \in v$, respectively. By the axiom on set construction every sentential formula $\alpha(x)$ singles out, in the set $\mathcal{R}$ of real numbers, the subset of those and only those real numbers x which satisfy the formula $\alpha(x)$; hence, all the cases covered by schema 707 follow from Axiom 707*. In the elementary theory of real numbers Schema 707 is adopted instead of Axiom 707* only for the reason that, as we do not have there the concept of set at our disposal, we are not able to formulate Axiom 707* as one sentence.

All classical mathematics can be formalized in terms of the theory of real numbers, as described above, a theory which contains in it the entire set theory.

The non-elementary arithmetic of natural numbers can be described in a similar way. Let the terms: 0, S, $\mathcal{N}$, Z, $\in$, which denote, respectively, the number zero, the successor function $S(x) = x+1$, the set of natural numbers $\mathcal{N}$, the concept of set, and the concept of elementhood, be the primitive terms of that theory. The axiom system includes Axioms 307–313 of set theory and the following axioms of the arithmetic of natural numbers:

801. $0 \in \mathcal{N}$,

802. $x \in \mathcal{N} \rightarrow S(x) \in \mathcal{N}$,

803. $x \in \mathcal{N} \rightarrow \sim (0 = S(x))$,

804. $S(x) = S(y) \rightarrow x = y$,

805. $Z(\mathcal{N})$,

806. $\bigwedge z\big((Z(z) \wedge 0 \in z \wedge \bigwedge u(u \in z \rightarrow S(u) \in z)) \rightarrow \bigwedge x(x \in \mathcal{N} \rightarrow x \in z)\big)$.

But set theory itself has the property, analysed in the Introduction, that we can construct in it both the theory of natural numbers and that of real numbers. Thus, if we adopt the axioms of set theory as a whole, the specific axioms of those theories are, strictly speaking, superfluous. The whole of mathematics can always be covered by an adequately strong system of set theory. But mathematicians are often interested in such problems as what minimum apparatus of set theory suffices to yield a group of important mathematical theorems that form a certain whole. From that point of view it seems interesting to single out the elementary theories, which do not utilize set theory at all, and next to make distinctions between the various levels and the various means of set construction.

In proceeding in this way many authors call certain weak versions of set theory *higher order logics*. The name: first order logic is accordingly reserved for the functional calculus, while the theory which can generally be described as consisting in the introduction, next to the individuals of the lowest order, of sets of such individuals, is called *second order logic*. Introduction of sets of those sets yields *third order logic*. Introduction of all sets of a finite order results in what is called *simple type theory*. All those theories may, however, be considered to be special fragments of set theory, formalizable as elementary theories with specific primitive terms. They are thus all based on the classical logical calculus or the functional calculus. Hence, these two deserve above all to be called logic, regarded as the instrument of all rational inference.

EXERCISES

1. Prove that the identity axioms entail the following extensionality schemata:
$$x = y \rightarrow (\alpha(x)_i \equiv \alpha(y)),$$
$$x = y \rightarrow \varphi(x) = \varphi(y).$$

2. What restrictions on the variables are to be made in the case of the first of the above schemata?

3. Demonstrate that the latter of these two schemata follows from the former.

9. CERTAIN SYNTACTIC PROPERTIES OF THE CLASSICAL LOGICAL CALCULUS

Below are some theorems of the logical calculus which are syntactic in nature, that is, refer only to the shape of the formulae which are the logical theorems.

Theorem on the normal form

When introducing the c.l.c. in its traditional interpretation (cf. Section 5) we pointed out the fact that Theorems L25 and L26 and De Morgan's laws make it possible to shift all the quantifiers so that they begin a given formula. The necessary theorems are not obtainable in a system of logic true in the empty domain. This can easily be grasped in view of the fact that in the empty domain every formula beginning with a universal quantifier is true, and every formula beginning with an existential quantifier is false. To bring out the role of the assumption of non-emptiness in the proof of the theorem on the normal form we shall now prove a number of consequences of the assumption of non-emptiness in the axiomatic interpretation of the predicate calculus. The numbers of the theorems referred to are those of theorems as proved in Section 6.

We first prove the general schema

0.7. $\bigwedge x\alpha \to \alpha,$

which is an extension of schema 2, since it covers also those formulae α which do note have any free variables.

PROOF. If α has a free variable x, then 0.7 coincides with axiom schema 2, assumed in the c.l.c. axiom system. If α has no free variable x, we reason as follows:

0.1. $\bigwedge x\Big(\alpha \to \big((P(x) \to P(x)) \to \alpha\big)\Big)$

    (from the sentential calculus),

0.2. $\bigwedge x\Big(\alpha \to \big((P(x) \to P(x)) \to \alpha\big)\Big) \to \Big(\bigwedge x\alpha \to \bigwedge x\big((P(x) \to$

    $P(x)) \to \alpha\big)\Big)$

    (axiom schema 1),

0.3. $\bigwedge x\alpha \to \bigwedge x\big((P(x) \to P(x)) \to \alpha\big)$

    (from 0.1, 0.2, by detachment),

0.4. $\bigwedge x\big((P(x) \to P(x)) \to \alpha\big) \to \Big(\bigvee x(P(x) \to P(x)) \to \alpha\Big)$

    (a special case of Theorem 40*),

0.5. $\bigwedge x\alpha \to \Big(\bigvee x(P(x) \to P(x)) \to \alpha\Big)$

    (from 0.3, 0.4, and the sentential calculus theorem

    $(p \to q) \to ((q \to r) \to (p \to r)))$,

0.6. $\bigvee x(P(x) \to P(x)) \to (\bigwedge x\alpha \to \alpha)$

    (from 0.5 and the sentential calculus theorem

    $(p \to (q \to r)) \to (q \to (p \to r)))$,

0.7. is obtained from 0,6 and 0 (p. 163) by detachment.

The following theorem is dual with respect to 0.7:

0.8. $\alpha \to \bigvee x\alpha.$

For an α which does not contain a free variable x this theorem is proved in a way similar to the proof of 0.7. Theorem 0.1 yields the following theorems (in addition to those derived from it above):

0.9. $\alpha \to \bigwedge x\bigl((P(x) \to P(x)) \to \alpha\bigr)$
    (from 0.1 and Law 22*),

0.10. $\bigwedge x\bigl((P(x) \to P(x)) \to \alpha\bigr) \to \bigl(\bigvee x(P(x) \to P(x)) \to \bigvee x\alpha\bigr)$
    (axiom schema 6),

0.11. $\alpha \to \bigl(\bigvee x(P(x) \to P(x)) \to \bigvee x\alpha\bigr)$
    (from 0.9 and 0.10),

0.12. $\bigvee x(P(x) \to P(x)) \to (\alpha \to \bigvee x\alpha)$
    (from 0.11 and c.l.c.).

0.8 is obtained from 0.12 and 0 by detachment.

Theorems 0.7 and 0.8 make it possible to strengthen Axioms 5 and 10 by changing them into equivalences; this means that it may be assumed that a quantifier which does not bind anything may always be added and may always be omitted. Theorems 0.7 and 0.8 also make it possible to drop the assumption, in the case of axiom schemata 2 and 7, that the variable x must occur in α as a free variable. Since the restriction stating that x must occur in α as a free variable was used only in axiom schemata 2 and 7 (in the case of axiom schemata 4 and 9 reference to variables was merely conditional), all the restrictions concerning the necessity of occurrence in logical formulae of certain variables do not apply to logic when we consider only non-empty domains. But the restrictions concerning non-occurrence of certain variables in certain situations still hold.

It can easily be seen that Axiom 0 can be derived from Theorems 0.7 and 0.8 and from group I axioms. In view of the transitivity of implication Theorems 0.7 and 0.8 yield

0.13. $\bigwedge x\alpha \to \bigvee x\alpha.$

216

Hence

0.14. $\bigwedge x\big(P(x) \to P(x)\big) \to \bigvee x\big(P(x) \to P(x)\big)$

is a special case of 0.13.

The sentence

0.15. $\bigwedge x\big(P(x) \to P(x)\big)$

which is a substitution for the sentential calculus theorem $p \to p$, belongs to the group I axioms. Axiom 0 is obtained from 0.14 and 0.15 by detachment. Hence, it can be concluded that a system consisting of schemata I of the former axiom system (substitutions for sentential calculus theorems preceded by universal quantifiers), schemata II (where schema II.5 is strengthened by being changed into an equivalence), and schemata III (where schema III.10 is also strengthened by being changed into an equivalence) forms an axiom system of the classical logical calculus which is equivalent to the axiom system described in Section 5.

This new axiom system makes it possible to obtain a number of c.l.c. theorems in an easy and obvious way. Thus, for instance, since axioms schema II.5 and III.10 are now equivalences:

5'. $\alpha \equiv \bigwedge x\alpha,$

10'. $\alpha \equiv \bigvee x\alpha$

(if x does not occur in α as a free variable), under the extensionality of logical formulae, we may always replace both $\bigwedge x\alpha$ and $\bigvee x\alpha$ by α, if x does not occur in α as a free variable. Accordingly, axiom schema II.1 immediately yields Theorem 15* on the shifting of the quantifier to the consequent, if x does not occur in the antecedent as a free variable. Likewise, now that the theorem

30. $\bigwedge x(\alpha \wedge \beta) \equiv \bigwedge x\alpha \wedge \bigwedge x\beta$

has been proved, on the assumption that x does not occur in α as a free variable, then, on the strength of equivalence 5', $\bigwedge x\alpha$ may be replaced by α, which yields the theorem

88. $\bigwedge x(\alpha \wedge \beta) \equiv \alpha \wedge \bigwedge x\beta.$

In an similar way, now that the theorem

50. $\bigvee x(\alpha \vee \beta) \equiv \bigvee x\alpha \vee \bigvee x\beta$

has been proved, on the assumption that x does not occur in α as a free variable, then $\bigvee x\alpha$ may immediately be replaced by α, which yields the theorem

89. $\bigvee x(\alpha \vee \beta) \equiv \alpha \vee \bigvee x\beta$.

Of the theorems mentioned above, Theorem 88 will now be proved in a rigorous way. The proof, however, will differ somewhat from that outlined above; this will be done to bring out the facts as to which implications are true in the empty domain and which are not.

Our key theorems will, as previously, be marked with an asterisk.

0.16. $\bigwedge x((\alpha \wedge \beta) \to \alpha)$
      (c.l.c.).

0.17. $\bigwedge x(\alpha \wedge \beta) \to \bigwedge x\alpha$
      (from 0.16, axiom schema 1).

0.18. $\bigwedge x((\alpha \wedge \beta) \to \beta)$
      (c.l.c.).

0.19. $\bigwedge x(\alpha \wedge \beta) \to \bigwedge x\beta$
      (from 0.18, axiom schema 1).

0.20. $\bigwedge x\beta \to \beta$
      (from 0.7).

0.21. $\bigwedge x(\alpha \wedge \beta) \to \beta$
      (from 0.19 and 0.20).

0.22* $\bigwedge x(\alpha \wedge \beta) \to (\bigwedge x\alpha \wedge \beta)$
      (from 0.17 and 0.21).

0.23. $\beta \to (\alpha \to (\alpha \wedge \beta))$
      (c.l.c.).

0.24. $\beta \to \bigwedge x(\alpha \to (\alpha \wedge \beta))$
      (from 0.23 and Theorem 22)
(if x does not occur in β as a free variable).

0.25. $\bigwedge x(\alpha \to (\alpha \wedge \beta)) \to (\bigwedge x\alpha \to \bigwedge x(\alpha \wedge \beta))$
      (axiom schema 1).

0.26. $\beta \to (\bigwedge x\alpha \to \bigwedge x(\alpha \wedge \beta))$
      (from 0.24 and 0.25)
(if x does not occur in β as a free variable).

0.27*. $\left(\bigwedge x\alpha \wedge \beta\right) \rightarrow \bigwedge x(\alpha \wedge \beta)$
(from 0.26)
(if x does not occur in β as a free variable).

0.28*. $\bigwedge x(\alpha \wedge \beta) \equiv \left(\bigwedge x\alpha \wedge \beta\right)$
(from 0.22* and 0.27*)
(if x does not occur in β as a free variable).

Some other consequences of Axiom 0:

0.29. $\bigwedge x(\alpha \wedge \sim \beta) \rightarrow \left(\bigwedge x\alpha \wedge \sim \beta\right)$
(from 0.22*).

0.30. $\left(\bigwedge x\alpha \rightarrow \beta\right) \rightarrow \sim \bigwedge x(\alpha \wedge \sim \beta)$
(from 0.29 and c.l.c.).

0.31*. $\left(\bigwedge x\alpha \rightarrow \beta\right) \rightarrow \bigvee x(\alpha \rightarrow \beta)$
(from 0.30 and De Morgan's laws).

0.32*. $\bigvee x(\alpha \rightarrow \beta) \equiv \left(\bigwedge x\alpha \rightarrow \beta\right)$
(from 0.31* and 59*)
(if x does not occur in β as a free variable).

0.33. $\bigwedge x(\sim \alpha \wedge \beta) \rightarrow \left(\bigwedge x \sim \alpha \wedge \beta\right)$
(from 0.29*).

0.34. $(\beta \rightarrow \sim \bigwedge x \sim \alpha) \rightarrow \sim \bigwedge x(\sim \alpha \wedge \beta)$
(from 0.33 and c.l.c.).

0.35*. $(\beta \rightarrow \bigvee x\alpha) \rightarrow \bigvee x(\beta \rightarrow \alpha)$
(from 0.34 and De Morgan's laws).

0.36. $(\beta \rightarrow \alpha) \rightarrow (\beta \rightarrow \bigvee x\alpha)$
(from 0.8 and c.l.c.).

0.37. $\bigvee x(\beta \rightarrow \alpha) \rightarrow \bigvee x(\beta \rightarrow \bigvee x\alpha)$
(from 0.36 and Axiom 6).

0.38*. $\bigvee x(\beta \rightarrow \alpha) \rightarrow (\beta \rightarrow \bigvee x\alpha)$
(from 0.37 and Axiom 10)
(if x does not occur in β as a free variable).

0.39*. $\bigvee x(\beta \rightarrow \alpha) \equiv (\beta \rightarrow \bigvee x\alpha)$
(from 0.35* and 0.38*)
(if x does not occur in β as a free variable).

The consequences of Axiom 0 are interesting in that the following theorem, known as the *theorem on the reduction to a normal form*, can be proved from them:

THEOREM 9. *For every formula A there exists a formula B, of a normal form, such that*

$$(119) \quad \ulcorner \bigwedge ...(A \equiv B) \urcorner \in L.$$

A formula is said to have a *normal form* if all the quantifiers it has occur at the beginning. Such a formula then has the form

$$(120) \quad ... \bigwedge x \bigvee y \bigwedge z \bigwedge v ... (...),$$

consisting of a finite sequence of quantifiers, followed by a formula which does not contain any more quantifiers. A formula B, constructed in this way, may, of course, contain free variables (the same which occur in A). Hence, in order to obtain a sentence, we must precede the equivalence $A \equiv B$ with a universal quantifier that would bind all the free variables occurring in A or in B.

PROOF. To simplify our reasoning it suffices to assume that the formula A contains only two sentential connectives: negation and implication. For negation and implication we have the following c.l.c. theorems:

$$(121) \quad \sim \bigvee x\alpha \equiv \bigwedge x \sim \alpha,$$

$$(122) \quad \sim \bigwedge x\alpha \equiv \bigvee x \sim \alpha,$$

$$(123) \quad (\alpha \rightarrow \bigwedge x\beta) \equiv \bigwedge x(\alpha \rightarrow \beta),$$

$$(124) \quad (\bigwedge x\alpha \rightarrow \beta) \equiv \bigvee x(\alpha \rightarrow \beta),$$

$$(125) \quad (\bigvee x\alpha \rightarrow \beta) \equiv \bigwedge x(\alpha \rightarrow \beta),$$

$$(126) \quad (\alpha \rightarrow \bigvee x\beta) \equiv \bigvee x(\alpha \rightarrow \beta),$$

under the assumption that in cases (123) and (126) x does not occur as a free variable in α, and in cases (124) and (125) x does not occur as a free variable in β.

The above schemata are just Theorems: 15*, 22*, 38*, 40*, 0.32*, 0.39*, 61*, 64*, and direct consequences of Theorems 62* and 65*. Theorem 0 is required not only for the proofs of (124) and (126), but also for the proofs of the formulae (121)–(126) in the form in which they have been given above, that is, with the admission of the possibility that the formulae α and β may not contain the variable x at all.

On a closer examination of Theorems (121)–(126) we can see easily that the quantifier may be shifted so as to precede negation ((121) or (122)), provided that it is changed into the opposite; that it may be shifted from the consequent, without being changed, so as to precede the entire implication ((123)–(126)); and that it may be shifted from the antecedent so as to precede the entire implication, provided that it is changed into its opposite ((124)–(125)). Thus the quantifiers may always be taken from the inside of a formula and made to precede it, though in some cases they have then to be changed into the opposite quantifier. Such transformations yield equivalent formulae. In this way iterated applications of Theorems (121)–(126) may shift all the quantifiers to positions preceding the entire formula in question, which yields a formula B that is equivalent to the original formula A and satisfies Theorem 9. The necessary condition of those transformations, which states that a variable which occurs bound in one part of an implication may not occur free in the other part of that implication can always be satisfied at the very outset by changing the shapes of all the bound variables so that they should not conflict with one another.

Abbreviated proofs of reduction to a normal form of a few simple formulae will be given below by way of example:

A. $\big(P(x) \vee \bigwedge yR(x, y)\big) \to \bigwedge yS(x, y)$.

We first differentiate variables by relettering and eliminate the disjunction symbol. These operations yield the following equivalent formula:

$$\big(\sim P(x) \to \bigwedge yR(x, y)\big) \to .\bigwedge zS(x, z).$$

We then shift the quantifiers:

$\bigwedge y\big(\sim P(x) \to R(x, y)\big) \to \bigwedge zS(x, z)$
    (application of (123)).

$\bigvee y\big|\big(\sim P(x) \to R(x, y)\big) \to \bigwedge zS(x, z)\big)$
    (from (124)).

B. $\bigvee y\bigwedge z\big|\big(\sim P(x) \to R(x, y)\big) \to S(x, z)\big)$
    (from (123)).

Thus formula B is equivalent to formula A and has all its quantifiers placed at the beginning. It can easily be seen that certain formulae are reducible to several different normal forms. For instance, formula A is also equivalent to formula B':

B'. $\bigwedge z \bigvee y\big((\sim P(x) \to R(x, y)) \to S(x, z)\big).$

Obviously, it is not necessary to express the remaining sentential connectives in terms of implication and negation. The laws of logic make it possible directly to shift a quantifier to a position preceding any other sentential connective. This is especially simple in the case of conjunction and disjunction. If the β part of a formula does not have x as a free variable, both the universal and the existential quantifier may be shifted without any other change so that it precedes the conjunction and the disjunction symbol alike. This may be done on the strength of the laws:

87. $(\bigwedge x\alpha \vee \beta) \equiv \bigwedge x(\alpha \vee \beta),$

0.28*. $(\bigwedge x\alpha \wedge \beta) \equiv \bigwedge x(\alpha \wedge \beta),$

89. $(\bigvee x\alpha \vee \beta) \equiv \bigvee x(\alpha \vee \beta),$

82. $(\bigvee x\alpha \wedge \beta) \equiv \bigvee x(\alpha \wedge \beta).$

We shall now reduce the following formula to a normal form:

C. $\bigwedge x(\bigwedge z R(x, z) \wedge \bigvee y R(x, y)) \vee \bigvee y S(y).$

We shall first differentiate the variables by relettering them:

$$\bigwedge x(\bigwedge z R(x, z) \wedge \bigvee y R(x, y)) \vee \bigvee v S(v).$$

Next, by applying 0.28* and 86 we obtain

$$\bigwedge x\bigwedge z\bigvee y(R(x, z) \vee R(x, y)) \vee \bigvee v S(v).$$

Finally, on applying 87 and 89, we obtain

C'. $\bigwedge x\bigwedge z\bigvee y\bigvee v\big((R(x, z) \wedge R(x, y)) \vee S(v)\big).$

C' has a normal form, and $(C \equiv C') \in L$.

Extensionality of logical formulae

This term refers to the property that any two equivalent formulae may, in mathematical theories, be replaced by one another.

THEOREM 10 (on the extensionality of logical formulae). *If $x_1, ..., x_n$, $y_1, ..., y_l$ are all the free variables occurring in the formulae α and β,* *and if C^α is any formula that contains α or a formula obtained from α by the substitution for the variables $x_1, \ldots, x_n$ of some other variables different from the bound variables occurring in the formulae α or β, and if C^β differs from C^α only in that in certain places (not necessarily all of them) in which in C^α there occurs α or a formula obtained from α by a substitution for the variables $x_1, \ldots, x_n$, in the corresponding places in C^β there occurs β or a formula obtained from β by an appropriate substitution, while the variables $y_1, \ldots, y_l$ are all the free variables in C^α and C^β, then the sentence:*

$$(127) \qquad \bigwedge \ldots y_1, \ldots, y_l \big(\bigwedge x_1, \ldots, x_n (\alpha \equiv \beta) \to (C^\alpha \equiv C^\beta) \big)$$

is a theorem in L.

PROOF. Any formula C^α which satisfies the conditions specified above is either just a result of a substitution in α or is obtained from such substitutions and from certain formulae that are not results of substitutions in α, connected by sentential connectives and having their variables bound by quantifiers.

For the formulae obtained by substitutions in α we have:

$$(128) \qquad \bigwedge \ldots \big(\bigwedge x_1, \ldots, x_n (\alpha \equiv \beta) \to \big(\alpha(x_i/z) \equiv \beta(x_i/z) \big) \big).$$

Theorem (128) is obtained from Axiom 2: $\bigwedge x_1, \ldots, x_n (\alpha \equiv \beta) \to (\alpha \equiv \beta)$, and possibly from Axiom 4 which allows us to identify or change variables, especially if in the formulae $\alpha(x_i/z)$ and $\beta(x_i/z)$ not one but several of the variables $x_1, \ldots, x_n$ are replaced by the variable z.

In connecting formulae with sentential connectives we may refer to the following sentential calculus theorems:

$$(\gamma \equiv \delta) \to (\sim \gamma \equiv \sim \delta),$$

$$(\gamma \equiv \delta \land \varepsilon \equiv \zeta) \to \big((\gamma \to \varepsilon) \equiv (\delta \to \zeta) \big),$$

$$(129) \qquad (\gamma \equiv \delta \land \varepsilon \equiv \zeta) \to \big((\gamma \lor \varepsilon) \equiv (\delta \lor \zeta) \big),$$

$$(\gamma \equiv \delta \land \varepsilon \equiv \zeta) \to \big((\gamma \land \varepsilon) \equiv (\delta \land \zeta) \big),$$

$$(\gamma \equiv \delta \land \varepsilon \equiv \zeta) \to \big((\gamma \equiv \varepsilon) \equiv (\delta \equiv \zeta) \big).$$

In binding variables with quantifiers we refer to the theorems:

$$(130) \qquad \bigwedge z(\gamma \equiv \delta) \to \big(\bigwedge z\gamma \equiv \bigwedge z\delta \big),$$

$$\bigwedge z(\gamma \equiv \delta) \to \big(\bigvee z\gamma \equiv \bigvee z\delta \big),$$

223

derivable from axioms schemata 1 and 6 and from laws of the sentential calculus. Since $\varepsilon \equiv \varepsilon$ is a sentential calculus tautology, further theorems are derivable from (129) by the replacement of ζ by ε and the cancelling of the premise $\varepsilon \equiv \varepsilon$ in the antecedent.

Formula (128) shows that under the assumption $\bigwedge x_1, \ldots, x_n(\alpha \equiv \beta)$ the corresponding component formulae of the sentences C^α and C^β are equivalent. Formulae (129) show that linking equivalent formulae with sentential connectives yields equivalent formulae, while formulae (130) show that preceding equivalent formulae with quantifiers yields equivalent formulae. Hence, C^α and C^β, as they have the same structure, must be equivalent. This is exactly what is stated by formula (127).

Theorem 10 is considered to be characteristic of the sentential connectives and quantifiers in the sense they have in logic. This is so because Theorem 10 intuitively true only for those formulae C^α and C^β which are constructed of sentential connectives and quantifiers, and does not hold for formulae C^α and C^β constructed of other phrases, used, for instance, in the humanities and the social sciences. The phrase: X knows that p, is an example of a phrase for which formula (127) does not hold. Formula (127) has its analogue, for instance, in the sentence:

$$(p \equiv q) \to (X \text{ knows that } p \equiv X \text{ knows that } q).$$

This sentence is intuitively false. If p is replaced by the sentence: *Kipling wrote The Jungle Book*, and q by: *Kipling wrote Kim*, we have $p \equiv q$. Yet there are people who know that p and do not know that q, and vice versa. Those statements for which formula (127) holds are called *extensional*, whereas those for which it does not hold are called *intensional*. Thus, Theorem 10 states that logical statements are all extensional.

It can easily be seen that the Theorem on Extensionality also holds for a system of logic that is true in all domains including the empty domain, and also for intuitionistic logic, since the proof refers only to theorems of those calculi.

Restrictibility of logical theorems true in the empty domain

Analysis will now be made of a property which holds only for logical theorems true in the empty domain and which consists in the fact that relativizations of logical theorems to arbitrarily selected predicates are logical theorems. *Relativization* to a predicate α consists in the replacement of all quantifiers by quantifiers restricted to the the predicate α. If α is a formula with one free variable, then the formulae which satisfy the following schemata:

(131)
$$\bigwedge x_{i\alpha}\beta \equiv \bigwedge x_i(\alpha' \to \beta),$$
$$\bigvee x_{i\alpha}\beta \equiv \bigvee x_i(\alpha' \wedge \beta),$$

are quantifier formulae relativized to α, where α' stands for the formula obtained from α by the substitution of the variable x_i for the variable free in α, if x_i does not occur in α as a bound variable. If x_i occurs in α as a bound variable, then we first replace it by another bound variable other than x_i and other than any other variable occurring in α, and only then substitute x_i for the variable free in α.

Schemata (131) are adopted as general definitional schemata in the c.l.c. which introduce restricted quantifiers for any formulae α and β. Quite often the predicate α has the form "$x \in Z$", and then we say briefly that *the quantifiers are restricted to the set* Z and we use the abbreviations:

$$\bigwedge x \in Z\beta \quad \text{instead of} \quad \bigwedge x(x \in Z \to \beta),$$
$$\bigvee x \in Z\beta \quad \text{instead of} \quad \bigvee x(x \in Z \wedge \beta).$$

If all the quantifiers occurring in a formula are restricted to a certain predicate (set), it is said that that formula is relativized to that predicate (set). If a formula A is transformed into a new formula B which differs from A only in having all its quantifiers restricted to a fixed predicate (set), it is said that formula B is derived from formula A by the *restriction of quantifiers* to that predicate (set).

THEOREM 11. *Relativization of a logical theorem which is true in the empty domain yields a logical theorem which is true in the empty domain.*

PROOF BY INDUCTION. I. It will be demonstrated first that relativized c.l.c. axioms, except for Axiom 0, are c.l.c. theorems true in the empty domain. The restricted axioms have the forms:

1. $\bigwedge x_\alpha(\beta \to \gamma) \to (\bigwedge x_\alpha\beta \to \bigwedge x_\alpha\gamma)$

   (derived from $(\alpha \to (\beta \to \gamma)) \to ((\alpha \to \beta) \to (\alpha \to \gamma))$ under the distribution of the implication symbol and Definition (131)).

225

2. $\bigwedge x_\alpha (\bigwedge x_\alpha \beta(x) \to \beta(x))$

> (derived from $\bigwedge x \big(\bigwedge x(\alpha \to \beta(x)) \to (\alpha \to \beta(x))\big)$ by the shifting of the last α to the beginning of the formula and the application of formula (131)).

3. $\bigwedge x_\alpha \bigwedge y_\alpha \beta \to \bigwedge y_\alpha \bigwedge x_\alpha \beta$

> (derived from $\bigwedge x \bigwedge y \big((\alpha(x_1/x) \wedge \alpha(x_1/y)) \to \beta\big) \to$
> $\bigwedge y \bigwedge x \big((\alpha(x_1/x) \wedge \alpha(x_1/y)) \to \beta\big)$ by transformations which are easy to find).

4. $\bigwedge x_\alpha \bigwedge y_\alpha \beta(x, y) \to \bigwedge x_\alpha \beta(x, x)$

> (derived from $\bigwedge x \bigwedge y \big((\alpha(x_1/x) \wedge \alpha(x_1/y)) \to \beta(x, y)\big) \to$
> $\bigwedge x \big((\alpha(x_1/x) \wedge \alpha(x_1/x)) \to \beta(x, x)\big)$ by transformations which are easy to find).

In the two cases above, in order to display the substitution under consideration, it is assumed that x_1 is a free variable in α.

5. $\beta \to \bigwedge x_\alpha^{\cdot \cdot} \beta$ (if x does not occur in β as a free variable)

> (derived from $\beta \to (\alpha \to \beta)$ by shifting the quantifier $\bigwedge x$ to the consequent: Theorem 16*).

6. $\bigwedge x_\alpha (\beta \to \gamma) \to (\bigvee x_\alpha \beta \to \bigvee x_\alpha \gamma)$

> (derived from $(\alpha \to (\beta \to \gamma)) \to ((\alpha \wedge \beta) \to (\alpha \wedge \gamma))$ by the application of axiom schemata 1 and 6).

7. $\bigwedge x_\alpha (\beta(x) \to \bigvee x_\alpha \beta(x))$

> (derived from $\alpha \to \big(\beta(x) \to (\alpha \wedge \beta(x))\big)$ and quantifier laws).

8. $\bigvee x_\alpha \bigvee y_\alpha \beta \to \bigvee y_\alpha \bigvee x_\alpha \beta$

> (derived from $(\alpha_1 \wedge \alpha_2 \wedge \beta) \to (\alpha_2 \wedge \alpha_1 \wedge \beta)$ and quantifier laws).

9. $\bigvee x_\alpha \beta(x, x) \to \bigvee x_\alpha \bigvee y_\alpha \beta(x, y)$

> (derived from axiom schema 9, like the restricted axiom schema 4 from axiom schema 4).

10. $\bigvee x_\alpha \beta \to \beta$

> (derived from $(\alpha \wedge \beta) \to \beta$ and quantifier laws).

More rigorous proofs of the above restricted formulae are left to the reader as a useful exercise in the predicate calculus. It can easily be seen that Axiom 0 is not needed in obtaining the appropriate restricted theorem.

Formula A restricted to a predicate α will be denoted by $A(\bigwedge/\alpha)$, while formula A restricted to a set X (that is, to the predicate: $x \in X$) will be denoted by $A(\bigwedge/X)$.

II. A restricted implication remains an implication: $\ulcorner A \to B \urcorner (\bigwedge/\alpha)$ is an implication of the form $A(\bigwedge/\alpha) \to B(\bigwedge/\alpha)$, and hence, if B is obtained from A and $A \to B$ by detachment, then $B(\bigwedge/\alpha)$ is obtained by detachment from $A(\bigwedge/\alpha)$ and $\ulcorner A \to B \urcorner (\bigwedge/\alpha)$.

Logic does not distinguish any extralogical concepts

The next property of the classical logical calculus with which we shall be concerned is the fact that, in the logical calculus, if something can be proved about one atomic predicate, the same can be proved about any other atomic predicate. No extralogical concept is distinguished. In logic, what can be proved about addition is the same as what can be proved about multiplication. If different things are asserted about two extralogical concepts this must always be due to some extralogical assumptions. This property of logic will be discussed in a number of subsections.

Logic does not distinguish any individual constant

THEOREM 12. *If a is a constant and $\alpha(x)$ is a sentential formula of one free variable x and does not contain any quantifier binding the variable x, and if $\alpha(a) \in L_1$, then $\bigwedge x\alpha(x) \in L_1$ (and the same holds for L^*).*

The meaning of this theorem can also be formulated in the following way: The axioms of logic are so universal that they do not single out any object from among others. Hence if a statement about an object a belongs to logic, then the same statement may be made about any other object.

PROOF. If $\alpha(a) \in L_1$, that is, if $\alpha(a) \in \mathrm{Cn}_0(\mathrm{Ax}\,L^* \cup \mathrm{Ax}\,\mathrm{Id})$, then there is a proof by detachment $D_1, \ldots, D_n$ of the sentence $\alpha(a) = D_n$ from the set $\mathrm{Ax}\,L^* \cup \mathrm{Ax}\,\mathrm{Id}$. It may be assumed about the sentences $D_1, \ldots, D_n$ that the variable x does not occur in them. At least some of the sentences $D_1, \ldots, D_n$ contain the constant a.

227

Let $D_i(x)$ stand for the formula derived from D_i by the replacement of the constant a by the variable x in all those places where the constant a occurs in D_i. If the constant a does not occur in D_i, then it is assumed that $D_i(x) = D_i$. It will be demonstrated that, for every $i \leqslant n$, $\bigwedge x D_i(x) \in L_1$. The proof will be by induction.

1. If D_i is one of the $\operatorname{Ax} L^*$ or $\operatorname{Ax} \operatorname{Id}$, then the sentence $\bigwedge x D_i(x)$ also is one of the $\operatorname{Ax} L^*$ or $\operatorname{Ax} \operatorname{Id}$. It is so because the sentence $\bigwedge x D_i(x)$ falls under the same axiom schema as does D_i.

2. Suppose now that $m \leqslant n$ and that for all $j, k < m$ it is true that $\bigwedge x D_j(x)$, $\bigwedge x D_k(x) \in L_1$; it will also be proved that $\bigwedge x D_m(x) \in L_1$. The sentence D_m either belongs to $\operatorname{Ax} L^* \cup \operatorname{Ax} \operatorname{Id}$ or is derived from earlier sentences by detachment. If D_m is an axiom, then, by 1 above, $\bigwedge x D_m(x) \in L_1$. If D_m is derived by detachment from earlier sentences D_j and D_k, where $k, j < m$, that is, if $D_k = D_j \to D_m$, then, by the inductive assumption, $\bigwedge x D_j(x) \in L_1$ and $\bigwedge x(D_j(x) \to D_m(x)) \in L_1$. Hence, by the axioms of II.1 on the distribution of quantifiers over implication we have: $\bigwedge x D_j(x) \to \bigwedge x D_m(x) \in L_1$. On the strength of the rule of detachment we accept the formula $\bigwedge x D_m(x) \in L_1$.

Thus all the premisses of the proof have the property under consideration. Consequently, $D_n = \alpha(a)$ also has that property. But $D_n(x) = \alpha(x)$, and hence $\bigwedge x \alpha(x) \in L_1$.

Logic does not distinguish any predicate symbol nor any function symbol

Reference will first be made to Theorem 12, which states that if a sentence $\alpha(a)$ is a theorem in the c.l.c., where a is a constant, then $\bigwedge x \alpha(x) \in L_1$. On the other hand, we know that the sentences

$$\bigwedge x \alpha(x) \to \alpha(b)$$

are particular cases of substitution for schema II.2 (where b also is any constant). This yields the formula

(132) $\alpha(a) \in L_1 \to \alpha(b) \in L_1$,

which states that if a sentence which states something about a constant is a logical theorem, then a sentence which states the same about any other constant is a logical theorem, too. It is so because in logic nothing special is assumed about any terms would not be assumed about all other terms. Similar theorems can be obtained for any predicates and functions.

THEOREM 13. *If a sentence A contains a predicate symbol P with n arguments, and if a sentence B differs from A only in that in those places in which the symbol P occurs in A, a certain other predicate symbol Q, also with n arguments, always occurs in B, then the implication $A \in L^* \to B \in L^*$ is true (if the symbol P is other than the identity symbol, then the implication $A \in L_1 \to B \in L_1$ is true, too).*

O u t l i n e   o f   p r o o f   b y   i n d u c t i o n.

1. The c.l.c. axioms $\mathrm{Ax}\,L^*$ and, likewise, the axioms of identity $\mathrm{Ax}\,\mathrm{Id}$ have been defined as arbitrary substitutions for certain general schemata. If a sentence which contains a predicate P is a correct substitution for a general schema, then a similar sentence, which instead of the predicate P contains a predicate Q, with the same number of arguments as P, is an equally correct substitution for that general schema.

2. Let $A(P)$ denote a sentence containing the predicate P, and $A(Q)$, a sentence derived from the sentence $A(P)$ by the replacement of the predicate P by the predicate Q. Thus, Part 1 of the proof states that if $A(P)$ is an axiom, then $A(Q)$ is an axiom, too. Note now that if $A(P)$ is an implication:

$$A(P) = \big(B(P) \to C(P)\big),$$

then also

$$A(Q) = \big(B(Q) \to C(Q)\big).$$

Hence, if the sentence $C(P)$ is obtained from the sentences $A(P)$ and $B(P)$ by detachment, then the sentence $C(Q)$ also is obtained from the sentences $A(Q)$ and $B(Q)$ by detachment.

Thus, if $A \in L_1$, that is, if A is derived from axioms by a finite number of detachments, then the sentence $B = A(Q)$ also is derived from axioms by the same number of detachments. Consequently, $B \in L_1$.

Thus, logical theorems do not distinguish any predicate. The same things are assumed about all predicates. If something can be proved in logic about a predicate P, the same holds for any other predicate Q.

The same reasoning leads to the conclusion that logical axioms do not single out any function. If something can be proved about a specific function, the same holds true for any other function as well. In logic, the same things can be proved about addition as can be proved about multiplication, the same things can be proved about the relation of *divisible by* as about the *less than* relation, etc.

THEOREM 14. *If a sentence A contains a function symbol F of n arguments, and if a sentence B differs from A only in that in those places in which the symbol F occurs in A, a function symbol G, also of n arguments, occurs in B, then the implication $A \in L_1 \rightarrow B \in L_1$ is true (and the same holds for L^*).*

The proof is analogical to that of Theorem 12.

Theorems 12, 13 and 14 taken together describe the general character of logical axioms. In logic, nothing more is asserted about one constant term than is asserted about any other constant term. The same applies to predicates and functions. These theorems, taken together, may accordingly be called *theorems on the fact that no constant symbols are distinguished in logic.*

The proof of the theorem that follows is left to the reader as an exercise:

THEOREM 15. *If $A(P) \in L_1$, then also $A(\alpha) \in L_1$, where the sentence $A(\alpha)$ differs from the sentence $A(P)$ only in that in all those places in which the atomic predicate $P(x_1, \ldots, x_n)$, or the result of some substitution in it, occurs in $A(P)$, the sentence $A(\alpha)$ has the predicate $\alpha(x_1, \ldots, x_n)$, not necessarily an atomic one, with all its variables $x_1, \ldots, x_n$ occurring as free, or the result of a substitution in that predicate, such that if any of the variables $x_1, \ldots, x_n$ is to be replaced by a variable which already occurs in α as a bound variable, then that variable bound in α is first changed in α into another variable, other than any variable occurring in α or in $A(P)$. (The same holds for L^* and for L.)*

The proof is by induction, analogical to the proofs of the preceding theorems.

Logic does not distinguish functions from univalent predicates

The following theorem on non-distinction states that in logic there is no difference between function symbols and those predicates for which statements on existence and uniqueness are assumed.

First of all we have to define what is meant by the statement that a sentence B states the same about a predicate $P(x, y)$ as a sentence

A states about a function $F(x)$. In the language of the theory L functions may be superimposed upon one another, and in order to define translation we shall define first when a sentence A' is a purification of a sentence A with respect to a functor F. Loosely speaking, a purification of the sentence A is a sentence in which the functor F occurs only in atomic sentences of the form: $F(\pi) = y$, so that $F(\pi)$ does not occur as an argument in any superpositions: $G(F(\pi))$. First of all, all the occurrences of the functor F in A are classified into pure and impure. An occurrence is *pure* if it has the form: $F(\pi) = y$, and any other occurrence is *impure*.

DEFINITION 26. *A' is a purification of a sentence A with respect to a functor F if and only if A' is derived from A by successive replacement of each impure occurrence of the functor F by its pure occurrence in accordance with the following schema*:

If $\beta(F(\pi))$ is the least sentential formula containing the functior F in its impure occurrence(s), then that formula is replaced by a formula of the form:

(133) $\bigvee y(\beta(y) \wedge F(\pi) = y)$,

in which the functor F occurs in a pure form and in which the added variable y differs from all other variables.

Since the equivalence

$$\beta(F(\pi)) \equiv \bigvee y(\beta(y) \wedge F(\pi) = y)$$

is a logical theorem by the theorem on the extensionality of logical formulae, we have the conclusion:

(134)   *If A' is a purification of A with respect to F, then $\ulcorner A \equiv A' \urcorner \in L_1$.*

A sentence B is a translation of a sentence A by the replacement of a functor F by a predicate P (in symbolic notation: $B = \mathrm{Tr}\, A(F/P)$) if and only if the sentence B is derived from a sentence A' (which is a purification of A with respect to F) by the replacement of every occurrence of the form: $F(\pi) = y$ by $P(\pi, y)$.

THEOREM 16. $A \in L_1 \equiv \ulcorner \mathrm{Funct}(P) \to \mathrm{Tr}\, A(F/P) \urcorner \in L_1$, *where*

(135) $\mathrm{Funct}\,(P) = \Big(\bigwedge x \bigvee y P(x, y) \wedge \bigwedge x, y, z\big((P(x,y) \wedge P(x,z)) \to$
$$y = z\big)\Big).$$

231

This theorem shall be called the *theorem on the non-distinction in logic between univalent predicates and functions*. Obviously, it also holds for predicates and functions of many arguments. For simplicity's sake we confine ourselves to two arguments.

PROOF. The implication ← is a direct consequence of Theorem 15. It follows that if a sentence containing an atomic predicate $P(x, y)$ is a logical theorem, then the sentence derived from that mentioned above by the replacement of the predicate $P(x, y)$ by the predicate $F(x) = y$ is a logical theorem, too. The last-named predicate evidently satisfies the conditions of existence and uniqueness. These conditions are consequences of the axioms of identity. By way of example, we shall prove the condition of existence for the predicate $F(x) = y$. We have

(a) $\bigwedge x_0 (x_0 = x_0 \rightarrow \bigvee y x_0 = y)$
   (from L1),

(b) $\bigwedge x_0 (x_0 = x_0 \rightarrow \bigvee y x_0 = y) \rightarrow (F(x) = F(x) \rightarrow \bigvee y F(x) = y)$
   (from L2),

(c) $F(x) = F(x) \rightarrow \bigvee y F(x) = y$
   ((a), (b), det.),

(d) $\bigwedge x_0 x_0 = x_0 \rightarrow F(x) = F(x)$
   (from L2),

(e) $F(x) = F(x)$
   ((d), Ax. Id 101, det.),

$\bigwedge x \bigvee y F(x) = y$
   ((c), (e), det., R3).

By detaching the conditions of existence and uniqueness for $F(x) = y$ we obtain the sentence A as a logical theorem.

The converse implication → is obtained by induction. If the sentence A falls under any one of the axiom schemata of system L^*, except for schemata II.2 and III.7, then the sentence $\mathrm{Tr}\, A(F/P)$ falls under the same axiom schema. Under the laws of the sentential calculus we may add the antecedent $\mathrm{Funct}(P)$ to obtain

$$\ulcorner \mathrm{Funct}(P) \rightarrow \mathrm{Tr}\, A(F/P) \urcorner \in L^*.$$

Axioms II.2 and III.7 will be discussed separately.

9. SYNTACTIC PROPERTIES

The special forms of those axioms, important for our considerations, are:

(136) $\quad \bigwedge \ldots \left(\bigwedge x \gamma(x) \to \gamma\left(\varrho(F(\pi))\right) \right),$

(137) $\quad \bigwedge \ldots \left(\gamma\left(\varrho(F(\pi))\right) \to \bigvee x \gamma(x) \right).$

The occurrences of F in the formula $\gamma(x)$ need not concern us, because the replacement of those occurrences by the corresponding formulae (133) takes place both in the antecedent and in the consequent and thus does not lead outside the axioms of schemata 2 and 7. On the contrary, the occurrence of F in ϱ, as shown in schemata (136) and (137), on the transformation Tr leads outside schemata of logical axioms.

Suppose that the variable x occurs in $\gamma(x)$ in k places. In each place that variable is an argument in an atomic sentential formula. Let those formulae be denoted by $R_1(x), \ldots, R_k(x)$.

(136) and (137) may thus be described as follows:

(136') $\quad \bigwedge \ldots \left(\bigwedge x \gamma(R_1(x), \ldots, R_k(x)) \to \gamma\left(R_1\left(\varrho(F(\pi))\right), \ldots, \right.\right.$
$\qquad R_k\left(\varrho(F(\pi))\right) \Big) \Big),$

(137') $\bigwedge \ldots \left(\gamma\left(R_1\left(\varrho(F(\pi))\right), \ldots, R_k\left(\varrho(F(\pi))\right)\right) \to \bigvee x \gamma(R_1(x), \ldots, R_k(x)) \right),$

Upon the transformation Tr these formulae change into

(138) $\quad \bigwedge \ldots \left(\left(\bigwedge x \gamma(R_1(x), \ldots, R_k(x)) \to \gamma\left(\bigvee y\left(R_1(\varrho(y)) \wedge P(\pi, y)\right), \right.\right.\right.$
$\qquad \ldots, \bigvee y\left(R_k(\varrho(y)) \wedge P(\pi, y)\right) \Big) \Big).$

(139) $\quad \bigvee \ldots \left(\gamma\left(\bigvee y\left(R_1(\varrho(y)) \wedge P(\pi, y)\right), \ldots, \bigvee y\left(R_k(\varrho(y)) \wedge \right.\right.\right.$
$\qquad P(\pi, y) \Big) \Big) \to \bigvee x \gamma(R_1(x), \ldots, R_k(x)) \Big).$

The above formulae are not axioms. They can be obtained in the following way. The sentences

(140) $\quad \bigwedge \ldots y\left(\bigwedge x \gamma(R_1(x), \ldots, R_k(x)) \to \gamma\left(R_1(\varrho(y)), \ldots, R_k\left((\varrho(y))\right)\right) \right).$

(141) $\quad \bigwedge \ldots y\left(\gamma\left(R_1(\varrho(y)), \ldots, R_k(\varrho(y))\right) \to \bigvee x \gamma(R_1(x), \ldots, R_k(x)) \right)$

fall under the axiom schemata of 2 and 7. On the other hand, the sentences

(142) $\bigwedge \ldots \bigvee yP(\pi, y)$,

(143) $\bigwedge \ldots y_1y_2\big((P(\pi, y_1) \wedge P(\pi, y_2)) \to y_1 = y_2\big)$

are consequences of Funct(P). Now (140) and (142), using the sentential calculus theorem $(\alpha \wedge (p \to q)) \to (p \to (\alpha \wedge q))$ and the quantifier theorem 0.38*, yield

(144) $\bigwedge \ldots \Big(\bigwedge x\gamma(R_1(x), \ldots, R_k(x)) \to$

$\bigvee y\big(P(\pi, y) \wedge \gamma(R_1(\varrho(y)), \ldots, R_k(\varrho(y)))\big)\Big)$.

Further, (141), under the sentential calculus theorem $(p \to q) \to ((\alpha \wedge p) \to q)$ and the quantifier theorem 40*, yields

(145) $\bigwedge \ldots \Big(\bigvee y\big(P(\pi, y) \wedge \gamma(R_1(\varrho(y)), \ldots, R_k(\varrho(y)))\big) \to$

$\bigvee x\gamma(R_1(x), \ldots, R_k(x))\Big)$.

The following schemata are consequences of (142) and (143):

(146) $\bigwedge \ldots \Big(\bigvee y\big(P(\pi, y) \wedge \sim \beta(\varrho(y))\big) \equiv$

$\sim \bigvee y\big(P(\pi, y) \wedge \beta(\varrho(y))\big)\Big)$,

(147) $\bigwedge \ldots \Big(\bigvee y\big(P(\pi, y) \wedge \beta_1(\varrho(y)) \wedge \beta_2(\varrho(y))\big) \equiv$

$\big(\bigvee y_1\big(P(\pi, y_1) \wedge \beta_1(\varrho(y_1))\big)\big) \wedge \bigvee y_2\big(P(\pi, y_2) \wedge$

$\beta_2(\varrho(y_2))\big)\Big)$,

(148) $\bigwedge \ldots \Big(\bigvee y\big(P(\pi, y) \wedge \bigvee z\beta(z, \varrho(y))\big) \equiv$

$\bigvee z\bigvee y\big(P(\pi, y) \wedge \beta(z, \varrho(y))\big)\Big)$,

where β, β_1 and β_2 are any sentential formulae in γ.

These schemata show that the quantifier $\bigvee y$, combined with the formula $P(\pi, y)$, may be shifted from a position preceding a negation symbol to that following that negation symbol, from a position before a quantifier to that within the scope of that quantifier; furthermore, if it stands before a conjunction it may be split so as to occur in the terms of that conjunction. Since all other sentential connectives can

234

be expressed in terms of conjunction and negation, and since the universal quantifier can be expressed in terms of the existential quantifier and the negation symbol, formulae (146)–(148) demonstrate that the formula $\bigvee y(P(\pi, y) \wedge \dots)$ may be shifted from a position preceding any quantifier to a position within the scope of that quantifier, and that from a position preceding any sentential connective it may be split so as to occur in the terms linked by that sentential connective, or may be shifted to one of the terms if it is not required to occur in the other term.

In this way, by Theorem 10 on extensionality, schemata (146)–(148) yield the equivalence:

$$(149) \qquad \bigvee y\Big(P(\pi, y) \wedge \gamma\big(R_1(\varrho(y)), \dots, R_k(\varrho(y))\big)\Big) \equiv$$
$$\gamma\Big(\bigvee y\big(P(\pi, y) \wedge R_1(\varrho(y))\big), \dots, \bigvee y\big(P(\pi, y) \wedge R_k(\varrho(y))\big)\Big).$$

Now, (149) together with (144) and (145) yield (138) and (139), which means that the translations of Axioms (136) and (137) belong to $Cn(Funct(P))$.

Apart from axioms schemata 2 and 7, discussed above, the function symbol F may also occur in Axiom on Extensionality Ex2:

$$x = z \to F(x) = F(z).$$

The purification of the above sentence has the form of the sentence:

$$x = z \to \bigvee y(F(x) = y \wedge F(z) = y),$$

and hence its translation, upon the replacement of F by P, forms the sentence

$$x = z \to \bigvee y(P(x, y) \wedge P(z, y)),$$

which is a direct consequence of the condition of existence, contained in $Funct(P)$.

Note now that

$$Tr(\ulcorner B \to A \urcorner)(F/P) = \ulcorner Tr B \ (F/P) \to Tr A(F/P) \urcorner;$$

hence, if A is derived by detachment from B and $C = B \to A$, then also $Tr A(F/P)$ is derived by detachment from $Tr B(F/P)$ and $Tr C(F/P)$. This, together with the previous analysis, yields that if $A \in L_1$, then $Tr A(F/P) \in Cn_1(Funct(P))$, and hence, by the Deduction Theorem, $\ulcorner Funct(P) \to Tr A(F/P) \urcorner \in L_1$.

Theorems 15 and 16 yield the following conclusion:

THEOREM 17. *If* $A \in L_1$, *then* $\ulcorner \text{Funct}(\alpha) \to A(\alpha) \urcorner \in L_1$, *where* Funct$(\alpha)$ *is a sentence which states that the non-atomic predicate* $\alpha(x, y)$ *is univalent and* $A(\alpha)$ *is a sentence derived from* Tr$A(F/P)$ *by the replacement of the atomic predicate* $P(x, y)$ *and its substitutions by the non-atomic predicate* $\alpha(x, y)$ *and its corresponding substitutions, and the bound variables occurring in* $\alpha(x, y)$ *are different from all the variables occurring in* Tr$A(F/P)$.

Theorem 17 may be summarized informally by saying that if in logic something can be proved about a function, then the same can be proved about any univalent predicate, not necessarily an atomic one.

EXERCISES

1. Prove that a restriction of an intuitionistic theorem true in the empty domain is an intuitionistic theorem.

2. Can the analogue of Theorem 16 in the intuitionistic functional calculus be proved? If so, how can that be done?

3. Prove that if a sentence A contains neither function symbols nor individual constant, then $A \in L \equiv A \in L^*$.

4. Prove that $A \in L_1 \equiv A'' \in L$, where A'' is a sentence obtained by the translation of all function symbols into the predicate language and by preceding that translation with appropriate conditions concerning those predicates and the concept of identity. This will be a theorem on a non-essential difference between the classical functional calculus and the classical predicate calculus. Formulate it in a rigorous form.

H i n t. The proof is based on Theorem 16.

10. ON DEFINITIONS

The preceding sections were concerned with certain examples of mathematical theories. When mathematical theories are actually being developed, the theorems which are proved in the process usually include, next to primitive concepts of the given theory, concepts which are introduced by definitions. It often happens that, when developing a mathematical theory, we are interested in relationships which could be described better, more briefly, and in a more interesting way by introducing a new concept. Such an addition of a new concept requires a characterization of the concept. Such a characterization can be made by reducing the meaning of the new concept to that of the concepts already occurring in the theory, in which case we say that the new concept is defined by means of concepts previously introduced, or it can be made by means of a new theory, which is an enlargement of the old one, if the meaning of the new concept is not reduced to that of concepts introduced previously. The meaning of the new concept is captured by the way that concept is used. Normally, the way of operation with the new concept is rigorously reduced to the way of using the old concepts by means of a special formula, which is called definition of the new concept. According to the type of the concept being introduced its definition takes on a certain form, but usually it amounts to an equivalence.

Distinction will be made below between several types of definitions. The first type to be discussed covers definitions of relations.

Definitions of relations

Definitions of relations as a rule take on the form of an equivalence.

For instance, when defining the *less than* relation by means of addition and multiplication in the arithmetic of real numbers, we may adopt the following definition:

$$x < y \equiv \bigvee z(z \neq 0 \wedge y = x + z^2).$$

The predicate to be defined occurs on one side of the equivalence symbol. The defining formula, formed of primitive terms or terms defined earlier, occurs on the other side. In this way relations of any number of arguments can be defined.

In general, when using logical formalism without free variables we may describe equivalence-type definitions of relations in the manner stated below:

DEFINITION 27. *A closed sentence, consisting of the equivalence* $\alpha \equiv \beta$ *preceded by universal quantifiers, is called an equivalence* (*or ordinary*) *definition if and only if*:

1. *The left term of the equivalence is an atomic predicate, that is, consists of a predicate symbol followed by parentheses with variables any two of which differ in shape* (cf. Definition 14).

2. *The right term of the equivalence is a* (*non-atomic*) *predicate con-*

237

taining the same free variables which occur in the atomic predicate on the left side of the equivalence symbols.

3. *The right term of the equivalence does not contain any predicate symbol equiform with the predicate symbol on the left side of the equivalence symbol.*

If an equivalence satisfies Definition 27, then the left term is called the *definiendum*, and the right term, the *definiens*. The predicate symbol occurring in the left term is called the *symbol (concept) being defined*.

Usually, when we develop a mathematical theory, we introduce many equivalence definitions which define new concepts whose properties we want to study. Definitions are added to mathematical theories as new axioms, whose truth is based on convention: in a definition the new atomic predicate is defined by convention as equivalent to a certain compound predicate. When adding new definitions to a theory which we develop we may treat them as new axioms true by convention only if we see to it that the term being defined does not occur earlier in the theory in question. Should it so occur, it would have a specific meaning in that theory and, hence, could not be given a new meaning in an arbitrary manner. If this principle is observed, then the equivalence definitions function merely is convenient notational and conceptual abbreviations, without essentially enriching a given mathematical theory.

The fact that definitions serve merely as abbreviations finds a rigorous formulation in two methodological theorems: the theorem on definitional replacement and the theorem on elimination of definitions from proofs. The former is a consequence of the theorem on the extensionality of logical formulae.

THEOREM 18 (on definitional replacement). *If a sentence C^P differs from a sentence C^α only in that in some places in which the formula $P(x, y)$ (or the result of substitution in it) occurs in C^P, C^α has the formula $\alpha(x, y)$ (or a corresponding result of substitution), the bound variables occurring in the formula $\alpha(x, y)$ being different from any variables occurring in C^P in the parentheses that follow the predicate symbol P, and if D is the definition:*

$$D: \bigwedge x, y\big(P(x, y) \equiv \alpha(x, y)\big),$$

then

$$(C^P \equiv C^\alpha) \in \mathrm{Cn}(D).$$

238

The proof follows from Theorem 10 and theorems on the properties of consequence. That theorem refers to the possibility of replacing the definiens by the definiendum, and conversely, in sentences constructed of sentential connectives and quantifiers.

The theorem on elimination of definitions from proofs will be preceded by the following explanation. The correct addition of a new definition obviously increases the number of theorems. For instance, a definition is not a theorem in a given system before it is added to the system. But the correct addition of a new definition does not change the set of those theorems which do not contain the term just defined. For instance, the addition to set theory of Definition (118) of the concept of equinumerosity does not change the set of those set-theoretical theorems which do not contain the concept of equinumerosity. Hence, the *theorem on elimination of definitions from proofs* may be formulated as follows:

THEOREM 19. *If A is an equivalence definition of a predicate symbol P which does not occur in any sentence of the set X, and if a sentence B does not contain the predicate P, then*

$$B \in \mathrm{Cn}(X \cup \{A\}) \to B \in \mathrm{Cn}(X).$$

PROOF. Suppose that a definition of the predicate symbol P has the form:

A. $\bigwedge x, y\big(P(x, y) \equiv \alpha(x, y)\big).$

It follows from the assumption that $B \in \mathrm{Cn}(X \cup \{A\})$ and from the Deduction Theorem that there exist sentences $E_1, \ldots, E_k$ such that $E_1, \ldots, E_k \in X$ and

$$\big((E_1 \wedge \ldots \wedge E_k \wedge A) \to B\big) \in L.$$

This and the theorem stating that logic does not distinguish any predicate constant (Theorem 13) yield that if in the logical theorem above the predicate $P(x, y)$ is replaced by any other predicate with the same free variables, then the result is a logical law, too. The predicate $P(x, y)$ by assumption occurs in the sentence A only, which is its definition. If in A the predicate $P(x, y)$ is replaced by a formula $\alpha(x, y)$, then A becomes the logical tautology:

A′. $\bigwedge x, y\big(\alpha(x, y) \equiv \alpha(x, y)\big).$

239

The theorem stating that logic does not privilege any constant yields accordingly that

$$((E_1 \wedge \ldots \wedge E_k \wedge A') \to B) \in L.$$

Since $A' \in L$, hence also

$$((E_1 \wedge \ldots \wedge E_k) \to B) \in L$$

so that $B \in \mathrm{Cn}(X)$.

Defining individual constants

We shall now be concerned with defining individual and functional constants. Before formulating a general theory of such definitions, we shall first show, by way of example, how to introduce an individual constant in group theory. In group theory, as formulated in the preceding section, it is possible to prove the existence of a zero-element with respect to the group operation. The theorem is:

$$\bigvee z \bigwedge y (y = y + z)$$
(existence of right-hand zero-element).

PROOF. We have

407. $\bigwedge x \bigvee z (x + z = x)$
    (from 405, axiom schema 4, combining of universal quantifiers),

408. $v + (x + z) = (v + x) + z$
    (from 404, change of variables),

409. $x + z = x \to v + (x + z) = v + x$
    (identity laws),

410. $v + (x + z) = v + x \to v + x = v + (x + z)$
    (identity laws),

411. $x + z = x \to v + x = v + (x + z)$
    (from 409, 410, transitivity of implication),

412. $\big(v + x = v + (x + z) \wedge v + (x + z) = (v + x) + z\big) \to$
    $v + x = (v + x) + z$
    (identity laws),

413. $v + (x + z) = (v + x) + z \to$
    $\big(v + x = v + (x + z) \to v + x = (v + x) + z\big)$
    (from 412 and a sentential calculus theorem),

414. $v+x = v+(x+z) \rightarrow v+x = (v+x)+z$
(from 413, 408, detachment),

415. $x+z = x \rightarrow v+x = (v+x)+z$
(from 411, 414, transitivity of implication),

416. $(v+x = y \wedge v+x = (v+x)+z) \rightarrow y = y+z$
(identity laws),

417. $v+x = y \rightarrow (x+z = x \rightarrow y = y+z)$

(from 415, 416, and the sentential calculus theorem

$(p \rightarrow q) \rightarrow ((r \wedge q \rightarrow s) \rightarrow (r \rightarrow (p \rightarrow s)))$),

418. $\bigvee v(v+x = y) \rightarrow (x+z = x \rightarrow y = y+z)$
(from 417, schema 40*),

419. $\bigwedge x, y \bigvee v(v+x = y) \rightarrow \bigwedge x, y(x+z = x \rightarrow y = y+z)$
(from 418, distribution of the quantifier over the implication),

420. $x+z = x \rightarrow y = y+z$
(from 404, 405, detachment),

421. $x+z = x \rightarrow \bigwedge y(y = y+z)$
(from 420, shifting of the universal quantifier $\bigwedge y$ to the consequent),

422. $\bigvee z(x+z = x) \rightarrow \bigvee z \bigwedge y(y = y+z)$
(from 421, distribution of the quantifier over the implication (axiom schema 6)),

423. $\bigwedge x \bigvee z(x+z = x) \rightarrow \bigwedge x \bigvee z \bigwedge y(y = y+z)$
(from 422, distribution of the quantifier over the implication (Axiom 1)),

424. $\bigwedge x \bigvee z \bigwedge y(y = y+z)$
(from 423, 407, detachment),

425. $\bigvee z \bigwedge y(y = y+z)$
(from 424, 0.7).

At the end of the proof we have an example of the elimination of the superflous quantifier $\bigwedge x$. It has been demonstrated about the element z (the existence of which was being proved) that it exists for every x, but the property which is being proved is independent of x and hence the quantifier $\bigwedge x$ ceases to bind any variable and may be disregarded on the strength of logical theorems for non-empty domains.

The above proof, even though it is not fully formalized, is very long. In an ordinary exposition of group theory it is, of course, outlined more briefly. We may, in an analogous way, prove the existence of a left-hand zero-element:

426. $\bigvee v \bigwedge y(v+y = y)$.

Next it can be proved easily that every left-hand zero-element is identical with every right-hand zero-element, so that there exists only one zero-element, which is both a right-hand and a left-hand one:

427. $\left(\bigwedge y(y = y+z) \wedge \bigwedge y(v+y = y) \right) \to z = v$.

For if $\bigwedge y(y = y+z)$, then in particular $v = v+z$, and if at the same time $\bigwedge y(v+y = y)$, then in particular $v+z = z$. Now, if $v = v+z$ and $v+z = z$, then $v = z$.

It follows from 426 and 427 that there exists only one right-hand zero-element, for should there be two such elements, then by 427 they would have to be identical with the left-hand zero-element, which exists under 426. Hence we have the theorem

428. $\bigwedge w, z\left(\bigwedge (y(y = y+z) \wedge \bigwedge y(y = y+w)) \to z = w \right)$.

We have thus proved about the right-hand zero-element that it exists (425) and that it is unique (428). In such a case we may introduce in a mathematical theory a separate constant symbol to be the name of that element, e.g., the symbol 0; concerning that symbol we adopt the following sentence as a new axiom:

429. $\bigwedge y(y = y+0)$.

The introduction, in group theory, of the term 0 and the adoption of Axiom 429 about that term is fully justified under the general principles of introducing constant terms in any theory based on the c.l.c. with identity. Those principles may be formulated with reference to any theory based on the c.l.c. if it is assumed, as was done in the preceding section, that the said system of the c.l.c. with identity and functions contains as predicate and function (term) symbols all the symbols (both primitive and introduced) of all the theories ever to be considered. Hence, by the Deduction Theorem, the c.l.c. contains, as implications, all the theorems of all the theories we are concerned with.

DEFINITION 28. *A name term a is correctly introduced in a theory T as a constant name of the only element that satisfies the formula α(x) if and only if*

1. *the sentences*

$$\bigvee x\alpha(x),$$

$$\bigwedge x, v\Big((\alpha(x)\wedge\alpha(v))\to x = v\Big)$$

are theorems in the theory T;

2. *the term a does not occur in those sentences and has not so far occurred in the theory T;*

3. *the term a and the axiom*

A. α(a)

are added to the theory T.

It may be said in accordance with the definition given above that the term 0 has been correctly introduced into the theory of the operation +, based on Axiom 404–406, as a constant name of the unique element which satisfies the formula

$$\alpha(x) = \bigwedge y(y = y+x).$$

Note that condition 3 above may be replaced by the following one:

3'. *the term a and the definition*

A'. $\bigwedge x(x = a \equiv \alpha(x))$

are added to the theory T.

The equivalence of the sentences A and A' in any theory which includes theorems on existence and uniqueness is obvious. The sentence A' resembles equivalence definitions of relations, but what is defined here is an equality (to *a*).

Intuitively, it seems quite understandable that if one and only one element of the domain under consideration by a given theory has a certain property, then that element may be given a separate name, and the sentence stating that the object bearing that name has the property in question may be accepted. New names are introduced very often in practical mathematical activity. Yet in metamathematical research every theory is usually considered as a theory of only its primitive terms, since this makes a description of that theory easier. Moreover, correctly introduced terms do not essentially increase the set of theorems of a given theory. As we shall see, a theorem on elimination, analogous to the theorem on elimination proved above for equivalence definitions, holds true.

Defining individual constants may, however, be treated as a special case of defining functional constants, an individual constant being a function of zero arguments. We shall accordingly examine now the definition of functional constants and shall prove for them a theorem on elimination, a special case of which will be the theorem on the elimination of individual constants.

Defining function constants

Let us return to group theory. The following theorems result from theorems on zero:

430. $\bigwedge x \bigvee z(x+z = 0)$
(from Axiom 405),

431. $\bigwedge x \bigwedge z, v((x+z = 0 \wedge x+v = 0) \rightarrow z = v)$.

PROOF. If $x+z = 0$ and $x+v = 0$, then $x+z = x+v$. By Axiom 406, for the element x there exists a w such that $w+x = 0$. The element w which has that property is now added to both sides of the equality $x+z = x+v$. This is a legitimate step under the Axiom Ex2 (p. 208). We thus obtain the equality $w+(x+z) = w+(x+v)$. This, by 404, yields $(w+x)+z = (w+x)+v$. But since $w+x = 0$, again by Ex2 we obtain that $0+z = 0+v$, and hence $z = v$.

Theorems 430 and 431 state that for every x there exists exactly one element which, when added to x, yields zero. That element is called the *complement of the element* x or the *element inverse to* x with respect to the operation $+$, and is denoted by $-x$. We thus introduce a new symbol, namely $-$, which is the function symbol for a function of one argument and is described by the equivalence

$$\bigwedge x, y(y = -x \equiv x+y = 0),$$

or by the shorter theorem:

$$x + (-x) = 0.$$

The defining of a function term in either way is legitimate under the principle of introduction of new functional terms, analogous to the principle of introduction of new constant terms, analysed above. The principle in question may formulated as a definition similar to Definition 28. It will be formulated for a function of n arguments, where n is an arbitrary natural number.

DEFINITION 29. *A function symbol F of n arguments is correctly defined in a theory T if and only if*:

1. *the sentences*

(150) $\quad \bigwedge x_1, \ldots, x_n \bigvee z \alpha(x_1, \ldots, x_n, z),$

(151) $\quad \bigwedge x_1, \ldots, x_n \bigwedge z, v\big((\alpha(x_1, \ldots, x_n, z) \wedge \alpha(x_1, \ldots, x_n, v)) \rightarrow$
$z = v\big)$

are theorems in the theory T;

2. *the term F does not occur in these sentences nor has it occurred so far in the theory T*;

3. *the term F and the definition*:

A. $\bigwedge x_1, \ldots, x_n, y\big(y = F(x_1, \ldots, x_n) \equiv \alpha(x_1, \ldots, x_n, y)\big)$

are joined to the theory T;

or:

3'. *the term F and the sentence*:

A'. $\bigwedge x_1, \ldots, x_n \alpha\big(x_1, \ldots, x_n, F(x_1, \ldots, x_n)\big)$

are added to the theory T.

It can easily be seen that in any theory T which includes the theorems on existence (150) and uniqueness (151) the sentences A and A' are equivalent, each is a consequence of the other, and, hence, a new functional constant may be introduced in either way, as has been illustrated above by the example of the definition of the inverse element.

THEOREM 20 (on the elimination of definitions of function constants). *If a function symbol F occurs neither in the sentences forming the set X, nor in sentences* (150) *and* (151), *nor in a sentence B, and if* $\{(150), (151)\} \subset$ $Cn_1(X)$, *then*

$$B \in Cn_1(X \cup \{A\}) \to B \in Cn_1(X).$$

PROOF. By the Deduction Theorem, there exists a finite set of sentences $E_1, \ldots, E_k \in X$ such that

$$\ulcorner (E_1 \wedge \ldots \wedge E_k \wedge A) \to B \urcorner \in L_1.|$$

By Theorem 17 (stating that logic does not distinguish functional constants from single-valued predicates), if in the theorem above the formula $y = F(x_1, \ldots, x_n)$ is replaced by $\alpha(x_1, \ldots, x_n)$, then that sentence is a consequence of the theorems on existence and uniqueness for the formula $\alpha(x_1, \ldots, x_n, y)$, that is,

(152) $\qquad \ulcorner ((E_1 \wedge \ldots \wedge E_k \wedge A) \to B)^\alpha_{\underline{A}} \urcorner \in Cn_1 \{(150), (151)\},$

where $\big((E_1 \wedge \ldots \wedge E_k \wedge A) \to B\big)^\alpha$ is the ¦sentence $(E_1 \wedge \ldots \wedge E_k \wedge A) \to B$ after the replacement of the formula $y = F(x_1, \ldots, x_n)$ by $\alpha(x_1, \ldots, x_n, y)$. But, by assumption, in these sentences only A contains the term F, and after the replacement mentioned above the sentence A becomes the following tautology:

$$\bigwedge x_1, \ldots, x_n, y \big(\alpha(x_1, \ldots, x_n, y) \equiv \alpha(x_1, \ldots, x_n, y)\big).$$

Hence

(153) $\qquad \ulcorner ((E_1 \wedge \ldots \wedge E_k \wedge A) \to B)^\alpha_{\underline{A}} \equiv ((E_1 \wedge \ldots \wedge E_k) \to B) \urcorner \in L_1.$

Formulae (152) and (153) yield

$$\ulcorner (E_1 \wedge \ldots \wedge E_k) \to B \urcorner \in Cn_1\big(\{(150), (151)\}\big).$$

Hence, in accordance with the ̇assumption of the theorem under consideration, $B \in Cn_1(X)$.

Thus, if the concepts being defined occur neither in conclusions nor in premises, they may always be eliminated from proofs.

On the occasion of discussing theorems on elimination it is worthwhile noting that sometimes it is stressed in mathematics that certain definitions are necessary for the proofs of certain theorems, even though the concepts being defined do not occur in those theorems. Strictly speaking, such cases ought to be explained differently. Usually, what is essential there is not so much a definition of a predicate as the use of a predicate in order, for instance, to single out in a set those elements which satisfy that predicate. Thus, the essential point is the use of a special case of the axiom on set construction or another axiom of a similar nature (the axiom on substitution in general set theory, the axiom on induction in the arithmetic of natural numbers, the axiom on continuity in the arithmetic of real numbers). The entire logical calculus can be formulated so that definitions play a creative role in it, but such an approach is both inconvenient in metalogical research and less common, and accordingly will not be discussed here.

In view of the theorems described above, definitions merely play a convenience in the construction of theories. This is why we shall avail ourselves of definitions in only those cases in which we are interested in the content of a theory and in which we carry out certain arguments in that theory. In those cases in which we are interested in a theory merely from the metamathematical point of view we shall usually consider that theory as containing no definitions. A rigorous description of a theory with definitions is much more complicated than a description of a theory without definitions. A theory with definitions must be treated as a sequence of increasingly comprehensive theories, such that every theory has one predicate symbol and one axiom more than the preceding one, the axiom in question being a definition of the given predicate symbol.

Conditional defining

In actual mathematical research definitions of functional constants often do not fall exactly under the schema presented above. Definitions of functions quite often are conditional definitions.

An important example of conditional definitions is provided by the theory of number fields. The field of real numbers is a special case of a field. The axioms of field theory yield the theorems on the conditional performability of division:

$$\bigwedge x, y(x \neq 0 \to \bigvee z(x \cdot z = y)),$$

and on uniqueness:

$$\bigwedge x, y, z, v((x \neq 0 \land x \cdot z = y \land x \cdot v = y) \to z = v).$$

Hence, we may correctly introduce, in a conditional way, the division symbol / and adopt the conditional definition axiom:

$$\bigwedge x, y\big(x \neq 0 \rightarrow (x \cdot (y/x) = y)\big),$$

or the conditional definition of the form:

$$\bigwedge z, x, y(x \neq 0 \rightarrow (z = y/x \equiv x \cdot z = y)).$$

This kind of definition falls under the following schema, which might call *conditional definition*:

DEFINITION 30. *A function symbol F of n arguments is correctly introduced in a conditional way into a theory T if and only if*:

1. *the sentences*

(154) $\quad \bigwedge x_1, ..., x_n(\beta(x_1, ..., x_n) \rightarrow \bigvee z\alpha(x_1, ..., x_n, z)),$

(155) $\quad \bigwedge x_1, ..., x_n \bigwedge z, v\big((\beta(x_1, ..., x_n) \wedge \alpha(x_1, ..., x_n, z) \wedge$

$\alpha(x_1, ..., x_n, v)) \rightarrow z = v\big)$

are theorems in the theory T;

2. *the term F does not occur in those sentences, nor has it occurred in the theory T so far*;

3. *the term F and the conditional definition*:

(156) $\quad \bigwedge x_1, ..., x_n, y\big(\beta(x_1, ..., x_n) \rightarrow$

$(y = F(x_1, ..., x_n) \equiv \alpha(x_1, ..., x_n, y))\big)$

or its equivalent sentence

(157) $\quad \bigwedge x_1, ..., x_n\big(\beta(x_1, ..., x_n) \rightarrow \alpha(x_1, ..., x_n, F(x_1, ..., x_n))\big)$

is added to the theory T.

As in the case of non-conditional definitions, the *theorem on elimination* can also be proved for conditional definitions:

THEOREM 21. *If a function symbol F occurs neither in the sentences belonging to a set X, nor in sentence* (154) *or* (155), *nor in a sentence B, and if* $\{(154), (155)\} \subset \mathrm{Cn}_1(X)$, *then*

$$B \in \mathrm{Cn}_1(X \cup \{(156)\}) \rightarrow B \in \mathrm{Cn}_1(X).$$

The proof consists in reducing this case to the theorem on elimination for non-conditional definitions.

We first extend the theory in question by adding to it a new constant a, about which nothing is assumed. Next we construct a definition of a new functional constant F'. Consider the sentential formula $\alpha'(X, y)$ which has the form:

(158) $\qquad \alpha'(X, y) = \big((\sim \beta(X) \wedge y = a) \vee (\beta(X) \wedge \alpha(X, y))\big)$,

where X denotes a complex of variables, $x_1, ..., x_n$.

The following sentences result from sentences (154) and (155):

(159) $\qquad \bigwedge X \bigvee y \alpha'(X, y)$,

(160) $\qquad \bigwedge X, y, z \big((\alpha'(X, y) \wedge \alpha'(X, z)) \to y = z\big)$.

We may accordingly, by Definition 29, correctly add to the theory under consideration both the new function term F' and the axiom

(161) $\qquad \bigwedge X, \alpha'(X, F'(X))$.

Note that, under (158), the sentence

(162) $\qquad \bigwedge X \big(\beta(X) \to \alpha(X, F'(X))\big)$,

which differs from (157) only by the replacement of F by F', is a consequence of (161).

If now $B \in \mathrm{Cn}_1(X \cup \{(157)\})$, then under the Deduction Theorem there exist $A_1, ..., A_k \in X$ such that

(163) $\qquad \big((A_1 \wedge ... \wedge A_k \wedge (157)) \to B\big) \in L_1$.

Hence, by Theorem 16 (on the fact that no difference between constants is made in logic),

(164) $\qquad \big((A_1 \wedge ... \wedge A_k \wedge (157)) \to B\big)' \in L_1$,

where the function $'$ means that all the occurrences of the symbol F have been replaced by the symbols F', also holds. But since neither the sentences belonging to the set X, nor the sentence B contain the symbol F, the function $'$ leaves them unchanged. But, in accordance with our remark concerning formula (162), we have $(157)' = (162)$. Hence (164) means the same as

(165) $\qquad \big((A_1 \wedge ... \wedge A_k \wedge (162)) \to B\big) \in L_1$.

Since $(162) \in \mathrm{Cn}_1(\{(161)\})$, hence it follows from (165) that

(166) $\qquad B \in \mathrm{Cn}_1(X \cup \{(161)\})$.

Since by Theorem 20 the term F' and its definition (161) may be eliminated from the proof of the sentence B, it results from (166) that $B \in \text{Cn}_1(X)$, which was to be proved.

In mathematics conditional definitions of functions are perhaps even more numerous than non-conditional definitions. The symbol of the concept is sometimes formulated with the statement that the symbol limit of a series lim $\{a_n\}$ in mathematical analysis is one of the most important mathematical symbols, defined conditionally. In the traditional approaches to mathematical analysis the conditional nature of that concept is sometimes formulated with the statement that the symbol lim $\{a_n\}$ may be meaningfully applied to convergent series only. By Definition 27, the concept of lim $\{a_n\}$ may be introduced correctly in the following way. First, for brevity's sake we adopt the convention that

$$\{a_n\} \Rightarrow a \equiv \bigwedge k \bigvee m \bigwedge n \left(n > m \to |a_n - a| < \frac{1}{k+1} \right).$$

Next we have to prove in analysis that

$$(\{a_n\} \Rightarrow a \wedge \{a_n\} \Rightarrow b) \to a = b.$$

When that is done, we may adopted the new term lim $\{a_n\}$ and to define it by the axiom

(167) $\bigwedge \{a_n\} (\bigvee a(\{a_n\} \Rightarrow a) \to (\{a_n\} \Rightarrow \lim \{a_n\}))$.

The condition $\beta(X)$ has here the form of the condition $\bigwedge a\{a_n\} \Rightarrow a$.

Theorem 21 states that the adding to mathematical analysis of sentence (167) which conditionally defines the term lim does not modify the set of those theorems of mathematical analysis which do not contain the term lim. The theorem on existence in this case has the form of the tautology

$$\bigvee a\{a_n\} \Rightarrow a \to \bigvee a\{a_n\} \Rightarrow a.$$

Instead of the conditional sentence (167) we may of course adopt a definition which has the form of an equivalence:

$$\bigwedge \{a_n\}, b(\bigvee a\{a_n\} \Rightarrow a \to (b = \lim \{a_n\} \equiv \{a_n\} \Rightarrow b)).$$

The situation is similar to the case of the definition of the operation of minimum in the arithmetic of natural numbers. There is at most one

250

least element in any set X, and if the set X is not empty, then a least element does exist. Hence we may adopt the conditional definition:

$$\bigwedge X, y\Big(\bigvee x\, x \in X \to \big(y = \mu x(x \in X) \equiv$$
$$(y \in X \wedge \bigwedge x(x \in X \to y \leqslant x))\big)\Big).$$

Definitions in set theory

Since set theory is being widely used as an instrument of extending mathematical theories, it is worthwhile considering certain types of definitions which occur in set theory. First of all, the concepts of power set 2^X, union $\bigcup X$, and intersection $\bigcap X$, mentioned in the Introduction, have conditional definitions. The axioms of the existence of power sets and of the existence of unions of sets are conditional axioms of existence. On the other hand, the axiom on extensionality provides for the uniqueness of these sets. Hence, in Zermelo's set theory we may adopt the following conditional definitions:

$$\bigwedge X\Big(Z(X) \to \big(Z(2^X) \wedge \bigwedge u(u \in 2^X \equiv Z(u) \wedge \bigwedge x(x \in u \to$$
$$x \in X))\big)\Big),$$

$$\bigwedge X\Big(\big(Z(X) \wedge \bigwedge y(y \in X \to Z(y))\big) \to \big(Z(\bigcup X) \wedge$$
$$\bigwedge u(u \in \bigcup X \equiv \bigvee v(u \in v \wedge v \in X))\big)\Big).$$

These definitions are of the type (157), and not of the type (156), although each of them contains an equivalence.

In set theory we often define constants which denote sets just by indicating which elements belong to a given set. The condition of existence is usually satisfied owing to the axiom on substitution or the axiom on set construction. On the other hand, uniqueness is guaranteed by the axiom of extensionality. The introduction of new constants by the axiom on set construction can be reduced to a general rule of introduction of defined constants by formulating the following theorem:

THEOREM 22. *If a is a constant in set theory such that the sentence $Z(a)$ is a theorem in set theory, then for every new function symbol F the following sentence*:

$$\bigwedge x, y, \ldots, z\Big(Z(F(y, \ldots, z)) \wedge \big(x \in F(y, \ldots, z) \equiv$$
$$(x \in a \wedge \alpha(x, y, \ldots, z))\big)\Big),$$

where $\alpha(x, y, ..., z)$ *is any formula which does not contain the symbol F being defined and hence may be added correctly to set theory as a definition.*

PROOF. If $Z(a)$ is a theorem in set theory, then under the axiom on subsets the following sentence also is a theorem in set theory:

$$(168) \qquad \bigvee u\Big(Z(u) \wedge \bigwedge x\big(x \in u \equiv (x \in a \wedge \alpha(x, y, ..., z))\big)\Big).$$

Let $\beta(y, ..., z, u)$ stand for the formula

$$\Big(Z(u) \wedge \bigwedge x\big(x \in u \equiv (x \in a \wedge \alpha(x, y, ..., z))\big)\Big).$$

Sentence (168) states that $\bigwedge x, y, ..., z \bigvee u \beta(x, y, ..., z, u)$. Thus, this sentence formulates the condition of existence. The condition of uniqueness follows from the axiom of extensionality in the following way. The formula $\beta(x, y, ..., z, u)$ falls under the schema

$$Z(u) \wedge \bigwedge x(x \in u \equiv \gamma).$$

Hence, the sentence

$$(\beta(x, y, ..., z, u) \wedge \beta(x, y, ..., z, w)) \rightarrow$$
$$\Big(\bigwedge x(x \in u \equiv \gamma) \wedge \bigwedge x(x \in w \equiv \gamma)\Big)$$

is a logical theorem, and so is

$$\Big(\bigwedge x(x \in u \equiv \gamma) \wedge \bigwedge x(x \in w \equiv \gamma)\Big) \rightarrow$$
$$\bigwedge x(x \in u \equiv x \in w).$$

On the other hand, the axiom on extensionality states that

$$\Big(Z(u) \wedge Z(w) \wedge \bigwedge x(x \in u \equiv x \in w)\Big) \rightarrow u = w.$$

Thus the c.l.c. theorems referred to and the axiom on extensionality taken together yield the sentence

$$\bigwedge x, y, ..., z, u, w\Big((\beta(x, y, ..., z, u) \wedge \beta(x, y, ..., z, w)) \rightarrow$$
$$u = w\Big),$$

which is the condition of uniqueness. By Definition 29, we may accordingly correctly introduce a new function symbol F by adopting the axiom

$$\bigwedge x, y, ..., z \beta(x, y, ..., F(x, y, ..., z)).$$

This axiom, which complies with the meaning of the formula β, is identical with that axiom referred to as added in Theorem 22.

10. ON DEFINITIONS

In view of the above, Definitions (40)–(42), (45), (46), (48), (49), (51), which occur in the Introduction, have been correctly added to set theory as presented in the Introduction.

Definitions by induction

Definitions by induction are a special case of definitions constructed by means of set theoretical concepts. From the set theoretical point of view, a definition by induction is a definition of the common part (intersection) of a family of sets. But definitions by induction are recorded in various ways. The basic variation occurs between the way in which definitions by induction are recorded in everyday practice of mathematicians and the way in which they are written down rigorously, in a formalized manner, using set-theoretical concepts. Ordinary, non-formalized definitions by induction have already been seen in the Introduction.

Every definition by induction is a definition of a set in which we can make a distinction between original elements and those obtained from the original ones later by repetition of certain operations. Hence, every definition by induction consists of two basic parts, which are:

1. initial conditions,
2. induction conditions.

The *initial conditions* state explicitly which objects belong to a given set. The *induction conditions* have a conditional form: on the assumption that certain objects $u_1, ..., u_n$ already belong to the set in question, they state what relation must exist between them and a new object x if that object also is to belong to that set. The elementhood of that new object may depend on the fact that a finite or an infinite number of some other objects belong to the set in question. We will be concerned below exclusively with definitions by induction of a special form, namely those in which the number of the objects $u_1, ..., u_n$ on which the elementhood of a new object x depends is finite.

The definition of finiteness as given in the Introduction was a definition by induction. Conditions 1 and 2 for the concept of finiteness may be formulated as follows:

1. The initial condition: *the empty set is finite*;
2. The induction condition: *if y is a finite set and if x differs from y by one element, then x is also a finite set.*

In Chapter I we have defined by induction the set of well-formed formulae. We shall soon return to that example. In the text below we shall also encounter many other definitions by induction.

The definition of finiteness, as given in the Introduction, was written down quite rigorously as one formula falling under the general schema

(169) $\qquad x \in F \equiv \bigwedge z\big((1(z) \wedge 2(z)) \rightarrow x \in z\big),$

where $1(z)$ stands for the original conditions, and $2(z)$, for the induction conditions. Schema (169) is a general set theoretical schema which covers all the rigorous formulations of those definitions by induction with which we shall now be concerned.

In order to introduce a term F in set theory very correctly we must use the axiom on set construction. For that purpose we must first have a theorem on the existence of a z such that $Z(z) \wedge 1(z) \wedge 2(z)$. If we have that theorem, then F can be singled out in z. Schema (169), as compared with formula (33) in the Introduction, states that F is a common part of the family of all those sets which satisfy the conditions $1(z)$ and $2(z)$. Should that family be empty, then it would be true for every x that it satisfies the implication describing elementhood in the set F, which obviously does not correspond to the intentions underlying definitions by induction.

Current formulations of definitions by induction usually state that

(170) $\qquad$ *A set F is the least set satisfying the original condition $1(z)$ and closed under the induction condition $2(z)$.*

We shall now analyse both formulations, (169) and (170), of definitions by induction in order to demonstrate that their exact meaning is the same. To simplify our reasoning we shall assume that the conditions 1 and 2 have a comparatively simple form:

1. The original condition specifies those objects which we assume are clearly given elements of a definite set F. The condition $1(z)$ in the simplest case has, thus, the form of a conjunction:

(171) $\qquad 1(z) \equiv a_1 \in z \wedge \ldots \wedge a_k \in z.$

2. The inductive condition $2(z)$ in the simplest case is an implication, e.g., of the form

(172) $\qquad 2(z) \equiv \bigwedge u_1, u_2\big(u_1, u_2 \in z \rightarrow s(u_1, u_2) \in z\big),$

where $s(u_1, u_2)$ is a function which defines a new element of the set F if u_1 and u_2 are elements of the set F.

The following theorem can easily be proved:

THEOREM 23. *If conditions* $1(z)$ *and* $2(z)$ *have the form of* (171) *and* (172), *respectively, then a definition which has the form of* (169) *is equivalent to the conjunction of the following three sentences*: $1(F)$, $2(F)$, *and* $\bigwedge z\big((1(z) \wedge 2(z)) \to F \subset z\big)$, *which state that the set* F *satisfies the conditions* 1 *and* 2 *and is included in every set* z *which satisfies conditions* 1 *and* 2.

Theorem 23 is considered a justification of the fact that formalized and ordinary inductive definitions are equivalent.

PROOF. We shall first demonstrate that sentences $1(F)$ and $2(F)$ are consequences of (169). In fact, on the assumption of (171) formula (169) has the form

$$x \in F \equiv \bigwedge z\big((a_1 \in z \wedge \ldots \wedge a_k \in z \wedge 2(z)) \to x \in z\big);$$

on substituting, e.g., a_1 for x we obtain an equivalence of the type

$$a_1 \in F \equiv \bigwedge z((a_1 \in z \wedge \ldots) \to a_1 \in z).$$

Its right side is a logical theorem; this proves the left side, and hence $a_1 \in F$. It is proved in a similar way that $a_2 \in F, \ldots, a_k \in F$, that is, by (171), $1(F)$. For proving the formula $2(F)$ we note that the sentence

$$(1(z) \wedge 2(z)) \to (u_1, u_2 \in z \to s(u_1, u_2) \in z),$$

derived from $(p \wedge q) \to q$ by the substitution $q/2(z)$ and by shifting the quantifier $\bigwedge u_1, u_2$ from the consequent to the position preceding the entire formula, is a logical theorem. This sentence and the theorem $(p \to (q \to r)) \to ((p \to q) \to (p \to r))$ yield

$$\big((1(z) \wedge 2(z)) \to u_1, u_2 \in z\big) \to \big((1(z) \wedge 2(z)) \to s(u_1, u_2) \in z\big).$$

Next we distribute the quantifier $\bigwedge z$

(173) $\bigwedge z\big((1(z) \wedge 2(z)) \to u_1, u_2 \in z\big) \to$

$\bigwedge z\big(1(z) \wedge 2(z) \to s(u_1, u_2) \in z\big),$

255

and then formulae (173) and (169) and the appropriate logical theorems yield the implication

$$u_1, u_2 \in F \to s(u_1, u_2) \in F,$$

which, by (172), means that $2(F)$.

Thus, if the conditions 1 and 2 have the form described above, then the formulae $1(F)$ and $2(F)$ are consequences of Definition (169).

The fact that the set F is included in every set z which satisfies the conditions $1(z)$ and $2(z)$ also follows immediately from Definition (169). It is so because the implication

(174) $\qquad \bigwedge x \big(x \in F \to \bigwedge z \big((1(z) \wedge 2(z)) \to x \in z \big) \big)$

is a consequence of (169). Now (174) may be transformed as follows:

$$\bigwedge z, x \big(x \in F \to \big((1(z) \wedge 2(z)) \to x \in z \big) \big),$$

(175) $\qquad \bigwedge z, x \big((1(z) \wedge 2(z)) \to (x \in F \to x \in z) \big),$

$$\bigwedge z \big((1(z) \wedge 2(z)) \to \bigwedge x (x \in F \to x \in z) \big).$$

The last sentence under (175) states that the set F is included in every set z that satisfies the conditions 1 and 2.

Note also that the sentence

$$\bigwedge z \big((1(z) \wedge 2(z)) \to x \in z \big) \to \big((1(F) \wedge 2(F)) \to x \in F \big)$$

is a logical theorem; hence, the same holds for the sentence

$$1(F) \wedge 2(F) \to \big(\bigwedge z \big((1(z) \wedge 2(z)) \to x \in z \big) \to x \in F \big),$$

which shows that the implication

(176) $\qquad \bigwedge z \big((1(z) \wedge 2(z)) \to x \in \bar{z} \big) \to x \in F,$

converse to (174), is a consequence of the sentences $1(F)$ and $2(F)$. As demonstrated above, (174) is equivalent to the last implication of (175). Implications (174) and (176) taken together yield equivalence (169). Hence, the statement that the set F satisfies the conditions 1 and 2 and is included in every set satisfying the conditions 1 and 2 is equivalent to the statement that the set F is defined in the manner of (169). Sentence (170), stating that the set F is the least of all the sets satisfying conditions 1 and 2, means exactly this that the set F itself satisfies the conditions 1 and 2 and is included in every set z satisfying those conditions. Thus, the ordinary formulations of inductive definitions of type (170) are equivalent to the formulations of the form (169). Hence, by Theorems 20 and 22, inductive definitions of type (170) do not increase the set of those theorems which do not contain the term defined by those definitions. Accordingly, such definitions may be used without the risk of contradiction, provided only that there exists a z such that $Z(z)$, $1(z)$, and $2(z)$.

The set-theoretical definition of the set of natural numbers provides a simple example of inductive definition. In the Introduction we have defined natural numbers without any direct reference to induction, but to define them we have used an inductively defined concept of finiteness. But it is possible to proceed in a different way. We may first define the number 0 and the successor relation, and next define the set of natural numbers as the least set containing 0 and closed under the successor operation. 0 and successor are defined as follows:

$$x \in 0 \equiv x \in 2^A \wedge x \text{ eqnum } A,$$
$$x \in S(u) \equiv \big(x \in 2^A \wedge u \in CN \wedge \bigvee y, z \big(y \in u \wedge z \in A \wedge$$
$$\sim (z \in y) \wedge (y \cup \{z\}) \text{ eqnum } x \big) \big).$$

The set A is an infinite set the existence of which has been assumed in the Introduction. 0 is the set of the empty subsets of the set A, and the successor $S(u)$ is the set of those subsets of A which have one element more than have the subsets which belong to the cardinal number u.

The inductive definition of the set $\mathcal{N}$ of natural numbers may accordingly take on the following form:

$$(177) \qquad x \in \mathcal{N} \equiv x \in 2^{2^A} \wedge \bigwedge z \big(\big(0 \in z \wedge \bigwedge u (u \in z \rightarrow S(u) \in z) \big) \rightarrow$$
$$x \in z \big).$$

As can be easily proved, this definition is equivalent to Definition (48) in the Introduction. It falls under Schema (169). Conditions 1 and 2 have a very simple structure:

$$1(z) \equiv 0 \in z,$$
$$2(z) \equiv \bigwedge u (u \in z \rightarrow S(u) \in z).$$

It follows from the general considerations presented above that Definition (177) has as consequences the sentences: $0 \in \mathcal{N}$, $\bigwedge u (u \in \mathcal{N} \rightarrow S(u) \in \mathcal{N})$, and

$$\bigwedge z \big(\big(0 \in z \wedge \bigwedge u (u \in z \rightarrow S(u) \in z) \big) \rightarrow \mathcal{N} \subset z \big).$$

257

Note that the above sentence is the axiom on induction which we have adopted in the non-elementary arithmetic of natural numbers as characteristic of the natural numbers. In set theory that axiom is a trivial conclusion from the inductive definition of the set of natural numbers. In many cases proofs by induction may be interpreted as applications of inductive definitions.

We shall now quote a few examples of inductive definitions drawn from non-elementary arithmetic. The *less than* relation as holding between natural numbers may be defined as follows. In set theory the *less than* relation is, of course, treated as a set of pairs. Since 0 is the least number, hence all the pairs $\langle 0, S(x) \rangle$ obviously belong to the *less than* relation. This property may be taken as the original condition of the definition by induction under consideration. The induction condition may be formulated as follows:

$$\langle x, y \rangle \in z \rightarrow \langle S(x), S(y) \rangle \in z.$$

For if $x < y$, then $S(x) < S(y)$. In its complete formulation the definition in question is as follows:

$$x < y \equiv \bigwedge z \Big(\big(\bigwedge x (\langle 0, S(x) \rangle \in z) \wedge \bigwedge x, y (\langle x, y \rangle \in z \rightarrow \langle S(x), S(y) \rangle \in z) \big) \rightarrow \langle x, y \rangle \in z \Big).$$

Under the general comments made in this section, this definition has the following sentences as its consequences:

$$\bigwedge x(0 < S(x)),$$

$$\bigwedge x, y(x < y \rightarrow S(x) < S(y)).$$

The *less than* relation as a set of pairs is the least relation that satisfies these two conditions.

In non-elementary arithmetic, the relations of addition and multiplication, are also defined by induction. The original condition states what the result is of adding the number 0 to a given number y. The induction condition states what the result is of adding the successor $S(x)$ if it is known what the result is of adding the predecessor x. The operation of addition may be interpreted set-theoretically as a relation of three arguments, that is, as the set of triples which satisfy the following conditions:

1. $\langle 0, y, y \rangle \in z$,

2. $\bigwedge x, u(\langle x, y, u \rangle \in z \rightarrow \langle S(x), y, S(u) \rangle \in z)$.

The mathematical meaning of these conditions is as follows:

1. $0 + y = y$,

2. $x + y = u \rightarrow S(x) + y = S(u)$.

Usually these conditions are written down more briefly:

(178)   1. $0 + y = y$,

        2. $S(x) + y = S(x + y)$.

Condition 2 of the brief notation is equivalent in the classical logical calculus with identity to condition 2 written in the form of an implication. The original condition 1 states that the addition of the number 0 does not change the number y to which 0 is added. The induction condition 2 states that the addition of a successor $S(x)$ increases the sum obtained by the addition of the predecessor x by one only. This defines the addition of any number x to a number y. The complete definition is as follows:

$$(179) \quad + (t, y, s) \equiv \bigwedge z \Big(\big(\langle 0, y, y \rangle \in z \wedge \bigwedge x, u(\langle x, y, u \rangle \in z \rightarrow$$
$$\langle S(x), y, S(u) \rangle \in z) \big) \rightarrow \langle t, y, s \rangle \in z \Big).$$

In accordance with the general comments on definitions by induction, (179) has as its consequences the sentences:

(180)   1. $+ (0, y, y)$,

        2. $+ (x, y, u) \rightarrow + (S(x), y, S(u))$.

(180) and Definition (179), by the axiom on mathematical induction, taking the entire domain to be the natural numbers, make it possible to deduce theorems on existence and uniqueness:

$$\bigwedge x, y \bigvee u + (x, y, u),$$
$$\bigwedge x, y, u, w \big((+ (x, y, u) \wedge + (x, y, w)) \rightarrow u = w \big).$$

We can, thus define addition correctly as a function, in accordance with the rule of defining functions:

$$\bigwedge x, y, u(u = x + y \equiv + (x, y, u)),$$

or more briefly:

(181) $\bigwedge x, y + (x, y, x+y)$.

Sentences (178), which are commonly treated as the definitional conditions describing addition by induction, can easily be deduced from sentences (180) and (181).

Likewise, multiplication is defined by induction by reference to addition:

$$\times (t, y, s) \equiv \bigwedge z \Big(\big(\langle 0, y, 0 \rangle \in z \wedge \bigwedge x, u (\langle x, y, u \rangle \in z \rightarrow$$
$$\langle S(x), y, u+y \rangle \in z) \big) \rightarrow \langle t, y, s \rangle \in z \Big).$$

On the introduction of a functional multiplication symbol this definition yields the sentences:

(182)
1. $0 \cdot y = 0$,
2. $S(x) \cdot y = (x \cdot y) + y$,

which are usually taken as the definitional conditions describing multiplication. In fact, condition 1 states what the result is of multiplying by 0, while condition 2 states what the result is of multiplying by a successor $S(x)$ if we know how to multiply by its predecessor.

Complete formal statements of inductive definitions always contain set-theoretical concepts.

It is worthwhile mentioning that if only the concept of the successor function is adopted as the original arithmetic concept, it is not possible to define addition and multiplication, without referring to set-theoretical concepts. Thus, in the elementary arithmetic of natural numbers the successor function alone does not suffice as the primitive concept. Nor do the successor function and the operation of addition make it possible to define the operation of multiplication. It is only the operations of multiplication and addition or the successor function and the concept of divisibility which make it possible to define all the functions usually analysed in arithmetic.

Quite often, in order to do without set-theoretical concepts we introduce into arithmetic a special schema of definition by induction. The schema states that

10. ON DEFINITIONS

(183)   *If g and h are functions in a given theory, then we may add to that theory a new function symbol f and assume the following axioms about it:*

1. $f(0, y_1, \ldots, y_n) = g(y_1, \ldots, y_n)$,
2. $f(S(x), y_1, \ldots, y_n) = h(x, y_1, \ldots, y_n, f(x, y_1, \ldots, y_n))$.

Formulae (178) and (182) are special cases of schema (183). Thus, in the arithmetic of natural numbers which contains only 0 and the successor function we can define addition, and then multiplication, by making use of the above schema of definition by induction. Addition and multiplication have already been defined above by formulae falling under schema (169). It can be proved in general that every function defined by schema (183) can also be defined by a set-theoretical definition falling under schema (169).

EXERCISES

1. Prove the equivalence, in set theory, of two definitions of natural numbers: Definition (177) in this section and Definition (48) in the Introduction.

2. Prove by induction, on the strength of formulae (180), the performability and the uniqueness of result of the operation of addition.

H i n t. Formulae (180) suffice to prove the existence of the result of that operation, but to prove the uniqueness of that result it is convenient first to define the set:

$$\langle t, y, s \rangle \in U \equiv \bigvee X \Big(\langle 0, y, y \rangle \in X \wedge \langle t, y, s \rangle \in X \wedge \bigwedge x \big(x < t \to$$
$$\bigwedge u (\langle x, y, u \rangle \in X \equiv \langle S(x), y, S(u) \rangle \in X) \big) \wedge$$
$$\bigwedge x, y, u, w (\langle x, y, u \rangle \in X \wedge \langle x, y, w \rangle \in X \to u = w) \Big).$$

The above is defining by induction, as it were, by successive stages. Next, it is proved that the set U satisfies conditions (180). This yields the implication: $+ (t, y, s) \to \langle t, y, s \rangle \in U$. The uniqueness of the result of the operation being defined follows easily.

3. Prove the same for multiplication and for any function f defined by schema (183). Make use of the method of succesive stages, which in the case of functions yields definitions equivalent to definitions that fall under schema (169).

4. Prove that the consequence function can be defined as follows:

$$x \in \mathrm{Cn}_0(X) \equiv \bigwedge z \Big(\big(X \subset z \wedge \bigwedge A, B((A \in z \wedge \ulcorner A \to B \urcorner \in z) \to$$
$$B \in z) \big) \to x \in z \Big).$$

Prove that Definition 20 in the preceding chapter is equivalent to the definition given above. Try to apply the above definition to inductive proofs of theorems on deduction, etc.

I. CLASSICAL LOGICAL CALCULUS

Formulated in an everyday language, the set of consequences is the least set containing the axioms and closed under the operation of detachment.

5. Demonstrate that the addition of the symbol Λ of the empty set and of the axiom: $Z(\Lambda) \wedge \bigwedge y \sim (y \in \Lambda)$, does not increase the set of those theorems of set theory which do not contain the symbol Λ.

Chapter II

MODELS OF AXIOMATIC THEORIES

The preceding chapter was concerned with the logical calculus, used by mathematicians in constructing various axiomatic theories.

Mathematical theories are constructed to serve various purposes. Some of them are constructed with the intention of grasping certain properties common to many mathematical domains. This applies, for instance, to group theory and to lattice theory. But there are many mathematical theories which have been constructed with the intention of describing one mathematical domain only. This applies, for instance, to the principal theories in classical mathematics: the arithmetic of natural numbers and the arithmetic of real numbers. When building an axiom system of the arithmetic of natural numbers we want it to express certain properties of natural numbers which we consider characteristic. We want the axioms of the arithmetic of natural number to be satisfied only by the domain of natural numbers. Likewise, when building an axiom system of the theory of real numbers we want to find axioms for addition and multiplication which would be satisfied only by addition and multiplication in the domain of real numbers. On the other hand, we have no such intention, for instance, when we construct group theory. The axioms of group theory are satisfied both by the addition of integers and by the addition of reals. They are also satisfied by the multiplication of rationals other than zero. They are further satisfied by many other operations whose definitions are much more intricate, such as the symmetric difference of two sets

$$A \div B = (A - B) \cup (B - A)$$

among subsets of a set; the same holds for the operation $\oplus$, defined by a matrix in Section 2 of the Introduction, in the set consisting of three numbers: 0, 1, 2. Each of these domains satisfies the axioms of group theory, that is a *model* of group theory.

That branch of mathematics which will be discussed now is accordingly called the *theory of models* and is concerned with relations between mathematical theories and their models, that is, with domains which satisfy the axioms of a given theory. The study of this branch of mathematics will begin with an analysis of the concepts of satisfaction and model and with rigorous definitions of these concepts. Next, the most important theorem in model theory, namely that which states that every consistent theory has a denumerable model, will be proved. Finally, it will be demonstrated that every theory has many quite different models. A theory constructed with the intention of describing one selected mathematical domain never fulfills that task. There are always many mathematical domains which satisfy the axioms of a given theory. Thus, the arithmetic of natural numbers does not describe the natural numbers only. The same applies to any classical mathematical theory. Hence the study of models sheds interesting light on the role and the epistemological value of mathematical theories.

1. THE CONCEPT OF SATISFACTION

The theory of satisfaction and models is sometimes called *logical semantics*. Semantics, in general, is the discipline which analyses the connections between a language and that reality to which the expressions of that language refer. In metamathematics, the relation between language and reality is interpreted as the connection between a mathematical theory and that domain which that theory describes. The basic relation which holds between the formulae of a theory and mathematical domains is that of *satisfaction*. The intuitive meaning of that relation can be inexactly, though convincingly, described as follows.

The intuitive concept of satisfaction and truth

If the formula $\alpha(x, y, ..., z)$ with the free variables $x, y, ..., z$ contains predicate symbols $P, Q, ..., R$ and function symbols $F, G, ..., H$, then it is said that the formula $\alpha(x, y, ..., z)$ *is satisfied* in the domain $\langle X, p, q, ..., r, f, g, ..., h \rangle$ by the elements $a, b, ..., c$—when the constants $P, Q, ..., R; F, G, ..., H$ are interpreted as the names of the relations $p, q, ..., r$ and the functions $f, g, ..., h$, respectively, if and only if the elements $a, b, ..., c$ are in the set X and if those elements, together with the relations $p, q, ..., r$ and the functions $f, g, ..., h$ *behave in the set X as the formula $\alpha(x, y, ..., z)$ states*, with the proviso that $x, y, ..., z$ are respectively names of the elements $a, b, ..., c$, the predicates $P, Q, ..., R$ are names of the relations $p, q, ..., r$, and the function symbols $F, G, ..., H$ are names of the functions $f, g, ..., h$.

For instance, the numbers 2, 3, 5 satisfy the formula $x \oplus y = z$ in the domain of natural numbers $\langle \mathcal{N}, + \rangle$, when the symbol $\oplus$ is interpreted as a name of the function of addition, since those numbers behave as the sentence $x \oplus y = z$ states, if it is assumed that x, y, z are, respectively, names of the numbers 2, 3, 5, and $\oplus$ is a name of addition. In fact, it is true that $2 + 3 = 5$. Likewise, 2, 3, 6 satisfy the same formula in the domain $\langle \mathcal{N}, \times \rangle$ if $\oplus$ is interpreted as a name of multiplication, since $2 \cdot 3 = 6$. The concept of satisfaction for more complex formulae is given in an analogous way. For instance, the formula $\bigwedge x(x > y \rightarrow x > z)$, with two free variables y and z, is satisfied by the objects 9 and 8 in the domain $\langle \mathcal{N}, > \rangle$, if the sign $>$ is interpreted as a symbol of the *greater than* relation, since it is true that, for every natural number x, if x is greater than 9, then x is greater than 8. Likewise, the formula $\bigvee x(y < x \wedge x < z)$, with the free variables y and z, is satisfied by the numbers 1 and 2 in the domain of real numbers $\langle \mathcal{R}, < \rangle$, if $<$ is interpreted as the sign of the *less than* relation, since the real numbers 1 and 2 behave as the formula $\bigvee x(y < x \vee x < z)$ states, namely, there exists a real number x such that $1 < x$ and $x < 2$. For instance, $1\frac{1}{2}$ is a number such that $1 < 1\frac{1}{2}$ and $1\frac{1}{2} < 2$. On the other hand, the same formula is not satisfied by the same numbers 1 and 2 in the domain of natural numbers $\langle \mathcal{N}, < \rangle$, since there is no natural number intermediate between the numbers 1 and 2.

If a formula α does not contain any free variables, then the concept of satisfaction may be applied to it as well. It suffices then to say that the formula α is satisfied in the domain $\mathfrak{M}$ when the constant symbols are interpreted as names of the corresponding predicates and functions, and there is no necessity to add which objects satisfy the formula α in that domain, since no variables occur in α as free. Thus, for instance, the formula $\bigwedge x, y(F(x, y) = F(y, x))$ is satisfied in the domain of natural numbers $\langle \mathcal{N}, \times \rangle$ when the symbol F is interpreted as a name of multiplication, since multiplication in that domain is a commutative operation. Likewise, the axioms of group theory, which are closed sentences, are satisfied in the domain $\langle \mathcal{R}, + \rangle$, because the addition of real numbers behaves as the axioms of group theory state. If a formula A does not contain any free variables and is satisfied in the domain

265

$\mathfrak{M}$, this fact is conveyed by the statement that the formula A *is true* in the domain $\mathfrak{M}$. Thus, the axioms of group theory are true for addition in the domain of real numbers. Likewise, the sentence

$$\bigwedge x, y \big(R(x, y) \rightarrow \bigvee z(R(x, z) \wedge R(z, y)) \big)$$

is true in the domain of real numbers $\langle \mathscr{R}, < \rangle$, when the symbol R is interpreted as a name of the *less than* relation, since it is true that

$$\bigwedge x, y \big(x, y \in \mathscr{R} \rightarrow (x < y \rightarrow$$
$$\bigvee z(z \in \mathscr{R} \wedge x < z \wedge z < y)) \big).$$

The concept of truth of a sentence in a domain was used in the foregoing chapter (Definition 15*), but no rigorous definition was given. The above examples show that truth is a special case of satisfaction, namely, it is identical with satisfaction when the formula under consideration is a sentence, that is, does not contain any free variables. This interpretation of the concept of truth and satisfaction agrees with the classical philosophical concept of truth as agreement of our opinions (statements) with reality, that is, with the state of things in a domain to which our opinions (statements) refer.

The language studied versus metalanguage

In order to develop a correct notion of the role of the concepts of satisfaction, truth, and model, which are soon to be defined rigorously, we have to make a distinction between the language studied and the language in which the study is being carried out, that is, the *metalanguage*.

The present chapter iš concerned with the classical logical calculus. Hence, the language studied will be the language of the c.l.c. as described in the preceding chapter. It is assumed that all constant mathematical symbols have been taken into account. Our intention is to formulate certain theorems about that language and about sentences formulated in it. Those theorems and their proofs must also be recorded as certain formulae in a more or less formalized language. The language in which we intend to formulate and prove various theorems concerning formulae of the c.l.c. shall be called the *metalanguage*. Some logicians call the language studied the *object language*, and the language in which research on the *object language* is done, the *subject language*. The subject language is the language used by a man (subject) who is studying some other language. In special cases the language studied may be the same language in which the research is being carried out.

The subject language, if it is to serve metalogical research, must most importantly contain names of formulae in the object language: names of single formulae, names of classes of formulae, names of relations between formulae. Terms pertaining to the formulae in the language studied form the first group of terms which must occur in the subject language. If, when using the subject language, we want to study not only the syntax of the language under consideration, that is, if we want not only to describe the shapes of and relations between the formulae in that language, but also want to study the relations between the shapes of formulae and that mathematical reality to which these formulae refer, then the subject language must also contain certain terms referring to that mathematical reality. If we want to discuss various mathematical domains, then the subject language must contain all the terms of set theory. If it does, it thereby contains the terms of the arithmetic of natural numbers, of the arithmetic of real numbers, and of the whole of classical mathematics. Thus, the second group of terms in the subject language is that of concepts of mathematics and set theory.

Although the terms classed as group one refer to somewhat vaguely defined things, namely formulae, these terms can be made quite precise. It may be assumed that apart from the formulae existing empirically as certain inscriptions on paper, on the blackboard, etc., there exist inscriptions as geometrical entities. The concepts of equiformity, part, and succession can refer to those ideal inscriptions with precision. As shown in the preceding chapter (relations (84)), these three concepts suffice to describe the various syntactic relations between formulae. The use of the concept of quotation marks can be dispensed with, since the formulae in the language studied may consist of a finite number of basic signs, with which all other single signs are equiform.

The second group consists of the rigorous apparatus of set theory. In this way all the analyses in the subject language may be formulated as a formalized theory which contains as its primitive terms the concepts (84) from Chapter I and the concepts of set theory. This theory will not be described in any rigorous way, since we shall not be concerned with its formalization. That theory is termed *metalogic* or *meta-*

mathematics. Analyses in terms of that theory will be done in a rather intuitive way.

Now that we have made a general distinction between the language studied and its metalanguage we shall consider a couple of examples. The sentence:

(1)　*The formula $\bigwedge x, y(x \oplus y = y \oplus x)$ is satisfied in the domain $\langle \mathcal{N}, + \rangle$ if the function symbol $\oplus$ is interpreted as a name of the function of addition $+$*

has the same meaning as the following sentence:

(2)　$\bigwedge x, y(x, y \in \mathcal{N} \to x + y = y + x).$

In the subject language, that is, the metalanguage, sentences (1) and (2) are equivalent. To put it more precisely, the equivalence of sentences (1) and (2) is a theorem in metalogic. Both sentences are formulated in metalogic. Sentence (1) contains a part which has the form

(3)　$\bigwedge x, y(x \oplus y = y \oplus x).$

Formula (3), contained in sentence (1), belongs to the language studied. It is a sentence in the language studied but in the metalanguage it is not a sentence, but a name of a sentence of the language studied. In order to bring out the fact that a formula equiform with (3) and contained in sentence (1) plays there only the role of a name of a formula we might place that formula in quotation marks. This, however, is not necessary since we have preceded it with the term *formula.* It is natural that the term *formula* is followed by a name of a formula or a description of a formula. In fact, sentence (1) is equivalent to the sentence:

(4)　*Formula (3) is satisfied in the domain $\langle \mathcal{N}, + \rangle$ if the symbol $\oplus$ is interpreted as a name of the function $+$.*

Sentence (4) contains a different name of the same formula (the same up to equiformity) than in sentence (1). The formula equiform with (3), which occurs in (1), plays the role of a name of itself. It is often assumed in metamathematics that certain formulae are names of themselves and of all other formulae which are equiform with them. If in a certain occurrence a formula is a name of itself, it is said that it occurs there *in material supposition.* The formulae which follow the nouns:

268

formula, sentence, symbol, expression, always occur in material supposition. The formulae which follow sentential connectives, the connective
that and the phrase *it is true that* belong to the subject language, are
sentences in the subject language. They occur, it is usually said, *in
simple supposition,* and not in material supposition. Likewise, the symbols which follow such nouns as: *function, relation, property* occur
in simple supposition, belong to the subject language and denote in it
certain function, relations, and properties, and not symbols. Thus, for
instance, in sentence (4) the symbol ⊕ occurs in material supposition
and denotes a symbol, namely itself and all symbols equiform with it.
On the contrary, the symbol +, which occurs after the term *function*
is used in simple supposition and denotes the function of addition.
The best way of checking whether we are dealing with the occurrence
of a symbol or formula in simple supposition or in material supposition is by trying to replace that symbol or formula by other formulae.
If an inscription, when replaced by the phrase: *equiform with …, a formula equiform with, …,* yields a meaningful sentence, then that inscription is certainly used in material supposition and denotes a formula
in the language studied. If such a replacement does not yield a meaningful sentence, the inscription occurs in simple supposition. For
instance, the phrase: *the symbol* ⊕ can in sentence (4) be meaningfully
replaced by the phrase: *a symbol equiform with the function symbol occurring in formula* (3). Hence ⊕ occurs in (4) in material supposition. Even
if formula (3), which occurs as part of sentence (1), were equiform with
sentence (2), it would occur in sentence (1) in material supposition,
and in sentence (2), in simple supposition.

In sentence (1), the part which is equiform with formula (3) may be
replaced by any term or term variable, since that part of (1), which
occurs in material supposition, is a name of a formula. By substituting
a term for another term we obtain a formula which also is meaningful,
even though it may be false. By substituting the variable A, which ranges
over the set of formulae, for the said part of (1) we obtain the predicate:

(5)   *The formula A is satisfied in the domain $\langle \mathcal{N}, + \rangle$ if the func-
        tion symbol ⊕ is interpreted as a name of the function of
        addition.*

Now (5) contains one free variable, is a predicate in metalanguage,
and defines a property of formulae of the c.l.c., that is, it designates

a subset in the set of all formulae. Definite substitutions in this predicate are equivalent to arithmetical statements. For instance, the following sentence, which is a substitution of predicate (5):

> The formula $\bigvee x(x \oplus x = x)$ is satisfied in the domain $\langle \mathcal{N}, + \rangle$ if the symbol $\oplus$ is interpreted as the sign of addition,

is equivalent to the arithmetical statement:

$$\bigvee x(x \in \mathcal{N} \wedge x+x = x).$$

Likewise, the metalogical statement (1) is equivalent to the arithmetical statement (2). On the other hand, the entire predicate (5) is not equivalent to any arithmetical predicate and has a metalogical meaning of its own; the property which it defines cannot be defined by any arithmetical formula, because in predicate (5) the free variable ranges over formulae, and in arithmetical statements free variables range over numbers. The relation of satisfaction links objects which are remote from one another; it is a relation between a formula in the language studied, on the one hand, and a mathematical domain and a sequence of objects from that domain, on the other.

A rigorous definition of the concept of satisfaction

We shall now proceed to formulate a rigorous definition of the concept of satisfaction. To describe it clearly we have to give more precision to the language of logic. For that purpose it is convenient to number both the variables and the predicate symbols. The following symbolism will be adopted:

Each variable consists of the letter x and a number of subscripted strokes:

$$x_{I}, x_{II}, x_{III}, \ldots;$$

these variables will respectively be abbreviated by:

$$x_1, x_2, x_3, \ldots.$$

Each predicate symbol consists of the letter P and a number of superscripted and subscripted strokes:

$$P_{I}', P_{II}', P_{III}', \ldots, P_{I}'', P_{II}'', P_{III}'', \ldots, P_{I}''', P_{II}''', \ldots;$$

these symbols will be abbreviated into:

$$P_1^1, P_2^1, P_3^1, ..., P_1^2, P_2^2, P_3^2, ..., P_1^3, P_2^3, ..., P_m^n, ...$$

Analysis will for the time being be confined to logic without functions.

The convention is adopted that the superscript indicates the number of arguments of a given predicate. This symbolism ensures great precision of formulation, but encumbers formulae with indices. It will, therefore, be set aside whenever there is no danger of misunderstanding.

A rigorous definition of a well-formed formula is as follows (formulated in more than one sentence because of its length).

DEFINITION 1. *The set W of well-formed formulae is the least of those sets Z which satisfy the following conditions*:

1. I n i t i a l   c o n d i t i o n. *Every atomic formula*
$$P_m^n(x_{k_1}, ..., x_{k_n}),$$

that is, one which consists of a predicate symbol P_m^n and parentheses containing the number of variables indicated by the superscript n on the letter P, is an element of the set Z.

2. I n d u c t i o n   c o n d i t i o n. *If A, B $\in$ Z, then the formulae*
$$(A \rightarrow B), \quad \sim A, \quad (A \lor B), \quad (A \land B), \quad (A \equiv B),$$
$$\bigwedge x_k A, \quad \bigvee x_k A,$$

also are in Z.

The concept of satisfaction applies to all well-formed formulae. Since formulae may contain different numbers of variables, in order to make our procedure uniform we shall associate with the sequence of all variables

$$x_1, x_2, x_3, ...$$

a sequence of objects

$$\{a_n\} = \{a_1, a_2, a_3, ...\}$$

which belong to the domain under consideration. Thus the variable x_n, if it occurs free in α, is considered to be a name of the object a_n. Instead of saying that the formula $\alpha(x_k, x_l, ..., x_m)$, with the free variables $x_k, x_l, ..., x_m$, is satisfied by the objects $a_k, a_l, ..., a_m$, we may say briefly: the formula A is satisfied by the sequence $\{a_n\}$. The sequence $\{a_n\}$ will be called a *valuation*. It associates the variables with objects from a certain domain, that is, it imparts definite values to the variables. If the domain in question is a set of numbers, then the sequence $\{a_n\}$ establishes a valuation of the variables by numbers. The formulations: a formula A is satisfied by a sequence $\{a_n\}$ in a domain $\mathfrak{M}$; a formula A is satisfied in a domain $\mathfrak{M}$ for a valuation $\{a_n\}$; a sequence $\{a_n\}$ satisfies A in $\mathfrak{M}$ — will be used alternately.

Now come the final explanations which make the formulation of the definition easier. Since for more complicated formulae satisfaction is defined by induction with reference to satisfaction as defined for the component formulae, and the component formulae do not contain more predicate constants than does the compound formula as a whole, it may be assumed that we define satisfaction for those formulae only which contain a certain fixed and small number of predicate symbols $P_1, ..., P_n$.

Thus it may also be assumed that we define satisfaction in a fixed mathematical domain $\langle X, p_1^{t_1}, ..., p_n^{t_n} \rangle$ and for a pre-established interpretation of the predicates $P_1, ..., P_n$ as names of the relations $p_1^{t_1}, ..., ..., p_n^{t_n}$. The index t_i indicates the number of the arguments of a given predicate P_i and of the corresponding relation p_i. An inductive definition defines satisfaction first for the atomic formulae (initial condition) which satisfy the above predicate symbols, and next for all other compound formulae (induction condition). For brevity, the fixed domain $\langle X, p_1^{t_1}, ..., p_n^{t_n} \rangle$ is symbolized by $\mathfrak{M}$. It is assumed about each valuation that it is a certain sequence of objects from the set X. The domain $\mathfrak{M}$ and the interpretation of the terms $P_1^{t_1}, ..., P_n^{t_n}$ as names of the relations $p_1^{t_1}, ..., p_n^{t_n}$ are assumed to be fixed throughout the present section.

DEFINITION 2 (Tarski [92]).

1. Initial condition. *An atomic formula*

(6) $\qquad P_i^t(x_{k_1}, ..., x_{k_t})$

is satisfied in the domain $\mathfrak{M}$ for the valuation $\{a_n\}$ if and only if

(7) $\qquad p_i^t(a_{k_1}, ..., a_{k_t})$.

2. Induction condition. *If for any sequence $\{a_n\}$ of objects from the set X the concept of satisfaction is defined for the formulae*

1. CONCEPT OF SATISFACTION

A and B in the domain $\mathfrak{M}$, *then satisfaction for more complex formulae is defined as follows: for any valuation* $\{a_n\}$

(a) $(A \to B)$ *is satisfied in* $\mathfrak{M}$ *by* $\{a_n\}$ *if and only if: if A is satisfied in* $\mathfrak{M}$ *by* $\{a_n\}$, *then B is satisfied in* $\mathfrak{M}$ *by* $\{a_n\}$;

(b) $\sim A$ *is satisfied in* $\mathfrak{M}$ *by* $\{a_n\}$ *if and only if it is not the case that A is satisfied in* $\mathfrak{M}$ *by* $\{a_n\}$;

(c) $(A \lor B)$ *is satisfied in* $\mathfrak{M}$ *by* $\{a_n\}$ *if and only if A is satisfied in* $\mathfrak{M}$ *by* $\{a_n\}$ *or B is satisfied in* $\mathfrak{M}$ *by* $\{a_n\}$;

(d) $(A \land B)$ *is satisfied in* $\mathfrak{M}$ *by* $\{a_n\}$ *if and only if A is satisfied in* $\mathfrak{M}$ *by* $\{a_n\}$ *and B is satisfied in* $\mathfrak{M}$ *by* $\{a_n\}$;

(e) $(A \equiv B)$ *is satisfied in* $\mathfrak{M}$ *by* $\{a_n\}$ *if and only if: A is satisfied in* $\mathfrak{M}$ *by* $\{a_n\}$ *if and only if B is satisfied in* $\mathfrak{M}$ *by* $\{a_n\}$;

(f) $\bigwedge x_k A$ *is satisfied in* $\mathfrak{M}$ *by* $\{a_n\}$ *if and only if, for every* $d \in X$, *A is satisfied in* $\mathfrak{M}$ *by the sequence* $\{a_1, a_2, ..., a_{k-1}, d, a_{k+1}, ...\}$ *(obtained from the sequence* $\{a_n\}$ *by the replacement of the element* a_k *by the element d)*;

(g) $\bigvee x_k A$ *is satisfied in* $\mathfrak{M}$ *by* $\{a_n\}$ *if and only if, for some* $d \in X$, *A is satisfied in* $\mathfrak{M}$ *by the sequence* $\{a_1, a_2, ..., ..., a_{k-1}, d, a_{k+1}, ...\}$ *(obtained from the sequence* $\{a_n\}$ *by the replacement of the element* a_k *by the element d)*.

In this way the concept of satisfaction is defined for any c.l.c. formulae without functions. This definition will not be written in a formalized way as a single sentence, since it would be lengthy and difficult to read. It is known from the previous analysis of inductive definitions that such a definition can easily be written in a formalized system of metalogic, that conditions 1 and 2 (a)–(g) are consequences of a rigorous statement of the definition of satisfaction, and that the relation of satisfaction is contained in every relation that complies with conditions 1 and 2 (a)–(g).

Consider now the structure of that definition. It may be said that it has established a way of translating formulae of the language studied into formulae of the subject language. It associates atomic formulae

(6) with atomic formulae (7), and next, in the induction condition, it associates implications with implications, negations with negations, disjunctions with disjunctions, etc. But if we want to so characterize the role of Definition 2, we have to add that the association of formulae of the language studied with formulae of the subject language takes place, in Definition 2, within the subject language itself, and this is why this association may not be interpreted as an ordinary translation of one theory into another. In an ordinary translation from the language of a theory to the language of another theory we establish a mapping between the two languages. Such a mapping is defined in a third language, which then is a metalanguage for both languages studied. We would have such a situation if we established the association in a sentence stating that

(8)   Formula $P_i^t(x_{k_1}, \ldots, x_{k_t})$ means the same as formula $p_i^t(a_{k_1}, \ldots, a_{k_t})$.

In sentence (8) the names of both formulae, (6) and (7), occur in the same supposition; on the other hand the entire sentence (8) is in that metalanguage in which we investigate both the language of formula (6) and that of formula (7). Sentence (8) differs from the sentence that is the inital condition of Definition 2 in that formula (7) in Definition 2 occurs in simple supposition whereas in sentence (8) it occurs in material supposition. This difference is essential. The difference between satisfaction and translation will appear still clearer below.

The examples discussed above will now be used to verify the fact that the rigorous definition of satisfaction corresponds to the intuitive one. For that purpose the examples must be adjusted to the conventions adopted. The variables x, y, z will be treated as different ways of writing the variables x_1, x_2, x_3. Moreover, we must say that given formulae are satisfied by certain sequences, and not by individuals. Hence, for instance, it follows from the initial condition 1 of Definition 2 that every sequence $\{a_n\}$ such that $a_1 = 2$, $a_2 = 3$, $a_3 = 5$, with further terms arbitrary, satisfies the formula $x \oplus y = z$ in the domain $\langle \mathcal{N}, + \rangle$ if the symbol $\oplus$ is interpreted as a name of addition, since the formula $x_1 \oplus x_2 = x_3$ may be treated as atomic (strictly: the formula used ought to be $\oplus(x_1, x_2, x_3)$, since Definition 2 is confined to formulae without function symbols), and in fact $a_1 + a_2 = a_3$.

Likewise, it follows from the inital condition of Definition 2 that the formula $x_1 > x_2$ is satisfied in the domain $\langle \mathcal{N}, > \rangle$ by every sequence $\{a_n\}$ in which $a_1 > a_2$. It follows from the above that

(9)   If $a_2 = 9$, then $\{a_n\}$ satisfies the formula $x_1 > x_2$ if and only if $a_1 > 9$.

The formula $x_1 > x_3$ also is satisfied in the same domain by every sequence $\{a_n\}$ in which $a_1 > a_3$. Hence, for instance,

(10)   If $a_3 = 8$, then $\{a_n\}$ satisfies the formula $x_1 > x_3$ if and only if $a_1 > 8$.

It follows from the inductive condition 2 (a) of Definition 2 that

(11)   A sequence $\{a_n\}$ satisfies the implication $x_1 > x_2 \rightarrow x_1 > x_3$ if and only if it is true that if $\{a_n\}$ satisfies $x_1 > x_2$, then $\{a_n\}$ also satisfies $x_1 > x_3$.

It is self-evident from sentences (9), (10), and (11) that

(12)   If $a_2 = 9$ and $a_3 = 8$, then a sequence $\{a_n\}$ satisfies the implication $x_1 > x_2 \rightarrow x_1 > x_3$ if and only if it is true that if $a_1 > 9$, then $a_1 > 8$.

Since it is known from arithmetic (which is contained in the metasystem under consideration) that, for any number $d \in \mathcal{N}$, if $d > 9$, then $d > 8$, hence the following statement may be deduced from sentence (12):

(13)   For any $d \in \mathcal{N}$, a sequence $\{a_n\}$ such that $a_1 = d$, $a_2 = 9$, $a_3 = 8$ satisfies the implication $x_1 > x_2 \rightarrow x_1 > x_3$.

By condition (f) in Definition 2, sentence (13) yields the statement that every sequence $\{a_n\}$ such that $a_2 = 9$ and $a_3 = 8$ satisfies the general formula $\bigwedge x_1(x_1 > x_2 \rightarrow x_1 > x_3)$ in the domain $\langle \mathcal{N}, > \rangle$; in other words, the numbers 9 and 8 satisfy, in the domain $\langle \mathcal{N}, > \rangle$, the formula $\bigwedge x(x > y \rightarrow x > z)$ with two free variables, y and z, if the symbol $>$ is interpreted as a name of the relation $>$. Thus, Definition 2 makes it possible to rigorously prove intuitively true statements about the relation of satisfaction.

In what follows it will sometimes be said that a finite sequence $\{a_1, ..., a_n\}$ of objects satisfies a formula A in $\mathfrak{M}$. This will mean that a formula A is satisfied in $\mathfrak{M}$ by any sequence whose initial terms are identical with $a_1, ..., a_n$.

Theorem on satisfaction by a finite sequence

To provide more information about the relation of satisfaction some theorems about it will be proved. Note first that the satisfaction of a formula A by a sequence $\{a_n\}$ depends only on a finite number of terms of that sequence, namely only on those terms $a_{k_1}, \ldots, a_{k_t}$ which are the ones corresponding to the free variables in the formula A. This property is rigorously formulated in the following

THEOREM 1. *If a finite set of variables* $x_{k_1}, \ldots, x_{k_t}$ *contains all the free variables occurring in a well-formed formula A, and if $\{a_n\}$ and $\{b_n\}$ are any sequences of objects from a set X in a domain $\mathfrak{M}$, which are identical in the places* $k_1, \ldots, k_t$ *(that is, if $a_{k_i} = b_{k_i}$ for $1 \leqslant i \leqslant t$), then*

(14) $\{a_n\}$ *satisfies A in* $\mathfrak{M}$ $\equiv$ $\{b_n\}$ *satisfies A in* $\mathfrak{M}$.

PROOF BY INDUCTION. It is to be demonstrated that every well-formed formula A has the property described by Theorem 1. Let that property be denoted by $W(A)$.

> $W(A) \equiv$ *If a sequence of variables* $x_{k_1}, \ldots, x_{k_t}$ *contains all the free variables of the formula A, and if $\{a_n\}$ and $\{b_n\}$ are any sequences of objects from a set X in the domain $\mathfrak{M}$, which are identical in the places* $k_1, \ldots, k_t$ *(that is, if $a_{k_i} = b_{k_i}$ for $1 \leqslant i \leqslant t$), then $\{a_n\}$ satisfies A in $\mathfrak{M}$ $\equiv$ $\{b_n\}$ satisfies A in $\mathfrak{M}$.*

1. Every atomic formula has the property W.

In fact, if A is an atomic formula of the form of (6), then by condition 1 of Definition 2

> *a sequence* $\{a_n\}$ *satisfies A in* $\mathfrak{M}$ $\equiv p_i^t(a_{k_1}, \ldots, a_{k_t})$,
>
> *a sequence* $\{b_n\}$ *satisfies A in* $\mathfrak{M}$ $\equiv p_i^t(b_{k_1}, \ldots, b_{k_t})$.

Since $a_{k_i} = b_{k_i}$ for $1 \leqslant i \leqslant t$, hence the right sides of these two equivalences are equivalent. Accordingly, their left sides are equivalent, too, so that formula (14) is satisfied. The atomic formulae satisfy the theorem under consideration, that is they have the property W referred to in that theorem.

2. If the component formulae have the property W, then the compound formulae have it, too.

(a)   *Suppose that $W(A)$ and $W(B)$; it will be proved that $W(A \rightarrow B)$.*

If $x_{k_1}, \ldots, x_{k_t}$ are all the free variables occurring in the formula $A \rightarrow B$, then among the variables $x_{k_1}, \ldots, x_{k_t}$ there are all the free variables occurring in the formula A and all those occurring in the formula B. Since $W(A)$ and $W(B)$, that is, since by assumption A and B satisfy the theorem under consideration, for any sequences $\{a_n\}$ and $\{b_n\}$, if $a_{k_i} = b_{k_i}$ for $1 \leqslant i \leqslant t$, then condition (14) holds, and

(15) $\{a_n\}$ *satisfies B in* $\mathfrak{M} \equiv \{b_n\}$ *satisfies B in* $\mathfrak{M}$.

It follows from condition (a) of Definition 2 that

(16) $\{a_n\}$ *satisfies* $(A \rightarrow B) \equiv (\{a_n\}$ *satisfies* $A \rightarrow \{a_n\}$ *satisfies* $B)$,

(17) $\{b_n\}$ *satisfies* $(A \rightarrow B) \equiv (\{b_n\}$ *satisfies* $A \rightarrow \{b_n\}$ *satisfies* $B)$.

It follows immediately from equivalences (14)–(17) that

$$\{a_n\} \text{ satisfies } (A \rightarrow B) \equiv \{b_n\} \text{ satisfies } (A \rightarrow B),$$

so that $W(A \rightarrow B)$ holds.

(b)   *If $W(A)$, then $W(\sim A)$.*

It follows from condition (b) of Definition 2 that

$$\{a_n\} \text{ satisfies } \ulcorner \sim A \urcorner \equiv \sim (\{a_n\} \text{ satisfies } A),$$
$$\{b_n\} \text{ satisfies } \ulcorner \sim A \urcorner \equiv \sim (\{b_n\} \text{ satisfies } A).$$

We shall hereafter use in formulae the symbols $\ulcorner \; \urcorner$ as a special kinds of parentheses which help us distinguish formulae of the language under investigation (object language) from those in metalanguage. The formulae in $\ulcorner \; \urcorner$ are always in object language. If the same symbolism is used in both object language and metalanguage, then the use of $\ulcorner \; \urcorner$ is convenient. Without them, for instance, the formula $\sim A \in Z$ might be interpreted in two ways: either the negation of the formula A is in the set Z, or it is not true that the formula A is in the set Z. On the contrary, the formula $\ulcorner \sim A \urcorner \in Z$ admits of the first interpretation only. These special brackets $\ulcorner \; \urcorner$ will not always be used; they will be used only in those cases in which there might be a risk of mis-understanding.

If $W(A)$, that is, if equivalence (14) holds, then the right sides of these equivalences are equivalent. Hence their left sides are equivalent, too, so that

$$\{a_n\} \text{ satisfies } \ulcorner \sim A \urcorner \equiv \{b_n\} \text{ satisfies } \ulcorner \sim A \urcorner.$$

Hence, the formula $\sim A$ has the property W.

Since the other sentential connectives are definable in terms of implication and negation, it has, thus, been proved that

$$If \ W(A) \ and \ W(B), \ then \ W(A \lor B), \ W(A \land B), \ W(A \equiv B).$$

(c) $If \ W(A), \ then \ W(\bigwedge x_{k_i} A).$

It follows from condition (f) of Definition 2 that

(18) $\{a_n\} \text{ satisfies } \ulcorner \bigwedge x_{k_i} A \urcorner \equiv$

$\bigwedge d(d \in X \rightarrow \{a_1, ..., a_{k_i-1}, d, a_{k_i+1}, ...\} \text{ satisfies } A),$

(19) $\{b_n\} \text{ satisfies } \ulcorner \bigwedge x_{k_i} A \urcorner \equiv$

$\bigwedge d(d \in X \rightarrow \{b_1, ..., b_{k_i-1}, d, b_{k_i+1}, ...\} \text{ satisfies } A).$

If the variable x_{k_i} is one of the free variables in the formula A, that is, $1 \leqslant i \leqslant t$, for instance, $i = t$, then $x_{k_1}, ..., x_{k_{t-1}}$ are all the free variables in the formula $\bigwedge x_{k_i} A$. It is to be proved that if the sequences $\{a_n\}$ and $\{b_n\}$ are identical in the places $k_1, ..., k_{t-1}$, then

(20) $\{a_n\} \text{ satisfies } \ulcorner \bigwedge x_{k_i} A \urcorner \equiv \{b_n\} \text{ satisfies } \ulcorner \bigwedge x_{k_i} A \urcorner.$

If the sequences $\{a_n\}$ and $\{b_n\}$ do not differ in the places $k_1, ..., k_{t-1}$, then the sequences $\{a_n'\} = \{a_1, ..., a_{k_i-1}, d, a_{k_i+1}, ...\}$, $\{b_n'\} = \{b_1, ..., b_{k_i-1}, d, b_{k_i+1}, ...\}$ are identical in all the places $k_1, ..., k_t$, since in the place k_t both sequences have the same element d. The sequences $\{a_n'\}$ and $\{b_n'\}$, thus, satisfy the antecedent of the condition $W(A)$, and hence also satisfy the consequent, so that

(21) $\{a_n'\} \text{ satisfies } A \equiv \{b_n'\} \text{ satisfies } A.$

Thus the right sides of equivalences (18) and (19) are equivalent to one another. Hence the left sides are equivalent, too, so that equivalence (20) holds. This proves that $W(\bigwedge x_{k_i} A)$. If a variable x_{k_i}, bound by the quantifier, is not free in A, for instance, if $i > t$, then the formulae A and $\bigwedge x_{k_i} A$ have the same free variables. If in that case the sequences $\{a_n\}$ and $\{b_n\}$ do not differ from one another in the places $k_1, \ldots, k_t$, then the corresponding sequences $\{a'_n\}$ and $\{b'_n\}$ also do not differ from one another in the places $k_1, \ldots, k_t$, since the new element d occurs in the place k_i, other than the places $k_1, \ldots, k_t$. Hence, in this case, too, the sequences $\{a'_n\}$ and $\{b'_n\}$ satisfy the condition $W(A)$. Consequently, equivalence (21) holds. Hence, as in the preceding case, it can be concluded that the right sides of equivalences (18) and (19) are equivalent. Hence, the left sides are equivalent, too, so that equivalence (20) holds, which proves that $W(\bigwedge x_{k_i} A)$.

Since, under De Morgan's laws, the existential quantifier is defined in terms of the universal quantifier and negation, it follows from the proofs of (b) and (c) that if $W(A)$, then $W(\bigvee x_{k_i} A)$.

To sum up parts 1 and 2 of the proof, part 1 states that the atomic formulae have the property W, and part 2 states that if the component formulae have the property W, then the compound formulae have it, too. Let Z stand for the set of those formulae which have the property W:

(22) $A \in Z \equiv W(A)$.

Part 1 states that the atomic formulae are in Z, and part 2 states that if $A, B \in Z$, then $\ulcorner A \to B \urcorner \in Z, \ulcorner \sim A \urcorner \in Z, \ulcorner A \vee B \urcorner \in Z, \ulcorner A \wedge B \urcorner \in Z,$ $\ulcorner A \equiv B \urcorner \in Z, \ulcorner \bigwedge x_i A \urcorner \in Z, \ulcorner \bigvee x_i A \urcorner \in Z$. Recall now the induction Definition 1. One of its consequences is that the set of well-formed formulae is included in any set Z which contains the atomic formulae and, for any two formulae A and B, also contains the formulae $A \to B$, $\sim A, A \vee B, A \wedge B, A \equiv B, \bigwedge x_i A, \bigvee x_i A$. The set Z, as defined by formula (22), satisfies those conditions. Hence the set of well-formed formulae is included in the set Z. Thus, every well-formed formula is in the set Z, that is, has the property W.

This concludes a proof by induction of the theorem stating that, for any A, if A is a well-formed formula, then $W(A)$. This shows that a proof by induction in fact consists in an application of a rigorous formulation of an inductive definition.

Since only those terms of a sequence $\{a_n\}$ which correspond to the free variables in A affect the satisfaction of A by that sequence, then, if the formula A has no free variables the satisfaction of A by $\{a_n\}$ does not depend on any terms of $\{a_n\}$. This property can be stated rigorously as follows:

COROLLARY. *If a formula A is a sentence (does not contain any free variables) and if $\{a_n\}$ and $\{b_n\}$ are any sequences of objects from a set X in a domain $\mathfrak{M}$, then*

$$\{a_n\} \text{ satisfies } A \text{ in } \mathfrak{M} \equiv \{b_n\} \text{ satisfies } A \text{ in } \mathfrak{M}.$$

PROOF. If A is a sentence, then the empty set of variables contains all the free variables in A. Theorem 1 assumes that the set of variables $\{x_{k_1}, \ldots, x_{k_t}\}$ may be arbitrary. Hence, in a special case it may be empty. If $\{a_n\}$ and $\{b_n\}$ are quite arbitrary sequences of objects from X in $\mathfrak{M}$, then it may be said that $\{a_n\}$ and $\{b_n\}$ do not differ from one another as to the empty set of the indices of terms, and, hence, as to the set of indices of terms which corresponds to the set of variables in A. Thus it follows from Theorem 1 that

$$\{a_n\} \text{ satisfies } A \text{ in } \mathfrak{M} \equiv \{b_n\} \text{ satisfies } A \text{ in } \mathfrak{M}.$$

THEOREM 2. *If a formula A is a sentence and a set X in a domain $\mathfrak{M}$ is not empty, then the sentence A is satisfied in $\mathfrak{M}$ for a valuation $\{a_n\} \equiv$ the sentence A is satisfied in $\mathfrak{M}$ for any valuation $\{a_n\}$ from the set X.*

PROOF. If $\{a_n\}$ satisfies A in $\mathfrak{M}$, then, by the above corollary to Theorem 1, any valuation $\{b_n\}$ satisfies A in $\mathfrak{M}$. Now if any valuation from the set X satisfies the sentence A, and if the set X is not empty, that is, there exists an element $a \in X$, then there exists a sequence $\{a_n\}$ such that $a_n = a$ for any n, and the sequence $\{a_n\}$ satisfies the sentence A.

In view of Theorem 2, if a formula A is a sentence and if it is satisfied by a sequence from a domain $\mathfrak{M}$, it is simply said that it is satisfied in the domain $\mathfrak{M}$; it is superfluous to specify the sequence, since all the sequences are interchangeable. The sentences satisfied in a domain $\mathfrak{M}$ are also called true in a domain $\mathfrak{M}$. This issue will soon be discussed more extensively.

The following practical conclusion may also be drawn from Theorem 1. Instead of referring to an entire infinite sequence we may, in the case of satisfaction, refer only to that finite part of the sequence in question which is associated with the free variables occurring in a formula under consideration. Hence, we may return to the formulations used at first, and in what follows we will sometimes say that a finite sequence $\{a_1, ..., a_n\}$ satisfies a formula A in $\mathfrak{M}$. This will mean that the formula A is satisfied in $\mathfrak{M}$ by any sequence in which the terms $\{a_1, ..., a_n\}$ occur and are associated with the free variables occurring in A.

Satisfaction versus translation

A further property of the relation of satisfaction will be described by comparing satisfaction with translation. A definition of *translation* from one formalized language into another will be introduced now. In doing so we shall confine ourselves to languages with a fixed structure, namely to the languages of theories formalized in the classical logical calculus without functions. The language of every theory formalized in the c.l.c. without functions is part of the language of the c.l.c. without functions, which is now under consideration. It is accordingly assumed that two such languages have the same variables, the same parentheses, the same sentential connectives, and the same quantifiers. Translation from one language into another thus reduces to the replacement of predicate symbols of one theory by predicate symbols of the theory. Hence instead of translation we shall simply refer to *replacement of constants*.

DEFINITION 3. *A formula A is obtained from a formula B through the replacement of the constants $P_1, ..., P_n$ by the constants $Q_1, ..., Q_n$ (in symbols: $A = B(P_1/Q_1, ..., P_n/Q_n)$) if and only if the formula A differs from the formula B only in having the symbols $Q_1, ..., Q_n$ in all those places in which the constant symbols $P_1, ..., P_n$ respectively occur in B.*

The above definition provides sufficient information about the meaning of the concept under consideration; a more rigorous formulation takes on the form of an inductive definition:

The relation of replacement of the constants $P_1, ..., P_n$ by the constants $Q_1, ..., Q_n$ is the least of the relations R which satisfy the following conditions:

1. F o r   t h e   a t o m i c   f o r m u l a e. A relation R holds between the formula $P_i(x_{k_1}, ..., x_{k_{t_i}})$ and the formula $Q_i(x_{k_1}, ..., x_{k_{t_i}})$ and R also holds between two formulae A and B which do not contain the constants $P_1, ..., P_n$ and $Q_1, ..., Q_n$ if these formulae are equiform

2. I n d u c t i o n   c o n d i t i o n. If a relation R holds between the formulae A and B and between C and D, then it also holds between: $A \rightarrow C$ and $B \rightarrow D$, $\sim A$ and $\sim B$, $A \vee C$ and $B \vee D$, $A \wedge C$ and $B \wedge D$, $A \equiv C$ and $B \equiv D$, $\bigwedge x_i A$ and $\bigwedge x_i B$, $\bigvee x_i A$ and $\bigvee x_i B$.

For instance, the following formula (23) is obtained from the formula (24) by the replacement of the constants Q_1, Q_2 by the constants P_1, P_2:

$$(23) \qquad \bigwedge x\big(P_1(x, y) \rightarrow \bigvee z \bigwedge u(P_1(x, u) \wedge P_2(u, z))\big),$$

$$(24) \qquad \bigwedge x\big(Q_1(x, y) \rightarrow \bigvee z \bigwedge u(Q_1(x, u) \wedge Q_2(u, z))\big).$$

The relationship between the relation of satisfaction and that of translation will be stated in a rigorous form in a theorem to be formulated in a theory one level higher than metalogic. Hence, metalogical formulae will be used in material supposition, and not in simple supposition. The analysis carried out in this part of the present section accordingly belongs to meta-metalogic.

It is assumed in meta-metalogic that metalogic is an axiomatic system which contains set theory and all the metalogical predicates and definitions mentioned so far, in particular the definition of satisfaction. It is assumed that metalogic contains the same logical constants which logic does. In meta-metalogic reference will be made both to well-formed formulae of the c.l.c. and to metalogical formulae.

THEOREM 3. *If A is a well-formed c.l.c. formula (without functions) which contains the predicate symbols $P_1, \ldots, P_n$ and the free variables $x_1, \ldots, x_k$, and if $p_1, \ldots, p_n$ are metalogical predicate symbols, then the following metalogical formula is a metalogical theorem*:

(25)   *For any sequence $\{a_n\}$ from any domain $\langle X, p_1, \ldots, p_n \rangle$ the sequence $\{a_n\}$ satisfies $\ulcorner A \urcorner$ in $\langle X, p_1, \ldots, p_n \rangle$ if the constants $P_1, \ldots, P_n$ are interpreted as names of the relations $p_1, \ldots, p_n$, if and only if*

$$A(P_1/p_1, \ldots, P_n/p_n, \bigwedge /X, x_1/a_1, \ldots, x_k/a_k).$$

In other words, for any definite c.l.c. formula A, the equivalence whose left part has the form: "the sequence $\{a_n\}$ satisfies A in $\langle X, p_1, \ldots, p_n \rangle$", and the right side is obtained by a translation of that formula into the language of metalogic and by the restriction of the quantifiers occurring in A to the set X, is a metalogical theorem. By $A(P_1/p_1, \ldots, P_n/p_n, \bigwedge /X, x_1/a_1, \ldots, x_k/a_k)$ we mean exactly the formula obtained from A by the simultaneous replacement of the constants $P_1, \ldots, P_n$ by the constants $p_1, \ldots, p_n$, the restriction of the quantifiers to the set X (that is, to the predicate $x \in X$), and the replacement of the variables $x_1, \ldots, x_k$ by the constants $a_1, \ldots, a_k$. Thus the following sentences are special cases of equivalence (25):

(26) *The sequence* $\{a_n\}$ *satisfies* $\ulcorner P_1(x_1, x_2) \vee P_2(x_1, x_2) \urcorner$ *in* $\langle X, p_1, p_2 \rangle$ *if and only if* $p_1(a_1, a_2) \vee p_2(a_1, a_2)$.

(27) *The sequence* $\{a_n\}$ *satisfies* $\ulcorner \bigwedge x_1 \bigvee x_2 (P_1(x_1, x_3) \to P_2(x_2, x_3)) \urcorner$ *in* $\langle X, p_1, p_2 \rangle$ *if and only if*
$\bigwedge x_1 \in X \bigvee x_2 \in X (p_1(x_1, a_3) \to p_2(x_2, a_3))$.

The equivalence (1) $\equiv$ (2) (p. 268), also is a special case of equivalence (25).

The above sentences are metalogical theorems. The formulae which occur on their right sides do so, in metalogic, in simple supposition. On the contrary, formula (25) is not a sentence in metalogic. It is merely a description, in meta-metalogic, of a class of definite metalogical sentences, a class which contains, among other things, sentences (26) and (27).

The proof of Theorem 3 is by induction. For the atomic c.l.c. formulae the equivalence falling under formula (25) is equiform with the initial condition of Definition 2. For the compound formulae the equivalence falling under formula (25) is obtained from the equivalences which correspond to the component formulae under the induction condition of the definition of satisfaction. Thus, for instance, assume that the component formulae A and B have the property referred to in the theorem under consideration, so that the following sentences are theorems in metalogic:

$\{a_n\}$ *satisfies* A *in* $\mathfrak{M} \equiv$
$A(P_1/p_1, \ldots, P_n/p_n, \bigwedge /X, x_1/a_1, \ldots, x_k/a_k)$,
$\{a_n\}$ *satisfies* B *in* $\mathfrak{M} \equiv$
$B(P_1/p_1, \ldots, P_n/p_n, \bigwedge /X, x_1/a_1, \ldots, x_k/a_k)$.

Under part (a) of the induction condition of Definition 2 the sentence:

$\{a_n\}$ *satisfies* $\ulcorner A \to B \urcorner$ *in* $\mathfrak{M} \equiv$
$(\{a_n\}$ *satisfies* A *in* $\mathfrak{M} \to \{a_n\}$ *satisfies* B *in* $\mathfrak{M})$

also is a metalogical theorem.

The above three equivalences yield the following theorem in metalogic:

$\{a_n\}$ *satisfies* $\ulcorner A \to B \urcorner$ *in* $\mathfrak{M} \equiv$

$((A \to B)(P_1/p_1, \ldots, P_n/p_n, \bigwedge/X, x_1/a_1, \ldots, x_k/a_k)),$

which is an application of schema (25) to the implication $A \to B$. Thus, the implication $A \to B$ has the property referred to in the theorem under consideration if its antecedent and its consequent have that property.

The proof is equally obvious in the case of disjunction, negation, conjunction, equivalence, and quantifiers. Hence, every well-formed c.l.c. formula has the property referred to in Theorem 3.

Theorem 3 is the best substantiation of the correctness of the definition of satisfaction. The definition of satisfaction has been constructed precisely so that Theorem 3 would hold.

Theorems on isomorphism

A theorem characteristic of the concept of satisfaction and having numerous applications will be proved now.

THEOREM 4. *If A is a formula which has the predicates $P_1, \ldots, P_n$ as its only extra-logical constants, and if a domain $\mathfrak{A} = \langle X, p_1, \ldots, p_n \rangle$ is isomorphic with a domain $\mathfrak{B} = \langle Y, r_1, \ldots, r_n \rangle$ under an isomorphism f, then the following equivalence holds*:

(28)   *a sequence $\{a_k\}$ from X satisfies A in $\mathfrak{A}$, $P_1, \ldots, P_n$ being interpreted as names of the relations $p_1, \ldots, p_n$, if and only if the sequence $\{f(a_k)\}$ satisfies A in $\mathfrak{B}$, $P_1, \ldots, P_n$ being interpreted as names of the relations $r_1, \ldots, r_n$.*

PROOF BY INDUCTION. We shall first obtain equivalence (28) for atomic sentences. It follows from the definition of satisfaction that

A sequence $\{a_k\}$ satisfies $P_i(x_{j_1}, \ldots, x_{j_l})$ in $\mathfrak{A} \equiv$
$p_i(a_{j_1}, \ldots, a_{j_l}),$
A sequence $\{f(a_k)\}$ satisfies $P_i(x_{j_1}, \ldots, x_{j_l})$ in $\mathfrak{B} \equiv$
$r_i(f(a_{j_1}), \ldots, f(a_{j_l})).$

And it follows immediately from the isomorphism that

$$p_i(a_{j_1}, \ldots, a_{j_l}) \equiv r_i(f(a_{j_1}), \ldots, f(a_{j_l})).$$

These three equivalences yield equivalence (28) for the atomic formulae.

284

1. CONCEPT OF SATISFACTION

Equivalence (28) for the atomic formulae entails a similar equivalence for compound formulae. In the case of the sentential connectives this transition is immediate. The case of a quantifier will now be described in greater detail. It is assumed that equivalence (28) is true for a formula A, and the formula $\bigvee x_t A$ is analysed. If that formula is satisfied by a sequence $\{a_k\}$, this means that, for some $d \in X$, the sequence $\{a_1, \ldots, a_{t-1}, d, a_{t+1}, \ldots\}$ satisfies A in $\mathfrak{A}$, and hence, by the induction assumption, the sequence $\{f(a_1), \ldots, f(a_{t-1}), f(d), f(a_{t+1}), \ldots\}$ satisfies A in $\mathfrak{B}$; hence, and by the definition of satisfaction, the sequence $\{f(a_k)\}$ satisfies $\bigvee x_t A$ in $\mathfrak{B}$. Conversely, if the sequence $\{f(a_k)\}$ satisfies $\bigvee x_t A$ in $\mathfrak{B}$, this means that, for a certain $b \in Y$, the sequence $\{f(a_1), \ldots, f(a_{t-1}), b, f(a_{t+1}), \ldots\}$ satisfies A in $\mathfrak{B}$, but since f is an isomorphism, then there exists some $d \in X$ such that $b = f(d)$. This and the induction assumption, yields that the sequence $\{a_1, \ldots, a_{t-1}, d, a_{t+1}, \ldots\}$ satisfies A in $\mathfrak{A}$, and, hence, on the strenght of the definition of satisfaction, the sequence $\{a_k\}$ satisfies $\bigvee x_t A$ in $\mathfrak{A}$. The same statement as that made by Theorem 4 is made by the following

COROLLARY. *If the domains $\langle X', r' \rangle$ and $\langle X'', r'' \rangle$ are isomorphic, and if the relations T' and T'' are defined as follows*:

$aT'b \equiv a$ *and b satisfy a formula A in the domain $\langle X', r' \rangle$, the term R being interpreted as a name of the relation r',*

$aT''b \equiv a$ *and b satisfy a formula A in the domain $\langle X'', r'' \rangle$, the term R being interpreted as a name of the relation r'',*

where A is a formula that contains only the predicate R, then the domains $\langle X', r', T' \rangle$ and $\langle X'', r'', T'' \rangle$ are isomorphic as well.

The proof follows directly from Theorem 4.

These theorems are very often used in mathematics, though they are not always formulated explicitly.

EXERCISE

For each of the formulae listed below indicate a mathematical domain and a way of interpreting terms, for which the formula is satisfied:

$$\bigvee x(\bigwedge y(xRy)), \quad \bigvee x(xRy \rightarrow \sim (xRz)),$$

$$\bigvee x, y\Big((\bigwedge z(S(x, z, x)) \wedge \bigwedge z(M(y, z, y))) \rightarrow xRy\Big).$$

2. THE CONCEPTS OF TRUTH AND MODEL. THE PROPERTIES OF THE SET OF SENTENCES TRUE IN A MODEL

The concept of truth

As has been mentioned in the preceding section, truth, roughly speaking, is a special case of satisfaction, namely, it is satisfaction restricted to formulae without free variables.

It is quite natural that in most cases we cannot decide whether a formula with free variables is true or not, since its truth depends on the interpretation of the free variables it contains. On the other hand, a formula without free variables states a complete idea, which is either true or false. Yet, since it is commonly accepted that formulae with free variables may occur as theorems of mathematical theories, the concept of truth, too, is defined for any well-formed formulae, which may contain free variables.

DEFINITION 4 (Tarski [92]). *A formula A is true in a domain* $\mathfrak{M} = \langle X, p_1, \ldots, p_n \rangle$ *(in symbols:* $A \in E(\mathfrak{M})$*), the predicates* $P_1, \ldots, P_n$ *being interpreted as names of the relations* $p_1, \ldots, p_n$*, if and only if every sequence of objects from the set X satisfies A in* $\mathfrak{M}$ *for the said interpretation of the predicates.*

The concept of truth has frequently occurred in the analyses made so far, but it has been used in an intuitive manner. Now that we have a rigorous definition of truth we shall try to demonstrate that it agrees with those intuitions. We shall strive to prove such properties as: every sentence is either true or false; the consequences of true sentences are true sentences; the theorems of logic are true in every domain.

To bring out the fact that the concept of truth is being mainly applied to closed sentences, let it be noted that

THEOREM 5. *If A is a formula with free variables* $x_1, \ldots, x_k$*, then*

$$A \text{ is true in } \mathfrak{M} \equiv \ulcorner \bigwedge x_1, \ldots, x_k A \urcorner \text{ is true in } \mathfrak{M}.$$

In this theorem and in the following one we are concerned with truth for a certain interpretation of terms, established once and for all. In order not to overload symbolism we shall not refer below to that established interpretation of terms.

PROOF. If A is true in $\mathfrak{M}$, then, by Definition 4, every sequence $\{a_n\}$ with any object $d \in X$ in its kth place satisfies the formula A. By part

2. CONCEPT OF TRUTH AND MODEL

(f) of Definition 2, it may be said that every sequence $\{a_n\}$ satisfies the formula $\bigwedge x_k A$. That sequence may have at its $(k-1)$-th place any object $d \in X$. Hence we similarly conclude that the formula $\bigwedge x_{k-1} \bigwedge x_k A$ is satisfied, etc. After k steps we arrive at the conclusion that every sequence $\{a_n\}$ of objects from the set X satisfies the formula $\bigwedge x_1, ..., x_k A$, and thus, the formula is true in $\mathfrak{M}$.

Conversely, if $\ulcorner \bigwedge x_1, ..., x_k A \urcorner$ is true in $\mathfrak{M}$, that is, if it is satisfied by every sequence $\{a_n\}$, then it is satisfied by any sequence that has in its first place any object $d \in X$. Thus, any sequence $\{a_n\}$ satisfies the formula $\bigwedge x_2, ..., x_k A$, which has one free variable x_1. After k steps we arrive at the conclusion that any sequence $\{a_n\}$ of objects from X satisfies the formula A, that is, that A is true in $\mathfrak{M}$.

Hereafter, the set $E(\mathfrak{M})$ shall be considered to contain sentences only.

A domain $\mathfrak{M} = \langle X, p_1, ..., p_n \rangle$ will be called *non-empty* if and only if the set X of that domain is non-empty.

THEOREM 6. *If A is a sentence and if $\mathfrak{M}$ is a non-empty domain, then $A \in E(\mathfrak{M}) \equiv$ there exists a sequence $\{a_n\}$ which satisfies A in $\mathfrak{M}$.*

PROOF. By Definition 4,

$$A \in E(\mathfrak{M}) \equiv \text{every sequence } \{a_n\} \text{ from } X \text{ satisfies } A \text{ in } \mathfrak{M}.$$

Since A is a sentence, and $\mathfrak{M}$ is a non-empty domain, Theorem 2 states that

*Every sequence $\{a_n\}$ satisfies A in $\mathfrak{M} \equiv$
some sequence $\{a_n\}$ satisfies A in $\mathfrak{M}$.*

These two equivalences yield the equivalence which occurs in Theorem 6.

THEOREM 7. *If A is a sentence, and if $\mathfrak{M}$ is a non-empty domain, then*

$$A \notin E(\mathfrak{M}) \equiv \ulcorner \sim A \urcorner \in E(\mathfrak{M}).$$

PROOF. Suppose that $A \notin E(\mathfrak{M})$, that is, A is not true in $\mathfrak{M}$. By Definition 4, this means that not every sequence from the set X satisfies A in $\mathfrak{M}$. By De Morgan's laws, this means that there exists a sequence $\{a_n\}$ from X which does not satisfy A in $\mathfrak{M}$. By part (b) of Definition 2, that sequence satisfies the negation of the sentence A. This yields that $\{a_n\}$ satisfies $\ulcorner \sim A \urcorner$ in $\mathfrak{M}$. The formula $\sim A$ is a sentence in view of the fact that A is a sentence. The domain $\mathfrak{M}$ is non-empty, since there exists a sequence $\{a_n\}$ of its elements. Hence Theorem 6 is applicable; by substituting in it $\sim A$ for A we immediately obtain the statement that $\ulcorner \sim A \urcorner \in E(\mathfrak{M})$.

Conversely, if $\ulcorner \sim A \urcorner \in E(\mathfrak{M})$, then by referring to the fact that the domain $\mathfrak{M}$ is non-empty we conclude, by Theorem 6, that there exists a sequence $\{a_n\}$ which satisfies $\ulcorner \sim A \urcorner$ in $\mathfrak{M}$, that is, one which does not satisfy A in $\mathfrak{M}$, under part (b) of Definition 2. Thus it is not true that every sequence satisfies A in $\mathfrak{M}$, that is, $A \notin E(\mathfrak{M})$, by Definition 4. The assumption of the non-emptiness of $\mathfrak{M}$ turns out to be required only in the proof of the converse implication.

If the sentence $\ulcorner \sim A \urcorner$ is true in a given domain, then we may say that A is false in that given domain. Hence, Theorem 7 may be interpreted as the statement:

Every sentence is either true in $\mathfrak{M}$, or false in $\mathfrak{M}$.

THEOREM 8. *The c.l.c. axioms, except for Axiom 0, are true in every domain $\mathfrak{M}$ for every interpretation of the terms. The c.l.c. axioms, including Axiom 0, are true in every non-empty domain for every interpretation of the terms* (in symbols: $\mathrm{Ax} L \subset E(\mathfrak{M})$).

PROOF. Theorem 8 is formulated in the language of metalogic. A correct proof of it should accordingly be formulated in metalogic. But it seems that more essential issues can be pointed out if the proof is carried out in a different way. We shall rise to the level of meta-metalogic and there prove the theorem stating that: *the formula above named Theorem 8 is a theorem in metalogic.*

Such a proof does not encounter any difficulties. It has been said in the preceding section that, in order to prove various theorems, we assume the whole classical logical calculus to be contained in metalogic. Thus any axiom A of the classical logical calculus is thereby a metalogical theorem. On the other hand, it is known from Theorem 3 that the following equivalence is a metalogical theorem:

$$\{a_n\} \text{ satisfies } \ulcorner A \urcorner \text{ in } \mathfrak{M} \equiv A(P_1/p_1, ..., P_n/p_n, \bigwedge /X)$$

(for any sequence $\{a_n\}$ from X). Since A is a logical axiom other than Axiom 0, hence $A(P_1/p_1, ..., P_n/p_n, \bigwedge /X)$ also is a logical theorem true in the empty domain; this is in accordance with Theorem 11 (Chapter I, p. 225), which states that a relativized logical theorem is a logical theorem if we confine ourselves to logic including in the empty domain. Since the right side of the above equivalence is a metalogical theorem, as a consequence of logical axioms adopted in metalogic, its left side is a metalogical theorem, too. Hence, for every axiom A of the c.l.c., except for Axiom 0, the following theorem can be proved in metalogic:

Every sequence $\{a_n\}$ of objects from X satisfies $\ulcorner A \urcorner$ in $\mathfrak{M}$,

so that, by Definition 4, it can be proved that every logical axiom, except for Axiom 0, is true in $\mathfrak{M}$.

Consider, by way of example, one of the c.l.c. axioms, for instance, the sentential calculus theorem $\bigwedge x(P(x) \vee \sim P(x))$ (the law of excluded middle). Since the whole of logic is assumed in metalogic, metalogic includes the theorem:

$$\bigwedge x\big(x \in X \to (p(x) \vee \sim p(x))\big).$$

This metalogical theorem is a relativization of the law of excluded middle to a set X. By Theorem 3, this metalogical theorem is equivalent to the theorem:

A sequence $\{a_n\}$ satisfies the law of excluded middle in $\mathfrak{M}$.

This holds for every sequence $\{a_n\}$, and hence the law of excluded middle is true in $\mathfrak{M}$.

If the domain $\mathfrak{M}$ is non-empty, then Axiom 0 is self-evidently satisfied in it, too; as is known, Axiom 0 states only that there exists at least one object.

The above argument brings out the fact that in metalogic one can prove the truth of a sentence A in a domain if any only if axioms are assumed in metalogic which yield the sentence A relativized to the domain under consideration. Thus, Theorem 8 does not have any essential cognitive value. It is merely a trivial consequence of the fact that in all reasoning, in particular in metalogic, the c.l.c. axioms are adopted as self-evident. Hence, Theorem 8 may not be treated as a criterion of truth, in the intuitive sense of the word, of the c.l.c. axioms. The only correct criterion is the intuitive analysis of the self-evidence of the c.l.c. axioms as outlined in the preceding chapter (Theorem 5). On the other hand, Theorem 8, and Theorem 3 to an even greater degree, testify to the correctness of Definition 4 as a definition of truth.

THEOREM 9. $\mathrm{Cn}_0\big(E(\mathfrak{M})\big) \subset E(\mathfrak{M})$. *(The set of sentences true in a domain $\mathfrak{M}$ is a theory.)*

PROOF. Suppose that the implication $A \to B$ and its antecedent A are true in a domain $\mathfrak{M}$. It will be proved that then the consequent B is also true in $\mathfrak{M}$. In fact, if $\ulcorner A \to B \urcorner$ is satisfied by every sequence $\{a_n\}$ from X, this means, by part (a) of Definition 2, that, for every sequence $\{a_n\}$ from X, if $\{a_n\}$ satisfies A, then $\{a_n\}$ satisfies B in $\mathfrak{M}$. Hence, under the law of the distribution of the quantifier over the implication, we have that if every sequence from X satisfies A in $\mathfrak{M}$, then every sequence from X satisfies B in $\mathfrak{M}$. By Definition 4, if it is assumed that $A \in E(\mathfrak{M})$, then every sequence from X satisfies A in $\mathfrak{M}$. Hence it is also true that every sequence from X satisfies B in $\mathfrak{M}$, so that $B \in E(\mathfrak{M})$.

The set $E(\mathfrak{M})$ is, thus, closed under the rule of detachment. Accordingly, by a definition from the preceding chapter, it forms a theory. But it may be a theory which does not in the least resemble ordinary mathematical theories, namely it may not be based on any ordinary (computable) system of axioms.

A more detailed explanation of this fact is to be found in the next chapter.

THEOREM 10. *All c.l.c. theorems are true in every non-empty domain.*

PROOF. It is known from Theorem 8 that the c.l.c. axioms are true in every non-empty domain. Theorem 9 shows that the rule of detachment does not lead outside sentences true in a given domain. Hence all the sentences obtained by detachment from c.l.c. axioms (i.e., all c.l.c. theorems) are true in every non-empty domain.

The comments that follow Theorem 8 show that Theorem 10 is a trivial consequence of the adoption in metalogic of the ordinary c.l.c. axioms. The theorem converse to Theorem 10, to be proved in the next section, is more interesting; it states that only the c.l.c. theorems have the property of being true in every domain. Other sentences, which are true in some domains, are not true in others. For instance, the sentence

$$\bigvee y \bigwedge x R(y, x)$$

is true in the domain $\langle \mathcal{N}, \leqslant \rangle$ if the term R is interpreted as a name of the relation: less than or equal to, since there exists a least natural number. On the other hand, this sentence is not true in the domain $\langle \mathscr{R}, \leqslant \rangle$ for the same interpretation of the term R, since there is no least real number. However, the sentence

$$\bigwedge x, y\Big((R(x, y) \wedge \sim R(y, x)) \to$$
$$\bigvee z\big(R(x, z) \wedge R(z, y) \wedge \sim (R(z, x)) \wedge \sim (R(y, z))\big)\Big)$$

is true in the domain $\langle \mathscr{R}, \leqslant \rangle$, but is not true in the domain $\langle \mathscr{N}, \leqslant \rangle$, for the interpretation of the term R as a name of the relation $\leqslant$, because the real numbers are densely ordered as to magnitude, whereas the natural numbers are ordered in an isolated way.

On the philosophical value of the above definition of truth see: Tarski [92], [95], and [89].

The concepts of consistency, completeness and model

The set $E(\mathfrak{M})$ of sentences true in a domain $\mathfrak{M}$ can serve as a good example of the application of certain important methodological concepts. The concepts of consistency and completeness are what we are primarily interested in. Logicians, mathematicians and philosophers have long pondered the problems of consistency and completeness of science. A theory is *consistent* if the answers it provides are not mutually exclusive. A theory is *complete* if it provides an answer to every problem formulated in its language. It is obvious that we want to have consistent knowledge, since inconsistent knowledge is no knowledge at all, as it negates on one occasion what it asserts on another. We would also like to have complete knowledge, so that we would be able to solve every problem. These reflections have given rise to definitions of consistency and completeness, applicable to rigorous mathematical theories.

DEFINITION 5. *A set X of sentences is inconsistent if and only if there exists a sentence A such that $A \in \mathrm{Cn}(X)$ and $\ulcorner \sim A \urcorner \in \mathrm{Cn}(X)$. If such a sentence does not exist, then the set X is called consistent.*

DEFINITION 6. *A set X is a complete set of sentences containing the constant terms $P_1, \ldots, P_n$ if and only if for every sentence A containing as constant symbols only symbols from the sequence $P_1, \ldots, P_n$ it is true that*

(29) $A \in X$ *or* $\ulcorner \sim A \urcorner \in X.$

A set X is called a *complete system of axioms* of a theory with constant terms $P_1, \ldots, P_n$ if and only if the set of sentences which contain the constant symbols $P_1, \ldots, P_n$ and are consequences, $\mathrm{Cn}(X)$, of the set X is a complete set.

If a set (theory) X is not a complete set, then it is called *incomplete*. If a theory is incomplete, then for some sentence A the negation of the sentence (29) is true, so that

$$A \notin X \quad \text{and} \quad \ulcorner \sim A \urcorner \notin X.$$

A sentence which is not in a theory and whose negation also is not in that theory is called *independent* of that theory. Thus, a theory is complete if and only if there are no sentences, formulated in its language, which are independent of it. Independent sentences exist for every incomplete theory.

If a set X under consideration is not a theory, then in defining the concept of independent sentence we have to refer to the set of consequences of X. In general the definition has the following form:

DEFINITION 7. *A sentence A is independent of a set X if and only if*

$$A \notin \mathrm{Cn}(X) \quad \text{and} \quad \ulcorner \sim A \urcorner \notin \mathrm{Cn}(X).$$

The concepts of independent sentence and the completeness and consistency of mathematical theories, defined rigorously as above, are subject matters of metalogical research. Some simple properties of those concepts will be given below.

THEOREM 11. *If $X \subset Y$ and Y is consistent, then X is consistent, too.*

PROOF. If $X \subset Y$, then $\mathrm{Cn}(X) \subset \mathrm{Cn}(Y)$. If there were a sentence A such that $A \in \mathrm{Cn}(X)$ and $\ulcorner \sim A \urcorner \in \mathrm{Cn}(X)$, then also $A \in \mathrm{Cn}(Y)$ and $\ulcorner \sim A \urcorner \in \mathrm{Cn}(Y)$.

THEOREM 12. *If $\ulcorner \sim A \urcorner \notin \mathrm{Cn}(X)$, then the set $X \cup \{A\}$ is consistent.*

PROOF. If the set $X \cup \{A\}$ were inconsistent, this would mean that for some sentence B

$$\ulcorner B \wedge \sim B \urcorner \in \mathrm{Cn}(X \cup \{A\}),$$

and, hence, by the Deduction Theorem,

$$\ulcorner A \to (B \wedge \sim B) \urcorner \in \mathrm{Cn}(X)$$

would hold. Since the consequences are based on the c.l.c.,

$$\ulcorner (A \to (B \wedge \sim B)) \to \sim A \urcorner \in \mathrm{Cn}(X).$$

Thus, under the rule of detachment, it would follow that

$$\ulcorner {\sim} A \urcorner \in \mathrm{Cn}(X),$$

which contradicts the assumption.

Every consistent and incomplete set of sentences may, thus, be expanded slightly so that a greater set, which is still consistent, is obtained. As will be seen, by repeating that procedure an infinite number of times one can finally arrive at a system which is complete and consistent.

THEOREM 13. *A consistent set X of sentences containing the constant symbols $P_1, \ldots, P_n$ is complete if and only if every set of sentences which contains the symbols $P_1, \ldots, P_n$ and is greater than X is inconsistent.*

PROOF. Since X is complete, then for every sentence A with the symbols $P_1, \ldots, P_n$ either $A \in X$ or $\ulcorner {\sim} A \urcorner \in X$. Hence, if $X \subset Y$ and $X \neq Y$, that is, if there exists an A such that $A \in Y$ and $A \notin X$, then it follows from the completeness of the set X that $\ulcorner {\sim} A \urcorner \in X$. But $X \subset Y$, so that $\ulcorner {\sim} A \urcorner \in Y$. Thus, the set Y is inconsistent.

The converse implication is obtained from Theorem 12. For should the set X be incomplete, there would be a set Y greater than X, $Y = X \cup \{A\}$, where $A \notin \mathrm{Cn}(X)$ and $\ulcorner {\sim} A \urcorner \notin \mathrm{Cn}(X)$, and the set Y would, by Theorem 12, be consistent.

The concepts of completeness and consistency are also applicable to quantifier-free calculi, e.g., to the sentential calculi. Definition 5 of consistency may be applied without modifications to any system that contains negation and to which a concept of consequence applies. On the contrary, the concept of completeness, as applied to systems in which sentences without free variables cannot be singled out, must be defined in a way different from that used in Definition 6. We usually avail ourselves of the concept of completeness, as formulated in Theorem 13, and adopt the convention that

A consistent set X is complete if and only if every set of formulae of the language under investigation which is greater than X is inconsistent.

This definition is more general, since it refers only to the concept of consequence and the concepts of set theory. It can accordingly be applied to quantifier-free calculi. It can easily be proved, e.g., that the classical sentential calculus is a consistent and complete theory. Both properties are consequences of the zero-and-one verification method for the theorems of the c.l.c. For if, by the theorem on the decidability of the c.l.c., those and only those formulae which yield 1 on the zero-and-one verification procedure are c.l.c. theorems, then if a formula α is a theorem, then the formula $\sim \alpha$ cannot be a theorem, because if α on verification yields 1, then $\sim \alpha$ yields 0, and, hence, is not a c.l.c. theorem. Thus, α and $\sim \alpha$ cannot both be theorems in the c.l.c., and, hence, the c.l.c. is a consistent theory. It can also easily be noticed that every greater set of sentential formulae is inconsistent. For if a formula α, which is not a c.l.c. theorem, is joined to the c.l.c. theorems, then—as α is not in the c.l.c.—for some substitution, for instance, $p = 1$, $q = 0$, $\alpha = 0$. Hence, some other formula β, obtained from α by the substitutions $p/p \to p$, $q/ \sim (p \to p)$ always yields 0, and hence its negation, $\sim \beta$, always yields 1, and, thus, is a c.l.c. theorem. Thus, by joining α to the c.l.c. we obtain an inconsistent system, which has as its consequences two mutually contradictory sentences, β and $\sim \beta$.

Thus, the c.l.c. is a consistent set of formulae such that it cannot be augmented while preserving its consistency. Hence the c.l.c. is a complete set.

Among the mathematical theories mentioned in Chapter I the elementary theory of the *less than* relation and the elementary theory of real numbers are complete.

THEOREM 14. *A set T of sentences is consistent if and only if there exists a sentence A such that*

$$A \notin \mathrm{Cn}(T).$$

In other words, a set is inconsistent if and only if every sentence is a consequence of it.

PROOF. If every sentence is a consequence of the set T, then this means that, for every sentence A, both the sentence A and the sentence $\sim A$ are consequences of the set T. By Definition 5, the set T is thus inconsistent.

Conversely, if T is an inconsistent set, then, for some A, both $A \in \mathrm{Cn}(T)$ and $\sim A \in \mathrm{Cn}(T)$. On the other hand, the sentence

$$\left(A \to (\sim A \to B)\right)$$

is a logical theorem for any sentence B. By applying the operation of detachment twice we obtain that $B \in \mathrm{Cn}(T)$.

294

Thus, an inconsistent theory is complete, and every incomplete theory is consistent.

THEOREM 15. *The set $E(\mathfrak{M})$ of all sentences true in a non-empty domain $\mathfrak{M}$, for a fixed interpretation of predicates, is complete and consistent.*

PROOF. 1. C o m p l e t e n e s s. Let A be any sentence with the constants $P_1, ..., P_n$. By Theorem 7, if $A \notin E(\mathfrak{M})$, then $\ulcorner \sim A \urcorner \in E(\mathfrak{M})$, so that $A \in E(\mathfrak{M})$ or else $\ulcorner \sim A \urcorner \in E(\mathfrak{M})$. The set $E(\mathfrak{M})$ is, thus, complete.

2. C o n s i s t e n c y. If the set $E(\mathfrak{M})$ were inconsistent, this would mean that, for a sentence A, both $A \in \mathrm{Cn}(E(\mathfrak{M}))$ and $\ulcorner \sim A \urcorner \in \mathrm{Cn}(E(\mathfrak{M}))$. By Theorems 8 and 9, the set $E(\mathfrak{M})$ is a theory, so that $\mathrm{Cn}(E(\mathfrak{M})) \subset E(\mathfrak{M})$. This would mean that both $A \in E(\mathfrak{M})$ and $\ulcorner \sim A \urcorner \in E(\mathfrak{M})$. But, by Theorem 7, if $\ulcorner \sim A \urcorner \in E(\mathfrak{M})$, then $A \notin E(\mathfrak{M})$, hence the set $E(\mathfrak{M})$ cannot be inconsistent.

Theorems 11 and 15 taken together provide a method of proving the consistency of a theory. If suffices for a theory X to be included in the set of all sentences true in a domain $\mathfrak{M}$ to be consistent. As will be seen soon, this is the only method of proving the consistency of a set of sentences. The domain $\mathfrak{M}$ is then called a *model* of the theory X.

DEFINITION 8. *A domain $\mathfrak{M} = \langle X, p_1, ..., p_n \rangle$ is called a model for a set Z of sentences with the constants $P_1, ..., P_n$ if and only if X is not empty and $Z \subset E(\mathfrak{M})$, that is, if all the sentences of the set Z are true in the domain $\mathfrak{M}$ for the interpretation of the terms $P_1, ..., P_n$ as names of the relations $p_1, ..., p_n$.*

If it is added that $\mathfrak{M}$ is a finite (denumerable or non-denumerable) model for the sentences of the set Z, this means that the set X in the model $\mathfrak{M}$ is a finite, denumerable, or non-denumerable set, respectively.

THEOREM 16. *If there exists a model for the sentences from a set Z, then the set Z is consistent.*

PROOF. By Theorem 15, the set $E(\mathfrak{M})$ is consistent. Hence, if $Z \subset E(\mathfrak{M})$, then, by Theorem 11, Z is consistent.

Thus, for instance, in our system of metalogic, which includes the whole of set theory, we can define various domains specified in the Introduction, and prove that they are models for certain mathematical theories. This will yield proofs of consistency. For instance, having

295

defined the rational numbers, the *less than* relation and the relation of identity between those numbers, as was done in the Introduction (cf. Section 8), we can easily prove that the domain $\langle \mathcal{W}, <, = \rangle$ is a model of the elementary theory of the *less than* relation as described in Section 7 of Chapter I. Having next defined the real numbers in the way outlined in the Introduction (Section 12) we can demonstrate that the domain $\langle \mathcal{R}, +, \cdot, 0, 1, <, = \rangle$ is a model of the elementary theory of real numbers as described in Section 8 of Chapter I.

The proofs of consistency of certain simple mathematical theories, as outlined above, are not absolutely valid from the intuitive point of view, since they are constructed in a metalogic which includes the whole of set theory, that is, in a theory which contains, as its small fragments, analogues of all the theories specified above. A proof of consistency of a theory is, thus, obtained in a theory which is much stronger than the theory being analysed. It can be demonstrated that this is the only possibility of proving the consistency of a theory. The inconsistency of a stronger theory cannot be proved in a theory that is weaker or of the same strength as the theory being investigated. The proof has an intuitive value if that stronger theory is considered to be consistent. In order to prove the consistency of that stronger theory in a rigorous manner we would have to resort to a still stronger theory. The belief in the consistency of the fundamental mathematical theories is, thus, based on the experience of many generations of mathematicians who have not encountered a contradiction, and also on an intuitive clarity of the fundamental mathematical concepts.

Extension of the concepts of satisfaction and truth to formulae with individual constants and function symbols

In order to formulate certain further properties of the set of sentences true in a given domain we have to enlarge the language of the theories under consideration. As formulated in the preceding section, the concept of satisfaction referred to formulae which contain neither individual constants nor function symbols. If new symbols are added to the language used, the concepts of satisfaction and truth must be re-defined.

We consider at first by itself the concept of adding a sequence $\{s_t\}_U$ of constant terms, since we shall often enlarge the language by adding to it only individual constants, without adding function symbols. The index t may range over any set U. In this case the concept of satisfaction easily reduces to the concept defined in Definition 2. The extended concept requires an additional condition. Since a new kind of term: constant term, is involved, it must be stated for which interpretation of these new terms the formula is considered to be satisfied. Thus, it does not suffice to say that a given formula is satisfied (true) for the interpretation of predicate terms as names of certain relations; the new condition must be formulated with a new phrase: for the interpretation of the constants from the sequence $\{s_t\}$ as names of objects from the sequence $\{b_t\}$.

It suffices to adopt the convention that:

A sequence $\{a_n\}$ *satisfies* a formula A with a constant s_v in a domain $\mathfrak{M}$ for the interpretation of the symbols $P_1, \ldots, P_n$ as names of the relations $p_1, \ldots, p_n$ respectively and for the interpretation of constant terms $\{s_t\}_U$ as names of respective terms of the sequence $\{b_t\}_U$ if and only if, on replacing in the formula A the constant s_v by a variable x_m which does not occur in A, we obtain a formula A' without the constant s_v and with the property that A' is satisfied in $\mathfrak{M}$, in the sense defined in the preceding section, by the sequence $\{a_1, \ldots, a_{m-1}, b_v, a_{m+1}, \ldots\}$.

The concept of satisfaction of formulae with constants for a definite interpretation of the constants as names of the objects associated with them can, of course, be defined by induction, without any reference to the concept of satisfaction as defined earlier. For such a definition by induction only the initial condition of Definition 2 must be modified. Namely, an atomic formula

$$P_i^t(x_{k_1}, \ldots, x_{k_{m-1}}, s_v, x_{k_{m+1}}, \ldots, x_{k_t})$$

containing a constant s is satisfied in $\mathfrak{M}$ for the valuation $\{a_n\}$ if and only if

$$p_i^t(a_{k_1}, \ldots, a_{k_{m-1}}, b_v, a_{k_{m+1}}, \ldots, a_{k_t}).$$

The induction condition is left unchanged.

The concept of truth is defined analogously: A formula A with constants $\{s_t\}_U$ is *true* in $\mathfrak{M}$ for the interpretation of the terms $P_1, \ldots, P_n$ as names of relations $p_1, \ldots, p_n$ and for the interpretation of the constants $\{s_t\}_U$ as names of objects $\{b_t\}_U$ if and only if the formula A is satisfied in the domain $\mathfrak{M}$ for the said interpretations of terms and constants by every valuation of the variables in the sense defined above.

All the previous theorems on satisfaction and truth remain valid for the definitions given above.

EXAMPLES

The sentence $\bigwedge x_1 s_0 R x_1$ is true in the domain of natural numbers for the interpretation of the predicate R as a name of the relation of "less than or equal to" and for the interpretation of the constant s_0 as a name of the number 0, since the formula $\bigwedge x_1 x_2 R x_1$ is satisfied in that domain by every sequence which has the term 0 in the place marked by the number 2.

The sentence $s_0 R s_1$ is also true in that domain for the interpretation of R as a name of the *less than* relation and for the interpretation of s_0 and s_1 as names of the numbers 0 and 1, respectively.

Analysis will now be extended so as to include formulae which contain function constants. The concept of satisfaction will be defined for the formulae of that type. That concept will be defined in a slightly different way than in the case of Definition 2. The method to be used here could have also been used in the case of Definition 2. The method of formulation can be easily grasped if we adopt two symbols: truth $= 1$, falsehood $= 0$. Instead of saying that a formula A is satisfied by a valuation $\{a_n\}$, we may say that the value of a formula A for a sequence $\{a_n\}$ is 1. Instead of saying that a formula A is not satisfied by a sequence $\{a_n\}$, we may say that the value of a formula A for a sequence $\{a_n\}$ is 0. Instead of defining the relation of satisfaction, we may, thus, define the *value function* V which, for the sentential formulae, takes on two values only: 0 or 1. The function V is defined not on sentential formulae alone, but on term formulae as well. (cf. Chapter II, Section 5, Definition 26). The values of the function V for term formulae are individuals from the set X of a domain $\mathfrak{M}$. The function V is restricted to a sequence $\{a_n\} \subset X$ and to a certain interpretation of terms.

DEFINITION 2*a. *Let $V(A, \{a_n\})$ stand for the value of a formula A for a sequence $\{a_n\}$, for the interpretation of the predicate constants $P_1, ..., P_n$ as names of the relations $p_1, ..., p_n$ and of the function constants $F_1, ..., F_k$ as names of the functions $f_1, ...,f_k$ defined in a set X,* *and for the interpretation of the individual constants $\{s_t\}$ as names of the corresponding terms of the sequence $\{b_t\} \subset X$.*

The inductive definition of value has the following form:

1. $V(x_k, \{a_n\}) = a_k,$
2. $V(F_i(A_1, ..., A_k), \{a_n\}) = f_i(V(A_1, \{a_n\}), ..., V(A_k, \{a_n\})),$
3. $V(s_t, \{a_n\}) = b_t,$
4. $V(P_i^k(A_1, ..., A_k), \{a_n\}) = 1 \equiv p_i(V(A_1, \{a_n\}), ..., V(A_k, \{a_n\})),$
5. $V(\ulcorner \sim A \urcorner, \{a_n\}) = (1 - V(A, \{a_n\})),$
6. $V(\ulcorner A \to B \urcorner, \{a_n\}) = (V(A, \{a_n\}) \to V(B, \{a_n\})),$
7. $V(\bigvee x_i A, \{a_n\}) = 1 \equiv$
   $\bigvee d \in X(V(A, \{a_1, ..., a_{i-1}, d, a_{i+1}, ...\}) = 1).$

The operation $1 - p$ for $p = 0$ or 1 is considered to be defined so as in arithmetic. Likewise, $p \to q = \max((1-p), q)$. In cases 4 and 7, if the condition on the right side is not satisfied, it is assumed that the function V equals 0. The matrix of the function $p \to q$ for 0 and 1 is, as can easily be verified, identical with the matrix for implication. The above definition defines values for any well-formed term and sentential formulae, since the remaining logical terms: $\vee$, $\wedge$, $\equiv$, $\bigwedge$ may be considered as definable in terms of $\to$, $\sim$, $\bigvee$.

DEFINITION 2*b. *A formula A is satisfied by a sequence $\{a_n\}$ in a domain $\mathfrak{M}$, for a certain interpretation of terms, if and only if, for that interpretation of terms, $V(A, \{a_n\}) = 1$.*

DEFINITION 4*. *A formula A is true in $\mathfrak{M}$, for the interpretation of the predicates $P_1, ..., P_n$ as names of the relations $p_1, ..., p_n$, of the function terms $F_1, ..., F_m$ as names of the functions $f_1, ..., f_m$, and of the constant terms $\{s_t\}$ as names of terms of a sequence $\{b_t\}$ if and only if every sequence of objects from the set X of the domain $\mathfrak{M}$ satisfies A in $\mathfrak{M}$ for the same interpretation of terms.*

It can easily be verified that the concepts of satisfaction and truth for formulae without individual constants and functions, now extended in the way described above, coincide with those introduced first, and that all the theorems proved so far remain valid for these extended concepts if the concepts Cn and L are replaced in those theorems by Cn* and L*, respectively.

II. MODELS

For the interpretation of Id as a name of the relation of equality, of F as a name of the successor function $x+1$, and of s_0 as a name of the number 0, the following formulae have the following values in the domain of natural numbers:

$$V\left(\ulcorner F\big(F(F(s_0))\big)\urcorner, \{a_n\}\right) = 3,$$

$$V\left(\ulcorner \mathrm{Id}\big(s_0, F(s_0)\big)\urcorner, \{a_n\}\right) = 0,$$

$$V\left(\ulcorner F(x_k)\urcorner, \{a_n\}\right) = a_k + 1.$$

Descriptive completeness

The next concept, which will be applied to the example of the set of true sentences, is that of *descriptive completeness*, which is an extension of the earlier concept of ω-completeness (Tarski [91], Henkin [22]).

Suppose that in investigating a mathematical domain $\langle X, p_1, ..., p_n \rangle$ we have not only names for the relations $p_1, ..., p_n$, but also have at our disposal individual names for all the objects from the set X. The following observation can then be made: If, for every $t \in X$, s_t is an individual name of the object t, and, for every $t \in X$, the sentence $A(s_t)$ is true, then the sentence $\bigwedge x A(x)$, which is a natural generalization of the former sentences, is also true in the domain $\langle X, p_1, ..., p_n \rangle$. As will be seen, the concept of descriptive completeness derives from this observation.

To have a definite linguistic name for every element of a domain is possible only in the case of finite domains.

A domain $\mathfrak{M} = \langle X, p_1, ..., p_n \rangle$ is called *denumerable* if the set X is denumerable; a set is called *denumerable* if it is equinumerous with the set of natural numbers. Infinite sets which are not denumerable are called *non-denumerable*.

If a domain is denumerable, it is possible to formulate a general principle for constructing names of its elements. For instance, there are definite principles for constructing names for the natural numbers, that is figures, e.g., in the decimal system. Even so, for very large natural numbers the possibility of writing down the corresponding figures becomes illusory. Likewise, for any other method of naming numbers we can quote numbers so large that it is, practically speaking, impossible to write names for them. Nevertheless, it is usually assumed in mathematics that speaking about a denumerable number of constants has a definite sense, provided that the language of a theory is to be understood in a more abstract way, not as a set of definite inscriptions that can be written down, but as a set of inscriptions which would be possible to write down if we had at our disposal infinite time, infinite space, etc. Following the same line of reasoning we could agree to speak about any non-denumerable number of constant terms which we add to the language of a theory, because the language of a theory can be identified with certain geometrical figures, and, hence, with certain continuous functions. The number of continuous functions on a plane is the size of the continuum. When passing to abstract spaces of higher powers we can imagine a language consisting of inscriptions in such a space, a language whose power would be arbitrarily high.

Yet, speaking about a non-denumerable number of constants changes the concept of mathematical theory radically. This is why speaking about theories with non-denumerable numbers of constants will be avoided. The definitions to be introduced below will be formulated very generally, for sets of any cardinality, but the definitions will be applied almost exclusively to denumerable sets. Only once shall we go beyond denumerability in an essential way, and then only in the proof of a theorem. In the theorems as such the concept of mathematical theory will always occur in the ordinary sense, as a denumerable set of inscriptions derived by continuing a procedure into infinity.

Certain subsets, usually termed sequences and symbolized as $\{s_t\}_U$, will be singled out among the constant terms. The set U, over which the indices t will range, will usually remain undefined. It may be assumed that it is a set which is either finite, or denumerable, or non-denumerable. But the most important application of the concept which we are about to introduce is the case when the sequence $\{s_t\}$ of constants is denumerable, so that t ranges over the set $\mathcal{N}$ of natural numbers. The symbol U of the set over which the indices range will, in important formulations, be added to the symbol of the sequence: $\{s_t\}_U$.

DEFINITION 9. *A theory T is descriptively complete (briefly: d-complete) with respect to a sequence $\{s_t\}_U$ of constant terms occurring in that theory if and only if, for any formula $A(x)$ with one free variable x, the following implication holds*:

(30)   *if* $\bigwedge t \in U \ulcorner A(x/s_t) \urcorner \in T$, *then* $\ulcorner \bigwedge x A(x) \urcorner \in T$;

301

that implication states that if, for every definite constant term from the sequence $\{s_t\}_U$, the substitution $A(x/s_t)$, in which the only variable x free in A is replaced by the constant s_t, is a theorem, then the general sentence $\bigwedge xA(x)$ is a theorem of the theory T.

Hereafter, when referring to substitutions of term constants for variables we shall often used the simpler notation: $A(s_t)$, instead of: $A(x/s_t)$.

If the sequence $\{s_t\}$ is denumerable, so that $U = \mathcal{N}$, then the theory T, descriptively complete with respect to the sequence $\{s_i\}_{\mathcal{N}}$, is called ω-complete.

Implication (30) could be strengthened by changing it into an equivalence, since the converse implication is always satisfied if the set T is a theory based on the c.l.c. Its correctness follows from axiom schema 2 of the c.l.c. Implication (30) is a natural version of the generalization refered to above, quite justified on the assumption that the terms of the sequence $\{s_t\}_U$ are names of all those objects to which the theory T applies. If, for every object s_t taken separately, it can be proved that it has a property A, then it seems natural to accept the general sentence $\bigwedge xA(x)$, stating that every object considered in the theory T has the property A.

Implication (30) would certainly be adopted as a good rule of inference if it could be applied effectively. But if the constants s_t include an infinite number of elements which differ from one another, then rule (30) is inapplicable, because in order to apply it we would have to have infinitely many previously proved theorems $A(x/s_1), A(x/s_2), \ldots$, yet in practice we always have only a finite number of theorems proved previously. Hence implication (30) plays only the role of a property of mathematical theories with respect to a sequence of constant terms. That property is called d-completeness, or ω-completeness if the sequence $\{s_n\}$ is denumerable.

The property of ω-completeness applies with special clarity to systems of the arithmetic of natural numbers. A system T of arithmetic is ω-complete with respect to the sequence of constants 0, 1, 2, 3, ... if and only if it satisfies the following condition:

If the sentences: $A(0), A(1), A(2), \ldots$, for all numerical constants, are theorems in a system T, then the sentence $\bigwedge xA(x)$ also is a theorem in T.

Unfortunately, as follows from the investigations to be described in the next chapter, the ordinary system of the arithmetic of natural numbers is not ω-complete. The property of ω-completeness is possessed by theories of theoretical rather than practical importance, such as the set of true sentences.

It will now be demonstrated that the set of sentences true in a domain $\mathfrak{M}$ is d-complete with respect to a sequence $\{s_t\}$ containing names of all objects from the set X of the domain $\mathfrak{M}$.

THEOREM 17. *If a set X of a domain $\mathfrak{M}$ consists of all terms of a sequence $\{b_t\}_U$, and of these only, then the set $E(\mathfrak{M})$ of the sentences true in $\mathfrak{M}$, for the interpretation of the constants $\{s_t\}_U$ as names of the terms of the sequence $\{b_t\}_U$, is d-complete with respect to the sequence of constants terms $\{s_t\}_U$.*

PROOF. If, for every t, $\ulcorner A(x_1/s_t)\urcorner \in E(\mathfrak{M})$, then, by Definitions 2* and 6*, every element $d \in X$ satisfies the formula $A(x_1)$ in $\mathfrak{M}$. To put it more strictly, every sequence which has as its first term any element $d \in X$ satisfies $A(x_1)$. Hence, $\ulcorner A(x_1)\urcorner \in E(\mathfrak{M})$, which, by Theorem 5, yields

$$\ulcorner \bigwedge x_1 A(x_1)\urcorner \in E(\mathfrak{M}).$$

The set of sentences of the elementary arithmetic of natural numbers, sentences which refer to addition and multiplication and are true in the domain $\langle \mathcal{N}, +, \times \rangle$, as defined in set theory, which in turn is assumed in metalogic, is, thus, ω-complete with respect to the sequence of constants $0, 1, 2 = 1+1, 3 = 1+1+1, \ldots$

The principal applications of the concept of d-completeness are to theories which are remote from mathematical practice. Yet, some examples of d-complete theories, fairly simple and apt to occur in practice, although devoid of any great significance, can be given. An example of such a simple ω-complete theory is provided, for instance, by the theory of the *less than* relation, described in Section 7 of Chapter I, if it is enriched with certain constants. It is those added constants which form the denumerable sequence of constants $\{s_n\}$, with respect to which that theory is ω-complete. It was mentioned in Section 7 of Chapter I that the elementary theory of the *less than* relation is satisfied in the domain of rational numbers. Now the added constants $\{s_n\}$ are names of all rational numbers. Rational numbers are denoted by fractions.

Such fractions may be confined to irreducible ones. Thus we add to the theory of the *less than* relation the constant 0 and all those irreducible fractions of the form k/n which have zero neither in the numerator nor in the denominator, and also fractions with the minus sign. Thus 0, 1/1, 2/1, 3/1, ..., 1/2, 3/2, ..., 1/3, 2/3, ..., $-1/1$, $-2/1$, ... are constants of the entended system of the *less than* relation. Infinitely many axioms are adopted concerning those constants, the axioms being sentences of the form

$$k/n < k_1/n_1 \quad \text{or} \quad \sim(k/n < k_1/n_1).$$

All true atomic sentences and negations of all false atomic sentences are adopted as axioms. Thus, for instance, the axioms include:

$$0 < 1/1, 2/1 < 3/1, 1/2 < 2/3,$$
$$\sim(3/1 < 3/2), \sim(0 < -1/3), ...$$

Ordinary axioms of the elementary theory of the *less than* relation also are axioms of the theory under consideration.

This theory is ω-complete, since, as will be proved later, it consists of all sentences true in the domain $\langle \mathcal{W}, <, = \rangle$ of rational numbers, for the interpretation of the constants described above as names of the corresponding rational numbers.

THEOREM 18. *If theories T and S are consistent and complete, contain the same terms $P_1, ..., P_n$, and both are d-complete with respect to the same sequence of constants $\{s_t\}_U$, and if the atomic theorems of the theories T and S are the same, then these theories are identical: $T = S$.*

By saying that the theories T and S coincide in their atomic theorems (containing constants $\{s_t\}$) we mean that, for every predicate P_i and for all n and m from the set U, the formula

(31) $$\ulcorner P_i(s_n, s_m) \urcorner \in T \equiv \ulcorner P_i(s_n, s_m) \urcorner \in S$$

holds if the predicate P_i has two arguments. If P_i is a predicate with more arguments than two, then an analogous formula with a greater number of constants holds. (To simplify our proof we shall assume that all predicates $P_1, ..., P_n$ have two arguments each.)

The proof is by induction with respect to the length of sentences. The shortest sentences are those which above have been called atomci. They are obtained from atomic predicates by the substitution for the variables of constant terms from the sequences $\{s_t\}$. The theories T and S coincide in their atomic sentences by assumption.

Longer formulae are obtained from shorter ones by combining shorter formulae with sentential connectives or by adding quantifiers. Those sentences which have fewer sentential connectives or quantifiers are considered shorter. It will be demonstrated that if the theories T and S coincide in their shorter sentences, they also coincide in their longer ones.

Suppose that theories T and S coincide in sentences A and B. This means that

(32) $\quad A \in T \equiv A \in S \quad$ and $\quad B \in T \equiv B \in S.$

Since the theories T and S are based on the c.l.c., equivalences (32) yield the equivalences

$$\ulcorner A \wedge B \urcorner \in T \equiv \ulcorner A \wedge B \urcorner \in S,$$
$$\ulcorner \sim A \urcorner \in T \equiv \ulcorner \sim A \urcorner \in S,$$

and similar equivalences for other sentential connectives.

The equivalence for conjunction is obtained immediately.

We shall examine the equivalence for negation. If $\ulcorner \sim A \urcorner \in T$ and T is consistent, then $A \notin T$, and hence, by (32), we obtain that $A \notin S$. And since S is complete and $A \notin S$, then $\ulcorner \sim A \urcorner \in S$. This proves the implication $\ulcorner \sim A \urcorner \in T \rightarrow \ulcorner \sim A \urcorner \in S$. The converse implication is proved in a parallel way. These equivalences yield similar equivalences for other sentential connectives.

Consider now a sentence with the universal quantifier, $\bigwedge xA(x)$, obtained by the application of the universal quantifier to the formula $A(x)$ with one free variable x. It follows from the assumption that the theories T and S coincide in their shorter sentences that the equivalence

(33) $\quad \ulcorner A(x/s_n) \urcorner \in T \equiv \ulcorner A(x/s_n) \urcorner \in S$

is satisfied for every $n \in U$.

If now the sentence $\bigwedge xA(x)$ is in T, then, by axiom schema 2 of the c.l.c., $\ulcorner A(x/s_n) \urcorner \in T$ for every $n \in U$. Hence, by (33), $\ulcorner A(x/s_n) \urcorner \in S$ for every n. Hence, since the theory S is d-complete, $\ulcorner \bigwedge xA(x) \urcorner \in S$. This proves the implication $\ulcorner \bigwedge xA(x) \urcorner \in T \rightarrow \ulcorner \bigwedge xA(x) \urcorner \in S$. The converse implication is obtained similarly. Since the existential quantifier can be defined by the universal quantifier and negation, the corresponding equivalence for the existantial quantifier is obtainable, too.

Thus, we conclude by induction that for every sentence A written in terms of the theory in question the equivalence $A \in T \equiv A \in S$ is true. In accordance with the axiom of extensionality this means that $T = S$.

We have so far been considering d-completeness with respect to any sequence of constants. Such constants may be single constant terms or may be compound ones. An important role is played by the set of all those constants which can be formed of certain initial single constants and all the function symbols considered in a given theory. That set is often called the *set of terms* of a given theory.

Individual constants (briefly: *terms*) formed of symbols $F_1, \ldots, F_k$ and of symbols from a sequence $\{s_t\}$ are any superpositions of symbols $F_1, \ldots, F_k$ and symbols $\{s_t\}$, such as $F_1\big(s_2, F_3\big(F_4(s_2, s_5)\big)\big)$, $F_2\big(F_1(s_3, s_5)$, $F_3(s_5)\big)$, etc. Some authors refer to all individual phrases as terms. A rigorous definition by induction of the constants formed of $F_1, \ldots, F_k$ and $\{s_t\}$ follows.

DEFINITION 10.

1. *Every symbol from a sequence $\{s_t\}$ belongs to the terms formed of $F_1, \ldots, F_k$ and the sequence $\{s_t\}$.*

2. *If $t_1, \ldots, t_m$ are constant terms, formed of $F_1, \ldots, F_k$ and the sequence $\{s_t\}$, and if a functor F_i $(1 \leqslant i \leqslant k)$ is a functor of m arguments, then the term*

$$F_i(t_1, \ldots, t_m)$$

is also a term constant formed of $F_1, \ldots, F_k$ and the sequence $\{s_t\}$.

The concept of descriptive completeness will hereafter be used so as if Definition 9 we had written "*a sequence $\{s_t\}_U$ of terms*" instead of "*a sequence $\{s_t\}_U$ of constant terms*". The same applies to Theorems 17 and 18.

THEOREM 19. *If a theory T is d-complete with respect to a subsequence of terms $\{r_v\}$, then it is d-complete with respect to the sequence of all terms.*

PROOF. Suppose that a theory T is d-complete with respect to a subsequence of terms $\{r_v\}$. If $\{t_v\}$ is the sequence of all the terms, and if for every term t_v it is true that $\ulcorner A(t_v) \urcorner \in T$, then obviously for every

r_v it is true that $\ulcorner A(r_v) \urcorner \in T$, since the sequence $\{r_v\}$ is a subsequence of the sequence $\{t_v\}$. It follows from the assumption that $\ulcorner \bigwedge x A(x) \urcorner \in T$. Hence the following implication is true:

$$\text{If } \ulcorner A(t_v) \urcorner \in T \text{ for every term } t_v, \text{ then } \ulcorner \bigwedge x A(x) \urcorner \in T,$$

which, by Definition 9, means that T is d-complete with respect to the sequence of all the terms.

If a theory T is d-complete with respect to a sequence $\{s_t\}$ of terms, then, as mentioned before, it is natural to assume that the terms s_t are names of all the objects considered in the theory T. In particular, it may be assumed, e.g. that the terms s_t are names of themselves, and to construct for the theory T a model $\mathfrak{M} = \langle X, p_1, \ldots, p_n \rangle$ in which the set X is the set of all the terms s_t. Such a model shall be called a *model constructed on the terms of the theory T*. This model will be described in detail by proving the following theorem:

THEOREM 20. *If T is a theory with the predicates $P_1, \ldots, P_n$, function constants $F_1, \ldots, F_k$, and individual constants $\{s_t\}_U$, and if T is complete, consistent, and d-complete with respect to the sequence of all the terms formed of the constants $\{s_t\}_U$ and the function constants $F_1, \ldots, F_k$, then the theory T is identical with the set of sentences true in a model $\mathfrak{M}$ constructed of terms.*

PROOF. The model $\mathfrak{M}$ is defined as follows:

$$\mathfrak{M} = \langle X, p_1, \ldots, p_n, f_1, \ldots, f_k, \{\ulcorner s_t \urcorner\}_U \rangle.$$

The set X consists of all the terms formed of the constants $\{s_t\}$ and of the function terms $F_1, \ldots, F_k$. Those terms, treated as individual elements of the set X, will be written in special brackets: $\ulcorner \ \urcorner$. The relations p_i and the functions f_i are defined on terms: If t_n, t_m are terms, and if the predicate P_i and the function F_i have, assume, two arguments each, then

$$(34) \quad \begin{aligned} & p_i(\ulcorner t_n \urcorner, \ulcorner t_m \urcorner) \equiv \ulcorner P_i(t_n, t_m) \urcorner \in T, \\ & f_i(\ulcorner t_n \urcorner, \ulcorner t_m \urcorner) = \ulcorner F_i(t_n, t_m) \urcorner. \end{aligned}$$

As can be seen, the functions f_i do not yield formulae other than terms.

It will be demonstrated that $T = E(\mathfrak{M})$ for the interpretation of the predicates $P_1, \ldots, P_n$ as names of the relations $p_1, \ldots, p_n$, the function constants $F_1, \ldots, F_k$ as names of the functions $f_1, \ldots, f_k$, and the individual constants $\{s_t\}$ as names of themselves.

By Theorems 9, 10, 15, and 17, the set $E(\mathfrak{M})$ is consistent, complete, and d-complete with respect to the sequence of all the terms, since the terms are names of all the elements of the set X in the domain $\mathfrak{M}$, because every term is considered to be a name of itself. Hence, in order to prove that $T = E(\mathfrak{M})$ it suffices, by Theorem 18, to demonstrate that the atomic theorems of the set T coincide with the atomic sentences of the set $E(\mathfrak{M})$. To do so we have to prove first that, for any sequence $\{a_n\}$ whose terms are in X and for any term t, it is true that

$$(35) \qquad V(t, \{a_n\}) = \ulcorner t \urcorner,$$

for the adopted interpretation of constant terms.

In fact, for those terms which are just constants in the sequence $\{s_t\}$ formula (35) is a direct consequence of condition 3 in Definition 2*a. And if formula (35) is true for the terms t_n, t_m, then, by Definition (34) and part 2 of Definition 2*a, it is also true for the compound term $F_i(t_n, t_m)$. Hence, by induction with respect to the complexity of terms, we have that formula (35) is true for any terms.

Formula (35), Definition (34) of the relation p_i, and Definition 2* yield that, for any sequence $\{a_n\}$ whose terms are in X and for the interpretation of terms under consideration,

$$V(\ulcorner P_i(t_n, t_m) \urcorner, \{a_n\}) = 1 \equiv p_i(\ulcorner t_n \urcorner, \ulcorner t_m \urcorner) \equiv P_i(t_n, t_m) \in T,$$

which, by Definition 4*, means that

$$P_i(t_n, t_m) \in E(\mathfrak{M}) \equiv P_i(t_n, t_m) \in T.$$

Now by applying Theorem 18 we have that $T = E(\mathfrak{M})$ for the above described interpretation of terms.

If the theory T does not include function constants, but only the predicates $P_1, \ldots, P_n$ and a sequence $\{s_t\}$ of constant terms, then the sequence $\{s_t\}$ obviously is the set of all the terms of that theory, and in order to define a model constructed on the constants $\{s_t\}$ it suffices to define the relations p_i by formula (34).

d-consistency and constructiveness of theories

The concept of d-completeness is associated with the concepts of descriptive consistency and constructiveness of theories.

2. CONCEPT OF TRUTH AND MODEL

DEFINITION 11.

a. *A theory T is constructive with respect to a sequence $\{s_t\}_U$ of terms if and only if, for every formula $A(x)$ with one free variable x, the following implication holds:*

(36)   *If $\bigvee xA(x) \in T$, then, for some index t, $\ulcorner A(x/s_t)\urcorner \in T$.*

A constructive theory contains a concrete case of every existential theorem.

b. *A theory T is d-consistent with respect to a sequence $\{s_t\}_U$ of terms if and only if, for any formula $A(x)$ with one free variable x, the following implication holds:*

(37)   *If $\bigwedge t \in U \ulcorner A(x/s_t)\urcorner \in T$, then $\ulcorner \bigvee x \sim A(x)\urcorner \notin T$;*

the implication states that if, for every term s_t, the sentence $A(x/s_t)$ is a theorem, then the sentence stating that there exists an element which does not have the property A cannot be a theorem.

If a theory T is d-consistent with respect to a denumerable sequence of constant terms, then it is called *ω-consistent*.

The concept of d-consistency originates from the same intuitive idea as does that of d-completeness. It is assumed that the terms $\{s_t\}$ are names of all the elements of the domain considered in a given theory T. Hence, if, for every term s_t, $A(x/s_t)$ can be proved, this means that all the elements of the domain in question have the property A. Hence, there should be no theorem stating that there exists an object that does not have the property A. d-consistency is a weaker condition than d-completeness: in the case of d-completeness, for the same assumption, it is required that the general sentence $\bigwedge xA(x)$ should be a theorem, whereas in the case of d-consistency it is required that the negation of that sentence should not be a theorem.

On converting condition (37) on the strength of the law of transposition we obtain another formulation of the definition of d-consistency:

A theory T is d-consistent if and only if, for any formula $A(x)$ with one free variable x, the following implication holds:

(38)   *If $\bigvee xA(x) \in T$, then, for some index t, $\ulcorner \sim A(x/s_t)\urcorner \notin T$.*

If an existential sentence is a theorem, then the addition to a theory

T of at least one concrete case of that sentence cannot result in a contradiction.

Comparison of Definition 11a, condition (36), and condition (38) shows that the concept of d-consistency is also weaker than that of constructiveness.

THEOREM 21.

a. *If a theory T is consistent and constructive (with respect to $\{s_t\}$), then it is d-consistent as well.*

b. *If a theory T is complete and d-consistent, then it is d-complete as well.*

c. *If a theory T is consistent, complete, and d-complete, then it is constructive as well.*

PROOF. a. If $\bigvee xA(x) \in T$, then, by Definition 11a, for some t, $\ulcorner A(x/s_t)\urcorner \in T$. If T is consistent, then $\ulcorner \sim A(x/s_t)\urcorner \notin T$. This satisfies condition (38), which is a necessary and sufficient condition of d-consistency.

b. If $\bigwedge t \in U \ulcorner A(x/s_t)\urcorner \in T$, then Definition 11b yields that if T is d-consistent, then $\ulcorner \bigvee x \sim A(x)\urcorner \notin T$. And if T is a complete theory, then the negation of the above sentence, that is, the sentence $\sim \bigvee x \sim A(x)$, must be in T. By De Morgan's laws, the formula $\bigwedge xA(x)$ must also be a theorem of T.

c. If $\ulcorner \bigvee xA(x)\urcorner \in T$, then, by De Morgan's laws, $\ulcorner \sim \bigwedge x \sim (A(x))\urcorner \in T$. And if T is consistent, then $\ulcorner \bigwedge x \sim A(x)\urcorner \notin T$. This and the definition of d-completeness yield, by the law of transposition, that there is a $t \in U$ such that $\ulcorner \sim A(x/s_t)\urcorner \notin T$; this and the the definition of completeness yield that $\ulcorner A(x/s_t)\urcorner \in T$.

On the other hand, it is possible to give examples of theories which are consistent, complete, and d-inconsistent, and, hence, are not d-complete. Further, a theory can be d-inconsistent with respect to a sequence of constant terms, and d-consistent with respect to another sequence of constant terms.

A set of sentences true in a domain is, thus, complete, consistent, d-complete, d-consistent, and constructive with respect to a sequence of term constants which includes names of all the objects from that domain. These properties form sufficient grounds to believe that the set of true sentences has been defined correctly.

Note also the following theorem on isomorphism:

THEOREM 22. *If $\mathfrak{M}$ ism $\mathfrak{N}$, then* $E(\mathfrak{M}) = E(\mathfrak{N})$.

The proof follows directly from Theorem 4 (cf. the preceding section). It follows from formula (28) that if every sequence from $\mathfrak{M}$ satisfies A in $\mathfrak{M}$, then every sequence from $\mathfrak{N}$ satisfies A in $\mathfrak{N}$. This is so because every sequence from $\mathfrak{N}$ is an image of a sequence from $\mathfrak{M}$. Hence, by Definition 4, if $A \in E(\mathfrak{M})$, then $A \in E(\mathfrak{N})$. Converse results are obtained analogically.

In Theorem 22, truth is defined in each case, of course, for the interpretation of terms as names of those relations in the two models which are isomorphic.

This theorem is used frequently in all branches of mathematics, although it is rarely formulated as a separate theorem. If we prove something about a domain, we assert the same about any other domain isomorphic with the former.

EXERCISES

1. Give an example of two domains which are not distinguishable from one another (in the sense that $E(\mathfrak{N}) = E(\mathfrak{M})$), but are not isomorphic.

H i n t. Consider an order of the type $\omega^* + \omega$ and its multiples.

2. Give an example of two non-isomorphic domains each of which is isomorphic with a subdomain of the other.

3. It is said that a domain $\mathfrak{N} = \langle X, p \rangle$ is an *elementary extension of a domain* $\mathfrak{M} = \langle Y, r \rangle$ if there is a function f such that $f(Y) \subset X$, and if for every formula Φ and every sequence $\{a_n\}$ from Y the following equivalence holds:

$$\{a_n\} \text{ satisfies } \Phi \text{ in } \mathfrak{M} \equiv \{f(a_n)\} \text{ satisfies } \Phi \text{ in } \mathfrak{N}.$$

Investigate the properties of the relation of being an elementary extension.

4. Prove that the relation of being an elementary extension is stronger than indistinguishability, and weaker than isomorphism.

3. EXISTENCE OF ω-COMPLETE EXTENSIONS AND DENUMERABLE MODELS

Existence of ω-complete extensions

Theorem 20 shows that the search for a model of a theory can be reduced to the search for a consistent, complete and d-complete extension of that theory.

If a set X of sentences is consistent, then there is no major problem in finding a complete and d-complete supertheory. It is simply the case that if the set X is not yet a complete theory, then there exist independent sentences, that is, sentences A such that neither A nor $\sim A$ is in $Cn(X)$. There is a simple method of adding some of the independent sentences to the set X so that the result is a theory T which contains the set X and is consistent, complete, and ω-complete. By Theorem 19, that theory determines a model for the set X.

The method we are referring to is the *completion method* due to A. Lindenbaum and consists, in fact, in multiple iteration of the same reasoning which led to the proof of Theorem 12. This method will now be described in greater detail in proving the following theorem:

THEOREM 23. *If T is a consistent set of sentences with terms $P_1, \ldots, P_n, F_1, \ldots, F_k$ and possibly certain constants $\{v_n\}$, then there exists a theory S such that it has the same constant terms that the theory T has, has in addition a denumerable number of new constant terms $\{s_n\}$, is consistent, complete, and ω-complete with respect to the constants $\{s_n\}$, and contains the set T.*

Thus the language of the theory S, in addition to the terms of the theory T, also contains the sequence $\{s_n\}$ of constants which do not occur in T. Apart from the constant terms from the sequence $\{s_n\}$ no other symbols are added. The identity term Id, if it occurs in X, is one of the terms $P_1, \ldots, P_n$ and is not treated in any special way.

PROOF. In order to properly describe the theorems of the system S we have to arrange all the sentences of the system S as one sequence. This can be done in many ways. Every sentence consists of a finite number of separate signs, symbols, or terms. The number of initial symbols in every ordinary theory is finite, too. The assumption that there are denumerably many distinct symbols does not make the proof any more difficult. If it is said that there are infinitely many constants or variables, then it is usually assumed that their form has been fixed in advance. For instance, according to the convention adopted in the preceding section, each variable consist of the letter x and a number of subscripted strokes, so that each variable is a graphic combination of two signs only. Hence, if we confine ourselves to sentences of definite length, for instance, of ten signs, then the number of such sentences is finite. There are, of course, numbers such that there are no sentences with such a number of words: for instance, in the theories here under consideration there are no one-word and two-word sentences. But there can be three-word sentences of the form $a = b$, if the identity symbol is written between the names of variables.

If the terms of a sequence $\{s_n\}$ have, for instance, the form of the letter a with a number of subscripted strokes, then the sentence $a_/ = a_/$ would be a five-word sentence, and the sentence $a_/ = a_{///}$, a seven-word sentence. The sentence $\sim P_/(a_/)$, and some others, are seven-word sentences, while the sentences: $\bigwedge x_/ P_/'(x_/)$, $\bigvee x_/ x_/ = x_{///}$, etc., are ten-word sentences.

Thus all the formulae can be numbered, beginning with all one-word sentences (if such exist), next all two-word sentences, next all three-word sentences, etc.

Among the formulae with the same number of words we may adopt any ordering, for instance the alphabetical one, which consists in adopting a fixed order among the basic signs, as is the case with letters in an alphabet, and next we order words first after the first sign, and those which have the same first sign, after the second sign, etc. Such an ordering is called *alphabetical* or *lexicographical*. It arranges all the formulae of a language under consideration into a single denumerable sequence, so that it associates a number with each formula, when it is said hereafter that a formula is earlier than another, or that a formula is the earliest of the formulae of a certain form, this will always mean earlier with respect to the order fixed by the numbering of formulae as described above.

We can now proceed to describe the axioms of a system S. That system S will be described so that practically all its theorems will be its axioms. In other words, we will not strive to find in S a relatively small system of axioms of S. The system S will be made to include all those sentences which are needed to construct a complete and ω-complete theory that would contain a theory T.

The set S is defined so that we examine one by one all the sentences written in the language of the system S. The sentences are examined in the order of their numbers (cf. the system of numbering described above). Let A_n stand for the nth sentence of the language in question, in the order described above. The numbering begins with A_0.

313

When we consider, one by one, all the sentences from the sequence $\{A_n\}$, either a sentence A_n or its negation is included in the system S. Hence the system S can best be defined as the sum of an increasing sequence $\{S_n\}$. That sequence is defined by induction:

$$S_0 = T.$$

$$(39) \quad S_{n+1} = \begin{cases} S_n \cup \{A_n\}, & \text{if } \ulcorner \sim A_n \urcorner \notin \text{Cn*}(S_n) \text{ and } A_n \text{ does not begin with the existential quantifier } \bigvee x_1; \\ S_n \cup \{A_n, \ulcorner B(s_m) \urcorner\}, & \text{if } \ulcorner \sim A_n \urcorner \notin \text{Cn*}(S_n) \text{ and } A_n = \bigvee x_1 B(x_1) \text{ and } s_m \text{ is the first constant in the sequence } \{s_n\} \text{ which has not occurred in the sentences } S_0, \ldots, S_n; \\ S_n \cup \{\ulcorner \sim A_n \urcorner\}, & \text{if } \ulcorner \sim A_n \urcorner \in \text{Cn*}(S_n). \end{cases}$$

$$(40) \quad S = \bigcup_n S_n.$$

In a verbal formulation: S contains the whole set T, and then, one by one, the sentences A_n or their negations. If the negation of a sentence A_n is a consequence of sentences already included in S, then it is included in the system S, and if it is not a consequence of sentences so included, then we include in S the sentence A_n itself. And if a sentence A_n is existential, then together with A_n we include in S a concrete case of that sentence. That concrete case is included, as can easily be guessed, in order to ensure the ω-consistency of the system S.

We shall now analyse the properties of the set S and the sequence $\{S_n\}$.

1. *The sequence $\{S_n\}$ is increasing*: $S_n \subset S_{n+1}$.

This follows directly from its definition.

2. *For every n, the set S_n is consistent.*

The proof is by induction. The set $T = S_0$ is consistent by definition. If a set S_n is consistent, then, by Theorem 12, the set S_{n+1} is consistent, too, since it derives from the set S_n by adding to the former either a consequence of the set S_n, or a sentence whose negation is not a consequence of S_n. Only the second case of the three occurring in formula (39) may give rise to doubts. But it can easily be seen that the sentence $\sim B(s_m)$ also is not a consequence of S_n if $\ulcorner \sim \bigvee x_1(B(x_1)) \urcorner \notin Cn^*(S_n)$. For suppose that $\ulcorner \sim B(s_m) \urcorner \in Cn^*(S_n)$. But s_m by assumption does not occur in the sentences of the set S_n. Hence, it follows from the properties of consequence that there is a finite number of sentences $D_1, \ldots, D_k \in S_n$ such that

$$\ulcorner (D_1 \wedge \ldots \wedge D_k) \to \sim B(s_m) \urcorner \in L^*.$$

As is known, the theorems of logic do not privilege any constants, and do not provide criteria from distinguishing constants from variables, hence

$$\ulcorner \bigwedge x((D_1 \wedge \ldots \wedge D_k) \to \sim B(x)) \urcorner \in L^*;$$

since $D_1, \ldots, D_k$ by assumption do not contain the constant s_m, also

$$\ulcorner (D_1 \wedge \ldots \wedge D_k) \to \bigwedge x \sim B(x) \urcorner \in L^*,$$

so that, in accordance with the properties of consequence, we have $\ulcorner \bigwedge x \sim B(x) \urcorner \in Cn^*(S_n)$. If the addition of $B(s_m)$ results in a contradiction, then the sentence $\bigvee x_1 B(x_1)$ would also have to result in a contradiction.

3. *The set S is consistent.*

In accordance with the finitistic interpretation of the concept of consequence, if a contradiction $\ulcorner B \wedge \sim B \urcorner$ is a consequence of the set S, it would be a consequence of a finite number of sentences $A_{k_1}, \ldots, A_{k_n}$. But since the sequence $\{S_n\}$ of sets is increasing and S is its sum, that finite number of sentences $A_{k_1}, \ldots, A_{k_n}$ would have to be included as a whole in some set S_n. Thus, the contradiction $\ulcorner B \wedge \sim B \urcorner$ would be a consequence of a set S_m, which is in contradiction with 2 above.

4. *The set S is complete.*

Any sentence A of the language under consideration has its number: $A = A_n$. By formula (39),

$$A_n \in S_{n+1} \vee \ulcorner \sim A_n \urcorner \in S_{n+1}.$$

This and (40) yield that

$$A_n \in S \vee \ulcorner \sim A_n \urcorner \in S,$$

so that S is a complete set.

5. *The set S is a theory.*

If $A \in \mathrm{Cn}^*(S)$, then if $\ulcorner \sim A \urcorner \in S$, then S would be inconsistent, contrary to 3 above. If S is consistent, then $\ulcorner \sim A \urcorner \notin S$. Since S is complete, $A \in S \vee \ulcorner \sim A \urcorner \in S$. Since $\ulcorner \sim A \urcorner \notin S$, $A \in S$. This yields

$$\mathrm{Cn}^*_{P_1, \ldots, P_n, F_1, \ldots, F_k, \{v_n\}, \{s_n\}}(S) \subset S.$$

6. *The set S is ω-constructive with respect to the sequence $\{s_n\}$.*

In fact, if the sentence $\bigvee x_1 B(x_1)$ is in S and has its number $n \colon A_n = \ulcorner \bigvee x_1 B(x_1) \urcorner$, then, by formula (39):

$$S_{n+1} = S_n \cup \{A_n, \ulcorner B(s_m) \urcorner\}.$$

Only this case is possible, for should $\ulcorner \sim A_n \urcorner \in S_{n+1}$, then the system S would include the contradiction $A_n \wedge \sim A_n$. Hence, $B(s_m) \in S_{n+1}$, so that the sentence $B(s_m)$ for a constant s_m is in S together with the sentence $\bigvee x_1 B(x_1)$.

The set S is a complete, consistent, and ω-constructive theory with respect to the sequence $\{s_n\}$ of constants. Hence, by Theorem 21, the set S also is ω-complete with respect to the sequence $\{s_n\}$.

Theorem on the semantic completeness of the c.l.c. and its consequences

THEOREM 24 (K. Gödel [16]). *Every consistent set T of sentences has a denumerable model.*

More precisely: *If a set T is consistent and if the sentences in T contain predicate terms $P_1, \ldots, P_n$, function terms $F_1, \ldots, F_k$, and constants $\{v_n\}$, and do not contain any other constants, then there exists a domain*

$$\mathfrak{M} = \langle x, p_1, \ldots, p_n, f_1, \ldots, f_k, \{b_n\} \rangle$$

in which X is an enumerable set, $\{b_n\}$ is a designated subsequence of the set X, $p_1, \ldots, p_n$ are relations in X, $f_1, \ldots, f_k$ are totally separable functions with all their values in X, and $\mathfrak{M}$ is a model for $T \colon T \subset E(\mathfrak{M})$ for the interpretation of $P_1, \ldots, P_n$; $F_1, \ldots, F_k$; and $\{v_n\}$, respectively, as names of the relations $p_1, \ldots, p_n$; of the functions $f_1, \ldots, f_k$; and of the elements of $\{b_n\}$.

PROOF. If T is a consistent set, then, by Theorem 23, there exists a theory S, consistent, complete, and ω-complete with respect to an added sequence $\{s_n\}$ of constants, and such that

$$T \subset S.$$

3. SEMANTICAL COMPLETENESS

The constants of the sequence $\{s_n\}$ are, of course, some of the terms of the theory S. Hence, by Theorem 19, if S is ω-complete with respect to $\{s_n\}$, then S is d-complete with respect to the set of all its terms. Hence, by Theorem 20, the theory S is identical with the set of all the sentences true in the model:

$$\mathfrak{M} = \langle X, p_1, \ldots, p_n, f_1, \ldots, f_k, \{\ulcorner v_n \urcorner\}, \{\ulcorner s_n \urcorner\}\rangle$$

defined on terms, a model in which the relations $p_1, \ldots, p_n$ and the functions $f_1, \ldots, f_k$ are defined by formulae (34). Since $S = E(\mathfrak{M})$ and $T \subset S$, hence $T \subset E(\mathfrak{M})$, so that $\mathfrak{M}$ is a model for T. The set X is denumerable since it is the set of all terms formed of the denumerable set of constant terms $\{v_n\}$ and $\{s_n\}$ by superposition on them of a finite number of functions $F_1, \ldots, F_k$. Since the constants $\{s_n\}$ do not occur in sentences of the set T, the elements $\{\ulcorner s_n \urcorner\}$ need not to be singled out in the model $\mathfrak{M}$, when $\mathfrak{M}$ is analysed as a model of the set T. It suffices then to single out the sequence $b_n = \ulcorner v_n \urcorner$.

The precise formulation of Theorem 24 included the phrase: $f_1, \ldots, f_k$ are totally separable functions, which has not been explained so far.

DEFINITION 12. *A system of function* $f_1, \ldots, f_n$ *is a system of functions totally separable, or completely free, in a set* X *if and only if:*

0. *The values of the functions* f_i *are in* X.

1. *Each function* f_i *is one-to-one for arguments from* X.

2. *The value of a function* f_i *for a fixed set of arguments* $x_1, \ldots, x_n$ *from* X *always differs from each of its arguments.*

3. *The images of the set* X *given by different functions* f_i *and* f_j *are disjoint.*

Thus, for instance, should the functions $f_1, \ldots, f_k$ have two arguments each, the property of total separability could be symbolized as follows:

$$O_{f_i}(X \times X) \subset X,$$

$$f_i(x, y) \neq x, \quad f_i(x, y) \neq y,$$

$$f_i(x, y) = f_j(z, v) \equiv (x = z \wedge y = v \wedge i = j)$$

for all $i, j \in [1, \ldots, k]$ and $x, y, z, v \in X$,

To prove that the functions $f_1, \ldots, f_n$ from the model $\mathfrak{M}$ as defined by formulae (34) satisfy the above definition we point to the trivial fact that any two distinct terms differ by at least one symbol. Two terms t_1 and t_2, obtained by superposition from the terms t_1' and t_2' and the function symbols F_i and F_j:

$$\ulcorner t_1 \urcorner = \ulcorner F_i(t_1') \urcorner,$$
$$\ulcorner t_2 \urcorner = \ulcorner F_j(t_2') \urcorner,$$

are equiform if and only if both the terms t_1' and t_2' are equiform and the function symbols F_i and F_j are equiform. This fact and Definition (34) yield immediately that

$$f_i(t_1') = f_j(t_2') \equiv (t_1' = t_2' \wedge i = j).$$

It is also self-evident that

$$\ulcorner f_i(t) \urcorner \neq \ulcorner t \urcorner.$$

The functions $f_1, \ldots, f_k$ thus form a totally separable system.

Since the set X is denumerable, it is equinumerous with the set $\mathcal{N}$ of natural numbers. This means that there is a one-to-one function $h(x)$ which maps the set $\mathcal{N}$ onto the entire set X. Hence, we can construct a model $\mathfrak{M}'$ of the theory T, consisting of the set $\mathcal{N}$ of natural numbers and the relations $p_1', \ldots, p_n'$ defined as follows:

$$p_i'(x, y) \equiv p_i\big(h(x), h(y)\big).$$

The domain $\mathfrak{M}'$ is isomorphic with the domain $\mathfrak{M}$. Hence, since $\mathfrak{M}$ is a model for T, $\mathfrak{M}'$ is also a model for T in accordance with Theorem 22 on isomorphism. Theorem 24 may, thus, be formulated in the following way:

Every consistent set of sentences has a model in the set of natural numbers.

In a freer formulation:

For every consistent set Z of sentences there exists an interpretation of its terms in the domain $\aleph_0$ of natural numbers such that for that interpretation of terms $Z \subset E(\mathfrak{N}_0)$.

Theorem 24 makes it possible for us to prove many implications converse to theorems proved previously, and thus to formulate necessary and sufficient conditions of certain logical relationships. Thus, for instance, Theorem 24 is converse to Theorem 16 and makes it possible to strengthen Theorem 16 by changing it into an equivalence:

THEOREM 25. *A set T of sentences is consistent if and only if there exists a model for the set T.*

The proof follows directly from Theorems 16 and 24.

THEOREM 26 (Th. Skolem [85]). *A set T of sentences has a model if and only if it has a denumerable model.*

The proof follows directly from Theorems 24 and 25.

THEOREM 27. *If a sentence A is not a theorem of L*, then there exists a non-empty domain $\mathfrak{M}$ such that $\ulcorner \sim A \urcorner \in E(\mathfrak{M})$.*

PROOF. Let T stand for the set of those c.l.c. theorems which have the same terms as the sentence A has. The set T is a theory with those terms which occur in the sentence A.

If A is not a theorem, then, by Theorem 12, the set $\{ \sim A \} \cup T$ is consistent and, by Theorem 25, it has a model $\mathfrak{M}$ such that $\{ \sim A \} \cup T \subset E(\mathfrak{M})$, and hence $\ulcorner \sim A \urcorner \in E(\mathfrak{M})$.

THEOREM 28. *If a sentence A is true in every non-empty domain $\mathfrak{M}$ and for every interpretation of terms, then A is a theorem in L*.*

PROOF. If a sentence A is true in every non-empty domain, then, on the strength of the consistency of the set of true sentences (Theorem 15), there cannot exist a domain in which the sentence $\sim A$ would be true, too. Hence, by a transposition of Theorem 27, we have that A is a c.l.c. theorem.

THEOREM 29. *A sentence A is a theorem in L* if and only if it is true in every non-empty domain and for every interpretation of the terms it contains.*

In symbols:

(41) $A \in L^* \equiv \bigwedge \mathfrak{M} \neq \Lambda \big(A \in E(\mathfrak{M})\big).$

The proof follows directly from Theorems 28 and 10.

THEOREM 30. *If A is a sentence and X is a set of sentences, then*

(42) $A \in \mathrm{Cn}^*(X) \equiv \bigwedge \mathfrak{M} \neq \Lambda \big(X \subset E(\mathfrak{M}) \to A \in E(\mathfrak{M})\big).$

In a verbal formulation: *$A \in \mathrm{Cn}^*(X)$ if and only if, for every non-empty domain $\mathfrak{M}$ and for every interpretation of the terms contained in the sentence A and in the set X, the following implication holds: If X consists of sentences true in $\mathfrak{M}$ for an interpretation of the terms occurring in the sentences of the set $X \cup \{A\}$, then A also is a sentence true in $\mathfrak{M}$ for the same interpretation of the terms.*

319

The symbolic formulation is not satisfactory in view of the fact that the interpretation of terms is not indicated in the symbol $E(\mathfrak{M})$. But the introduction of more complicated notation would make the symbolism unclear.

PROOF. If $A \in \mathrm{Cn}^*(X)$, then, for a finite set of sentences $B_1, \ldots, B_n \in X$, $((B_1 \wedge \ldots \wedge B_n) \to A) \in L^*$, and, hence, by (41), for every domain $\mathfrak{M}$

(43) $\qquad \{(B_1 \wedge \ldots \wedge B_n) \to A\} \in E(\mathfrak{M})$.

By Theorem 9, the set $E(\mathfrak{M})$ is a theory, and, hence, by (43) and $X \subset E(\mathfrak{M})$, we have $\ulcorner(B_1 \wedge \ldots \wedge B_n)\urcorner \in E(\mathfrak{M})$, then also $A \in E(\mathfrak{M})$.

Conversely, suppose $A \notin \mathrm{Cn}^*(X)$; by Theorem 12, the set $\{\ulcorner \sim A \urcorner\} \cup X$ is consistent and, by Theorem 25, if has a non-empty model $\mathfrak{M}$.

Since $\{\ulcorner \sim A \urcorner\} \cup X \subset E(\mathfrak{M})$, hence $\ulcorner \sim A \urcorner \in E(\mathfrak{M})$ and $X \subset E(\mathfrak{M})$. As stated by Theorem 15, the set $E(\mathfrak{M})$ is consistent. Now, since $\ulcorner \sim A \urcorner \in E(\mathfrak{M})$, hence $A \notin E(\mathfrak{M})$. Consequently, there exists an $\mathfrak{M}$ such that $X \subset E(\mathfrak{M})$ and $A \notin E(\mathfrak{M})$, which means that the second part of equivalence (42) is not true if the the first is not true. This by transpositition yields the implication $\leftarrow$.

If we confine ourselves to formulae without term constants and function symbols, then in Theorems 27–30 the symbols L^* and Cn^* may be replaced by L and Cn. It can also easily be seen that for those sentences which contain neither term constants nor function symbols Theorems 31 and 32 may be formulated without the restriction that the domain in question be non-empty, namely:

(44) $\qquad \begin{aligned} &A \in L' \equiv \bigwedge \mathfrak{M} \, A \in E(\mathfrak{M}), \\ &A \in \mathrm{Cn}'(X) \equiv \bigwedge \mathfrak{M}(X \subset E(\mathfrak{M}) \to A \in E(\mathfrak{M})), \end{aligned}$

if L' stands for the set of the consequences by detachment of all logical axioms except for Axiom 0, which states that the domain in question is not empty, and if it is assumed that $\mathrm{Cn}'(X) = \mathrm{Cn}_0(L' \cup X)$.

Theorem 30 provides a general method of proving that a sentence is not a consequence of a set of sentences, because formula (42) yields, by negation, the following one:

$$A \notin \mathrm{Cn}^*(X) \equiv \bigwedge \mathfrak{M} \ne \Lambda(X \subset E(\mathfrak{M}) \wedge A \notin E(\mathfrak{M})).$$

For instance, if we want to demonstrate that the commutativity condition $x+y = y+x$ is not a consequence of Axioms 404–406 of group theory, it is necessary and sufficient to indicate a domain in which the axioms of identity and those of group theory are true, and the commutativity condition is not true. Such a domain does in fact exist (cf. exercise 4, p. 324).

The theorems just proved are important for understanding the classical logical calculus. When we adopted the axioms of the c.l.c., we demonstrated that the c.l.c. axioms are true, in an intuitive sense, in every domain. The same property was next ascribed to all c.l.c. theorems. At that time an intuitive concept of truth was being used. That concept has been formulated with more precision in the present chapter. It has turned out that the above property is preserved following that reformulation, and that the converse implication can be proved, too. The classical axiomatic logical calculus contains exactly all those sentences which are true in every domain, that is, it contains all logical tautologies, and logical tautologies only, in accordance with Definition 16* of logical tautology, as given is Section 6 of the preceding chapter. Theorem 28 is the converse implication of Theorem 6, intuitively arrived at in Section 6 of the preceding chapter. Hence, the axiom system of the c.l.c. has been chosen correctly, since it makes it possible to carry out all those reasonings which are universally valid, i.e., are valid in every domain.

This property of the c.l.c. might be termed *semantic completeness* of the axiomatic logical calculus, since that calculus includes the complete set of logical theorems true in all domains. This property is often called *completeness*, but this is obviously a property quite different from that defined by Definition 6 in the preceding section. This is why a different term had better be used. The c.l.c. includes the complete set of universally valid logical laws, but is not a complete theory in the sense of Definition 6. For instance, sentences of the type

$$\bigvee x, y \sim (y = x),$$

$$\bigvee x, y, z \big(\sim (x = y) \wedge \sim (x = z) \wedge \sim (y = z) \big), \quad \text{etc.,}$$

stating that there exist at least 2 different objects, at least 3 different objects, etc., are independent of the c.l.c., since they are not true in less numerous domains: the first is not true in one-element domains, the second is not true in two-element domains, etc. The negations of those sentences also are not universally valid, since they are not true in greater domains. Hence, those sentences are independent of the set of sentences true in all domains.

Theorem 29 is a definition of the classical logical calculus. It is a semantic definition, since it refers to models, that is, to the reality to which that calculus applies. This definition is independent of any system of axioms and rules. Reference to a given set of axioms and rules may seem arbitrary, and, hence, the semantic definition of the c.l.c. is reputed to be the best, as it is in a sense absolute. [But the result is, perhaps, not as far-reaching as it would seem.] The equivalence of the axiomatic system of calculus with the calculus defined semantically is arrived at because of the adoption in the metasystem of ordinary classical logic. Thus, the starting point is the intuitive adoption of certain laws of logic. Hence, it may be concluded that the theorems discussed in this section refer rather to the correctness of the concept of truth than to the correctness of the axiomatic system of logic. At any rate, the agreement of the concept of truth with the adoption of an axiomatic system of logic seems to be a non-trivial conclusion.

Theorems 25, 29, and 30 describe, by reference to the concept of truth, the meaning of such purely formal concepts as consistency, theorems of the axiomatic logical calculus, and logical consequence.

The meaning of Theorem 26 is somewhat different. It is often believed to be paradoxical, as it shows that ordinary methods of inference do not make it possible to single out non-denumerable domains. The theory of real numbers, if we do not go beyond ordinary logical methods and ordinary symbolism, always has a denumerable model, even though it may include very strong axioms. Likewise, set theory refers to sets of arbitrarily high powers, and yet has a denumerable model. Of course, those denumerable models of set theory or the theory of natural numbers are very strange and do not comply with the intentions underlying those theories, though they satisfy the axioms. The paradox is called the *Skolem–Löwenheim paradox*.

The theories mentioned above of course also have non-denumerable models. The non-denumerable set of real numbers is, of course, a model of the theory of real numbers. It is even possible to prove a stronger theorem, stating that every consistent set of sentences which has a denumerable model also has models of arbitrarily high power. Such models satisfy an additional condition, to be discussed in the next section, and this is why the proof of this property is postponed until the next section.

Note also a strengthening of Theorems 28–30, which will be required below and which follows directly from Theorem 24.

THEOREM 31. *If sentences from the set X and a sentence A contain function terms* $F_1, ..., F_n$, *then*

1. *if A is true in every non-empty domain* $\mathfrak{M}$ *and for every interpretation of the terms* $F_1, ..., F_n$ *as names of totally disjoint functions* $f_1, ..., f_n$, *then* $A \in L^*$,

2. *if, for every non-empty domain* $\mathfrak{M}$ *and for every interpretation of the terms* $F_1, ..., F_n$ *as names of totally disjoint functions* $f_1, ..., f_n$, *the following consequence holds*: *if* $X \subset E(\mathfrak{M})$, *then* $A \in E(\mathfrak{M})$, *then* $A \in Cn^*(X)$.

PROOF. Should $A \notin L^*$, that is, $A \notin Cn(AxL^*)$, then the set $AxL^* \cup \{\sim A\}$ would be consistent and, by Theorem 24, it would have a model $\mathfrak{M}$ in which the terms $F_1, ..., F_n$ would be interpreted as names of certain totally disjoint functions $f_1, ..., f_n$. Hence, in a some model $\mathfrak{M}$ both A and $\sim A$ would be true, which is not possible.

In the case 2 the proof is analogical.

The theorems discussed in this section have been the subject matter of various investigations. Various partly independent proofs have been given, in addition to the authors mentioned above, by H. Rasiowa and R. Sikorski [68], L. Henkin [21], L. Rieger [69], J. Łoś [47]. See also E. W. Beth [4].

EXERCISES

1. Prove that if there exist sentences independent of a theory T, then the theory T has at least two different (not isomorphic) denumerable models.

2. Prove the general theorem stating that the sum of the increasing sequence $T_1 \subset T_2 \subset T_3 \subset ...$ of consistent theories also is a consistent theory.

3. Referring to the following geometrical model:

$$r \nearrow^2 \searrow r$$
$$1 \xleftarrow{r} 3$$

describe arithmetically the domain $\langle \{1, 2, 3\}, r \rangle$, in which the following sentences are true:

$$\bigwedge x, y(xRy \rightarrow \sim (yRx)),$$
$$\bigwedge x, y(xRy \lor yRx \lor x = y),$$

323

and the sentence stating that the relation R is transitive:

$$\bigwedge x, y, z((xRy \wedge yRz) \rightarrow xRz)$$

is false.

4. Demonstrate that the six-element group of permutations of the 3 numbers $\{0, 1, 2\}$ is not commutative.

5. Prove in an analogical way that none of sentences 404–406 follow from the remaining ones and the axioms of identity; that none of the axioms of the theory of the *less than* relation 106–111 follow from the remaining ones; and that no pair: 611a–b, 612a–b, 613a–b follows from the remaining axioms of Boolean algebra.

6. Prove that if a set Z of sentences has the property that every finite subset of the set Z is consistent, then the whole set Z is consistent, too.

7. Deduce formulae (44).

8. Prove that if $\{v_n\}$ is a sequence of constants which univocally denote elements from a set X, then a system of functions $f_1, \dots, f_n$ is totally free if and only if, for the domain $\mathfrak{M} = \langle X, f_1, \dots, f_n, \{b_n\} \rangle$, where $\{b_n\}$ is a sequence of elements of the set X associated respectively with names of the sequence $\{v_n\}$, it is true that for any two terms t and t' (formed of $F_1, \dots, F_n$ and the sequence $\{v_n\}$)

$$\text{if } t \neq t', \text{ then } V(t, \{a_n\}) \neq V(t', \{a_n\}).$$

4. SOME OTHER CONCEPTS AND RESULTS IN MODEL THEORY

Model theory is now a highly developed branch of mathematics, and the present book presents only some of its results, which are more closely connected with logic.

The method of constructing a model for any consistent set of sentences, as described in the preceding section, may prove unsatisfactory. It is true that it yields a model, but, as has been noted at the end of the section, such a model will lack certain properties which may be essential for the domains we want to describe in the theory under investigation. The models discussed in the preceding section are always denumerable, whereas a theory may be intended to refer to non-denumerable sets. A model obtained by a mechanical completion of the set of theorems up to an ω-complete theory may, thus, not comply with the intentions of the theory in question. This has given rise to the notion of making a distinction between those models which comply with the intentions underlying a given theory and those models which distort the concepts used in a theory (*non-absolute, non-standard* models according to fairly common terminology).

A general definition of standardness and non-standardness will not be given here. It is possible to give such a general definition; it states more or less that a relation is a *standard* one if it is isomorphic with a relation which in the metasystem is considered to be a correctly defined version of a concept under consideration. We shall be concerned with two special cases of standardness. The first refers to the concept of identity.

Absolute models for the concept of identity

So far we have not singled out the identity predicate from among the other predicates of a theory under consideration. The identity predicate Id or = could be one of the predicates $P_0, ..., P_n$ as discussed in the preceding section. In the construction of a model out of constants it was interpreted as one of the relations $p_0, ..., p_n$. Now that we want to single it out, it will be assumed that in the formalized theories under consideration the identity predicate has the form Id, while in the metasystem in which theose theories are investigated, it is expressed by the equality symbol =.

DEFINITION 13. *A domain* $\mathfrak{M} = \langle X, =, p_1, ..., p_n \rangle$ *is a model of a theory T with the terms* Id, $P_1, ..., P_n$, *absolute with respect to the concept of identity (or for an absolute or standard interpretation of the concept of identity) if and only if every sentence of the theory T is true in the domain* $\mathfrak{M}$ *for the interpretation of the terms* Id, $P_1, ..., P_n$ *as names of the relations* $=, p_1, ..., p_n$, *it being assumed that the theory T is based on the ordinary axioms of identity which state that* Id *is an equivalence relation that preserves all the predicates* $P_1, ..., P_n$.

The requirement that in a model of any theory the term Id be always interpreted as a name of the identity relation seems rather natural. Thus Definition 13 formulates some of those requirements which might be associated with the concept of a "good" or "adequate" model, or a model which "correctly interprets" certain concepts. A model which interprets the term Id not as identity, of course, does not interpret it in accordance with the intentions underlying a given theory. By adding to a system T a denumerable number of constants $\{s_n\}$, as was done in the proof of Theorem 23, we usually obtain a model $\mathfrak{M}$, in which

325

the term Id (if it occurs among the terms $P_0, ..., P_n$ of the theory T, e.g., as the predicate P_0) is not interpreted absolutely. This means that in the model $\mathfrak{M} = \langle X, p_0, ..., p_n \rangle$ of the theory T there are two individuals $x, y \in X$ such that $x \neq y$, but $x p_0 y$, where p_0 is a relation which interprets the term Id occurring in the theory T. As will soon be proved, the relation p_0 must be an equivalence relation which preserves the remaining relations $p_1, ..., p_n$. Obviously, not every equivalence which preserves certain relations must be identity. But the case of identity is fairly simple. Every model with a non-absolute interpretation of identity can easily be changed into one with an absolute interpretation of identity; to do so we have to form a quotient domain with respect to the congruence p_0.

THEOREM 32. *If in a theory T it is assumed about the identity term* Id *that it is an equivalence which preserves the predicates $P_1, ..., P_n$, and if* $\mathfrak{M} = \langle X, p_0, p_1, ..., p_n \rangle$ *is a model of the theory T for the interpretation of the terms* Id, $P_1, ..., P_n$ *occurring in the theory T as names of the relations $p_0, ..., p_n$, then p_0 is a congruence in* $\mathfrak{M}$, *and the domain* $\mathfrak{M}/p_0 = \langle \mathrm{Abs} X_{p_0}, =, \bar{p}_1^{p_0}, ..., \bar{p}_n^{p_0} \rangle$ *is a model of the theory T with an absolute concept of identity.*

PROOF. If p_0 is the designatum of the term Id, then p_0 is a congruence in the set X which preserves the relations $p_1, ..., p_n$. This can be verified on an example. Suppose that P_1 is a predicate of two arguments. Since it is always assumed about identity that it preserves operations, the sentence

$$\bigwedge x, y, z, v\big(\big(\mathrm{Id}(x, y) \wedge \mathrm{Id}(z, v) \wedge P_1(x, z)\big) \to P_1(y, v)\big)$$

is an assumption of the theory T. This sentence is true by assumption in the domain $\mathfrak{M}$ for the interpretation of the terms Id and P_1 as names of relations p_0 and p_1. Hence, by Theorem 3, it is true that

$$\bigwedge x, y, z, v \in X\big((x p_0 y \wedge z p_0 v \wedge x p_1 z) \to y p_1 v\big),$$

which means that p_0 preserves the relation p_1. It can easily be proved in a similar way that p_0 preserves the remaining relations $p_2, ..., p_n$ and that it is an equivalence. Thus, p_0 is a congruence in $\mathfrak{M}$. All proofs refer to Theorem 3 and proceed in the same way. It follows from the definition of the quotient domain that, for any $x, y \in X$ and $1 \leqslant i \leqslant n$,

(45) $x p_i y \equiv [x]_{p_0} \bar{p}_i^{p_0} [y]_{p_0}.$

It is also true that

(46) $\qquad x p_0 y \equiv [x]_{p_0} = [y]_{p_0}.$

Concerning the construction of quotients see the Introduction, Section 9.

These equivalences show that if B is an atomic formula, then $\{a_n\}$ satisfies B in $\mathfrak{M}$ for the interpretation of Id, $P_1, \ldots, P_n$ as names of of the relations $p_0, p_1, \ldots, p_n$ if and only if $\{[a_n]_{p_0}\}$ satisfies B in $\mathfrak{M}/p_0$ for the interpretation of Id, $P_1, \ldots, P_n$ as names of the relations $= , \bar{p}_1^{p_0}, \ldots, \bar{p}_n^{p_0}.$

This property shall be written briefly as the formula

(47) $\qquad \{a_n\}$ satisfies B in $\mathfrak{M} \equiv \{[a_n]_{p_0}\}$ satisfies B in $\mathfrak{M}/p_0.$

It will be proved that all formulae correctly formed of the terms Id, $P_1, \ldots, P_n$ have property (47). In fact, as follows from formulae (45) and (46), all atomic formulae have it, and if the component formulae have it, then the compound formulae have it, too. By way of example we shall analyse the cases of negation and the universal quantifier.

By Definition 2 (part (b)), equivalence (47) immediately yields the equivalence

$\{a_n\}$ satisfies $\ulcorner \sim B \urcorner$ in $\mathfrak{M} \equiv$

$\{[a_n]_{p_0}\}$ satisfies $\ulcorner \sim B \urcorner$ in $\mathfrak{M}/p_0,$

which shows that the property under consideration is preserved in the case of negation.

Part (f) of Definition 2 states that

$\{a_n\}$ satisfies $\bigwedge x_k B$ in $\mathfrak{M} \equiv$

$\bigwedge d (d \in X \to \{a_1, \ldots, a_{k-1}, d, a_{k+1}, \ldots\}$ satisfies B in $\mathfrak{M}),$

$\{[a_n]\}$ satisfies $\bigwedge x_k B$ in $\mathfrak{M}/p_0 \equiv$

$\bigwedge d (d \in X \to \{[a_1], \ldots, [a_{k-1}], [d], [a_{k+1}], \ldots\}$ satisfies B in $\mathfrak{M}/p_0).$

These equivalences, together with (47), yield the formula

$\{a_n\}$ satisfies $\bigwedge x_k B$ in $\mathfrak{M} \equiv \{[a_n]\}$ satisfies $\bigwedge x_k B$ in $\mathfrak{M}/p_0,$

which shows that property (47) is preserved when a formula is preceded by the universal quantifier. These two examples show that in all other cases property (47) is preserved when it comes to compound formulae. Hence it is concluded by induction that property (47) is an attribute of all formulae of the theory T, in particular of sentences. For sentences, by Theorem 6, formula (47) means that

$$B \in E(\mathfrak{M}) \equiv B \in E(\mathfrak{M}/p_0)$$

for the appropriate interpretations of terms. Hence, if $\mathfrak{M}$ is a model for T, then $\mathfrak{M}/p_0$ is, too.

The following theorem is a simple consequence of Theorems 24 and 32:

THEOREM 33. *Every consistent theory with identity has a denumerable or finite model with an absolute concept of identity.*

PROOF. By Theorem 24, if T is consistent, then it has a denumerable model $\mathfrak{M}$, in which the term Id of the theory T is interpreted as the congruence relation p_0. Hence, by Theorem 32, $\mathfrak{M}/p_0$ is a model of the theory T with an absolute concept of identity. As the set Abs X_{p_0} is not more numerous than the set X, it is at most denumerable if the set X is denumerable.

For every theory T the following disjunction holds: either every model with an absolute concept of identity is finite, of a power not greater than a fixed natural number, or the theory T has infinite models, with an absolute concept of identity, of arbitrarily high powers. To prove this disjunction we have to prove two theorems.

THEOREM 34. *If a theory T has models of arbitrarily high finite powers with an absolute concept of identity, then it also has an infinite model with an absolute concept of identity.*

PROOF. If a theory T has absolute finite models of arbitrarily high powers, then for every n there exists a model $\mathfrak{M}_n$ of the theory T, which has at least n elements, that is, such that a sentence V_n of the form

$$(48) \qquad \bigvee x_1, \ldots, x_n [\sim (\mathrm{Id}(x_1, x_2)) \wedge \ldots \wedge \sim (\mathrm{Id}(x_1, x_n)) \wedge$$
$$\sim (\mathrm{Id}(x_2, x_3)) \wedge \ldots \wedge \sim (\mathrm{Id}(x_{n-1}, x_n))]$$

is true in it for the interpretation of the term Id as a name of the relation of equality, because this sentence states that at least n different objects exist. Since for every n the set $T \cup \{V_n\}$ has a model, that set is consistent. Hence also consistent is the set of sentences

$$(49) \qquad T' = T \cup \{V_1\} \cup \{V_2\} \cup \{V_3\} \cup \ldots$$

obtained by adding to T all the sentences of the sequence $\{V_n\}$. For if an inconsistency resulted from T', then in accordance with the properties of consequence it would have to be a consequence of a finite subset of the set T', and thus, for instance, it would have to be a consequence of the set $T \cup \{V_1\} \cup \{V_2\} \cup \ldots \cup \{V_n\}$. However, $\ulcorner(V_{i+1} \to V_i)\urcorner \in L$, for the implication $V_{i+1} \to V_i$ states that if there exist $i+1$ different objects, then there also exist i different objects. Hence

$$T \cup \{V_1\} \cup \{V_2\} \cup \ldots \cup \{V_n\} \subset Cn(T \cup \{V_n\}).$$

All sentences V_i for $i < n$ follow successively from V_n. Hence, if T' is inconsistent, $T \cup \{V_n\}$ would have to be inconsistent, too, and this, as has been demonstrated above, is impossible.

Since the theory T' is consistent, by Theorem 32, it has a model $\mathfrak{M}$ with an absolute concept of identity. In that model, for every n the sentence V_n is true, since it belongs to T'. Hence, that model cannot be finite, for if it were finite and had, e.g., k elements, then for every $n > k$ the sentence V_n would be false in it for an absolute interpretation of the term Id, since, for $n > k$, the sentence V_n would state that the model $\mathfrak{M}$ has more than k different elements, whereas that model would have exactly k elements. Hence, the model $\mathfrak{M}$ is infinite.

THEOREM 35. *If a theory T has an infinite model with an absolute concept of identity, then the theory T has infinite models with an absolute concept of identity of arbitrarily high powers.*

PROOF. Theorem 35 is the only one in whose proof the concept of d-completeness will be used in its full extent. The proof of Theorem 35 resembles that of Theorem 23, but for a much greater number of constant terms. The theory T will be gradually extended to a complete and d-complete theory. This extension will be carried out in four stages.

S t a g e I. A new theory T' is formed by adding to the theorems of the theory T of a denumerable set of sentences $\{V_n\}$ of the form of (48). The theory T' is consistent, for if the theory T has an infinite model $\mathfrak{M}$ with an absolute concept of identity, then all the sentences of the sequence $\{V_n\}$ are true in that model.

Each sentence V_n which is true in $\mathfrak{M}$ states that there exist at least n different objects. Hence, it is true in $\mathfrak{M}$, since the set X of the model $\mathfrak{M}$ has infinitely many objects.

329

S t a g e II. In stage II we go beyond the usual understanding of a mathematical theory. In the usual understanding of a mathematical theory there is only a denumerable number of sentences, and, hence, also at most a denumerable number of symbols. But in stage II we add to the language of the theory T a set $\{s_t\}_U$ of constant terms of an arbitrary infinite power $\mathfrak{m} = \overline{\overline{U}}$, and we form a theory T'' by adding to the theorems of the theory T' all the sentences of the form

$$(50) \qquad \sim \big(\mathrm{Id}(s_t, s_{t'})\big)$$

for any $t, t' \in U$ different from one another.

The set T'' is consistent, for should it be inconsistent, the inconsistency would have to result from a finite subset $Y \subset T''$. That finite subset would include a finite number of sentences of the form (50) and of the set T'. Suppose that the sentences of the form (50) in the set Y are the sentences

$$\sim \big(\mathrm{Id}(s_1, s_2)\big), \quad \sim \big(\mathrm{Id}(s_1, s_3)\big), \quad \ldots, \quad \sim \big(\mathrm{Id}(s_{n-1}, s_n)\big);$$

let K stand for the conjunction of these sentences, and Q, for the conjunction of the remaining sentences of the set Y, that is, sentences belonging to T'. Since the contradiction $A \wedge \sim A$ is a consequence of the set Y, this means that

$$\ulcorner A \wedge \sim A \urcorner \in \mathrm{Cn}(\{\ulcorner K \wedge Q \urcorner\}),$$

so that, in accordance with the Deduction Theorem,

$$\ulcorner (K \wedge Q) \rightarrow (A \wedge \sim A) \urcorner \in L,$$

and hence

$$\ulcorner Q \rightarrow \sim K \urcorner \in L.$$

The sentence $\sim K$ contains a number of added constants which by assumption do not occur in the sentences from the set T', and hence do not occur in the sentence Q. Hence, on the strenght of the theorem stating that in logic no distinction is made between constants and variables, we have

$$\ulcorner Q \rightarrow \bigwedge x_1, \ldots, x_n \big(\sim K(s_1/x_1, \ldots, s_n/x_n)\big) \urcorner \in L.$$

Thus, by De Morgan's laws and the rule of detachment,

$$\ulcorner \sim \big(\bigvee x_1, \ldots, x_n K(s_1/x_1, \ldots, s_n/x_n)\big) \urcorner \in \mathrm{Cn}(\{Q\}).$$

The sentence K is a conjunction of negations of identity. Hence the sentence $\bigvee x_1, \ldots, x_n K(s_1/x_1, \ldots, s_n/x_n)$ has the form (48), and, thus, is equiform with V_n. This yields that

$$\ulcorner \sim V_n \urcorner \in \mathrm{Cn}(\{Q\}).$$

Q is a conjunction of sentences belonging to T'. Since $Q \in \mathrm{Cn}(T')$, $\ulcorner \sim V_n \urcorner \in \mathrm{Cn}(T')$. It follows from the definition of the set T' that $V_n \in T'$. The set T' would thus be inconsistent. Since the set T' is consistent, T'' must be consistent, too.

Stage III. Guaranteeing descriptive constructiveness of the system. The proof of Theorem 35 differs from that of Theorem 25 in that we first guarantee descriptive constructiveness. This is achieved by adding, for every formula with one free variable x_0: $\Phi(x_0)$, an axiom of the form:

$$(\alpha) \qquad \bigvee x_0 \Phi(x_0) \rightarrow \Phi(t_\Phi),$$

in which t_Φ is a new individual constant added especially for the formula Φ. Since, however, the addition of a new constant increases the number of well-formed formulae, this procedure must be repeated a denumerable number of times. Hence this procedure can be described rigorously as follows. The sequence F of sets of formulae is defined as:

$F_0 =$ the set of well-formed formulae consisting of Id, $P_1, \ldots, P_n$, and the constants $\{s_t\}_U$.

$F_{n+1} =$ the set of well-formed formulae consisting of Id, $P_1, \ldots, P_n$, the constants $\{s_t\}_U$, and the constants from the sequence $\{t_\Phi\}$ for $\Phi \in F_n$.

When constructing formulae for the set F_{n+1}, we, thus, add new constants with indices which are formulae from the preceding set F_n. Next we define the set G of formulae as the sum of all the sets F_n:

$$G = \bigcup_n F_n.$$

The set G is, thus, the set of all well-formed formulae consisting of Id, $P_1, \ldots, P_n$, the constants $\{s_t\}_U$, and all the constants added in the process of forming the sets $\{F_n\}$, that is, the constants $\{t_\Phi\}_G$. Every well-formed formula Ψ, constructed as described above, contains a finite number of constants. Hence, there exists an n_0 such that the formula Ψ contains only constants t_Φ such that $\Phi \in F_n$ for $n < n_0$. Then $\Psi \in F_{n_0}$. The converse inclusion is self-evident.

The set T''' of sentences is now defined as follows:

$$A \in T''' \underset{\mathrm{df}}{\equiv} \{A \in T'' \vee A \text{ has the form } (\alpha) \text{ for a } \Phi \in G\}.$$

The set T''' is consistent.

PROOF. Should the set T''' be inconsistent, a finite subset Z, containing a finite number of sentences $A_1, ..., A_n$, falling under the schema (α), would be inconsistent, too. The set $T'' \cup \{A_1, ..., A_n\}$ would be inconsistent, too. Assume that the sentences $A_1, ..., A_n$ are numbered according to the sets of the sequence $\{F_n\}$, of which they are elements; hence, if $A_i \in F_k$, then, for $j < i$, also $A_j \in F_k$. This is possible, because the sets $\{F_n\}$ form an increasing sequence. Among sentences belonging to the same F_n the order may be arbitrary. With such an ordering it is true that if A_i has the form (α), then the constant t_Φ which occurs in it occurs neither in the sentences of the set T'' nor in the sentences $A_1, ..., A_{i-1}$. Now, the following theorem can easily be proved:

(51) *If a set X is consistent and if the constant t_Φ does not occur in the sentences of the set X, then the set $X \cup \{A\}$, where A has the form (α), is consistent, too.*

In fact, should the set $X \cup \{A\}$ be inconsistent, that would mean that $\ulcorner \sim A \urcorner \in \mathrm{Cn}(X)$. In accordance with (α) this means that

(52) $\ulcorner \bigvee x_0 \, \Phi(x_0) \urcorner \in \mathrm{Cn}(X),$

(53) $\ulcorner \sim \Phi(t_\Phi) \urcorner \in \mathrm{Cn}(X).$

It follows from (53) that there exists a finite subset $\{B_1, ..., B_k\} \subset X$ such that

(54) $\ulcorner \sim \Phi(t_\Phi) \urcorner \in \mathrm{Cn}(\{B_1, ..., B_k\}),$

so that, by the Deduction Theorem,

(55) $\ulcorner (B_1 \wedge ... \wedge B_k) \to \sim \Phi(t_\Phi) \urcorner \in L.$

In accordance with the theorem stating that no constants are privileged in logic it follows from (55) that

(56) $\ulcorner \bigwedge x((B_1 \wedge ... \wedge B_k) \to \sim \Phi(x)) \urcorner \in L.$

The constant t_Φ does not occur in the sentences $B_1, ..., B_k$ by assumption, and hence it follows from (56) that

(57) $\qquad \ulcorner (B_1 \wedge ... \wedge B_k) \to \bigwedge x \sim \Phi(x) \urcorner \in L,$

and hence, in accordance with the properties of the concept of consequence,

$$\ulcorner \bigwedge x \sim \Phi(x) \urcorner \in Cn(X),$$

which, taken together with (52), would mean that the set X is inconsistent. Thus, by Theorem (51), if the set T'' is consistent, then the set $T'' \cup \{A_0\}$ is consistent, too. Likewise, if the set $T'' \cup \{A_0, ..., A_{i-1}\}$ is consistent, then the set $T'' \cup \{A_0, ..., A_{i-1}, A_i\}$ is consistent, too. Thus, for every n, the set $T'' \cup \{A_0, ..., A_n\}$ is consistent. Hence, the set T''' is consistent.

Stage IV. The complete theory S. (A good understanding of this stage of the proof requires the knowledge of well-orderings on the level of popular handbooks of set theory.) We shall now construct a complete theory containing T'''. We shall do so by gradually extending the set T''' by an even simpler procedure than in the proo of Theorem 23. All the sentences belonging to the set G may form a non-denumerable set, and hence, instead of being ordered as a denumerable sequence, they must be ordered as a transfinite sequence. By Zermelo's theorem, which belongs to set theory and states that every set can be well-ordered, the set of the sentences belonging to the set G can form a transfinite sequence $\{A_\alpha\}_{\alpha < \gamma}$, in which the indices α are ordinal numbers less than γ. The theory S will be defined by induction of the γ order:

$$S_0 = T''',$$

$$S_{\alpha+1} = \begin{cases} S_\alpha \cup \{A_\alpha\}, & \text{if } \ulcorner \sim A_\alpha \urcorner \notin Cn(S_\alpha), \\ S_\alpha \cup \{\ulcorner \sim A_\alpha \urcorner\}, & \text{if } \ulcorner \sim A_\alpha \urcorner \in Cn(S_\alpha), \end{cases}$$

$$S_\lambda = \bigcup_{\xi < \gamma} S_\xi \quad \text{for limit ordinals } \lambda,$$

$$S = \bigcup_{\alpha < \gamma} S_\alpha.$$

It can now easily be proved that:

1. The set S is consistent. This is because every S_α is consistent on the strength of the same reasoning as that used in the proof of Theorem 23.

2. The set S is complete, for the sentence A_α is in $S_{\alpha+1}$, or else $\ulcorner \sim A_\alpha \urcorner \in S_{\alpha+1}$.

3. The set S is a theory, for if $A \in \mathrm{Cn}(S)$ and $A \notin S$, then $\ulcorner \sim A \urcorner \in S$ in accordance with the property of the concept of completeness, and, hence, S would be inconsistent.

4. S is descriptively complete with respect to the individual constants $\{t_\Phi\}_G$. S is constructive with respect to the individual constants $\{t_\Phi\}_G$. Since S is a theory and contains T''' and, hence, contains sentences of the form (α), if $\ulcorner \bigvee x_0 \Phi(x_0) \urcorner \in S$, then also $\ulcorner \Phi(t_\Phi) \urcorner \in S$. Thus, by Theorem 21, the theory S is also descriptively complete with respect to $\{t_\Phi\}_G$.

The theory S, being consistent, complete, and descriptively complete, determines a model on its constants by Theorem 20. That model can be partitioned by the congruence:

$$t_\alpha \simeq t_\beta \equiv \ulcorner t_\alpha = t_\beta \urcorner \in S;$$

this yields, as in the proof of Theorem 32, a model with the concept of identity interpreted absolutely.

In formulating the theory S we could also use the same construction of a descriptively complete theory as in the proof of Theorem 23. Here we just wanted to show a different method of constructing a descriptively complete theory.

Since at Stage II we added to T $\mathfrak{m}$ constants and assumed also that they all differed from one another, the number of the abstraction classes for the relation $\simeq$ is at least $\mathfrak{m}$. It can be proved that their number is exactly $\mathfrak{m}$. New constants were added, but their number also was only $\mathfrak{m}$. Transition from F_n to F_{n+1} was accompanied by the addition of constants for all the formulae $\Phi(x_0)$. There are as many such formulae as many there are finite combinations of $\mathfrak{m}$ elements. In accordance with theorems of set theory, the number of such combinations is $\mathfrak{m}$. Thus, when passing from F_0 to F_1 we add $\mathfrak{m}$ new constants. Yet, since $\mathfrak{m} + \mathfrak{m} = \mathfrak{m}$, the number of constants does not increase. A denumerable summing of sets of the power $\mathfrak{m}$ also yields only a set of the power $\mathfrak{m}$. Thus, the number of all constants is $\mathfrak{m}$. Hence, the number of abstraction classes cannot be greater than $\mathfrak{m}$, and, thus, it is exactly $\mathfrak{m}$. The model obtained by partition by the congruance $\simeq$ accordingly has exactly $\mathfrak{m}$ elements.

THEOREM 36. *For every theory T, either there exists a natural number k such that all models of the theory T with the absolute concept of identity have at most k elements, or the theory T has infinite models of arbitrarily high powers with the absolute concept of identity.*

The proof follows from Theorems 34 and 35.

In this section we shall hereafter consider only models with the absolute interpretation of the concept of identity. This is why in describing a model we shall not mention the identity relation at all.

We note that for models with an absolute concept of identity Theorems 29 and 30 can be reformulated to obtain the equivalences:

$$A \in L_1 \equiv \bigwedge \mathfrak{M} \neq \Lambda \big(A \in E(\mathfrak{M})\big),$$

$$A \in \mathrm{Cn}_1(X) \equiv \bigwedge \mathfrak{M} \neq \Lambda \big(X \subset E(\mathfrak{M}) \rightarrow A \in E(\mathfrak{M})\big),$$

where the symbol $\mathfrak{M} \neq \Lambda$ means that $\mathfrak{M}$ is a non-empty domain, and where the set $E(\mathfrak{M})$ is interpreted as the set of sentences true in $\mathfrak{M}$ for the interpretation of the term Id as a name of identity and for some interpretation of the other terms (cf. the remarks made in connection with Theorems 29 and 30).

Standard and non-standard models of arithmetic

We shall now discuss the problem what models of the elementary arithmetic of natural numbers may be taken as standard models. Consider the basic system of the elementary arithmetic of natural numbers, called *Peano's elementary system*. Peano's elementary system P in its classical form is presented as based on the L_1 system of logic and containing only 0 and S as the initial arithmetical terms. The following sentences are axioms in P:

1. $0 \neq Sx$,
2. $Sx = Sy \rightarrow x = y$,

and the infinite set of sentences falling under the induction schema:

3. $\bigwedge u_1, ..., u_n\big((A(x/0) \wedge \bigwedge x(A(x) \rightarrow A(x/Sx))) \rightarrow \bigwedge xA(x)\big)$.

In the above schema $A(x)$ is any formula containing the free variable x and any set $u_1, ..., u_n$ of other variables playing the role of parameters. This schema of induction corresponds to the non-elementary axiom 806 on induction (p. 213).

Moreover, in the classical interpretation of the system P the rule of introducing new functions by the inductive definition schema (183)

(p. 261), was adopted. In this sense, the theory P in its classical interpretation has infinitely many primitive terms, and every term n, except for 0 and S, is described by two equations falling under schema (183). Induction schema 3 applies to formulae $A(x)$ which contain not only the constants 0 and S, but also any constants introduced by inductive definition schema (183).

The theory P in its classical interpretation, thus, contains the symbols of addition and multiplication, described by the following induction equations:

4. $x+0 = x$,
5. $x+Sy = S(x+y)$,
6. $x \cdot 0 = 0$,
7. $x \cdot Sy = (x \cdot y)+x$,

and also many other symbols introduced analogically. In connection with what has been said about reducing inductive definitions to ordinary definitions (cf. Section 10 of the preceding chapter) it is worthwhile noting that in the theory P such a reduction cannot be made since the theory P does not contain the set-theoretical concept $\in$, used to carry out that reduction. Yet, closer investigations of the theory P have shown that the irreducibility of inductive definitions to ordinary ones applies to the concepts $+$ and $\cdot$ only. All other concepts defined by induction in arithmetic by recourse to schema (183) can also be defined by the concepts of addition and multiplication with the use of ordinary definitions only (cf. pp. 502, 503). Thus, the theory P may be described as a theory which does not contain the rule of inductive definition, but contains the following four primitive concepts: 0, S, $+$, $\cdot$, and Axioms 1–7. Presented in this way the theory P is just an ordinary mathematical theory based on the rule of detachment alone, as was the case of other theories analysed as examples.

The *less than* predicate can, of course, be defined in Peano's arithmetic P:

8. $x < y \equiv \bigvee z\, y = x+Sz$.

It can be proved that this predicate establishes an order among the individuals considered in that theory:

9. $x < y \rightarrow \sim(y < x)$,

10. $(x < y \wedge y < z) \rightarrow x < z,$

11. $x < y \vee y < x \vee x = y.$

Sentences 9–11 are theorems in P. Concerning the relation $<$ it can be proved that

12. $0 < Sx,$

13. $x < y \rightarrow Sx < Sy,$

14. $x < Sx,$

15. $\sim \bigvee z(x < z \wedge z < Sx),$

16. $x = 0 \vee \bigvee yx = Sy.$

Sentences 14 and 15 state that the ordering $<$ is isolated: for every element x there exists an element Sx such that there are no intermediate natural numbers between x and Sx. Sentences 12 and 16 taken together state that 0 is the first element in the ordering $<$, since every element other than 0 is a successor.

It can easily be demonstrated that if the term $<$ is adopted in the theory P as a new primitive term described by 9, 11, 12, and 13 as axioms, then equivalence 8 is a theorem in P.

Analysis of the absolute nature of models of the system described above will be carried out in metalogic. It has been assumed about metalogic that it includes set theory, and, hence, it includes definitions of natural numbers $\mathcal{N}$, the number 0, and the relations $S, <, +, \times,$ as given in the Introduction. Thus, we have in metalogic the defined domain

$$\mathfrak{N}_0 = \langle \mathcal{N}, 0, S, <, +, \times \rangle,$$

concerning which it can easily be proved in the same metalogic that $\mathfrak{N}_0$ is a model of the arithmetic P described above. The model $\mathfrak{N}_0$ will hereafter be called the *classical model of arithmetic*. That model is considered to be "good". In fact, it is obtained as a result of a very natural construction. The natural numbers $\mathcal{N}$ have been defined as powers of finite sets, which fully complies with the ordinary intuitive interpretation of the concept of natural number. Thus, it is assumed about the model $\mathfrak{N}_0$ that the primitive concepts of the arithmetic P are interpreted in it properly, that is, in an absolute manner. This assumption makes it possible to extend, in a simple manner, the concept of absoluteness so as to cover other models. Namely it is assumed that any model $\mathfrak{N}$ isomorphic with $\mathfrak{N}_0$ is absolute.

II. MODELS

DEFINITION 14. *A domain* $\mathfrak{N} = \langle X, 0, S', <', +', \times' \rangle$ *is a standard model of the system P of arithmetic, as described above, if and only if the domain* $\mathfrak{N}$ *is isomorphic with the domain* $\mathfrak{N}_0$.

It can easily be proved that it suffices, for instance, for the relation $<'$ in the model $\mathfrak{N}$ under consideration to be isomorphic with the relation $<$ in the model $\mathfrak{N}_0$ for all the remaining operations in the model $\mathfrak{N}$ to be isomorphic with the remaining operations in the model $\mathfrak{N}_0$.

Referring to Axioms 1 and 2, it is easy to prove by induction that the sentences

$$\sim \big(\underbrace{SS \ldots S(0)}_{n} = \underbrace{SS \ldots S(0)}_{m} \big),$$

for any n and m such that $n \neq m$, are theorems in P. Hence, every model of P must contain an infinite number of objects. By Theorem 36, the theory P has models of arbitrarily high powers, and, hence, also has non-denumerable models. A non-denumerable model cannot be isomorphic with a denumerable one, since isomorphism implies equinumerosity The set $\mathcal{N}$ of natural numbers is, of course, denumerable. Hence, there exist for the theory P models which are not isomorphic with the model $\mathfrak{N}_0$, and, hence, are non-standard (non-absolute). It is worthwhile demonstrating that non-standard models are also to be found among the denumerable models of the arithmetic P. The following theorem, which is even stronger, can also be proved.

THEOREM 37. *The set* $E(\mathfrak{N}_0)$ *of sentences true in the classical model of arithmetic, which contains only the terms* 0, S, $<$, $+$, $\times$, *has a denumerable non-standard model.*

PROOF. The proposed model is constructed by the extension of the theory $E(\mathfrak{N}_0)$ by a new individual constant ε, concerning which the sentences guaranteeing the non-classical nature of every model are assumed. The proof of the existence of the model is obtained from Theorem 33.

Infinitely many assumptions of the form:

$$0 < \varepsilon,$$
$$S(0) < \varepsilon,$$
$$SS(0) < \varepsilon,$$
$$\cdots \cdots \cdots$$

are adopted about the constant ε.

338

In general, let Z_n stand for the sentence

(58) $\underbrace{SS \ldots S(0)}_{n} < \varepsilon,$

and let the set Z be the set of all the sentences of the sequence $\{Z_n\}$. It will be proved that the set $E(\mathfrak{N}_0) \cup Z$ is consistent. An even stronger property will be proved outright. Let $P(x)$ be a predicate recorded by means of the constants $0, S, <, +, \times$ and having the property that there are infinitely many natural numbers which satisfy the formula $P(x)$ in the domain $\mathfrak{N}_0$. Thus, the predicate $P(x)$ can, for instance, have the form of the formula

$$\bigvee y (x = SS(0) \times y)$$

or the formula

$$\bigwedge y, z \Big(x = y \times z \to \big((y = x \land z = S(0)) \lor (z = x \land$$

$$y = S(0)) \big) \Big).$$

The former states that x is divisible by 2, that is, is even. The latter states that x is divisible only by itself or by 1, that is, is a prime number. Both the even and the prime numbers are infinitely many. If the predicate $P(x)$ has the property that it is satisfied by infinitely many numbers, then the set

(59) $E(\mathfrak{N}_0) \cup Z \cup \{P(x/\varepsilon)\}$

is consistent.

In fact, if that set were inconsistent, the inconsistency would have to result from a finite subset of that set. But no finite subset of that set is inconsistent. For let Y be a finite set and

$$Y \subset E(\mathfrak{N}_0) \cup Z \cup \{P(x/\varepsilon)\},$$

so Y contains only a finite number of sentences from the sequence $\{Z_n\}$. The sentences of the sequence $\{Z_n\}$ are ordered by the consequence relation, namely

$$\ulcorner Z_{n+1} \to Z_n \urcorner \in P,$$

since if ε is greater than the number $n+1$, then it is also greater than n. Let Z_{n_0} be a sentence from the sequence $\{Z_n\}$ with the greatest number out of those sentences of that sequence which are in Y. Hence,

$Y \subset \mathrm{Cn}(Y')$, where $Y' = \left(E(\mathfrak{N}_0) \cup \{\ulcorner Z_{n_0} \urcorner, \ulcorner P(\varepsilon) \urcorner\}\right)$. The sentences from Y' assume that ε is greater than n_0 and satisfies $P(x)$. Thus, the constant term ε may be considered a name of any number greater than n_0 which satisfies $P(x)$. The set Y' is, thus, consistent, since the domain $\mathfrak{N}_0$ is a model for it. It was assumed that in the domain $\mathfrak{N}_0$ there are infinitely many numbers which satisfy the predicate $P(x)$. Let q be the least number in $\mathcal{N}$, greater than n_0, which satisfies $P(x)$. It can easily be seen that the sentences from Y' are true in the domain $\mathfrak{N}_0$ for the ordinary interpretation of the terms 0, S, $<$, $+$, $\times$ and for the interpretation of the term constant ε as a name of the number q. There can be doubts only as to the sentences Z_{n_0} and $P(x/\varepsilon)$, since only these two sentences contain the term ε. The remaining sentences from Y' do not contain the term ε and are by assumption true in $\mathfrak{N}_0$. By Theorem 7, the sentence Z_{n_0} is true in $\mathfrak{N}_0$ for the interpretation of the constant ε as a name of the number q if and only if $n_0 < q$; likewise, the sentence $P(x/\varepsilon)$ is true in $\mathfrak{N}_0$ for the same interpretation of ε if and only if q satisfies the predicate $P(x)$. Since q has just been selected so as to be greater than n_0 and to satisfy $P(x)$, the sentences from Y' are true in $\mathfrak{N}_0$ and the set Y is consistent.

Set (59) is, thus, consistent and, by Theorem 33, it has the denumerable model $\mathfrak{N} = \langle X, 0', S', <', +', \times', e \rangle$ with an absolute concept of identity. This model has an element e whose name is just the term ε. That element must satisfy in the model $\mathfrak{N}$ all sentences of the form (58). Thus the element e is greater, in the sense of the relation $<'$, than any natural number $0'$, $S'(0')$, $S'S'(0')$, etc., that is, it is greater than any natural number in the model $\mathfrak{N}$, obtained from the number $0'$ by successive addition of 1. Let the set of those elements be denoted by X_0:

$$X_0 = \{0', S'(0'), S'S'(0'), \ldots\}.$$

The element e is not in the set X_0. All those elements of the model $\mathfrak{N}$ which are greater than e also are not in X_0, nor are the elements defined by the formulae $e-1$, $e-2$, ..., $[e/2]$, $[e/3]$, ... Hence, the model $\mathfrak{N}$ in addition to its classical part X_0 contains infinitely many non-classical elements.

The domain $\mathfrak{N}$ is a model for the set (59), and, hence, it is a model for the set $E(\mathfrak{N}_0)$. Yet $\mathfrak{N}$ is not isomorphic with $\mathfrak{N}_0$: The *less than* relation in the model $\mathfrak{N}$ is not isomorphic with the *less than* relation in the model $\mathfrak{N}_0$. $<'$ and $<$ are the respective *less than* relations in $\mathfrak{N}$ and $\mathfrak{N}_0$. Let g be a single-valued function which maps the set X of the model $\mathfrak{N}$ onto the whole set $\mathcal{N}$ of the model $\mathfrak{N}_0$. It will be proved that the function g does not transform the relation $<'$ into $<$. If g transformed $<'$ into $<$, this would mean that

$$\bigwedge x, y \in X\big(x <' y \to g(x) < g(y)\big).$$

In particular, for $y = e$ and for any $x \in X_0$ we would have

$$\bigwedge x \in X_0\big(x <' e \to g(x) < g(e)\big).$$

The set X_0 is infinite. For every $x \in X_0$ it is true that $x <' e$. Hence for every $x \in X_0$ it is true that $g(x) < g(e)$. If the function g is single-valued, then the image $O_g(X_0)$ of the set X_0 is equinumerous with the set X_0, and, thus, it is infinite, too, since the set X_0 is infinite. Hence, the set $\mathcal{N}$ includes an infinite subset $O_g(X_0)$ such that

$$\bigwedge y\big(y \in O_g(X_0) \to y < g(e)\big).$$

But since $g(e)$ is a natural number: $g(e) \in \mathcal{N}$, there exists only a finite number of natural numbers $y \in \mathcal{N}$ such that $y < g(e)$. In fact, the number of numbers less than $g(e)$ exactly equals $g(e)$. Hence, the infinite set $O_g(X_0)$ would have to be included in the finite set of numbers less than $g(e)$, which is impossible. Hence $\mathfrak{N}$ is not isomorphic with $\mathfrak{N}_0$.

The concept of indistinguishability of domains

The domains $\mathfrak{N}$ and $\mathfrak{N}_0$ can be used to illustrate another important concept, namely that of indistinguishability of domains. The domain $\mathfrak{N}$ has been constructed as a model of the set of sentences

$$E(\mathfrak{N}_0) \cup Z \cup \{P(x/\varepsilon)\},$$

where $E(\mathfrak{N}_0)$ stands for the set of those sentences which are true in $\mathfrak{N}_0$, contain the terms 0, S, $<$, $+$, $\times$, and do not contain the additional term ε, which occurs in sentences from the set $Z \cup \{P(x/\varepsilon)\}$. The sentences $E(\mathfrak{N}_0)$ are, thus, also true in the model $\mathfrak{N}$ and do not contain any terms other than 0, S, $<$, $+$, $\times$. Since $E(\mathfrak{N}_0)$ is a complete set, by Theorem 15, $E(\mathfrak{N}_0)$ is the set of all those sentences which are true in $\mathfrak{N}$ and contain only the terms 0, S, $<$, $+$, $\times$. Thus, the domains

$\mathfrak{N}$ and $\mathfrak{N}_0$ have the property that if we consider only those sentences which have as their terms only 0, S, $<$, $+$, $\times$, then the set of the sentences with those terms true in $\mathfrak{N}_0$ is identical with the set of sentences with those terms true in $\mathfrak{N}$.

The symbol $E(\mathfrak{A})$ is not unambiguous. To be rigorous we would have to indicate which terms are admitted in the sentences belonging to the set $E(\mathfrak{A})$. We have not been doing so thus far, because the set of the terms under consideration was being fixed in advance and there was no risk of a misunderstanding.

In those cases where there is a risk of a misunderstanding we shall use a more cumbersome symbolism:

$E_{\langle a, b, ..., c \rangle}(\mathfrak{A})$ = *the set of those sentences which contain the terms* $a, b, ..., c$ *and are true in the domain* $\mathfrak{A}$ *(for a given interpretation of those terms).*

This symbolism can be used to express the relationship between $\mathfrak{N}$ and $\mathfrak{N}_0$ with the formula

$$E_{\langle 0, S, <, +, \times \rangle}(\mathfrak{N}) = E_{\langle 0, S, <, +, \times \rangle}(\mathfrak{N}_0).$$

The domains $\mathfrak{N}$ and $\mathfrak{N}_0$, though not isomorphic, cannot be distinguished from one another by means of sentences recorded in terms specified above. Every sentence recorded in those terms and true in one domain is also true in the other. Such domains are called *indistinguishable* by specified terms.

DEFINITION 15 (Tarski [97]). *Two domains, $\mathfrak{A}$ and $\mathfrak{B}$, are called indistinguishable by the terms $a, b, ..., c$ (for their accepted interpretation) if and only if*

$$E_{\langle a, b, ..., c \rangle}(\mathfrak{A}) = E_{\langle a, b, ..., c \rangle}(\mathfrak{B}).$$

The domains $\mathfrak{N}$ and $\mathfrak{N}_0$ are accordingly indistinguishable by the terms 0, S, $<$, $+$, $\times$.

Domains which are undistinguishable by a set of concepts can prove to be distinguishable by a richer set of concepts. For instance, by adding to the terms of arithmetic under consideration the term fin which is a predicate of one argument and has the following meaning:

$\text{fin}(x) \equiv$ *there exist finitely many elements y which satisfy the formula $y < x$,* we can distinguish $\mathfrak{N}$ from $\mathfrak{N}_0$: the sentences $\bigwedge x \mathrm{fin}(x)$ is true in $\mathfrak{N}_0$ but it is false in $\mathfrak{N}$, since the element e does not satisfy $\mathrm{fin}(x)$. Likewise, by adding the following concepts of set theory: that of an arbitrary subset of the set of natural numbers and that of elementhood we can easily distinguish $\mathfrak{N}$ from $\mathfrak{N}_0$.

It can easily be seen that a theory is complete if and only if every two models of it are indistinguishable by the terms of that theory.

The indistinguishability relation is an equivalence relation. The abstraction classes of that relation are called *arithmetic classes of models*. There are many interesting studies concerned with undistinguishability and arithmetic classes.

The concept of categoricity of theories

Many mathematical theories have been constructed with the intention that each of them describe one definite mathematical domain as seen by mathematicians. The arithmetic of natural numbers was, for instance, such a theory. When constructing the arithmetic of natural numbers we want to include in it the characteristic properties of natural numbers. In doing so we want, if possible, to find those characteristic properties of natural numbers which are attributes of natural numbers only. This restriction, obviously, means "up to isomorphism". Thus, we would like a well selected axiom system to have the property that any two of its models are isomorphic. This goal had been formulated rather early, even before the concept of model received its present rigorous definition. It has been decided to call those theories which have that property *categorical*; while no rigorous definition of a model was available, proofs were being advanced that both arithmetic and geometry are categorical theories. Those proofs can be given some meaning if the concept of model is made more specialized, which will be discussed in the next part of this section. If the concept of model is left unchanged, then it turns out that it is impossible to find an axiom system for the said theories such that any two models are isomorphic. As is known from the theorems proved so far, every theory has models of arbitrarily high powers, and hence it has models which are non-equinumerous and as such are non-isomorphic. Even if we confine ourselves to models with an absolute concept of identity, Theorem 36 states that, when it comes to the most interesting theories, namely those which have infinite models, there is no theory all of whose models are isomorphic. Only a complete theory which describes a finite domain can be categorical in the sense that any two of its models with an absolute concept of identity are isomorphic.

This naturally gives rise to the problem whether, if we confine ourselves to models of a given power $\mathfrak{m}$, there exist theories for which any two models of the power $\mathfrak{m}$ are isomorphic. This has, accordingly, given rise to the concept of categoricity in power.

DEFINITION 16 (Łoś [46], Vaught [101]). *A theory T is called categorical in a power $\mathfrak{m}$ if and only if*

1. *it has a model of the power $\mathfrak{m}$ with an absolute concept of identity,*

2. *any two models of the theory with an absolute concept of identity and of the power equal to the cardinal number $\mathfrak{m}$ are isomorphic.*

This concept is not empty. There are theories which are categorical in certain powers. The arithmetic of natural numbers, as shown by Theorem 37, has proved to be non-categorical in any power, even a denumerable one. By Definition 16, there are many theories which are categorical in finite powers. Every operation defined in a finite set of n elements can be characterized in a complete way. A complete theory which characterizes it is categorical in the finite power n. But theories categorical in infinite powers are more interesting.

THEOREM 38. *The elementary theory of the less than relation* (cf. Chapter I, Section 7) *is categorical in the denumerable power.*

The proof is presented in popular books on set theory as a proof of the theorem that any two denumerable orders which are dense and have neither a first nor a last element are similar. The proof will be outlined here as it is usually presented.

PROOF. Let $\langle X, < \rangle$ and $\langle X', <' \rangle$ be two denumerable models of the theory of the *less than* relation. The sets X and X' are, thus, denumerable. The relations $<$ and $<'$ order the sets X and X', respectively, densely and without a first or a last element. The relation g, which establishes an isomorphism between the models, is defined as follows. The sets X and X', being denumerable, can be numbered, that is, arranged as sequences

$$X = \{x_1, x_2, x_3, ...\},$$
$$X' = \{y_1, y_2, y_3, ...\}.$$

The element x_1 of X is associated with the element y_1 of $X':g(x_1) = y_1$. Next, the element x_2 of X is associated with the first element of the sequence $\{y_n\}$ which bears the same relation to y_1 as x_2 bears to x_1. Thus, if $x_1 < x_2$, then we select as $g(x_2)$ the earliest element y_n in the sequence $\{y_n\}$ (the element with the least number n) such that $y_1 <' y_n$. Such an element does exist, because the set X' has no last element with respect to the relation $<'$, and if such an element did not exist, that would mean that y_1 is the last element of X' with respect to the relation $<'$. Likewise, if $x_2 < x_1$, then we select as $g(x_2)$ the earliest (as to the numbering) element of $\{y_n\}$ such that $y_n <' y_1$. Let $g(x_2) = y_{10}$. We now consider the earliest element of $\{y_n\}$ which is not associated with any element of $\{x_n\}$. This is the element y_2. We investigate the relations it bears to the elements y_1 and y_{10}, which are already associated with the elements x_1 and x_2. If it turns out, for instance, that $y_1 <' y_2$ and $y_2 <' y_{10}$, then we select as the element $g^{-1}(y_2)$ the earliest element x_n in $\{x_n\}$ such that it bears the same relations to x_1 and x_2 as y_2 does to y_1 and y_{10}, so that $x_1 < x_n$ and $x_n < x_2$. Such an element does exist since the relation $<$ is a dense ordering, and, hence, there are infinitely many elements of X between x_1 and x_2, and we can select out of them that element which has the least number. Suppose that element is x_5. Accordingly, we take $y_2 = g(x_5)$. Next we return to the sequence $\{x_n\}$ and consider in turn the earliest element in $\{x_n\}$ which has not yet been associated with any element of $\{y_n\}$. In our case that is x_3. We now select as $g(x_3)$ the earliest (as to the numbering) element of $\{y_n\}$ which bears the same relations to y_1, y_{10}, and y_2, as x_3 does to x_1, x_2, and x_5, respectively. Such an element must exist, since between any two elements of $\{y_n\}$, and also before and after each element of that sequence, there are infinitely many elements of $\{y_n\}$. Thus, for instance, if $x_3 < x_1 < x_5 < x_2$, then we select as $g(x_3)$ the earliest element y_n such that $y_n <' y_1 <' y_2 <' y_{10}$. Now we are in turn concerned with that earliest element of $\{y_n\}$ which has not yet been associated with any element of $\{x_n\}$ and we associate it with the earliest element of $\{x_n\}$ which bears the same relations to the elements already considered as the element of $\{y_n\}$ in question does to their counterparts. Then we again return to $\{x_n\}$ and in this way alternately find those elements which have not yet been associated with any element and we associate them with those earliest elements which bear the same relations to others. By selecting each time one element alternately from $\{x_n\}$ and $\{y_n\}$ we finally define the one-to-one mapping of the set X onto the whole set X'. That mapping is selected so as to satisfy the condition of isomorphism

$$\bigwedge x, x' \in X(x < x' \equiv g(x) <' g(x')).$$

Thus, the final result is that $\langle X, < \rangle$ is$_g$ $\langle X', <' \rangle$. On the other hand it is known that the theory of the *less than* relation has a denumerable model, namely the domain $\langle \mathscr{W}, < \rangle$, consisting of the set of rational numbers and the *less than* relation between them.

Of the theories described in the preceding chapter the elementary theory of the relation od "lying between" is also categorical in the denumerable power.

THEOREM 39. *If every model of a theory T is infinite and if T is categorical in an infinite power* $\mathfrak{m}$, *then T is a complete theory.*

PROOF. If the theory T were incomplete, there would exist an independent sentence A. The sets $T \cup \{A\}$ and $T \cup \{\ulcorner \sim A \urcorner\}$ would then each be consistent. Every model of each of these two sets is a model of the theory T, hence each such model is infinite. By Theorem 36, the sets $T \cup \{A\}$ and $T \cup \{\ulcorner \sim A \urcorner\}$ have models of arbitrary infinite powers. Hence, there exist two models $\mathfrak{A}$ and $\mathfrak{B}$ of a power $\mathfrak{m}$ such that $\mathfrak{A}$ is a model for $T \cup \{A\}$ and $\mathfrak{B}$ is a model for $T \cup \{\ulcorner \sim A \urcorner\}$; accordingly, $\mathfrak{A}$ and $\mathfrak{B}$ are models of the theory T and by assumption must be isomorphic. If they are isomorphic, then, by Theorem 22, every sentence which is true in one of them must be true in the other. But the sentence A is true in $\mathfrak{A}$ and false in $\mathfrak{B}$. Thus, there cannot exist sentences independent of T; hence, T is complete.

COROLLARY. *The elementary theory of the less than relation with the added assumption of non-emptiness is complete.*

The proof follows directly from Theorems 38 and 39. Existence of two objects follows from the assumption of non-emptiness by the axiom stating that for every element there exists another element which is greater than the former. The theorem stating that there exist two objects results, by the axiom of dense ordering, in the existence of infinitely many objects between them. Hence, the theory has no finite absolute models for the concept of identity. Thus, the conditions specified in

Theorem 39 are satisfied. In this way the concept of model can be used to prove such a formal concept as the completeness of a mathematical theory.

Special models without absolutely interpreted set-theoretical concepts and the classical concept of categoricity

As mentioned above, the concept of categoricity had been formulated before the concept of model received its rigorous definition. The categoricity of the arithmetics of natural numbers, rational numbers, and real numbers, and plane and three dimensional geometries, both Euclidean and hyperbolic, was being proved in the early inexact sense of the concept. Proofs of the categoricity of those theories can even now be found in handbooks of arithmetic and geometry. Those proofs are often not precise enough, but if a special concept of model is introduced, those proofs of categoricity, as presented in the traditional approaches to arithmetic and geometry, can be given a strictly defined sense. We shall, however, first describe those theories which are to be examined.

We shall analyse non-elementary theories, such as those mentioned in Chapter I, Section 8: a non-elementary theory of real numbers and a non-elementary arithmetic of natural numbers, that is, theories which contain set theory (complete or reduced, since sets of the first order suffice to formulate the non-elementary Axioms 707* and 806).

The primitive terms of non-elementary arithmetic are, for instance, 0, S, N, Z, $\in$. A model in the ordinary sense of the term should accordingly include interpretations of all those concepts. The models to be considered now will be partial models. Non-elementary arithmetic will have its special partial model in the system $\langle N, o, s \rangle$, where N is a set, o, a designated element in that set, and s, a function in that set. Models of this kind do not fall under Definition 8. Likewise, the theory of real numbers will have its model in the new sense in the system $\langle \mathscr{R}, 0, 1, <, +, \times \rangle$. This shows that in the present interpretation of the concept of model not all primitive concepts are covered by it: the concepts of set theory are left uninterpreted and remain outside the model. We would define a partial model for geometry in the same way. We shall now describe in general terms what kind of theory is being investigated, and what models are being considered. The theories in question will be called *theories with a set-theoretical superstructure*.

II. MODELS

DEFINITION 17. *T is a theory with a set-theoretical superstructure if the primitive concepts of T are*: $X, F_1, ..., F_k, Z, \in$, *and if the axioms of T state that*:

1. *X is the set of all those individuals which are not sets, and*

2. $F_1, ..., F_k$ *are functions in the set X, and*

3. *they include the axioms of set theory for the terms Z and $\in$, or some restricted formulations of those axioms, and*

4. *the remaining axioms of T are sentences such that every quantifier is restricted either to the set X, or to the set 2^X of all the subsets of X, or to the set 2^{2^X} of all the subsets of the set 2^X, etc.*

As can easily be seen, the non-elementary arithmetics under consideration can be given a formulation which falls under the above definition.

Throughout this section it is assumed that the symbol of elementhood in metalogic has the form ε; the former symbolism will be restored later.

DEFINITION 18. *The domain $\langle X, f_1, ..., f_k \rangle$ is a special partial model of a theory T with a set-theoretical superstructure if the domain $\langle M, X, f_1, ..., f_k, \mathrm{set}_M, \varepsilon_M \rangle$ is a model of the theory T in the ordinary sense and if the predicates $\mathrm{set}(x)$ and $x\varepsilon y$ are set-theoretical predicates (set and elementhood) as they occur in the language of metalogic, and set_M and ε_M are restrictions of those predicates to the elements of the set M. M is a set which includes the set X and such that with every set A which it includes, it also includes the set 2^A.*

As can be seen, the special models under consideration here are models in which the concepts of set theory are interpreted in a fixed way and absolutely, that is, as true elementhood and true concept of set, namely that which is used in our metalanguage. Abstraction is made from that interpretation in the very concept of special model, a special model being treated as partial.

It can easily be demonstrated that two such special partial models are isomorphic if it is the non-elementary arithmetic of natural numbers, or the non-elementary arithmetic of real numbers, or non-elementary geometry which is the theory under consideration. The proofs are not difficult, but the theorems are not now ascribed any profound meaning.

By way of example we shall prove the theorem on classical categoricity of the arithmetic of natural numbers.

THEOREM 40. *Any two special partial models of the non-elementary arithmetic of natural numbers are isomorphic.*

PROOF. Let $\mathfrak{N} = \langle N, o, s \rangle$ and $\mathfrak{N}' = \langle N', o', s' \rangle$ be two special partial models of the non-elementary arithmetic described in Chapter I (Section 8). By Definition 18, there exist sets M and M' which include N and N', respectively, and are closed under the power set operation, and such that

$\mathfrak{M} = \langle M, N, o, s, \text{set}_M, \varepsilon_M \rangle$ is a model of that arithmetic, and

$\mathfrak{M}' = \langle M', N', o', s', \text{set}_{M'}, \varepsilon_{M'} \rangle$ is a model of that arithmetic.

The following function is defined on the set N:

$$\varphi(o) = o',$$

$$\varphi\big(s(x)\big) = s'\big(\varphi(x)\big) \quad \text{for} \quad x \varepsilon N.$$

The possibility of recursive definitions in the model requires, of course, appropriate lemmas about the set N in that model. That function will prove to be an isomorphism of the domains $\mathfrak{N}$ and $\mathfrak{N}'$. We demonstrate first that the function φ is a mapping onto the whole set N'. Reference is made to the truth of the Axiom of induction 806 in the model $\mathfrak{M}'$. We define the set $U = \{y : \bigvee x \varepsilon N y = \varphi(x)\}$. The common part $Y = U \cap N'$, being a subset of N', belongs to M', so that we may apply to it the Axiom of induction, true in the model $\mathfrak{M}'$. The very definition of the mapping φ shows that $o' \varepsilon Y$, and then that if $y \varepsilon Y$, that is, for an $x \varepsilon N$, $y = \varphi(x)$, then $\varphi\big(s(x)\big) = s'\big(\varphi(x)\big) = s'(y)$, so that $s'(y) \varepsilon Y$. Thus it follows from the truth of the Axiom of induction that $N' \subset Y$, that is, $N' \subset U$.

It only remains to demonstrate that φ is a one-to-one function. To do so we refer to the truth of the Axiom of induction 806 in the model $\mathfrak{M}$. We define the set

$$X = \big\{ x : x \varepsilon N \wedge \bigwedge z \varepsilon N\big(\varphi(x) = \varphi(z) \to x = z\big) \big\}.$$

The set X by definition is a subset of N, and, hence, is in M; we may accordingly apply to it the Axiom on induction true in $\mathfrak{M}$. We prove first that $o \varepsilon X$, so that

$$\bigwedge z \varepsilon N\big(z \neq o \to \varphi(z) \neq \varphi(o)\big).$$

349

In fact, if $z \neq o$, then, for some $z \varepsilon N$, $z = s(y)$ and $\varphi(z) = \varphi(s(y)) = s'(\varphi(y))$, whereas $\varphi(o) = o'$. It follows from the truth of Axiom 803 in $\mathfrak{M}'$ that $o' \neq s'(u)$ for any $u \varepsilon N'$. Hence, $\varphi(z) \neq \varphi(o)$. Next we prove the inductive step: $x \varepsilon X \to s(x) \varepsilon X$. Thus, the induction assumption now is that

$$\bigwedge z \varepsilon N(\varphi(x) = \varphi(z) \to x = z).$$

We have to prove that

$$\bigwedge w \varepsilon N\Big(\varphi(s(x)) = \varphi(w) \to s(x) = w\Big).$$

Note first that if $\varphi(s(x)) = \varphi(w)$, then $w \neq o$. Should $w = o$, we would have $\varphi(w) = \varphi(o) = o'$, whereas in fact $\varphi(s(x)) = s'(\varphi(x))$, and we know from the truth of Axiom 803 in $\mathfrak{M}'$ that $o' \neq s'(\varphi(x))$. If $w \neq o$, then, for some $z \varepsilon N$, $w = s'(z)$. Hence, if $\varphi(s(x)) = \varphi(w)$, that is, $\varphi(s(x)) = \varphi(s(z))$, then, since the definition of the function φ yields $\varphi(s(x)) = s'(\varphi(x))$ and $\varphi(s(z)) = s'(\varphi(z))$, we obtain that $s'(\varphi(x)) = s'(\varphi(z))$. This and the truth of Axiom 804 in the model $\mathfrak{M}'$ yield that $\varphi(x) = \varphi(z)$. Hence, and by the induction assumption, we have that $x = z$, and accordingly also $s(x) = s(z) = w$.

Since φ is a one-to-one function, it is an isomorphism, because by definition it transforms o into o' and s into s'.

In the above proof we have resorted several times to the fact that the sentence:

$$(x \in N \wedge x \neq 0) \to \bigvee y \in N x = S(y)$$

is an arithmetical theorem. It can easily be arrived at by induction.

Modern models theory is now a separate branch of metamathematics, very involved. (Cf. Bell and Slomson [1], Robinson [73], Sacks [81] or Shoenfield [83].)

EXERCISES

1. Prove the classical categoricity of the non-elementary arithmetic of real numbers

H i n t. First single out the natural numbers in both models, and extend their isomorphism so as to cover the rational numbers first and the real numbers next.

2. The integral part of a fraction $[x/y]$ is the greatest natural number not greater than the fraction x/y. Prove that, in the non-standard model $\mathfrak{N}$ for P, if $x \in X_0$, then $x <' [\varepsilon/2] <' \varepsilon$.

3. Prove that the relation $<'$ in the non-standard model $\mathfrak{N}$ occurring in the proof of Theorem 37 has the order type $\omega + (\omega^* + \omega)\eta$. (The meaning of these symbols is to be found in any handbook on set theory.)

4. Find a sentence, involving the symbol < for ordinal numbers as the only non-logical symbol which makes a distinction between the order type $\omega+\omega$ and $\omega+\omega+\omega$.

5. Find a sentence, recorded by means of the symbols < and + for the ordinal numbers, which makes a distinction between the order type $\omega \cdot \omega$ and $\omega \cdot \omega \cdot \omega$. Next, find a sentence which makes a distinction between the same two types and is recorded by means of the symbol < alone.

6. Verify by reference to examples that there exist ordinal numbers which are greater than or equal to ω^ω and are indistinguishable by means of the relation < as holding between ordinal numbers.

7. Find a sentence, using the symbols < and + for the ordinal numbers, which makes a distinction between the type ω^ω and $\omega^{\omega\omega}$.

8. Prove the following stronger versions of the theorems on the existence of a denumerable model and the existence of models of arbitrary infinite powers.

For every infinite domain $\mathfrak{M}$ there exists a denumerable domain $\mathfrak{N}$ such that $\mathfrak{M}$ is an elementary extension of $\mathfrak{N}$ and, for every denumerable domain $\mathfrak{N}$ there exists a domain $\mathfrak{M}$ of an arbitrarily high power such that $\mathfrak{M}$ is an elementary extension of $\mathfrak{N}$.

N o t e. The definition of elementary extension is to be found in Exercise 3, p. 311.

5. SKOLEM'S ELIMINATION OF QUANTIFIERS, CONSISTENCY OF COMPOUND THEORIES AND INTERPOLATION THEOREMS

An axiom system of the theory of the 'less that' relation without existential quantifiers

The concept of model can be used to prove certain fairly important theorems about the classical logical calculus, formulated in terms of syntax only. We shall consider such theorems in the present section. In so doing we shall return to an analysis of models which are not necessarily absolute with respect to the concept of identity. The term Id will accordingly be treated as one of the predicates of a given theory, which may be interpreted in various ways. The starting point of the present analysis is the introduction of what is called the *Skolem function*. That construction, arising from Skolem's combinatorial investigations, is concerned with the role of quantifiers in c.l.c. theorems or in axioms of any theory. It turns out that if we use the c.l.c. as a basis, then the quantifiers can be replaced by functions. This fact will first be examined from an intuitive point of view. Recall the theory of the *less than* relation (Chapter I, Section 7). Axioms 109, 110, and 111, which occur there, can be written as follows:

109. $\bigwedge x, y \bigvee z(x < y \rightarrow (x < z \wedge z < y))$,

110. $\bigwedge x \bigvee y(y < x)$,

111. $\bigwedge x \bigvee y(x < y)$.

Axiom 109 states that there exists an element intermediate between x and y. Axiom 110 states that there exists an element less than x, and Axiom 111 states that there exists an element greater than x. The theory of the *less than* relation has a model, for instance, in the domain $\langle \mathscr{W}, < \rangle$, which consists of rational numbers and the *less than* relation between them. Axiom 109, as applied to that domain, states that for any two rational numbers x, y there exists an intermediate rational number z. The number $z = \frac{1}{2}(x+y)$, which is the arithmetical mean of those two numbers, can always be taken as that intermediate number, because the following formula for the arithmetical mean is true:

$$(60) \qquad \bigwedge x, y \in \mathscr{W} \left\{ x < y \rightarrow \left(x < \frac{x+y}{2} \wedge \frac{x+y}{2} < y \right) \right\}.$$

Axiom 110 states that a smaller number always exists. There are infinitely many such numbers, but for Axiom 110 to be satisfied it always suffices to indicate a number less by one, for it is true that

$$(61) \qquad \bigwedge x \in \mathscr{W}(x-1 < x).$$

Likewise, it suffices to indicate as a greater number, a number greater by one, since

$$(62) \qquad \bigwedge x \in \mathscr{W}(x < x+1).$$

If we now add to the language of the theory of the *less than* relation the function symbols S, F_1, F_2, and adopt the following axioms about them:

109'. $\bigwedge x, y \big(x < y \rightarrow (x < S(x, y) \wedge S(x, y) < y) \big)$,

110'. $\bigwedge x(F_1(x) < x)$,

111'. $\bigwedge x(x < F_2(x))$,

then it follows from sentences (60)–(62) that Axioms 109–111 are true in the domain $\langle \mathscr{W}, \; < \rangle$ for the interpretation of the function symbols S, F_1, F_2 as denoting, respectively, the functions: $\frac{1}{2}(x+y)$, $x-1$, $x+1$, defined in the set $\mathscr{W}$.

Axioms 109′–111′ have only universal quantifiers standing at the beginning of each formula. The existential quantifiers occurring in Axioms 109–111 have been replaced in Axioms 109′–111′ by the functions S, F_1, F_2, which indicate those elements whose existence is postulated in Axioms 109–111. Axioms 109–111 are consequences of Axioms 109′–111′. This will be demonstrated by taking Axiom 110 as an example. The sentence

(63) $\bigwedge x\big(F_1(x) < x \rightarrow \bigvee y(y < x)\big),$

is an axiom of the c.l.c. with functions. Sentence (63) yields, by the distribution of the universal quantifier over implication, the sentence

(64) $\bigwedge x\big(F_1(x) < x\big) \rightarrow \bigwedge x \bigvee y(y < x).$

Sentences (64) and 110′ yield sentence 110 by detachment. Sentences 109 and 111 follows from 109′ and 111′ in a similar way.

The above discussion shows that sentences 101–108 and 109′–111′ could be adopted as axioms of set theory. As has been demonstrated above, the original axiom system of the theory of the *less than* relation follows from this new axiom system. Further, the new axiom system is satisfied in every model in which the former axiom system is satisfied, as has been demonstrated by the example of the model $\langle \mathscr{W}, \; < \rangle$. As compared with the former system, the new axiom system is simpler by not having any quantifiers except for universal quantifiers standing at the beginning of formulae, but is more complicated by having as many new primitive function symbols as the former axiom system had existential quantifiers. But the increase in the number of the primitive terms is not significant complication from the logical point of view, since it does not change essentially the logical means necessary for the construction of the theory. On the contrary, the disappearance of the existential quantifiers essentially reduces the logical means required for proving certain theorems of the theory under consideration. This is why the axiom system free from existential quantifiers will be given considerable attention here. The operation itself will be called *elimination of the existential quantifiers by Skolem's method.*

353

Skolem equivalents of any set of sentences

It will now be demonstrated that the elimination of the existential quantifiers by Skolem's method can be applied to the axioms of any theory based on the c.l.c. Each axiom can be reduced to its normal form with all the quantifiers grouped at the beginning (Theorem 11, Chapter I, Section 9). If an axiom thereby becomes

$$(65) \qquad \bigwedge x \bigvee y A(x, y),$$

then we can imitate the theory of the *less than* relation; such was the case of Axioms 109–111. That y which exists for any x can namely be treated as a function of x: $F(x)$. We can thus adopt the new primitive term F and the new axiom

$$(66) \qquad \bigwedge x A(x, F(x)).$$

This is how we have obtained Axioms 109′–111′ from 109–111. But more complicated sentences, on being reduced to a normal form, may contain more quantifiers than two. In such cases the above process can be iterated. Consider, by way of example, a case of four quantifiers. Our observations will then be generalized. If an axiom in its normal form is

$$(67) \qquad \bigwedge x \bigvee y \bigwedge z \bigvee v B(x, y, z, v),$$

then this sentence may also be considered as falling under the schema (65) with a more complex inner part A. It may be assumed that (67) has the form of (65), where A has the form $\bigwedge z \bigvee v B(x, y, z, v)$. Hence the element y is treated as a function of x and denoted by $F(x)$. (67) is, thus, replaced by (66).

Since A is complex, (66) has the form

$$(68) \qquad \bigwedge x \bigwedge z \bigvee v B(x, F(x), z, v).$$

But (68) also resembles schema (65), the only difference being that v depends on two variables, x, and z, as it did in the case of Axiom 9 of the theory of the *less than* relation. The variable v is accordingly replaced by a function of two variables $G(x, z)$, and then (68) yields

$$(69) \qquad \bigwedge x \bigwedge z B(x, F(x), z, G(x, z)).$$

In this way sentence (67) with two existential quantifiers has been reduced to a sentence without existential quantifiers, two function constants being added in the process. If an axiom begins with the existential quantifier, e.g., $\bigvee y\,C(y)$, then y is not a function of any x, and hence in every model we may select one fixed y, so that y may be considered constant in every model. We may accordingly replace it by a new constant a which has not occurred in the theory so far. Thus the axiom $\bigvee y\,C(y)$ will have its equivalent in the axiom $C(a)$. The equivalents of sentences from a set Z, which do not contain existential quantifiers and have been obtained by the method described above, shall be called *Skolem equivalents* or *Skolem forms of the set Z*. The general method of forming Skolem equivalents can be formulated as the following definition:

DEFINITION 19. *A sentence A is a Skolem equivalent of a sentence B if and only if the sentence B has a normal form and the sentence A is obtained from the sentence B as the last element of a finite sequence of transformations*

$$(70) \qquad B_0, B_1, B_2, ..., B_n = A,$$

which satisfies the following conditions:

1. $B_0 = B$.

2. *Either the sentence B_1 is equiform with the sentence B_0, if B_0 does not begin with an existential quantifier, or, if B_0 begins with existential quantifiers, that is, if*

$$(71) \qquad B_0 = \bigvee x_1 ... \bigvee x_k C(x_1, ..., x_k),$$

where C does not have any existential quantifier at the beginning, then

$$(72) \qquad B_1 = C(x_1/a_1, ..., x_k/a_k),$$

where $a_1, ..., a_k$ are distinct constant terms other than any constants occurring in B_0.

3. *For all sentences B_i, $0 < i < n$, the transition from B_i to B_{i+1} can be described in a general manner, since the sentences B_i, $0 < i < n$, always begin in a similar way, namely with universal quantifiers. This transition is described as follows: If B_i has the form*

$$(73) \qquad \bigwedge x_1, ..., x_k \bigvee y_1, ..., y_t C(x_1, ..., x_k, y_1, ..., y_t),$$

355

where C does not begin with an existential quantifier, then B_{i+1} has the form

(74) $\quad \bigwedge x_1, \ldots, x_k \, C(x_1, \ldots, x_k, y_1/F_1(x_1, \ldots, x_k), \ldots,$

$\qquad y_t/F_t(x_1, \ldots, x_k)),$

where $F_1, \ldots, F_t$ are distinct function symbols other any constant symbols occurring in B_i. Those added function symbols $F_1, \ldots, F_t$ are called Skolem functions.

4. *The sentence $A = B_n$ is the first element in the sequence* (70) *which has a normal form and does not contain any existential quantifiers.*

DEFINITION 20. *A set X of sentences is a Skolem equivalent of a set Y of sentences in normal forms $(X = \mathrm{Skl}(Y))$ if and only if:*

1. *Every sentence from the set X is a Skolem equivalent of a sentence from the set Y.*

2. *For every sentence in Y there exists one and only one Skolem equivalent in X.*

3. *If a function term F_i or a constant a_i is a term added in the construction of the Skolem equivalent of the sentence $A \in Y$ (that is, if F_i and a_i do not occur in A, but do occur in the Skolem equivalent of A, which is in X), then that term may not occur in any other sentence from the set X. In other words, when constructing a Skolem equivalent of a set Y we must add different function terms F_i to different sentences.*

The concept of Skolem equivalent, as described above, is not unique and, strictly speaking, we should not use the term Skl as a function term. But in the actual cases to be analysed below this notation does not result in any misunderstanding, for it is always possible to assume that, for given sets, the Skolem equivalents are *ad hoc* defined uniquely (cf. Exercise 4).

THEOREM 41. $X \subset \mathrm{Cn}^*(\mathrm{Skl}(X))$.

PROOF. It will be demonstrated first that

(75) $\quad$ *If A is a Skolem equivalent of a sentence B, then*
$\qquad \ulcorner A \to B \urcorner \in L^*.$

Since $B = B_0$, $\ulcorner B_0 \to B \urcorner \in L^*$. A sentence B_1, if it differs from B_0, has the form (72). The sentence

$$C(x_1/a_1, \ldots, x_k/a_k) \to \bigvee x_1, \ldots, x_k \, C(x_1, \ldots, x_k)$$

356

is an axiom of L^*, hence, by (71) and (72), $\ulcorner B_1 \rightarrow B_0 \urcorner \in L^*$. A similar argument shows that, for $0 < i < n$, $\ulcorner B_{i+1} \rightarrow B_i \urcorner \in L^*$. The sentence

$$\bigwedge x_1, \ldots, x_k \big(C(x_1, \ldots, x_k, y_1/F_1(x_1, \ldots, x_k), \ldots,$$

$$y_t/F_t(x_1, \ldots, x_k)) \rightarrow \bigvee y_1, \ldots, y_t\, C(x_1, \ldots, x_k, y_1, \ldots, y_t) \big)$$

is an axiom of L^*. By applying to that sentence the axiom on the distri-bution of the quantifier over implication we obtain

$$\ulcorner (74) \rightarrow (73) \urcorner \in L^*,$$

and this, by Definition 19, means that $\ulcorner B_{i+1} \rightarrow B_i \urcorner \in L^*$. Thus, in the entire sequence (70) a given term is a logical consequence of the next one. This, in view of the transitivity of implication, yields $\ulcorner A \rightarrow B \urcorner \in L^*$.

Definition 20 and sentence (75) immediately yield the inclusion

$$X \subset \mathrm{Cn}^*(\mathrm{Skl}(X)).$$

Hereafter, in order to record theorems more briefly, we shall treat term constants $a_1, \ldots, a_k$ as function constants of 0 arguments; de-signated elements of a given domain will likewise be treated as functions of 0 arguments.

THEOREM 42. *If symbols $\{F_n\}$ are the added function symbols which occur in $\mathrm{Skl}(X)$, and do not occur in sentences from the set X, then for every non-empty domain $\mathfrak{M}$ and for every interpretation of the terms occurring in sentences from X the following equivalence holds*:

$$X \subset E(\mathfrak{M}) \equiv \text{*for some functions* } \{f_n\}, \ \mathrm{Skl}(X) \subset E(\mathfrak{M}) \text{ *for*}$$
the interpretation of $\{F_n\}$ as names of the functions $\{f_n\}$ defined on the set of the indi-viduals on the domain $\mathfrak{M}$.

PROOF. If $X \subset E(\mathfrak{M})$, then, for any $B \in X$, B is true in $\mathfrak{M}$. Suppose that A is a Skolem equivalent of B, and $F_1, \ldots, F_n$ are the added function symbols which occur in A and do not occur in B. Moreover, let A derive from B by the sequence (70) of transformations: $B = B_0, B_1, \ldots, B_n = A$, which satisfy the conditions laid down by Definition 19 of Chapter I. Of course, B_0 is true in $\mathfrak{M}$, since $B = B_0$. If B_0 has the form of (71) and is true in $\mathfrak{M}$, then this means that there exist in $\mathfrak{M}$ individuals $a'_1, \ldots, a'_k$ such that the sequence $\{a'_n\}$ beginning with those individuals satisfies in $\mathfrak{M}$ the formula $C(x_1, \ldots, x_k)$. If these individuals are denoted by

357

$a_1, \dots, a_k$, then, by the definition of truth extended so as to cover formuale with constant terms, the sentence B_1 of the form (72) is true in $\mathfrak{M}$ for the interpretation of the constants $a_1, \dots, a_k$ as names of the elements $a'_1, \dots, a'_k$.

Likewise, for any B_i, if B_i has the form of (73) and is true in a domain $\mathfrak{M}$ with a set Y, this means that

$$\bigwedge x'_1, \dots, x'_k \in Y \bigvee y'_1, \dots, y'_t \in Y(x'_1, \dots, x'_k, y'_1, \dots, y'_t$$

satisfy C in $\mathfrak{M}$).

For every combination of the objects $x'_1, \dots, x'_k$ from Y there may exist various combinations of $y'_1, \dots, y'_t \in Y$ which together with $x'_1, \dots, x'_k$ satisfy the formula C in $\mathfrak{M}$. By the axiom of choice, which we adopt in metalogic, those combinations can be well-ordered, and the earliest from them, $\langle y'_{i_1}, \dots, y'_{i_t} \rangle$, can be selected. For every combination $x'_1, \dots, x'_k \in Y$, the earliest combination $\langle y'_{i_1}, \dots, y'_{i_t} \rangle$ which satisfies C is thus uniquely associated with the combination $x'_1, \dots, x'_k$. Thus, there exists a function which with every combination $x'_1, \dots, x'_k \in Y$ associates the earliest combination $\langle y'_1, \dots, y'_t \rangle$ which together with $x'_1, \dots, x'_k$ satisfies C. Since in the combination $\langle y'_1, \dots, y'_t \rangle$ every element is uniquely defined by its position, there exist functions $f_1, \dots, f_t$ such that the combination $\langle f_1(x_1, \dots, x_k), \dots, f_t(x_1, \dots, x_k) \rangle$ is identical with the combination $\langle y'_{i_1}, \dots, y'_{i_t} \rangle$, associated with the combination $x_1, \dots, x_k$. Hence, there exist functions $f_1, \dots, f_t$, defined in the set Y, such that a sentence B_{i+1} of the form (74) is true in $\mathfrak{M}$ for the interpretation of the constants $F_1, \dots, F_t$ as names of the functions $f_1, \dots, f_t$. Accordingly, we obtain by induction that for every $B \in X$ there exists a combination of functions defined on the set X of the domain $\mathfrak{M}$ such that $B_n = A$ is true in $\mathfrak{M}$ for the interpretation of $F_1, \dots, F_k$ as names of the functions $f_1, \dots, f_k$.

This proof consists in the application of the same reasoning which underlies the very idea of replacing the existential quantifiers by Skolem functions. In mathematical analysis, theorems of the type

For every number x there exists a number y such that A(x, y)

have long been considered equivalent to theorems of the form

There exists a function f such that $A(x, f(x))$.

For if for every x there exists a corresponding y, then for a given x we can select one y such that $A(x, y)$, and we may consider it as associated with the number x. Thus, there exists a mapping, that is, a function. If a function exists, then it yields a $y = f(x)$ such that $A(x, y)$.

The converse implication is a consequence of Theorem 42. That theorem and Theorem 30 yield that if $\mathrm{Skl}(X) \subset E(\mathfrak{M})$, then also $X \subset E(\mathfrak{M})$ for the established interpretation of the terms which occur in sentences from the set X.

THEOREM 43. *For any set Z of sentences in a normal form the following equivalence holds*:

$$Z \text{ is consistent} \equiv \mathrm{Skl}(Z) \text{ is consistent.}$$

PROOF. If a set Z is consistent, then, by Theorem 24, it has a non-empty model $\mathfrak{M}$, $Z \subset E(\mathfrak{M})$. Hence, by Theorem 42, $\mathrm{Skl}(Z) \subset E(\mathfrak{M})$, and so the set $\mathrm{Skl}(Z)$, since it has a model, is consistent.

The converse implication follows from Theorem 41. Since $Z \subset \mathrm{Cn}^*(\mathrm{Skl}(Z))$, if the set Z were inconsistent, the set $\mathrm{Skl}(Z)$ would have to be inconsistent, too.

THEOREM 44. *If a sentence A does not contain function symbols added in $\mathrm{Skl}(X)$, and if $A \subset \mathrm{Cn}^*(\mathrm{Skl}(X))$, then $A \subset \mathrm{Cn}^*(X)$.*

PROOF. If $\mathfrak{M}$ is any non-empty model of the set X for a certain interpretation of the terms from the set X, then, by Theorem 42, there exist functions $\{f_n\}$ such that $\mathrm{Skl}(X) \subset E(\mathfrak{M})$ for the interpretation of the added function terms as names of the functions $\{f_n\}$ and for the originally fixed interpretation of the terms from the set X. This fact and Theorem 30 yield that $A \in E(\mathfrak{M})$ for the same interpretation of the terms occurring both in $\mathrm{Skl}(X)$ and in the sentence A, and for the interpretation of those terms which occur in A but do not occur in $\mathrm{Skl}(X)$. Since A does not contain added function terms, $A \in E(\mathfrak{M})$ for the original interpretation of those terms which occur both in X and in A, and for any interpretation of the terms occurring in A, but not in X.

Therefore, for any non-empty model $\mathfrak{M}$ and for any interpretation of the terms occurring in the sentences from the set X and in a sentence A, if $X \subset E(\mathfrak{M})$, then $A \in E(\mathfrak{M})$. By Theorem 30, this means that $A \in \mathrm{Cn}^*(X)$.

Reduction of the consistency of any set to the consistency of a set of sentences without variables

Theorem 43 reduces the consistency of any set of sentences to the consistency of a set of sentences which have a normal form and do not contain existential quantifiers. Such sentences are called *general sentences*.

DEFINITION 21. *A sentence A is a general sentence if all its quantifiers are at the beginning and it does not contain any existential quantifiers.*

Thus all Skolem equivalents are general sentences. The consistency of any set of general sentences will now be reduced to the consistency of a set of sentences which do not contain variables. A set of variable-free sentences which corresponds to a set X of general sentences will be called a *description* corresponding to the set X.

DEFINITION 22. *A set Z is a description, with respect to the constants $\{s_n\}$ and the function symbols $F_1, ..., F_k$, corresponding to a general sentence A if and only if the set Z consists of all those sentences which:*

1. *are substitutions of the sentence A, formed by means of constant terms composed of symbols from the sequence $\{s_n\}$ and of the function symbols $F_1, ..., F_k$;*

2. *do not contain variables (since certain constant terms composed of symbols from the sequence $\{s_n\}$ and of the function symbols $F_1, ..., F_k$ have been substituted for all the variables).*

Thus, for instance, the description of the general sentence $\bigwedge x \bigwedge y$ $R(x, y, F(x, y))$, with respect to the constant terms $0, 1, 2, ...$ and the function symbol F, contains, among other things, the sentences

$$R(0, 0, F(0, 0)),$$
$$R(0, 1, F(0, 1)),$$
$$R(1, 0, F(1, 0)),$$
$$R(0, 2, F(0, 2)),$$
$$R(1, F(0, 0), F(1, F(0, 0))),$$
$$R(1, F(0, 1), F(1, F(0, 1))),$$
$$R(F(0, 0), F(0, 0), F(F(0, 0), F(0, 0))),$$
$$R(F(F(0, 1), 1), 0, F(F(F(0, 1), 1), 0)), \quad \text{etc.}$$

360

5. SKOLEM FORMS

A description with respect to the constants from the sequence $0, 1, 2, \ldots$ will be called an *arithmetic description*. Descriptions corresponding to the sentences from a set Z will be called a *description corresponding to the whole set* Z, and will be denoted by $\mathrm{Descr}(Z)$.

All the logical concepts occurring in this section are restricted to the logic L^*; thus, for instance, the concept of consistency, which was used in Theorem 43 and will be used in the next theorem, has the following sense: a set Z is consistent if, on the strength of the logic L^*, it is not possible to deduce from it a pair of contradictory sentences, that is, if there is no sentence A such that

$$A \in \mathrm{Cn}^*(Z) \quad \text{and} \quad \ulcorner {\sim} A \urcorner \in \mathrm{Cn}^*(Z).$$

THEOREM 45. *If Z is a set of general sentences, then*

$$Z \text{ is consistent} \equiv \text{an arithmetic description corresponding to the set } Z \text{ is consistent.}$$

PROOF. Every sentence from an arithmetic description corresponding to a sentence $A \in Z$ is a consequence of A: it is obtained from A by the application of Axiom 2 of the c.l.c. Hence, if the set Z is consistent, then an arithmetic description corresponding to Z, being contained in $\mathrm{Cn}^*(Z)$, is also consistent.

Conversely, if an arithmetic description is consistent, then we can easily construct a model for the set Z. Consider a sequence of terms $\{t_n\}$ consisting of all the terms formed of symbols from a sequence $\{s_n\}$ and of the function symbols $F_1, \ldots, F_k$. If a description of the set Z is consistent, then, by Theorem 26, there exists a non-empty model $\mathfrak{M} = \langle X, p_1, \ldots, p_n, f_1, \ldots, f_k, \{b_n\}\rangle$ and every sentence from that description is true in that model for the interpretation of the terms $P_1, \ldots, P_n, F_1, \ldots, F_k$ as names of the relations $p_1, \ldots, p_n$ and the functions $f_1, \ldots, f_k$, respectively, and for the interpretation of the constants $\{s_n\}$ as names of the elements of the sequence $\{b_n\}$.

Since the sentences from that description do not contain variables, they refer only to those elements of the set X, whose names t_n occur in the sentences of the description. Thus, every sentence of the description is true in a narrower model $\mathfrak{M}' = \langle X', p_1, \ldots, p_n, f_1, \ldots, f_k, \{b_n\}\rangle$, where X' is the set of all and only those elements of the set X which have names in $\{t_n\}$. As stated by Theorem 20, the set $E(\mathfrak{M}')$ of the

361

sentences true in $\mathfrak{M}'$ is ω-complete with respect to the sequence of names $\{t_n\}$ that contains names of all elements of the set X'. Hence $E(\mathfrak{M}')$ must contain all the sentences of the set Z, since they are generalizations of their substitutions with respect to all the constant names in the sequence $\{t_n\}$. Hence $\mathfrak{M}'$ is a non-empty model for the set Z.

THEOREM 46. *For every set Z of sentences in a normal form*

$$Z \text{ is consistent} \equiv \text{an arithmetic description of } (\text{Skl}(Z)) \text{ is consistent.}$$

The proof follows directly from Theorems 43 and 45.

The consistency of any set of sentences thus reduces to the consistency of a set of sentences which does not contain any variables.

Variable-free sentences and logical tautologies

The reduction of the consistency of any set to a set of sentences without variables suggests a closer analysis of the logical relations among variable-free sentences.

It can easily be surmised that among those sentences of that type which do not contain the identity predicate Id only substitutions of theorems of the sentential calculus can be logically true. In order to be able to formulate that theorem in a rigorous form, to prove it, and to formulate auxiliary theorems, we shall introduce certain definitions. Note first that an atomic sentential formula is a sentential formula no part of which is a sentential formula. If in an atomic formula all variables are replaced by constant terms, then we obtain a constant atomic formula, which is a sentence. Sentences of this type are called *atomic sentences*. Thus the sentences

$$P_1^2(0, 3),$$

$$P_1^1(8),$$

$$P_7^2\big(F_1'\big(F_2'(3)\big), 3\big), \quad \text{etc.,}$$

are atomic sentences. Thus, every sentence from an arithmetical description consists of atomic sentences linked by sentential connectives.

The set of atomic sentences which form parts of sentences from a set Z will be denoted by at(Z). The set at(Z) will briefly be called the *set of atoms* of the set Z.

5. SKOLEM FORMS

DEFINITION 23. *Every function f which associates with the atomic sentences from a set Z the numbers* 0 *or* 1

(76) $\bigwedge A \in \text{at}(Z)\,(f(A) = 0 \lor f(A) = 1)$

is called a zero-one valuation of the atomic sentences from the set Z.

DEFINITION 24. *The value of a sentence A (which does not contain variables) for a valuation f is the number (zero or one) computed by means of the truth-tables for the c.l.c. sentential connectives from the values associated by the function f with the atomic sentences which are components of the sentence A.*

Thus, the rigorous definition of values for variable-free formulae is inductive and similar to Definition 2* of values for formulae with variables. Note the abbreviation:

$W(f, A) =$ *the value of the sentence A for the valuation f*,

1. (The initial condition) *If A is an atomic sentence, then*

$W(f, A) = f(A);$

2. (Induction conditions)

$W(f, \sim A) = 1 - W(f, A),$

$W(f, \ulcorner A \to B \urcorner) = W(f, A) \to W(f, B).$

The functions $1 - p$ and $p \to q$ have been discussed in connection with Definition 2*. The above definition defines the value for all sentential connectives, since the remaining connectives may be treated as defined by means of $\to$ and $\sim$. The conditions:

$W(f, \ulcorner A \lor B \urcorner) = W(f, \ulcorner \sim A \to B \urcorner),$

$W(f, \ulcorner A \land B \urcorner) = W(f, \ulcorner \sim (\sim A \lor \sim B) \urcorner)$

could possibly be included in Definition 24.

THEOREM 47. *If Z is a set of variable-free sentences, if $\mathfrak{M}$ is any non-empty domain, and if f is a valuation of atomic sentences* at(Z) *such that for every atomic sentence $A \in \text{at}(Z)$ it is true that, for an adopted interpretation of terms,*

(77) $A \in E(\mathfrak{M}) \equiv f(A) = 1,$

then for every sentence $B \in Z$ it is true that

(78) $B \in E(\mathfrak{M}) \equiv W(f, B) = 1,$

for the same interpretation of terms.

PROOF BY INDUCTION. By Definitions 4* and 2*b, if A is variable-free, then for any sequence $\{a_n\}$ of objects from $\mathfrak{M}$ it is true that

(79) $A \in E(\mathfrak{M}) \equiv V(A, \{a_n\}) = 1$.

This, (77), and the initial condition of Definition 24 yield that, for $A \in \text{at}(Z)$,

(80) $V(A, \{a_n\}) = 1 \equiv W(f, A) = 1$.

Both functions, W and V, take on for sentences the values 0 or 1 only. This fact, and equivalence (80), yield for $A \in \text{at}(Z)$ the equation

(81) $V(A, \{a_n\}) = W(f, A)$.

On comparing the induction condition from Definition 24 with conditions 5 and 6 from Definition 2*a we obtain the implications:

$$V(A, \{a_n\}) = W(f, A) \to V(\ulcorner \sim A \urcorner, \{a_n\}) = W(f, \ulcorner \sim A \urcorner),$$

$$\big(V(A, \{a_n\}) = W(f, A) \wedge V(B, \{a_n\}) = W(f, B)\big) \to$$

$$V(\ulcorner A \to B \urcorner, \{a_n\}) = W(f, \ulcorner A \to B \urcorner).$$

We conclude immediately from these implications that if equation (81) holds for the atomic sentences, then it holds for the sentence obtained from atomic sentences linked by sentential connectives, and hence, in particular, equation (81) must hold for all the sentences of the set Z. This and equivalence (79) yield equivalence (78).

THEOREM 48. *If Z is a set of variable-free sentences, and if $\mathfrak{M}$ is any non-empty domain, then for a fixed interpretation of terms it is true that*

$$Z \subset E(\mathfrak{M}) \equiv \textit{there exists a valuation } f \textit{ of atomic sentences}$$

at(Z) such that

1. for every atomic sentence A it is true that
$A \in E(\mathfrak{M}) \equiv f(A) = 1$;

2. for every sentence $B \in Z$ it is true that
$W(f, B) = 1$.

PROOF. If $Z \subset E(\mathfrak{M})$, then as the designated valuation we adopt the function f which is defined on the atomic sentences $A \in \text{at}(Z)$ in the following way:

$$f(A) = \begin{cases} 1, & \text{when} \quad A \in E(\mathfrak{M}), \\ 0, & \text{when} \quad A \notin E(\mathfrak{M}). \end{cases}$$

In other words, a function f for atoms is defined as the value V: $f(A) = V(A, \{a_n\})$ for any sequence $\{a_n\}$ of elements from a domain $\mathfrak{M}$.

This function by definition satisfies condition (77), and hence, by Theorem 47, it satisfies condition (78). Hence, since $Z \in E(\mathfrak{M})$, for every $B \in Z$, $W(f, B) = 1$, f satisfies condition 2.

Conversely, if f satisfies condition (77), then, by Theorem 47, it satisfies condition (78). Since in addition it satisfies condition 2, $W(f, B) = 1$ for every $B \in Z$, hence, also $B \in E(\mathfrak{M})$ for every $B \in Z$. Consequently, $Z \subset E(\mathfrak{M})$.

THEOREM 49. *If Z is a set of variable-free sentences, then*

> *The set Z has a model $\equiv$ for a certain valuation of the atomic sentences at (Z) every sentence from the set Z has the value 1.*

PROOF. If a set Z has a model $\mathfrak{M}$, that is, if $Z \subset E(\mathfrak{M})$, then, by Theorem 48, there exists a valuation of the atomic sentences which satisfies condition 2, that is, such that $W(f, B) = 1$ for $B \in Z$.

Conversely, if there exists a valuation f of the atomic sentences at(Z) such that $W(f, B) = 1$ for $B \in Z$, then the model

$$\mathfrak{M} = \langle X, p_1, \ldots, p_n, f_1, \ldots, f_k, \{\ulcorner s_n \urcorner\} \rangle$$

for the sentences from the set Z is constructed of terms in the manner of the model described in Theorem 23. The set X of objects from the domain $\mathfrak{M}$ is the set of all terms formed of individual constants from the sequence $\{s_n\}$ and of the function terms $F_1, \ldots, F_k$ occurring in sentences from the set Z. The constants s_n are interpreted as names of themselves, and the functions $f_1, \ldots, f_k$ and the relations $p_1, \ldots, p_n$ are defined as follows. Suppose that those functions and relations are of two arguments each, and let t_1 and t_2 stand for any two terms. Then we set

$$(82) \quad \begin{aligned} f_i(\ulcorner t_1 \urcorner, \ulcorner t_2 \urcorner) &= \ulcorner F_i(t_1, t_2) \urcorner, \\ p_i(\ulcorner t_1 \urcorner, \ulcorner t_2 \urcorner) &\equiv f(\ulcorner P_i(t_1, t_2) \urcorner) = 1. \end{aligned}$$

It follows immediately from the very definition of the domain $\mathfrak{M}$ that for the sentences of the form $P_i(t_1, t_2)$, where t_1 and t_2 are any two terms, it is true that, for the interpretation of the predicates $P_1, \ldots, P_n$

365

as names of the relations $p_1, ..., p_n$, for the interpretation of $F_1, ..., F_k$ as names of the functions $f_1. ..., f_k$, and for the interpretation of the constants s_n as names of themselves, the following equivalence holds:

(83) $\quad \ulcorner P_i(t_1, t_2) \urcorner \in E(\mathfrak{M}) \equiv f(\ulcorner P_i(t_1, t_2) \urcorner) = 1.$

The sentences of the form $P_i(t_1, t_2)$ are atomic. Hence, for $A \in \text{at}(Z)$ the equivalence

$$A \in E(\mathfrak{M}) \equiv f(A) = 1$$

is true.

This and Theorem 48 yield that $Z \subset E(\mathfrak{M})$, that is, that $\mathfrak{M}$ is a model for the set Z.

THEOREM 50. *If Z is a set of variable-free sentences then*

> *Z is consistent $\equiv$ for some valuation of the atomic sentences at(Z) the value of every sentence from the set Z is 1.*

The proof follows directly from Theorems 49 and 27.

THEOREM 51. *If A is a variable-free sentence, then*

> *A is a substitution in a c.l.c. tautology $\equiv$ for every valuation f of atomic sentences $W(f, A) = 1$.*

PROOF. Theorem 51 is, strictly speaking, a reformulation, in a new terminology, of Theorem 1 (Chapter I) on the decidability of the set of the c.l.c. tautologies. It is so because, by that theorem, every sentential schema which for every substitution of zeros and ones (that is, for every valuation) yields one as the value of the entire formula is a c.l.c. tautology.

THEOREM 52. *If A is a variable-free sentence, then*

> *$A \in L^* \equiv A$ is a substitution in a c.l.c. tautology.*

PROOF. If A is a substitution in a c.l.c. tautology, then, by the definitions of the sets $A \times L$, $A \times L^*$, and L^*, $A \in L^*$, because all substitutions in the c.l.c. tautologies are logical theorems.

Conversely, suppose that A is not a substitution in any c.l.c. tautology. Hence, by Theorem 51, it follows that there exists a valuation f of atomic sentences such that for that valuation $W(f, A) = 0$, so that

$W(f, \sim A) = 1$. This and Theorem 50 yield that the sentence $\sim A$ has a non-empty model $\mathfrak{M}$, so that $\sim A \in E(\mathfrak{M})$. This and the consistency of the set $E(\mathfrak{M})$ yield that $A \notin E(\mathfrak{M})$. Thus the sentence A is not true in every non-empty model, and, hence, by Theorem 29, does not belong to L^*.

THEOREM 53. *If A is a general sentence, then*

$A \in L^* \equiv A$ *is the result of a substitution in a c.l.c. tautology prefaced by quantifiers which bind all the variables in A.*

PROOF. The implication ← follows from the definition of the c.l.c., since the logical calculus contains all the substitutions of the c.l.c. tautologies. The implication → is proved by *reductio ad absurdum*. Consider a substitution A' of the sentence A, in which each variable x_n is replaced respectively by the constant n. Let A have the form $\bigwedge x_1, ..., x_k B(x_1, ..., x_k)$. Then $A' = B(x_1/1, ..., x_k/k)$. The sentences A and A' obviously fall under one and the same sentential schema α. A is derived from the schema α when certain atomic formulae are substituted in it for the sentential variables and the whole is prefaced by universal quantifiers; A' is derived from the schema α when certain atomic sentences which are the corresponding substitutions of the atomic formulae occurring in A are substituted in it for the sentential variables. If at one place the schema α has a sentential variable p, and if that variable is replaced in the sentence A by a formula $P_n^2(x_i, x_j)$, then in the sentence A' it is replaced by the sentence $P_n^2(i, j)$.

Suppose that the schema α is not a c.l.c. tautology, which means that for some substitution of zeros and ones for the sentential variables it yields zero. This means in turn that, by Theorem 51, there exists a valuation f of the atomic sentences occurring in A' such that $W(f, A') = 0$, that is, $W(f, \sim A') = 1$. Hence, by Theorem 49, the sentence $\sim A'$ has a model $\mathfrak{M}$: $\ulcorner \sim A' \urcorner \in E(\mathfrak{M})$.

It follows from the definition of the sentence A' and axiom schema 2 of the c.l.c. that $\ulcorner A \to A' \urcorner \in L^*$. Hence, if $A \in L^*$, then also $A' \in L^*$, so that A' is true in every domain, and consequently $A' \in E(\mathfrak{M})$. Thus, the assumption that the implication → does not hold yields that the set $E(\mathfrak{M})$ is inconsistent. Yet, by Theorem 15, the set $E(\mathfrak{M})$ is consistent.

The interpolation theorems for the sentential calculus

The concepts developed in this section will be used to prove fairly important theorems on the existence of an intermediate formula and on the consistency of compound theories. These theorems will first be proved for the sets of variable-free sentences, and later will be extended so as to cover any theories; this will be done by reference to arithmetical descripitons and Skolem equivalents.

THEOREM 54 (Craig [11]). *If A and B are variable-free sentences such that* $at(\{A\}) \cap at(\{B\}) \neq \Lambda$ *and* $\ulcorner A \to B \urcorner \in L^*$, *then there exists a sentence C such that* $at(\{C\}) \subset at(\{A\}) \cap at(\{B\})$ *and*

$$\ulcorner A \to C \urcorner \in L^* \quad \text{and} \quad \ulcorner C \to B \urcorner \in L^*.$$

PROOF. To simplify the proof suppose that A has only two atoms, p and q, and that B also has only two atoms, p and r. To prove the theorem put $A = A(p, q)$, $B = B(p, r)$. Let 0 and 1 (interpreted as logical values) denote, respectively, the following sentences:

$$1 = (p \to p),$$
$$0 = \ \sim (p \to p).$$

If

(84) $\quad \ulcorner A(p, q) \to B(p, r) \urcorner \in L^*,$

then the following formula $C(p)$

(85) $\quad C(p) = \ulcorner (B(p, 0) \wedge B(p, 1)) \urcorner$

is a possible intermediate formula.

In fact, if (84) holds, then, by Theorem 52, the implication $A(p, q) \to B(p, r)$ is a substitution instance of a c.l.c. tautology. The atoms p, q, r may accordingly be interpreted as sentential variables. And if (84) is a sentential calculus tautology for p, q, r as sentential variables, then (84) remains a tautology if certain compound formulae are substituted for certain variables. Thus, for instance, we may substitute 0 or 1 for the sentential variable r. Hence,

$$\ulcorner A(p, q) \to B(p, 0) \urcorner \in L^*,$$
$$\ulcorner A(p, q) \to B(p, 1) \urcorner \in L^*.$$

It follows from these formulae that

(86) $\quad \ulcorner A(p, q) \to (B(p, 0) \wedge B(p, 1)) \urcorner \in L^*.$

On the other hand, it is true that

(87) $\ulcorner (B(p, 1) \wedge B(p, 0)) \rightarrow B(p, r) \urcorner \in L^*$.

The above formula is a tautology. It would be false only if, for some substitution of 0 or 1 for r, the consequent were false while the antecedent were true. But then the antecedent would prove false too, and so the implication would remain true. Formulae (86) and (87) show that a formula $C(p)$ of the form (85) in fact satisfies the conditions laid down in the theorem. If B contains atoms other than those shared with A, $C(p)$ would have to be a conjunction of more elements. For instance, if B contains the atoms $p, r, s : B = B(p, r, s)$, then $C(p)$ would have the form

$$B(p, 0, 0) \wedge B(p, 0, 1) \wedge B(p, 1, 0) \wedge B(p, 1, 1).$$

Theorem 54 (called *Craig's interpolation theorem*) can be strengthened if we point out that not only must atoms which occur in only one element not occur in the intermediate formula, but also other atoms, which occur in both parts of the implication may not occur in the intermediate formula.

THEOREM 55 (Lyndon [45]). *If A and B are variable-free sentences having some common atoms, and if $\ulcorner A \rightarrow B \urcorner \in L^*$, then there exists an intermediate sentence C such that*

$$\ulcorner A \rightarrow C \urcorner \in L^* \quad and \quad \ulcorner C \rightarrow B \urcorner \in L^*,$$

and such that the atoms of C satisfy the condition

(α)   *If an atom a occurs affirmatively in the normal form of the formula C, then the atom a also occurs affirmatively in a place of the normal form of the formula A and in a place of the normal form of the formula B; if, on the contrary, the atom a occurs negated in the normal form of the formula C, then it also occurs negated in a place of the formula A and, also negated, in a place of the normal form of the formula B.*

By *affirmative occurrence* is meant an occurrence of the atom in the normal form without a preceding negation symbol, and by *negated occurrence* is meant an occurrence in which the atom in question is preceded by the negation symbol. By the normal form is meant here the disjunctive-conjunctive normal form as described in Section 3 of Chapter I.

PROOF. The form of the intermediate formula will depend on the normal form of the formulae A and B. Suppose that the sentence A is a disjunction of the formulae A_k:

$$A = A_1 \vee \ldots \vee A_K,$$

and each A_k $(1 \leqslant k \leqslant K)$ is a conjunction of the formulae A_{k_n}:

$$A_k = A_{k_1} \wedge \ldots \wedge A_{k_N},$$

such that each A_{k_n} $(1 \leqslant n \leqslant N)$ is either an atomic sentence or a negation of an atomic sentence. The same holds for the sentence B: it is a disjunction of the formulae B_t $(1 \leqslant t \leqslant T)$:

$$B = B_1 \vee \ldots \vee B_T,$$

and each B_t is a conjunction of the formulae B_{t_s} $(1 \leqslant s \leqslant S)$:

$$B_t = B_{t_1} \wedge \ldots \wedge B_{t_S},$$

such that the formulae B_{t_s} are either atomic or negations of atoms. The formula C is defined as follows: C is a disjunction of as many elements as A has:

$$C = C_1 \vee \ldots \vee C_K.$$

Further, each C_k $(1 \leqslant k \leqslant K)$ is defined as follows: either $C_k = 0$, if A_k is inconsistent in the sense that the same atom occurs in A_k both in the affirmative and in the negative form (which means that, for some n and n', $A_{k_n} = \sim A_{k_{n'}}$), or C_k is a conjunction of as many elements as the corresponding A_k has:

$$C_k = C_{k_1} \wedge \ldots \wedge C_{k_N},$$

and the elements C_{k_n} are defined as follows:

$$(88) \qquad C_{k_n} = \begin{cases} A_{k_n}, & \text{if } A_{k_n} \text{ occurs in the normal form of the formula } B, \text{ so that, for some } t \text{ and } s: A_{k_n} = B_{t_s}, \\ 1 & \text{if } A_{k_n} \text{ does not occur in the normal form of the formula } B. \end{cases}$$

Since C_{k_n} is an atom, or a negation of an atom, or 1, hence by the very definition of C_{k_n} it follows that condition (α) is satisfied. If an atom occurs in C and is not contained in 1 or 0, then it occurs in A in the same form (affirmative, if it is affirmative in C, negated, if it is negated in C), since that atom or its negation is identical with C_{k_n}, and by definition $C_{k_n} = A_{k_n}$. Furthermore, if that atom occurs in C and is not contained in 1 or 0, that is, if it is identical with a C_{k_n}, or C_{k_n} is its negation, which means that atom does not reduce to 1, so that the upper condition of the two specified in the definition of C_{k_n} holds, then there exist t and s such that $C_{k_n} = A_{k_n} = B_{ts}$, and hence that atom occurs in B in the same form as in C (affirmative if it is affirmative in C, negated, if it is negated in C).

It remains to show that C is in fact an intermediate formula, that is, that

$$\ulcorner A \to C \urcorner \in L^* \quad \text{and} \quad \ulcorner C \to B \urcorner \in L^*.$$

The first implication is self-evident. Since, by (88), either $C_{k_n} = A_{k_n}$, or $C_{k_n} = 1$, then in both cases it is true that $\ulcorner A_{k_n} \to C_{k_n} \urcorner \in L^*$, and, hence, under the sentential calculus laws of multiplying implications, $\ulcorner A_k \to C_k \urcorner \in L^*$. If A_k is inconsistent, then also $\ulcorner A_k \to C_k \urcorner \in L^*$. By the laws of adding implications, $\ulcorner A \to C \urcorner \in L^*$.

The second implication can be obtained as follows. Consider the negation of B: $\sim B$. It can, of course, also be reduced to a normal form: $\sim B \equiv D_1 \vee \ldots \vee D_U$, such that every D_u $(1 \leqslant u \leqslant U)$ is a conjunction: $D_u = D_{u_1} \wedge \ldots \wedge D_{u_W}$ whose elements are atoms or negations of atoms. Of course, the atoms of the formula $\sim B$ are the same as the atoms of the formula B, they can occur different numbers of times, but an atom occurs in a normal form of $\sim B$ if and only if it occurs in a normal form of the formula B. It can easily be noticed that

(89)   *An atom occurs in the normal form of the formula B affirmatively if and only if it occurs negated in the normal form of the formula $\sim$ B, and it occurs negated in the normal form of the formula B if and only if it occurs affirmatively in the normal form of the formula $\sim$ B.*

Definition (88) of the element C_{k_n} can accordingly be written as

(90) $$C_{k_n} = \begin{cases} A_{k_n} & \text{if there exist } u \text{ and } w \text{ such that } A_{k_n} = \sim D_{u_w}, \\ & \text{or } \sim A_{k_n} = D_{u_w}, \\ 1 & \text{otherwise.} \end{cases}$$

Further reasoning is as follows. If $\ulcorner A \to B \urcorner \in L^*$, then $\ulcorner \sim (A \wedge \sim B) \urcorner \in L^*$, and, hence, for all k and u,

(91) $\ulcorner \sim (A_k \wedge D_u) \urcorner \in L^*$.

Use is made here of the distributivity of conjunction with respect to disjunction and De Morgan's laws in the sentential calculus. Next, from

$$\ulcorner \sim (A_k \wedge D_u) \urcorner \in L^*,$$

we wish to conclude that $\ulcorner \sim (C_k \wedge D_u) \urcorner \in L^*$.

If A_k is inconsistent, then by definition $C_k = 0$, and, hence, obviously $\ulcorner \sim (C_k \wedge D_u) \urcorner \in L^*$. If A_k is consistent and identical with C_k, then $\ulcorner \sim (C_k \wedge D_u) \urcorner \in L^*$ immediately by (91). If A_k is consistent and not identical with C_k, then in the transition from A_k to C_k, there is an A_{k_n} which is, in accordance with (90), replaced by 1. This is done when there do not exist u and w such that $A_{k_n} = \sim D_{u_w}$ or $\sim A_{k_n} = D_{u_w}$. A_{k_n} and D_{u_w} are atoms or negations of atoms. Hence, it follows from (90) that for all u and w

(92)   *If $C_{k_n} = 1$ and if the atom contained in A_{k_n} is identical with that contained in D_{u_w}, then $A_{k_n}' = D_{u_w}$.*

If A_k is consistent, then the transition from A_k to C_k can be imagined as a finite sequence of steps such that at each step only one of the A_{k_n} is replaced by 1. Thus, there exists a sequence $A_k^1, \ldots, A_k^M$ of formulae such that $A_k^1 = A_k$, $A_k^M = C_k$, and for every m $(1 \leqslant m \leqslant M)$ the formula A_k^{m+1} is derived from A_k^m by the replacement of a single A_{k_n} contained in A_k^m by the symbol 1.

It will be demonstrated that for each m $(1 \leqslant m \leqslant M)$

(93) $\ulcorner \sim (A_k^m \wedge D_u) \urcorner \in L^*$.

The proof is by induction with respect to m. For $m = 1$ the proof is in formula (91). We take now (93) as the induction assumption and deduce from it that

(94) $\ulcorner \sim (A_k^{m+1} \wedge D_u) \urcorner \in L^*$.

Formula A_k^m can be presented as the conjunction

$$A_k^m = A_{k_1} \wedge \ldots \wedge A_{k_n} \wedge \ldots \wedge A_{k_N},$$

while A_k^{m+1} would then be a conjunction of the form

$$A_k^{m+1} = A_{k_1} \wedge \ldots \wedge 1 \wedge \ldots \wedge A_{k_N}.$$

The formula A_{k_n} has been replaced in it by the symbol 1.

Consider two cases now. The atom contained in A_{k_n} either occurs in D_u, or it does not.

If it does, then it occurs in a D_{u_w}, and then, by (92), $A_{k_n} = D_{u_w} = q$. Hence, denoting all the remaining elements of A_k^m by A_q and all the remaining elements of D_u by D_q we may, in accordance with the laws of associativity and commutativity of conjuction, present the induction assumption (93) as

$$\ulcorner \sim (A_q \wedge q \wedge D_q \wedge q)\urcorner \in L^*.$$

Theorem (94) would then have the form

$$\ulcorner \sim (A_q \wedge 1 \wedge D_q \wedge q)\urcorner \in L^*.$$

The transition from the induction assumption to the theorem is, thus, obtained on the strength of the following sentential calculus law:

$$\sim (p \wedge q \wedge r \wedge q) \to \sim (p \wedge 1 \wedge r \wedge q).$$

If the atom contained in A_{k_n} does not occur in D_u, then two cases of other occurrences of the atom are considered. The first is that there exists another element of the conjunction A_k which contains the same atom, but with the opposite sign (that is, negative if it is positive in A_{k_n}, or positive if it is negative in A_{k_n}). But then A_k would be inconsistent, and that case has been considered previously. Hence, the only remaining possibility is that if the atom contained in A_{k_n} occurs in some other element of the conjunction A_k, then it occurs in the same way as it does in A_{k_n}, so that A_{k_n} simply occurs more than once in the conjunction A_k. The conjuction A_k^m is then equivalent to a sentence in which the element A_{k_n} occurs only once. Hence, if the conjunction of all the remaining elements of A_k^m is denoted by A_q and if A_{k_n} is denoted by q, then

(95) $\qquad \ulcorner A_k^m \equiv (A_q \wedge q)\urcorner \in L^* \quad$ and $\quad \ulcorner A_k^{m+1} \to (A_q \wedge 1)\urcorner \in L^*,$

which, together with the induction assumption (93), yields that

$$\ulcorner \sim (A_q \wedge q \wedge D_u)\urcorner \in L^*,$$

where neither A_q nor D_u contains the atom contained in q. If $\ulcorner \sim (A_q \wedge q \wedge D_u) \urcorner \in L^*$, this means that for every valuation f of the atoms

$$W(f, \ulcorner (A_q \wedge q \wedge D_u) \urcorner) = 0.$$

Hence, we infer that for every valuation of atoms also:

(96) $\qquad W(f, \ulcorner (A_q \wedge 1 \wedge D_u) \urcorner) = 0.$

For should, for some valuation of atoms, $W(f, \ulcorner (A_q \wedge 1 \wedge D_u) \urcorner) = 1$, then we could define the following valuation of atoms:

$$g(a) = \begin{cases} f(a) & \text{for the atoms contained in } A_q \text{ or in } D_u, \\ 0 & \text{if } a \text{ is an atom contained in } q \text{ and } q = \sim a, \\ 1 & \text{if } a = q \text{ and } q \text{ is an atom.} \end{cases}$$

The above formula defines correctly a valuation of the atoms occurring in the formula $(A_q \wedge q \wedge D_u)$, since an atom contained in q occurs neither in A_q nor in D_u. This and the definition of the function g show that $W(g, q) = 1$ and $W(g, A_q) = W(f, A_q)$ and $W(g, D_u) = W(f, D_u)$, and hence

$$W(g, (A_q \wedge q \wedge D_u)) = W(g, (A_q \wedge 1 \wedge D_u)) =$$
$$W(f, (A_q \wedge 1 \wedge D_u)) = 1.$$

Thus, there would exist a valuation for which $W(g, (A_q \wedge q \wedge D_u)) = 1$.

We have, thus, proved that formula (96) holds for all valuations f of atoms, and this, by the theorem on the decidability of the sentential calculus (Theorem 51), yields that $\ulcorner \sim (A_q \wedge 1 \wedge D_u) \urcorner \in L^*$. This and (95) yield that $\ulcorner \sim (A_k^{m+1} \wedge D_u) \urcorner \in L^*$ also when the atom contained in A_{k_n} does not occur in D_u.

Formula (93) has, thus, been proved by induction. Hence, in the special case of $m = M$ we have that $\ulcorner \sim (C_k \wedge D_u) \urcorner \in L^*$. Thus, both when A_k is inconsistent and when it is not we have that $\ulcorner \sim (C_k \wedge D_u) \urcorner \in L^*$ for all k and u. Hence, under De Morgan's laws and the laws of distributivity of conjunction with respect to disjunction, it follows that $\ulcorner \sim (C \wedge \sim B) \urcorner \in L^*$, so that $\ulcorner C \to B \urcorner \in L^*$.

Note that Theorem 55 implies Theorem 54. It follows from condition (α) that if an atom a occurs in an intermediate formula C, then it must occur in formulae A and B. C, thus, contains only those atoms which are common to A and B. But Theorem 55 is much stronger: it follows from it that if an atom occurs only affirmatively in the antecedent and only negatively in the consequent, then there exists an intermediate formula C in which it does not occur at all.

The same observation as in Theorem 54 is formulated in the following

THEOREM 56. *If Z is an inconsistent set of variable-free sentences and if Z is a set-theoretical sum of two consistent sets X and Y*

$$Z = X \cup Y,$$

then there exist atoms which are common to the sets X and Y: $\text{at}(X) \cap \text{at}(Y) \neq \Lambda$, and there exists a sentence C consisting of atoms common to the sets X and Y: $\text{at}(\{C\}) \subset \text{at}(X) \cap \text{at}(Y)$ such that

$$C \in \text{Cn}^*(X) \quad and \quad \ulcorner \sim C \urcorner \in \text{Cn}^*(Y).$$

PROOF. If Z is an inconsistent set, that is, $\ulcorner (P \wedge \sim P) \urcorner \in \text{Cn}(Z)$ for a sentence P, then it follows from the finitistic nature of consequence that there exists a finite subset Z' of the set Z, which is inconsistent, that is, $\ulcorner P \wedge \sim P \urcorner \in \text{Cn}(Z')$. Let there be two conjunctions:

$A = $ *the conjunction of those sentences of Z' which are in X,*
$B = $ *the conjunction of those sentences of Z' which are in Y.*

The conjunction $(A \wedge B)$ is thus inconsistent: $\ulcorner P \wedge \sim P \urcorner \subset \text{Cn}(\{A \wedge B\})$. On the other hand, neither A nor B taken separately is inconsistent, since the sets X and Y taken separately are not inconsistent. Since A and B are consistent, then, by Theorem 50, there exist valuations g and h of the atoms $\text{at}(\{A\})$ and $\text{at}(\{B\})$ such that $W(g, A) = 1$ and $W(h, B) = 1$. If the sets $\text{at}(\{A\})$ and $\text{at}(\{B\})$ were disjoint, then the valuations g and h would define a common valuation f for all the atoms $D \in \text{at}(\{A \wedge B\}) = \text{at}(\{A\}) \cup \text{at}(\{B\})$ on the assumption that

$$f(D) = \begin{cases} g(D) & \text{when} \quad D \in \text{at}(\{A\}), \\ h(D) & \text{when} \quad D \in \text{at}(\{B\}). \end{cases}$$

It follows self-evidently from the definition of the valuation f that $W(f, A) = W(g, A) = 1, W(f, B) = W(h, B) = 1$. Hence $W(f, A \wedge B) = 1$, so that, by Theorem 50, the sentence $A \wedge B$ would be consistent. But since $A \wedge B$ is inconsistent, then the sets $\text{at}(\{A\})$ and $\text{at}(\{B\})$ cannot be disjoint. Since $A \in X$ and $B \in Y$, $\text{at}(\{A\}) \subset \text{at}(X)$, $\text{at}(\{B\}) \subset \text{at}(Y)$, and $\mathrm{at}(X) \cap \mathrm{at}(Y) \neq \varLambda$. Since the conjunction $A \wedge B$ is inconsistent, we have

(97) $\ulcorner A \to \,\sim B \urcorner \in L^*$.

As demonstrated above, the sentences A and B have certain atoms in common, hence, by Theorem 54, there exists an intermediate sentence C such that

$$\mathrm{at}(\{C\}) \subset \mathrm{at}(\{A\}) \cap \mathrm{at}(\{B\})$$

and

(98) $\ulcorner A \to C \urcorner \in L^*$ and $\ulcorner C \to \,\sim B \urcorner \in L^*$.

It follows from (98) that

(99) $\ulcorner A \to C \urcorner \in L^*$ and $\ulcorner B \to \,\sim C \urcorner \in L^*$,

and since $A \in X$ and $B \in Y$, hence (99) yields the desired formulae

$$C \in \mathrm{Cn}^*(X) \quad \text{and} \quad \ulcorner \sim C \urcorner \in \mathrm{Cn}^*(Y).$$

Theorem 56 is very intuitive. In a free formulation it states that if the union of two consistent sets of constant sentences is inconsistent, then that inconsistency must be expressed in terms common to both sets. The following is a generalization of Theorem 56 covering arbitrary theories:

THEOREM 57 (Robinson [72]). *If X and Y are any consistent and disjoint sets of closed sentences, if the sentences of the set X have $P, \ldots, Q$, $R, \ldots, S$ as the only extralogical terms, and if the sentences of the set Y have $P, \ldots, Q, U, \ldots, W$ as the only extralogical terms, and if the set $X \cup Y$ is inconsistent, then there exists a sentence C, which has as extralogical terms only the terms $P, \ldots, Q$, common to the sets X and Y and such that*

$$C \in \mathrm{Cn}^*(X) \quad and \quad \ulcorner \sim C \urcorner \in \mathrm{Cn}^*(Y).$$

PROOF. Consider the arithmetical description associated with the Skolem equivalent of the set $X \cup Y$: arith des $(\mathrm{Skl}(X \cup Y))$. That description may be taken to be the union of two descriptions that correspond to the sets X and Y:

(100)   arith des $(\mathrm{Skl}(X \cup Y)) =$
         arith des $(\mathrm{Skl}(X)) \cup$ arith des $(\mathrm{Skl}(Y))$.

If the set $X \cup Y$ is inconsistent, then, by Theorem 46, arith des $(Skl(X \cup Y))$ is inconsistent, too. Hence the sum of the descriptions is inconsistent. Theorem 56 accordingly yields that there exists a sentence A such that

(101) $\quad \text{at}(\{A\}) \subset \text{at}\left(\text{arith des}(Skl(X))\right) \cap \text{at}\left(\text{arith des}(Skl(Y))\right)$

and

(102) $\quad$ $A \in \text{Cn}^*\left(\text{arith des}(Skl(X))\right),$
$\ulcorner \sim A \urcorner \in \text{Cn}^* \left(\text{arith des}(Skl(Y))\right).$

Since arith des $(Skl(X)) \subset \text{Cn}^*(Skl(X))$, and since the corresponding formula holds for Y, (102) yields that

(103) $\quad$ $A \in \text{Cn}^*(Skl(X)),$
$\ulcorner \sim A \urcorner \in \text{Cn}^*(Skl(Y)).$

It follows from (101) that the sentence A contains only the predicates $P, \ldots, Q$, that is, those which are common to the sets X and Y, and added Skolem functions which occur in both $Skl(X)$ and $Skl(Y)$. Furthermore, it contains the constants $0, 1, 2, \ldots$, which occur neither in $Skl(X)$ nor in $Skl(Y)$. Starting with the sentence A it is possible to construct a sentence C, which contains neither the constants $0, 1, 2, \ldots$ nor any Skolem functions, and which satisfies the conditions laid down in Theorem 57.

Transition from A to C can be described as a finite sequence of transformations that follow one another: $A = A_0, A_1, \ldots, A_{n+1} = C$. The initial sentence $A = A_0$ contains neither variables nor quantifiers, the terminal sentence $A_{n+1} = C$ does contain variables and quantifiers, but contains neither any constants from the sequence $0, 1, 2, \ldots$ nor any Skolem functions. The sentence A may, for instance, have the form of:

(104) $\quad A\left(0, 1, F_1(0), F_1(G_1(1)), F_2\left(G_1(F_1(0)), 3\right), G_1\left(F_1(0)\right)\right).$

Let F_j stand for the Skolem functions occurring in $Skl(X)$, and G_j, for the Skolem functions occurring in $Skl(Y)$. All the terms occurring in A are formed of the constants $0, 1, 2, \ldots$ and the functions F_j and G_j.

Concerning the terms contained in A let a distinction be made between terms of the first kind and those of the second kind. *Terms of the first kind* are terms $t_1, \ldots, t_i$ which are arguments in atomic sentences of the form $P(t_1, \ldots, t_i), \ldots, Q(t_1, \ldots, t_i)$, occurring in A. *Terms of the second kind* are those which are arguments in term formulae $F_j(t_1, \ldots, t_i)$ or $G_j(t_1, \ldots, t_i)$, contained in A. Terms of the second kind are parts of other terms. The classification is not disjoint, since a term may occur in one place as a term of the first kind, and in another place may be part of another term and, thus, occur as a term of the second kind. Let the terms listed in formula (104) be all the terms of the first kind. Some of them, e.g., $F_1(0)$ and $G_1(F_1(0))$, are also terms of the second kind.

All those terms which occur in A, be it in one place only, as terms of the first kind are ordered as a finite sequence $\{t_0, \ldots, t_n\}$ beginning with the greatest ones (that is, those consisting of the greatest number of symbols) and coming down to the smallest ones (those consisting of one symbol each). Among those terms which have the same number of symbols a certain conventional, essentially arbitrary ordering given in advance is adopted. The terms occurring in formula (104), thus, form the following sequence:

$$F_2\big(G_1(F_1(0)), 3\big), \quad F_1\big(G_1(1)\big), \quad G_1\big(F_1(0)\big), \quad F_1(0), \quad 0, \quad 1,$$
$$t_0, \qquad\qquad t_1, \qquad\qquad t_2, \qquad\qquad t_3, \qquad t_4, \quad t_5.$$

The sequence of successive transformations $\{A_0, \ldots, A_{n+1}\}$ is defined as follows:

$$A_0 = A;$$

for $0 \leqslant i \leqslant n$:

$$(105) \qquad A_{i+1} = \begin{cases} \bigvee x_i A_i(t_i/x_i) & \text{when } t_i \text{ begins with a function } F_j; \\ \bigwedge x_i A_i(t_i/x_i) & \text{otherwise.} \end{cases}$$

Note that in forming $A_i(t_i/x_i)$ we replace the term t_i by the variable x_i in all those places in which t_i occurs in A_i.

A_{i+1} is thus derived by the replacement in A_i of the term t_i by the variable x_i and by preceding the whole by the existential quantifier when t_i begins with a Skolem function F_j, and by the universal quantifier when that term begins with a function G_j or is a numerical constant.

The last element in the sequence, $A_{n+1} = C$, obviously does not contain any terms, but has $n+1$ quantifiers at the beginning. In the case of

(104), the last element of the sequence of transformations, $A_6 = C$, has the form

$$\bigwedge x_5 \bigwedge x_4 \bigvee x_3 \bigwedge x_2 \bigvee x_1 \bigvee x_0 \, A(x_4, x_5, x_3, x_1, x_0, x_2).$$

It will now be proved that $C \in \mathrm{Cn}^* \, (\mathrm{Skl}(X))$, and $\ulcorner \sim C \urcorner \in \mathrm{Cn}^*(\mathrm{Skl}(Y))$. It will be a semantic proof, based on a semantic characteristic of consequence, resulting from Theorem 31.

The basic step in the proof will be the application of the following lemma:

(γ)   *If $A\big(H(t_0)\big)$ is a sentence in which the term $H(t_0)$ is not part of any other term that begins with the functor H, if the variables occurring in the sentence $A\big(H(t_0)\big)$ are not arguments in any term formulae that begin with the functor H, and if X is a set of sentences which do not contain the functor H, and if the variable x_i does not occur in $A\big(H(t_0)\big)$, then, if $A\big(H(t_0)\big) \in \mathrm{Cn}^*(X)$, then also $\ulcorner \bigwedge x_i A(x_i) \urcorner \in \mathrm{Cn}^*(X)$.*

PROOF OF THE LEMMA. To demonstrate that $\ulcorner \bigwedge x_i \, A(x_i) \urcorner \in \mathrm{Cn}^*(X)$ consider any model $\mathfrak{M}$, in which all those functions which are interpretations of the function terms occurring in sentences from X and in the sentence A form a totally free combination (are fully separable). It is assumed that

(106) $X \subset E(\mathfrak{M})$,

and it will be proved that $\ulcorner \bigwedge x_i A(x_i) \urcorner \in E(\mathfrak{M})$. For that purpose consider any element a from the domain $\mathfrak{M}$ and a function h which interprets the functor H in the domain $\mathfrak{M}$. A new function h' is defined on the elements of $\mathfrak{M}$:

(107) $h'(u) = \begin{cases} a & \text{when } u = V(t_0, \{a_n\}) \text{ for a sequence } \{a_n\} \\ & \text{from } \mathfrak{M}, \\ h(u) & \text{otherwise.} \end{cases}$

As is known from Definition 2*a, in the above formula the value of the constant term $V(t_0, \{a_n\})$, of course, does not depend on the sequence $\{a_n\}$.

Note also that Lemma (γ) is true if t_0 is interpreted not as a single term, but as a complex of terms $t_1, \ldots, t_k$. In such a case, if the functor

H has k arguments, its interpretations h and h' also must be functions of k arguments, and h' is then defined as follows:

$$h'(u_1, \ldots, u_k) = \begin{cases} a & \text{when } u_1 = V(t_1, \{a_n\}) \text{ and} \\ & \ldots \text{ and } u_k = V(t_k, \{a_n\}), \\ h(u_1, \ldots, u_k) & \text{otherwise.} \end{cases}$$

Consider now the domain $\mathfrak{M}'$ obtained from $\mathfrak{M}$ through the replacement of the function h by the function h'. Thus, if $\mathfrak{M} = \langle M, r_1, \ldots, r_s, f_1, \ldots, f_u, h \rangle$, then $\mathfrak{M}' = \langle M, r_1, \ldots, r_s, f_1, \ldots, f_u, h' \rangle$. Consider also the valuation of formulae in the latter domain. Since both valuations, in $\mathfrak{M}$ and in $\mathfrak{M}'$, will be considered simultaneously, the domain to which a given value refers must be indicated. Let then $V(\Phi, \{a_n\})$ stand as before for a value in $\mathfrak{M}$, and $V'(\Phi, \{a_n\})$, for a value (of course also defined by Definition 2*a) in $\mathfrak{M}'$. It is self-evident that for those formulae Φ which do not contain the function term H these values coincide, that is, $V(\Phi, \{a_n\}) = V'(\Phi, \{a_n\})$, since these domains differ only in the interpretation of the function term H, and all the remaining terms, both functional $F_1, \ldots, F_u$ and relational $R_1, \ldots, R_s$, are in both models interpreted by the same functions and relations $f_1, \ldots, f_u$, $r_1, \ldots, r_s$. (It is assumed that $H, F_1, \ldots, F_u, R_1, \ldots, R_s$ are the only extralogical terms which occur in the sentences from X and in $A(H(t_0))$.) Since it has been assumed in the lemma that the sentences from X do not contain the function constant H, the values of the sentences from X are the same in both models. Accordingly, since by (106) the sentences from X are true in $\mathfrak{M}$, they are also true in $\mathfrak{M}'$:

(108) $X \subset E(\mathfrak{M}')$.

Since also $A(H(t_0)) \in Cn^*(X)$, hence, by Theorem 9, it follows from (108) that

(109) $A(H(t_0)) \in E(\mathfrak{M}')$.

Consider now a formula $A(x_i)$ such that the formula $A(H(t_0))$ is derived from $A(x_i)$ by the substitution of the constant term $H(t_0)$ for the free variable x_i; consider also sequences $\{a_n\}$ for which $a_i = a$. It will be proved that for such sequences

(110) $V(A(x_i), \{a_n\}) = V'(A(H(t_0)), \{a_n\})$.

The proof will be by induction with respect to the subformulae of the formula $A(H(t_0))$. If B is any subformula of the formula $A(H(t_0))$, then let B^x stand for a formula which differs from B only in having the variable x_i in those places in which B has the term $H(t_0)$. If B does not contain $H(t_0)$, then $B^x = B$. Thus it will be proved by induction, for the sequences $\{a_n\}$ under consideration, that is, those for which $a_i = a$, that

(111) $V(B^x, \{a_n\}) = V'(B, \{a_n\})$.

If B is a variable or an individual constant, then it does not contain the functor H, and, hence, not the term $H(t_0)$ either. Consequently, $B^x = B$, and as the term H does not occur in B, the values of the formula B are the same in both models:

$$V(B^x, \{a_n\}) = V(B, \{a_n\}) = V'(B, \{a_n\}).$$

Thus equation (111) is true for variables and individual constants. Now assume that equation (111) holds for simpler formulae B; it will be proved that if it holds for the simpler ones, it also holds for the more complex ones. Consider first the case when a compound formula is a term formula. A term formula is obtained from simpler ones when these are placed in parentheses and preceded by a function symbol. If that function symbol differs from the functor H, then

$$V(F_j(B^x), \{a_n\}) = f_j(V(B^x, \{a_n\})) = f_j(V'(B, \{a_n\})) =$$
$$V'(F_j(B), \{a_n\}),$$

because those function terms which are other than H have the same interpretation in both models. Since also $(F_j(B))^x = F_j(B^x)$,

$$V((F_j(B))^x, \{a_n\}) = V'(F_j(B), \{a_n\}),$$

which thus yields formula (111) for a compound formula. If F_j has more arguments, the method of the proof is the same.

If H is the initial function symbol of a compound formula, then in investigating $H(B)$ note first that, by the conditions of the lemma above, the formula B cannot be a variable. Thus, B is a constant term. Consider two cases: either $B = t_0$ or $B \neq t_0$.

If $B = t_0$, then $H(B) = H(t_0)$. Then $(H(B))^x = (H(t_0))^x = x_i$, because the transformation B^x just transforms the formula $H(t_0)$ into the variable x_i. Since we only consider the sequence $\{a_n\}$, for which

$a_i = a$, by the definition of value, by the induction assumption (111), and by the definition (107) of the function h', we obtain the equations

$$V(H(B)^x, \{a_n\}) = V(H(t_0)^x, \{a_n\}) = V(x_i, \{a_n\}) = a_i = a,$$

$$V'(H(B), \{a_n\}) = h'(V'(B, \{a_n\})) = h'(V(B^x, \{a_n\})) =$$
$$h'(V(t_0, \{a_n\})) = a.$$

It follows from these equations that

(112) $V(H(B)^x, \{a_n\}) = V'(H(B), \{a_n\})$.

If $B \neq t_0$, then it follows by the assumption of the lemma above that B cannot contain the formula $H(t_0)$, for it has been assumed in the lemma that the formula $H(t_0)$ in no place in the sentence $A(H(t_0))$ is part of a larger term that begins with the functor H. Hence, it is concluded that $B^x = B$ and $H(B)^x = H(B)$. Since $B \neq t_0$, we also have

(113) $V(B, \{a_n\}) \neq V(t_0, \{a_n\})$.

Formula (113) is a consequence of the fact that it has been assumed about the model $\mathfrak{M}$ that the functions which interpret function terms are totally separable in it. This consequence may even be considered a definition of the total freedom of a combination of functions. A combination of functions is *totally separable* if and only if the values of two different terms are different (cf. Exercise 8 on p. 324). We can now compute the value:

$$V(H(B)^x, \{a_n\}) = V(H(B), \{a_n\}) = h(V(B, \{a_n\})) =$$
$$h'(V(B, \{a_n\})) = h'(V(B^x, \{a_n\})) =$$
$$h'(V'(B, \{a_n\})) = V'(H(B), \{a_n\}).$$

In this computation use is made of the second formula for h' ($h'(u) = h(u)$), in view of the fact that (113) holds. Moreover, use is made of the definition of value, the induction assumption (111), and the equations $B^x = B$ and $H(B)^x = H(B)$.

Thus, formula (112) is obtained both when $B = t_0$ and when $B \neq t_0$. In this way all the ways in which a compound term formula can be formed have been analysed. Formula (111) is thus true for all term formulae B contained in the formula A. Since in both models predicates are interpreted as names of the same relations, formula (111) is transferred from term formulae to all atomic sentential formulae:

$$V(R_j(B_1, B_2)^x, \{a_n\}) = V(R_j(B_1^x, B_2^x), \{a_n\}) = 1 \equiv$$
$$r_j(V(B_1^x, \{a_n\}), V(B_2^x, \{a_n\})) \equiv$$
$$r_j(V'(B_1, \{a_n\}), V'(B_2, \{a_n\})) \equiv$$
$$V'(R_j(B_1, B_2), \{a_n\}) = 1.$$

Formula (111) is thus true for all atomic formulae, and, thus, also for all subformulae of formula $A(H(t_0))$:

$$V((\sim B)^x, \{a_n\}) = V(\sim (B^x), \{a_n\}) = 1 - V(B^x, \{a_n\}) =$$
$$1 - V'(B, \{a_n\}) = V'(\sim B, \{a_n\}),$$

$$V((B \wedge C)^x, \{a_n\}) = V(B^x \wedge C^x, \{a_n\}) =$$
$$V(B^x, \{a_n\}) \cdot V(C^x, \{a_n\}) =$$
$$V'(B, \{a_n\}) \cdot V'(C, \{a_n\}) =$$
$$V'(B \wedge C, \{a_n\}),$$

$$V((\bigvee x_j B)^x, \{a_n\}) = V(\bigvee x_j B^x, \{a_n\}) =$$
$$V(B^x, \{a_n(a_j/d)\}) = V'(B, \{a_n(a_j/d)\}) =$$
$$V'(\bigvee x_j B, \{a_n\}).$$

In the computation of the value of the formula with the quantifier use is made of the assumption that the variable x_i does not occur in $A(H(t_0))$, so that the bound variable x_j differs from x_i, that is, $j \neq i$, and, hence, the sequence $\{a_n(a_j/d)\}$, obtained by the replacement of a_j by some other element d, which satisfies the formula B, continues to be such that $a_i = a$, and, thus, induction assumption (111) is applicable to it.

Now that we have formula (111) for all the subformulae of the formula $A(H(t_0))$ we obtain formula (110) as its special case. Since, by (109), $V'(A(H(t_0)), \{a_n\}) = 1$, we obtain from (110) and (109) that $V(A(x_i), \{a_n\}) = 1$ for every sequence such that $a_i = a$. Since the element a has been chosen arbitrarily, this, by Definitions 2*a and 2*b, proves that the sentence $\bigwedge x_i A(x_i)$ is true in the model $\mathfrak{M}$.

It has thus been demonstrated that for every model $\mathfrak{M}$ with a totally free combination of those functions which interpret the function terms occurring in the sentences from the set X and in the sentence $A(H(t_0))$ the following implication holds:

$$X \subset E(\mathfrak{M}) \rightarrow \bigwedge x_i A(x_i) \in E(\mathfrak{M}).$$

By Theorem 31, this means that

$$\bigwedge x_i A(x_i) \in \text{Cn}^*(X).$$

Now that Lemma (γ) has been proved, we return to the main course of the proof. The lemma will be used to demonstrate that the following formulae

(114) $A_i \in \text{Cn}^*(\text{Skl}(X))$ and $\ulcorner \sim A_i \urcorner \in \text{Cn}^*(\text{Skl}(Y))$

hold for every $i \leqslant n+1$. For $i = 0$ formula (114) follows from (103), because $A_0 = A$. Assume now (114) as the induction assumption. It will be proved that

(115) $A_{i+1} \in \text{Cn}^*(\text{Skl}(X))$ and $\ulcorner \sim A_{i+1} \urcorner \in \text{Cn}^*(\text{Skl}(Y))$.

Since in (105) A_{i+1} is defined for two cases, in the inductive proof, too, a distinction is made between two cases.

C a s e I. The term t_i begins with a function F_j. By (105):

(116) $A_{i+1} = \bigvee x_i A_i(t_i/x_i).$

Since

$$\ulcorner A_i(t_i) \to \bigvee x_i A_i(t_i/x_i) \urcorner \in L^*,$$

that is, $\ulcorner A_i \to A_{i+1} \urcorner \in L^*$, the law on the transitivity of implication and the induction assumption (114) yield that $A_{i+1} \in \text{Cn}^*(\text{Skl}(X))$. On the other hand, in this case the sentence $\sim A_{i+1}$, by (116) and De Morgan's laws, satisfies the equivalence

(117) $\sim A_{i+1} \equiv \bigwedge x_i \sim A_i(t_i/x_i).$

Consider the sentences $A_i = A_i(t_i)$ and $\bigwedge x_i \sim A_i(t_i/x_i)$. The term t_i begins with a functor F_j, which was added when $\text{Skl}(X)$ was formed, and thus does not occur in the sentences from $\text{Skl}(Y)$. Moreover, t_i is the longest of all the terms contained in $\sim A_i$, and, thus, is not part of any other term that begins with the same functor F_j. The variable x_i does not occur in $\sim A_i$. The other variables occurring in $\sim A_i$ are not arguments in any term formulae that begin with F_j, since the variables occurring in A_i replace constant terms, if a term formula $F_j(x_i)$ occurred in A_i, this would mean that x_i had replaced a term t_i which is longer than, or at least as long as, the term t_i, but then the term $F_j(t_i)$ would be still longer and should have replaced by a variable earlier. In this way the sentences $\sim A_i$ and $\bigwedge x_i \sim A_i(t_i/x_i)$ satisfy the assumptions of Lemma (γ). Hence if $\sim A_i \in \mathrm{Cn}^*(\mathrm{Skl}(Y))$, then also

$$\ulcorner\bigwedge x_i \sim A_i(t_i/x_i)\urcorner \in \mathrm{Cn}^*(\mathrm{Skl}(Y)).$$

Thus, in accordance with (117), the conditions laid down in (115) have been deduced from (114) for case I.

C a s e II. The term t_i does not begin with a function F_j. Then by (105),

$$(118) \qquad A_{i+1} = \bigwedge x_i A(t_i/x_i).$$

If t_i is an arithmetical constant, then Theorem 14 (Chapter I, Section 9) yields, by the Deduction Theorem, that

$$\ulcorner\bigwedge x_i A_i(t_i/x_i)\urcorner \in \mathrm{Cn}^*(\mathrm{Skl}(X)),$$

since arithmetical constants do not occur in $\mathrm{Skl}(X)$. The argument is that, if $\ulcorner X' \rightarrow A_i(t_i)\urcorner \in L^*$, then also

$$\ulcorner\bigwedge x_i(X' \rightarrow A_i(t_i/x_i))\urcorner \in L^*,$$

and, hence, also $\ulcorner X' \rightarrow \bigwedge x_i A_i(t_i/x_i)\urcorner \in L^*$, where X' is a finite conjunction of sentences from $\mathrm{Skl}(X)$.

If t_i is not an arithmetical constant and does not begin with a functor F_j, then it must begin with a functor G_j which was added when $\mathrm{Skl}(Y)$ was formed, and, hence does not occur in $\mathrm{Skl}(X)$. Moreover, t_i is one of the longest terms in A_i which has not yet been replaced by a variable, and as such it is not an argument in any other term beginning with G_j, and as all longer terms have already been replaced by variables, no variable occurring in A_i, that is, derived from a term that is longer than, or as long as, t_i, can be an argument in any term formula beginning with G_j. Thus, all the conditions laid down in Lemma (γ) are satisfied; hence, if, by the inductive assumption, $A_i \in \mathrm{Cn}^*(\mathrm{Skl}(X))$, then also

$$\ulcorner\bigwedge x_i A_i(t_i/x_i)\urcorner \in \mathrm{Cn}^*(\mathrm{Skl}(X)).$$

In the other part, of the case now under consideration the negation $\sim A_{i+1}$ satisfies the equivalence

$$(119) \qquad \sim A_{i+1} \equiv \bigvee x_i \sim A_i(t_i/x_i);$$

since, under the induction assumption, $\sim A_i(t_i) \in \mathrm{Cn}^*(\mathrm{Skl}(Y))$, then $\ulcorner\bigvee x_i \sim A_i(t_i/x_i)\urcorner \in \mathrm{Cn}^*(\mathrm{Skl}(Y))$. Thus, in this case, too, the conditions laid down in (115) have been satisfied.

385

As a special case of (114) for $i = n+1$ we have that

$$C \in \text{Cn}^*(\text{Skl}(X)) \quad \text{and} \quad \ulcorner \sim C \urcorner \in \text{Cn}^*(\text{Skl}(Y));$$

which together with Theorem 44 yield that

$$C \in \text{Cn}^*(X) \quad \text{and} \quad \ulcorner \sim C \urcorner \in \text{Cn}^*(Y).$$

By (101), the atoms of the sentence A can contain symbols of only those predicates which are common to X and Y. Hence, the sentence C contains only the predicates $P, ..., Q$, which are common to sentences from X and Y.

An analogue of Theorem 54 on an intermediate formula, but which covers any formulae with variables and predicates, will be formulated now.

As an abbreviation, symbols of the type $A_{P,Q,R}$ will be used to stand for formulae containing only the terms P, Q, R.

THEOREM 58 (Craig [11]). *If* $\ulcorner A_{P,Q} \to B_{P,R} \urcorner \in L^*$, *then there exists a formula* C *such that*

$$\ulcorner A_{P,Q} \to C_P \urcorner \in L^* \quad \text{and} \quad \ulcorner C_P \to B_{P,R} \urcorner \in L^*.$$

PROOF. If $\ulcorner A_{P,Q} \to B_{P,R} \urcorner \in L^*$, then the set consisting of two sentences, $A_{P,Q}$ and $\ulcorner \sim B_{P,R} \urcorner$, results in a contradiction. On substituting $X = \{A_{P,Q}\}$ and $Y = \{\ulcorner \sim B_{P,R} \urcorner\}$, in Theorem 57 we conclude that there exists a sentence C_P such that

$$C_P \in \text{Cn}^*(\{A_{P,Q}\}) \quad \text{and} \quad \ulcorner \sim C_P \urcorner \in \text{Cn}^*(\{\ulcorner \sim B_{P,R} \urcorner\}).$$

Hence, by the Deduction Theorem, we conclude that

$$\ulcorner A_{P,Q} \to C_P \urcorner \in L^* \quad \text{and} \quad \ulcorner \sim B_{P,R} \to \sim C_P \urcorner \in L^*.$$

The theorem remains valid if P, Q, R are interpreted not as single predicates, but as any finite sets of predicates.

The proof of Theorem 57 also leads to a conclusion parallel to Lyndon's Theorem 55. For if the variable-free formula A from the proof of Theorem 57 satisfies conditions (α) from Theorem 55, then the formula C satisfies analogical conditions for predicates. In this way we obtain

THEOREM 59 (Lyndon [45]). *If* $\ulcorner A_{P,Q} \to B_{P,R} \urcorner \in L^*$, *then there exists an intermediate formula* C_P *such that if a predicate* P *occurs in it affirmatively, then it also occurs affirmatively in* $A_{P,Q}$ *and in* $B_{P,R}$; *the same holds for negated occurrences.*

Above theorems (57, 58 and 59) have been proved in principle for predicates only. The proof of their analogues for functional constants does not differ essentially, though a more complicated symbolism must be used.

EXERCISES

1. Write down the Skolem equivalent of the following set of two sentences:

$$\bigwedge x \bigvee y \bigwedge z \bigvee u\, P(x, y) \vee Q(y, z, u),$$

$$\bigwedge v \bigvee x \bigwedge y \sim P(x, v) \wedge \sim Q(v, x, y).$$

2. Demonstrate that the arithmetical description of the above set is inconsistent, which means that the set itself is inconsistent.

3. Define the concept of Skolem equivalent so that two sentences different in form should always contain different added Skolem functions. For instance, order all the formulae and to the formula with the number n add the symbols $F_{2^n 2i+1}$ as Skolem functions.

4. Extend the proof of Theorem 57 so as to cover the sets Z of sentences containing function terms.

5. Let $A_{P,Q}(x_1, ..., x_n)$ stand for any formula with the terms P, Q and the variables $x_1, ..., x_n$. Prove the following generalization of Theorem 58:

If $\ulcorner \bigwedge x, y(A_{P,Q}(x, y) \to B_{P,R}(x, y)) \urcorner \in L^*$, then there exists a formula $C_P(x, y)$ such that

$$\ulcorner \bigwedge x, y(A_{P,Q}(x, y) \to C_P(x, y)) \urcorner \in L^*$$

and

$$\ulcorner \bigwedge x, y(C_P(x, y) \to B_{P,R}(x, y)) \urcorner \in L^*.$$

H i n t. Refer to Theorem 58 and Theorem 14 from Chapter I, which state that no constant term is privileged in logic.

6. Demonstrate that, for any sets Z and Z' of sentences in normal forms, if the terms added in $\text{Skl}(Z')$ are other than those occurring in Z and Z', then the following implication holds:

$$\text{Skl}(Z) \subset \text{Cn}^*(\text{Skl}(Z')) \to Z \subset \text{Cn}^*(Z').$$

H i n t. Use Theorem 30 twice and Theorem 42.

7. Prove that if Z is a set of variable-free sentences and if A is a variable-free sentence, then

$$A \in \text{Cn}(Z) \equiv A \in \text{Cn}_0(Z \cup \text{Ax} LI).$$

8. Note that Theorem 56 is also true if the term Cn^* be interpreted as a consequence based on intuitionistic logic. In fact, if $\ulcorner X \to C \urcorner \in L$, then the implication $\ulcorner X \to \sim \sim C \urcorner$ is a theorem in the intuitionistic sentential calculus, as follows from Theorem 73 (Chapter I, Section 3, Exercise 5). On the other hand, Theorem 57 is not known in intuitionistic logic and seems to pose more difficult problems.

387

II. MODELS

6. DEFINABILITY

So far definitions have been discussed from one viewpoint only, namely as a method of introducing new terms into a theory. Introduction of new terms has, moreover, been treated as a purely technical procedure which facilitates reasoning by providing convenient abbreviations. But defining also has another aspect, which is somewhat more general. It happens that in a theory we use from the outset a number of concepts and we are interested in finding out whether some of them can be defined by means of others. Defining some concepts by means of others sheds light upon a certain ordering of concepts of a given theory. Those concepts which can be used to define others are, as it were, "stronger". Relationships between concepts as to their definability resemble relationships between sentences of a theory as to their deducibility (consequence). The set of theorems of a theory is characterized by axioms as the initial theorems, from which other theorems are deduced, and the set of concepts of a theory is likewise characterized by the initial concepts of the theory (primitive concepts), which are used to define other concepts of that theory by application of certain methods of correct introduction of new concepts. In examining the axioms of a theory we try to find out whether some of them are not superfluous in the sense of being deducible from the remaining ones, and in examining the primitive concepts of a theory we likewise try to find out whether some of them are not superfluous in the sense of being definable by remaining ones. The concept of definability is, thus, in a way similar to that of consequence. Like the definition of consequence, the rigorous definition of definability is syntactic in nature. We shall confine ourselves here to analysing definability of predicate symbols.

DEFINITION 25. *A predicate symbol P_e^n is (equivalentially) definable in a theory T by symbols $Q_1, ..., Q_m$ if and only if an equivalence of the form*:

$$(120) \qquad \bigwedge x_1, ..., x_n \big(P_e^n(x_1, ..., x_n) \equiv A\big),$$

where the formula A contains $Q_1, ..., Q_m$ as its only predicate symbols and at most the variables $x_1, ..., x_n$ as free variables, is a theorem in the theory T.

In a very abbreviated formulation: a term is definable in a theory if a definition of it is a theorem in that theory. This definition is purely syntactic. As in the case of the concept of consequence, which, though defined syntactically at first, was next characterized semantically in Theorem 32, the concept of definability will also be characterized semantically in Theorem 63.

Consider a few examples first. The arithmetic P, as described in Section 4, may be treated as a theory which has five primitive terms, namely the symbols 0, S, $<$, $+$, $\cdot$. It can easily be demonstrated that some of those symbols can be defined in the theory P by means of remaining ones. The symbols of successor is definable by the symbol of the *less than* relation since the following theorem which falls under schema (120) can easily be deduced from the axioms of P:

$$\bigwedge x, y\Big(S(x, y) \equiv \big(x < y \wedge \bigwedge z(z < y \rightarrow (z = x \vee z < x))\big)\Big).$$

(Since Definition 25 refers to predicate symbols the successor symbol has been changed from a function symbol into a predicate symbol: $S(x, y) \equiv y = S(x)$.)

Likewise, the term $<$ is definable by the addition and successor symbols, since the axioms of P easily yield the theorem:

$$x < y \equiv \Big(\bigvee z(x + S(z) = y)\Big).$$

On the contrary, the multiplication symbol is not definable by the remaining terms of the theory P.

Definability of the term $<$ in the elementary theory of real numbers (cf. Chapter I, Section 10, p. 237) provides a much less trivial example of definability. The elementary arithmetic of real numbers as presented there has the primitive terms: $0, 1, +, <, \cdot$, and $=$ as a logical term. The schema of continuity, adapted in that theory, can be used to prove that the following sentence:

$$\bigwedge x, y\Big(x < y \equiv \bigvee z(z \neq 0 \wedge y = x + (z \cdot z))\Big)$$

is a theorem in that theory. Thus the symbol $<$ is definable in that theory by means of remaining terms, even though in presentations of the arithmetic of real numbers the symbol $<$ is treated as a primitive term: this is done so in order to bring out role of the axiom of continuity. On the other hand, the symbol of multiplication is not definable by means of remaining primitive terms in the elementary arithmetic of real numbers.

389

The question arises as to what the general method of proving non-definability of a term by other terms is. That method is given by the theorems which are to be proved below. For that purpose suppose that we are concerned with a theory T whose primitive terms are P_e^n and $Q_1, ..., Q_m$ and that we are interested in the definability of the term P_e^n by the terms $Q_1, ..., Q_m$.

THEOREM 60. *If a term P_e^n is definable in a theory T by terms $Q_1, ..., Q_m$, then for every two models $\mathfrak{M} = \langle X, p, q_1, ..., q_m \rangle$, $\mathfrak{N} = \langle Y, p', q_1', ..., q_m' \rangle$ of the theory T and for every function establishing an isomorphism between them it is true that if the submodels $\langle X, q_1, ..., q_m \rangle$ and $\langle Y, q_1', ..., q_m' \rangle$ are isomorphic, then that isomorphism also covers the relations p and p', that is, can be extended so as to cover the entire models $\mathfrak{M}$ and $\mathfrak{N}$.*

By using the symbolism adopted in the Introduction we can write down the assertion of the above theorem in the following way:

(121)   *For any function f, if $\langle X, q_1, ..., q_m \rangle$ ism$_f$ $\langle Y, q_1', ..., q_m' \rangle$, then $\bigwedge t_1, ..., t_n \in X \big(p(t_1, ..., t_n) \equiv p'(f(t_1), ..., f(t_n)) \big)$, that is, $\mathfrak{M}$ ism$_f$ $\mathfrak{N}$.*

The proof is an easy application of Theorem 4 from Section 1. If P_e^n is definable by remaining terms, then under the definition of definability there exists a formula A such that a definition of the form (120) is a theorem in T. Hence, (120) is true in both models $\mathfrak{M}$ and $\mathfrak{N}$ which are models of T. By the definition of truth, (120) is, thus, satisfied by any sequences selected, respectively, from X and Y, so that the following equivalences hold for $\{a_n\} \subset X$ and $\{b_n\} \subset Y$:

(122) $\{a_n\}$ satisfies $P_e^n(x_1, ..., x_n)$ in $\mathfrak{M} \equiv \{a_n\}$ satisfies A in $\mathfrak{M}$,

$\{b_n\}$ satisfies $P_e^n(x_1, ..., x_n)$ in $\mathfrak{N} \equiv \{b_n\}$ satisfies A in $\mathfrak{N}$.

Under the definition of satisfaction we have

$\{a_n\}$ satisfies $P(x_1, ..., x_n)$ in $\mathfrak{M} \equiv p(a_1, ..., a_n)$,

and similarly for $\{b_n\}$. Hence formulae (122) yield the formulae

(123) $p(a_1, ..., a_n) \equiv \{a_n\}$ satisfies A in $\mathfrak{M}$,

$p'(b_1, ..., b_n) \equiv \{b_n\}$ satisfies A in $\mathfrak{N}$.

It follows from Theorem 4 (Section 1, p. 284) that $\{a_n\}$ satisfies A in $\mathfrak{M}$

if and only if $\{f(a_n)\}$ satisfies A in $\mathfrak{N}$. This and (123), on the substitution of $f(a_i)$ for b_i, yield that

$$p(a_1, ..., a_n) \equiv p'(f(a_1), ..., f(a_n)).$$

Theorem 60 can clearly be restricted to the case in which the relation establishing isomorphism is that of identity.

THEOREM 61. *If X is a set, if p, p', $q_1, ..., q_m$ are relations in that set, and if a term P is definable by $Q_1, ..., Q_m$ in a theory T which is true in a model $\langle X, p, q_1, ..., q_m \rangle$ for the interpretation of $P, Q_1, ..., Q_m$ as names of the relations $p, q_1, ..., q_m$, and which is true in a model $\langle X, p', q_1, ..., q_m \rangle$ for the interpretation of $P, Q_1, ..., Q_m$ as names of the relations $p', q_1, ..., q_m$, then $p = p'$.* (The relation p is the same relation as p' since the identity symbol $=$ for relations is interpreted so that relations are identical if they hold between the same objects:

$$p = p' \equiv \bigwedge \alpha_1, ..., \alpha_n(p(\alpha_1, ..., \alpha_n) \equiv p'(\alpha_1, ..., \alpha_n)).)$$

The proof follows from Theorem 60 if f is interpreted as the identity function.

By transposition Theorem 61 yields the simplest method of proving non-definability of terms. If we want to prove by this method that a term P is not definable by the remaining terms $Q_1, ..., Q_m$, we have to find a domain in which, for the same interpretation of the terms $Q_1, ..., Q_m$, we can interpret the term P in two ways, both as a relation p and as a relation p', different from the former.

As an example of applying Theorem 61 we shall prove the non-definability of the multiplication symbol by the remaining terms of the theory P in its weaker form. Consider the non-standard model $\mathfrak{N}$ as defined in Section 4 (Theorem 37). In that model we define multiplication 2 by means of multiplication 1:

$$x \times '' y = \begin{cases} x \times ' y & \text{when} \quad x \in X_0 \text{ or } y \in X_0, \\ S'(x \times ' y) & \text{when} \quad x \notin X_0 \text{ and } y \notin X_0. \end{cases}$$

Multiplication $\times ''$ coincides with multiplication $\times '$ in the classical part of the model $\mathfrak{N}$ (on the set X_0), and differs from it outside the classical part. It can easily be verified that both $\langle X, 0', S', <', +', \times ' \rangle$ and $\langle X, 0', S', <', +', \times '' \rangle$ are models of the theory P examined in Section 4, if the induction schema 3 is removed from that theory. Thus

391

the multiplication symbol is not definable in P by the symbols of zero, successor, *less than* relation and addition without schema 3. Nor is it definable by those symbols in the set of all sentences true in $\mathfrak{N}_0$. The proof of the latter fact is, however, more difficult.

We shall now demonstrate that the method outlined above is general in nature.

THEOREM 62 (Beth [4]). *If for any domain* $\mathfrak{M} = \langle X, p, p', q_1, ..., q_m \rangle$ *the following implication*:

(124) $T \subset E(\langle X, p, q_1, ..., q_m \rangle) \wedge T \subset E(\langle X, p', q_1, ..., q_m \rangle) \rightarrow$
          $p = p'$

holds, then the term P *is definable by* $Q_1, ..., Q_m$.

PROOF. Note first that in the antecedent of implication (124) reference is first made to truth for the interpretation of P, Q_1, ..., Q_m as names of the relations $p, q_1, ..., q_m$, and then to truth for the interpretation of $P, Q_1, ..., Q_m$ as names of the relations $p', q_1, ..., q_m$. To prove the theorem under consideration we shall take into account an auxiliary theory S to be defined as follows: its terms are $P, P', Q_1, ..., Q_m$; further

(125) $S = T \cup T'$,

where T' is the set of sentences obtained from T by the replacement of the symbol P by the symbol P', without any change in the remaining symbols. Note now that implication (124) implies the following implication:

(126)   *If* $S \subset E(\mathfrak{M})$ *for the interpretation of* P, P', $Q_1, ..., Q_m$
          as names of the relations p, p', $q_1, ..., q_m$, *then* $p = p'$.

For if $S \subset E(\mathfrak{M})$, then it follows from the definition of S that $T \subset E(\mathfrak{M})$ and $T' \subset E(\mathfrak{M})$ for the interpretation of terms stated in implication (126). If $T' \subset E(\mathfrak{M})$ for the interpretation of P' as a name of the relation p', and if T' is derived from T by the replacement of P by P', then of course also $T \subset E(\langle X, p', q_1, ..., q_m \rangle)$ for the interpretation of P as a name of the relation p'. The assumptions of implication (124) are thus satisfied, and accordingly we may accept that $p = p'$.

We shall now interpret the equation $p = p'$ by reference to the concept of satisfaction. The equation $p = p'$ means that the equivalence

$p(a_1, ..., a_n) \equiv p'(a_1, ..., a_n)$ holds in the model $\mathfrak{M}$ for any objects $a_1, ..., a_n$; hence the sentence

(127) $\qquad \bigwedge x_1, ..., x_n\big(P(x_1, ..., x_n) \equiv P'(x_1, ..., x_n)\big)$

is true for the interpretation of P and P' as names of the relations p and p'. Hence implication (126) may be transformed into

(128) $\qquad$ *If $S \subset E(\mathfrak{M})$, then sentence (127) is in $E(\mathfrak{M})$.*

Since assumption (124) refers to any $\mathfrak{M}$, formula (128) also holds for any $\mathfrak{M}$, and this, in accordance with the model characteristics of the concept of consequence (Theorem 30, Section 3), implies that

(129) $\qquad$ (127) $\in \mathrm{Cn}(S)$.

Thus, in accordance with the properties of the concept of consequence, there exist finite subsets $T_0 \subset T$ and $T_0' \subset T'$ such that (127) $\in \mathrm{Cn}(T_0 \cup T_0')$. We may also add to the sets T_0 and T_0' finite numbers of sentences so that the sets preserve the same relation between them as holds between T and T', that is, that T_0' is derived from T_0 through the replacement of P by P'. Next we may combine all the sentences of T_0 into the conjunction $A_{P,Q_1,...,Q_m}$, and likewise, all the sentences of T_0' into the conjunction $A'_{P',Q_1,...,Q_m}$ so that A' is derived from A through the replacement of every occurrence of the predicate P by the predicate P'. Thus ultimately we may conclude from (129) and (127) that

(130) $\qquad \ulcorner (A_{P,Q_1,...,Q_m} \wedge A'_{P',Q_1,...,Q_m}) \rightarrow$
$\qquad \bigwedge x_1, ..., x_n\big(P(x_1, ..., x_n) \equiv P'(x_1, ..., x_n)\big)\urcorner \in L^*,$

where

(131) $\qquad A_{P,Q_1,...,Q_m} \in \mathrm{Cn}(T)$.

A simple logical transformation of (130) yields

(132) $\qquad \ulcorner \bigwedge x_1, ..., x_n\big((A_{P,Q_1,...,Q_m} \wedge P(x_1, ..., x_n)) \rightarrow$
$\qquad (A'_{P',Q_1,...,Q_m} \rightarrow P'(x_1, ..., x_n))\big)\urcorner \in L^*.$

The antecedent of this implication does not contain any occurrences of P', while the consequent does not contain any occurrence of P. By Theorem 58 on the existence of an intermediate formula, there exists a formula $C_{Q_1,...,Q_m}(x_1, ..., x_n)$ which contains neither P nor P', and such that

(133) $\ulcorner \bigwedge x_1, \ldots, x_n \big((A_{P,Q_1,\ldots,Q_m} \wedge P(x_1, \ldots, x_n)) \to$
$C_{Q_1,\ldots,Q_m}(x_1, \ldots, x_n) \big) \urcorner \in L^*,$

(134) $\ulcorner \bigwedge x_1, \ldots, x_n \big(C_{Q_1,\ldots,Q_m}(x_1, \ldots, x_n) \to$
$(A'_{P',Q_1,\ldots,Q_m} \to P'(x_1, \ldots, x_n)) \big) \urcorner \in L^*.$

It is true that Theorem 58 refers to sentences, but it can easily be given the form required in the case now under consideration (cf. Exercise 5 in the preceding section). We may substitute, in formula (132), certain constants for variables; for instance, they may be numbers x_i/i, under the assumption that those constants do not occur in (132). In this way (132) becomes an implication without any quantifiers at the beginning. Theorem 58 can be applied to it directly, which yields an intermediate sentence C' which contains constants common to the antecedent and the subsequent, that is, the constants $Q_1, \ldots, Q_m$ and $1, \ldots, n$. The intermediate sentence would, thus, satisfy the implications:

(135) $\ulcorner (A \wedge P(1, \ldots, n)) \to C'_{Q_1,\ldots,Q_m}(1, \ldots, n) \urcorner \in L^*,$
$\ulcorner C'_{Q_1,\ldots,Q_m}(1, \ldots, n) \to (A' \to P'(1, \ldots, n)) \urcorner \in L^*.$

But, in view of Theorem 14 (Chapter I), which states that no distinction is made between constants and variables, the constants $1, \ldots, n$ in implications (135) may be replaced by the variables $x_1, \ldots, x_n$. The sentence C' thus yields a formula with variables

$$C_{Q_1,\ldots,Q_m}(x_1, \ldots, x_n) \equiv C'_{Q_1,\ldots,Q_m}(1/x_1, \ldots, n/x_n),$$

which satisfies formulae (133) and (134).

Theorem 13 (Chapter I), which states that no extra-logical terms are privileged in logic, is then applied to formula (134). It follows from that theorem that the sentence derived from (134) by the replacement of all occurrences of P' by P must be a logical theorem.

Thus, we have that

(136) $\ulcorner \bigwedge x_1, \ldots, x_n \big(C_{Q_1,\ldots,Q_m}(x_1, \ldots, x_n) \to$
$(A_{P,Q_1,\ldots,Q_m} \to P(x_1, \ldots, x_n)) \big) \urcorner \in L^*.$

Formulae (133) and (136) yield

(137) $\ulcorner A_{P,Q_1,\ldots,Q_m} \to \bigwedge x_1, \ldots, x_n (P(x_1, \ldots, x_n) \equiv$
$C_{Q_1,\ldots,Q_m}(x_1, \ldots, x_n)) \urcorner \in L^*.$

Now since, by (131), $A_{P,Q_1,...,Q_m} \in \text{Cn}^*(T)$, hence (137) yields

$$(138) \quad \ulcorner \bigwedge x_1, ..., x_n \big(P(x_1, ..., x_n) \equiv C_{Q_1,...,Q_m}(x_1, ..., x_n)\big) \urcorner \in \text{Cn}^*(T).$$

The formula $C_{Q_1,...,Q_m}$ does not contain the predicate P. Equivalence (138) is thus a definition, in the theory T, of the term P by means of the terms $Q_1, ..., Q_m$.

THEOREM 63 (Craig [12]). *A term P is definable in T by the remaining terms $Q_1, ..., Q_m$ of T if and only if for every domain $\langle X, p, p', q_1, ..., q_m \rangle$ it is true that*

$$(139) \quad \textit{if } T \subset E(\langle X, p, q_1, ..., q_m \rangle), \textit{ and } T \subset E(\langle X, p', q_1, ..., q_m \rangle),$$
$$\textit{then } p = p'.$$

The proof follows directly from Theorems 61 and 62.

Theorem 63 is a model theoretical characterization of the concept of definability, like the model theoretical characterization of the concepts of consequence, consistency, and logical tautology, obtained earlier in this chapter.

EXERCISES

1. Prove the theorem which is the converse of Theorem 60.

2. Verify that addition is definable by means of multiplication and the successor operation

$$a+b = c \equiv S(a \cdot c) \cdot S(b \cdot c) = S\big(c \cdot c \cdot S(a \cdot b)\big).$$

3. Try to formulate definitions and theorems in this section so that they would be applicable to the defining of function terms.

4. Define the ordinal number $\omega \cdot \omega + \omega + 3$ by means of the relation $<$ between ordinal numbers (that is, describe a property $P(x)$ which is possessed only by the number in question).

5. Define the relation $<$ between integers only by means of $=$, $+$, $\times$ as operations on integers.

H i n t. Use the theorem that every natural number is the sum of four squares.

Chapter III

LOGICAL HIERARCHY OF CONCEPTS

Modern metalogical research falls, broadly, into two areas. The first, namely that concerned with models of axiomatic theories, has been described in the preceding chapter. The present chapter will be dedicated to the other area of metalogic, one aspect of which is indicated by the above heading. The other aspect might be formulated as: research on effectiveness of logical constructions. Both aspects are closely inter-connected, since logical hierarchy of predicates consists mainly in sin-gling out certain degrees of constructiveness or effectiveness of logical constructions. Moreover, analysis of effectiveness of logical constructions results in singling out those which are most effective. Such constructions are termed computable (general recursive or algorithmic). Hence, this branch of metalogic is mainly concerned with the investigations into such constructions. The present chapter will accordingly begin with such investigations.

1. THE CONCEPT OF EFFECTIVENESS IN ARITHMETIC

The intuitive concept of computability and decidability

Let us begin with the most important example of the application of the concept of effectiveness, namely its application to logical con-structions in the arithmetic of natural numbers. The concept of effec-tiveness in that field has been rigorously formulated as a definition of what are called general recursive (computable) functions or relations. Before we proceed to analyse that concept, a few preliminary comments are in order.

The concept of effectiveness has its analogues in other sciences; the same holds for the related concept of decidability. These concepts from other sciences pertain to empirical procedures rather than to theoretical thinking.

When various scientific problems are being solved, the problem often arises of finding a method which would make it possible to solve the problem in question by means of a finite, possibly small number of easy tests. For instance, in analytic chemistry we want to have methods which make it possible to find out the chemical composition of a substance by means of a finite number of successive tests. In chemical analysis such methods have been worked out in detail, and there are standard rules for carrying out that kind of research, which is indispensable to industry.

In mathematics the situation is often similar. We want to have a method of making sure, in a finite number of simple operations, whether a formula is true, whether a proof is correct, whether a number n has a property W, or what the value of a function for given arguments is. Many such methods are known, and they are called effective. A general, though not precise definition of an effective method and of the concept of decidability is as follows:

A method is *effective* if it makes it possible to solve, in a finite number of simple operations, problems to which it is applicable.

A general problem (a class of definite questions) is *decidable* if there exists an effective method of solving every definite question in that class.

Chapter I (Section 2, Theorem 1) described the zero-one method for finding out which sentential schemata are, and which are not c.l.c. tautologies. The theorem stating the existence of that method was accordingly termed the theorem on the decidability of the c.l.c.

It is known from elementary arithmetic that there exists an easy method of finding out whether a natural number greater than 1 is or is not a prime number, that is, a number divisible without remainder only by 1 and by itself. Hence, the problem whether a natural number x greater than 1 is or is not a prime number is decidable. Such a problem can be decided by a finite number of operations of division or multiplication: we can write out all the natural numbers greater than 1 and less than x (the number of such numbers is finite) and divide x by each of them. If x is not divisible without remainder by any of them, then, obviously, under the definition of prime number, we have to conclude that x is a prime number, and if x is divisible without remainder by at least one of the said numbers, then we conclude that x is not a prime number. This procedure gives an effective method by which to decide, in a finite number of steps, whether x is or is not a prime number.

In mathematics there are very many methods of this kind and very many decidable problems; these methods are very useful, since they make it possible to automate a number of arguments. Arithmetical formulae usually refer to certain methods of computation, and, thereby, of deciding certain problems.

For instance, if we have a polynomial of the second degree

$$ax^2 + bx + c, \quad \text{where} \quad a \neq 0,$$

with integral coefficients, and if we want to find out whether it has real square roots, that is, whether there exists at least one real number x such that $ax^2 + bx + c = 0$, then, as is well-known, it suffices to compute what is called the discriminant:

$$\Delta = b^2 - 4ac.$$

If Δ is negative, then there are no roots which are real numbers; if Δ is not negative, then there exist square roots which are real numbers and are computed by a well-known formula.

Thus, the problem of the existence of real roots of a second degree polynomial with integral or rational coefficients is decidable.

If for a mathematical theory T the general problem, whether an arbitrary formula is or is not a theorem of T is decidable, then that theory is called *decidable*.

Thus, the classical sentential calculus is a decidable theory. On the other hand, as will be shown soon, the axiomatic elementary arithmetic of natural numbers is not a decidable theory.

It is worthwhile noting that there is no essential difference between deciding whether some substance in a test tube contains, for instance, chlorine or does not contain it, and deciding whether a given natural number is a prime or is not, because every given natural number is given in a system of notation, and dividing it by lesser numbers is a graphic operation which consists in transforming certain inscriptions into some other inscriptions in accordance with certain general rules. Likewise, any other effective method in mathematics reduces to certain empirical operations on inscriptions. Thus, the concept of decidability in mathematics is actually just an idealization and a narrowing of the concept of empirical decidability.

Therefore, generally speaking, when considering any domain we can ask ourselves which objects from that domain are empirically known to us, and for which properties we can decide whether they occur or not, and which mental operations lead from decidable properties to decidable properties. Let us begin with the domain of natural numbers. The various natural numbers are empirically given to us in written form, for example, in decimal notation. Certain properties, relations, and functions are also empirically given by the methods for determining or computing them, and certain mental operations do not lead beyond those functions or relations which are empirically given. Those functions and relations which are empirically given or determinable are called *general recursive* or *computable*. Before proceeding to formulate a rigorous mathematical definition of those functions, we shall reflect on the consequences of the following intuitive and provisional description:

A function f is *computable* (belongs to the class of computable or general recursive functions) if and only if there exists an effective method of computing the values of that function, that is, a method which makes it possible, for every natural number n, to compute the value $y = f(n)$ by means of a finite number of simple calculations.

In this sense the function $y = x^2$ is computable, since there exists a method of computing the square of any natural number, a method which consists in multiplying that number by itself. Obviously, the two-argument functions of addition and multiplication, $y = x+z$ and $y = x \cdot z$, respectively, are computable, as there exist simple rules of addition and multiplication in decimal notation. The functions defined by simple formulae which refer only to addition and multiplication are also computable, for instance, polynomials such as

$$f_1(x) = 2x^4 + 3x^3,$$

$$f_2(x, y) = xy^2 + 7y^5x^3, \text{ etc.}$$

For any two given natural numbers, for example, 3 and 8, the value of $f_2(3, 8)$ can be computed by a finite number of simple operations of addition and multiplication.

But combining known operations is not the only method of defining computable functions. For instance, the function

the n-th successive prime number

is a computable function which is not defined by combinations of basic operations.

As is proved in arithmetic, there are infinitely many prime numbers. Hence, we can imagine that they are arranged in an infinite sequence according to magnitude

$$2, \ 3, \ 5, \ 7, \ 11, \ 13, \ 17, \ 19, \ ...$$

We can further imagine that they are numbered by successive natural numbers beginning with the number 0, so that we use all the natural numbers in that enumeration. Thus,

> 2 *is the prime number bearing the number* 0,
>
> 3 *is the prime number bearing the number* 1, etc.

This function is abbreviated

> p_n *is the prime number bearing the number n*

($p_0 = 2, p_1 = 3, p_2 = 5, p_{10} = 29$). The function p_n is computable in an intuitive sense. The method of its computation can be described easily. In order to find the prime number which bears the number n we must consider, one by one, all the natural numbers 0, 1, 2, 3, 4, ..., and to test each of them for the property of being a prime number. As has been noted above, such a test can be carried out in a finite number of operations. While testing in this way all the natural numbers one by one, we can number those which prove to be primes. Since there are infinitely many prime numbers, at some time we must encounter the n-th successive prime number. In this way we can determine the value of the function p_n.

All the principal methods of defining computable functions will be examined in the next section.

Not every computable function is a polynomial, but, as we shall see, every such function can be defined by certain formulae. Hence each such function is definable in the sense of Definition 25 (Chapter II, Section 6).

The intuitive concepts of a computable relation holding between natural numbers and of a computable set of natural numbers will now be introduced:

A relation R is *computable* (general recursive) if and only if there exists an effective method of determining for every pair of natural numbers x, y whether the relation R does or does not hold between them.

1. EFFECTIVENESS

A set Z is *computable* (general recursive) if and only if there exists a general effective method of determining for every natural numer whether it is or is not an element of the set Z.

The concepts: *general recursive* and *computable* will always be treated as equivalent. *General recursive* will often be abbreviated to *recursive*.

For instance, the set of even numbers is computable, since it suffices to find out whether a given natural number, when recorded in the decimal notation, ends in one of the following figures: 0, 2, 4, 6, 8. If it does, the number in question is even. The set of numbers divisible by 3 is computable. The easiest method of finding out whether a given number is divisible by 3 is to find out whether the sum of its digits is divisible by 3. As seen above, the set of prime numbers is computable.

Likewise, the set of numbers divisible by any selected number z is computable, for we can find out by trying whether a number x is or is not divisible by z. This means simply that the relation

$$x \text{ divides by } z \text{ without remainder}$$

is computable.

In the text below we shall often make use of what are called characteristic functions.

Every function f such that $f(x) = 0 \equiv x \in Z$ will be called a *characteristic function of the set Z*.

Every function f such that

$$R(x_1, ..., x_n) \equiv f(x_1, ..., x_n) = 0$$

will be called a *characteristic function of the relation R*.

There may be many functions characteristic of a given set Z. If a number x is not in Z, then a function f can take on any value other than zero. (I deviate here from the custom according to which a characteristic function takes only two values, 0 and 1.) The same holds for relations. For instance the functions

$$f_1(x, y) = (x-y) + (5y - 5x),$$
$$f_2(x, y) = (2x - 2y) + (3y - 3x)$$

both are functions characteristic of the relation of equality.

By making use of the provisional definitions introduced so far it can easily be seen that

A set Z is computable if and only if there exists at least one computable characteristic function for the set Z.

In fact, if a set Z is computable and a function f is defined as follows:

$$f(x) = \begin{cases} 0 & \text{if } x \text{ is in } Z, \\ 1 & \text{if } x \text{ is not in } Z, \end{cases}$$

then the function f is a characteristic function of the set Z; the function f is also computable, since, as we have a method of finding out whether x is or is not in Z, we also have, in view of the definition above, a method of establishing the value of the function f.

Conversely, if a function g is computable and characteristic for a set Z, that is, the equivalence

$$x \in Z \equiv g(x) = 0$$

holds, then we have a method of deciding for any natural number whether it is or is not in Z, because, in accordance with the equivalence above, it suffices to compute the value of the function $g(x)$ and to see whether it does or does not equal zero, in order to find out whether the number x is or is not in Z.

Likewise, it is true that *a relation R is computable if and only if there exists at least one computable characteristic function for the relation R.*

Logical constructions which do not lead outside computable functions

Without, for the time being, giving a rigorous definition of the concept of computability, let us consider which mental constructions do not lead outside computable functions in the intuitive sense of the term. The basic ways of defining functions in the arithmetic of natural numbers will be analysed.

A. DEFINITION BY SUPERPOSITION. It can easily be seen that if we can compute the values of some functions

$$f(\ldots, x, \ldots, z, \ldots) \quad \text{and} \quad g(\ldots, y, \ldots),$$

then we can also compute the values of a function h which is defined by the superposition of the functions f and g:

$$h(\ldots, x, \ldots, y, \ldots) = f(\ldots, x, \ldots, g(\ldots, y, \ldots), \ldots).$$

The value of the function h can be computed if the value of the function $z = g(\ldots, y, \ldots)$ is computed first and then substituted for z in the formula of the function $f(\ldots, x, \ldots, z, \ldots)$.

If the functions f and g are computable, then so is the function h. This property can also be described as follows: Superposition as an operation used in defining new functions does not lead outside the class of computable functions (it leads from computable functions to computable functions), or, in other words: the class of functions computable in the intuitive sense of the term is closed under the operation of superposition.

As has been seen above, all the polynomials with integral coefficients are computable functions, since they are obtained by superposition of operations of multiplication and addition.

The operation of identifying variables is also interpreted as an operation of superposition. It leads from a function $f(x_1, \ldots, x_j, \ldots, x_k, \ldots, x_n)$ to a new function of a smaller number of arguments, because for two different variables x_j, x_k we may substitute one and the same function of the same variable, in particular the identity function, that is, the same variable x_t, and, thus, obtain the function $f(x_1, \ldots, x_t, \ldots, x_t, \ldots, \ldots, x_n)$. For instance, the function $x^2 = x \cdot x$ is obtained by the identification of both arguments of the two-argument function of multiplication $x \cdot y$.

B. DEFINITIONS BY CASES. In some cases, when defining a function, we make distinctions among various cases and in each case define the function in question by a different formula. For instance, the function $\text{sgn}(x)$ is defined for natural numbers as follows:

$$\text{sgn}(x) = \begin{cases} 0 & \text{if } x = 0, \\ 1 & \text{if } x > 0. \end{cases}$$

For such a definition to be correct the conditions must be mutually exclusive and must cover all the possible cases for natural numbers.

Two examples more:

$$(1) \qquad x \doteq y = \begin{cases} 0 & \text{if } x \leqslant y, \\ x - y & \text{if } x > y. \end{cases}$$

$$[x/2] = \begin{cases} x/2 & \text{if } x \text{ is even}, \\ (x-1)/2 & \text{if } x \text{ is odd}. \end{cases}$$

The function $[x/2]$ denotes the integral part of the result of dividing the number x by 2. By definition, the values of this functions are always natural numbers.

The above definitions fall under the general schema:

$$(2) \qquad f(x_1, ..., x_n) = \begin{cases} g(x_1, ..., x_n) \\ \quad \text{if the condition } R(x_1, ..., x_n) \text{ holds,} \\ h(x_1, ..., x_n) \\ \quad \text{if the condition } R(x_1, ..., x_n) \text{ does not} \\ \quad \text{hold.} \end{cases}$$

Functions defined by a distinguishing more than two cases can be easily imagined. For instance, the function $[x/3]$ (the integral part of the fraction with a natural number x in the numerator and the number 3 in the denominator) can be defined for all natural numbers x by the following three conditions:

$$[x/3] = \begin{cases} x/3 & \text{if } x \text{ divides by 3 without remainder,} \\ (x-1)/3 & \text{if } x \text{ divided by 3 leaves 1 as remainder,} \\ (x-2)/3 & \text{if } x \text{ divided by 3 leaves 2 as remainder.} \end{cases}$$

It can easily be seen that if the conditions are computable relations and if the functions g and h, which define a given function under specified conditions, are computable, then the function which is defined conditionally is computable; for in order to compute its value it suffices to find out which condition holds; this can be done when the conditions are computable; next we compute the value of the function in question by that method of computation which defines the function associated with the given condition. It can easily be seen that in all the definitions given above the functions and the defining conditions are computable. Strictly speaking, in all the definitions given above except for (1) the defining functions do not take on only natural numbers as values but they are used in those cases only in which they take on natural numbers as values.

The above discussion shows that definition by cases with computable conditions and computable defining functions does not lead outside functions computable in the intuitive sense.

C. INDUCTIVE DEFINITION OF FUNCTIONS. A general schema (169) of inductive definitions was given in Chapter I, Section 10. When analysing inductive definitions of functions, we gave there also a general schema (183) for inductively defining some functions by other functions. This was accompanied by the comment that any function defined by schema (183) can also be defined by the initial induction schema (169). Two fairly general induction schemata, which are restrictions of schema (183) to functions of one and two arguments, will now be given:

(3)
$$f(0) = k,$$
$$f(n+1) = h(n, f(n)),$$

(4)
$$\text{(a) } f(0, x) = g(x),$$
$$\text{(b) } f(n+1, x) = h(x, n, f(n, x)).$$

Schema (3) covers, for instance, the inductive definition of the function $n!$ (factorial) by multiplication:

(5)
$$0! = 1,$$
$$(n+1)! = (n+1) \cdot n!.$$

Schema (4) covers the inductive definitions of the functions: addition (by reference to the successor function), multiplication (by reference to addition), exponentiation (by reference to multiplication). They will be recorded below, the symbol S being used for the successor function:

(6)
$$0+x = x,$$
$$S(n)+x = S(x+n).$$

(7)
$$0 \cdot x = 0,$$
$$S(n) \cdot x = x+(n \cdot x).$$

(8)
$$x^0 = 1,$$
$$x^{S(n)} = x^n \cdot x.$$

Each of these definitions consists of two equations. The first equation is the initial condition and establishes the value of the function for $n = 0$ by reference to known functions or to constant numbers. The second equation is the induction condition, in which the value of the function for the successor $n+1$ is established by reference to the value of the function for the preceding number n and by reference to other known functions. Schema (183) is not the most general schema of inductive definition of functions in arithmetic. It is called the *schema of simple recursion* or *recursion with respect to a single variable*. As shown by its applications, in this schema only one variable changes from n to $n+1$. There are schemata of simultaneous recursion with respect to several variables; such schemata are not reducible to the schema of recursion on a single variable. It can easily be seen that the methods of definition which fall under the schema of simple recursion do not lead outside functions computable in the intuitive sense of the term.

We can argue as follows. If we can compute the function g and the function h, then we can easily compute the function $f(n, u)$, defined by the general schema (4), for any n and u. But to do so we must compute the function h n times. We have to first compute $f(0, u)$, which we can do by referring to equation (a), since we know how to compute the function g. Next we compute $f(1, u) = h(0, u, f(0, u))$ by referring to the value of $f(0, u)$ already obtained. Next, by referring to $f(1, u)$ we compute $f(2, u)$, then $f(3, u)$, etc. After n steps, by using formula (b) each time, we compute $f(n, u)$.

By repeating n times the addition of a number x we can, by (7), compute the product $n \cdot x$. Likewise, by repeating n times multiplication by x we obtain, by (8), x^n, that is, the n-th power of x. This is exactly how in practice multiplication and exponentiation are carried out in an elementary way. The schema of simple recursion makes it possible to define a large number of computable functions encountered in mathematical practice. This schema will be used now to define two other simple functions, which will be used later on, namely the predecessor function $P(x)$ and subtraction of natural numbers from natural numbers, previously defined by conditional formula (1):

$$(9) \quad \begin{aligned} &P(0) = 0, \\ &P(n+1) = n. \end{aligned}$$

$$(10) \quad \begin{aligned} &x \mathbin{\dot-} 0 = x, \\ &x \mathbin{\dot-} (n+1) = P(x \mathbin{\dot-} n). \end{aligned}$$

D. THE MINIMIZATION OPERATION. The fourth method of defining functions to be discussed here is definition by the *minimization operation*. This operation is applied to any relation $R(..., x, ...)$; as a result it yields a number or a function symbolized by

$$(11) \quad (\mu x)[R(..., x, ...)].$$

This formula is read: the least x such that the condition $R(..., x, ...)$ holds. For instance,

$$4 = (\mu x)[x > 3],$$
$$6 = (\mu x)[x = 12 \text{ or } x = 8 \text{ or } x = 6].$$

The first of these two equations states that the number 4 is the least of all natural numbers greater than 3. The second states that the number 6 is the least of the three numbers: 12, 8, and 6. In a general case the convention is that formula (11) means

1) the least of those numbers for which the condition $R(..., x, ...)$ is satisfied, if such numbers exist,

2) zero, if such numbers do not exist.

For instance,

$$0 = (\mu x) [x = x]$$

on the strength of condition 1), but

$$0 = (\mu x)[x \neq x],$$
$$0 = (\mu x)[x < x]$$

on the strength of condition 2), since there are no numbers x such for which $x \neq x$ or $x < x$.

The minimization operation makes it possible, of course, to define functions as well. If we have any relation $R(x, z)$ we can use the minimization operation to define a function $f(z)$ as follows:

$$(12) \qquad f(z) = (\mu x)[R(x, z)].$$

For instance,

$$(13) \qquad [\sqrt{z}] = (\mu x)[(x+1)^2 > z].$$

The function $[\sqrt{z}]$ stands for the integral part of the square root of z. The symbols of the functions $[\sqrt{z}]$ and $[x/y]$ consist of two symbols each, but in the arithmetic of natural numbers these symbols are treated as inseparable, because none of the functions denoted by component symbols may be treated as a function with the argument and the value in the set of natural numbers. For instance, $[\sqrt{1}] = 1, [\sqrt{2}] = 1, [\sqrt{3}] = 1,$ $[\sqrt{4}] = 2, [\sqrt{5}] = 2, [\sqrt{6}] = 2, [\sqrt{9}] = 3, [\sqrt{10}] = 3.$

The function

(14) $[z/y] = (\mu x)\,[(x+1)\cdot y > z]$

stands for the integer part of the fraction z/y, for example, $[1/1] = 1$, $[1/2] = 0$, $[5/2] = 2$. The examples show that the functions defined by formulae (13) and (14) correspond to what their symbols stand for. The ordinary methods of extracting the square root and division make it possible to compute those functions. The integral part of a square root is the greatest of those numbers x whose square is less than or equal to z. Thus $(x+1)^2$ is greater than z, but x is the least of those numbers for which $(x+1)^2 > z$. The integral part of the fraction z/y is the greatest integer x which is less than or equal to a given fraction, that is $x \leqslant z/y$, but $x+1$ is greater than the fraction z/y: $(x+1) > z/y$; hence, by multiplying both sides of the above inequality by y we obtain the relation $(x+1)\cdot y > z$, which occurs in the definition.

A distinction will now be made between two cases of applying the minimization operation to defining a new function. The first case is called *effective*. The minimization operation is applied effectively to a relation $R(x, z)$ if a function $f(x)$ is defined by formula (12) and the following condition is satisfied:

(15)   *For every natural number z there exists a natural number x such that the relation R(x, z) holds.*

In a symbolic notation:

$$\bigwedge z \bigvee x R(x,z).$$

This is called the *condition of effective application of the minimization operation*. For instance, in the case of formula (13) the minimization operation is applied effectively, because the effectiveness condition is satisfied, as it is true that

$$\bigwedge z \bigvee x\big((x+1)^2 > z\big).$$

It suffices to substitute z for x, as it is true that $(z+1)^2 > z$ for every natural number z. Hence such numbers x for which $(x+1)^2 > z$ do exist, and the minimization operation selects the least of them.

On the other hand, if the effectiveness condition is not satisfied, then the application of the minimization operation is called *ineffective*. For instance, in the case of formula (14) the minimization operation is applied ineffectively, since for $y = 0$ and for any z there does not exist any x such that $(x+1)\cdot 0 > z$.

The function $[z/y]$ can, however, be defined by the minimization operation in an effective way very easily, for instance, by the formula

(16) $[z/y] = (\mu x)[y = 0$ or $(x+1) \cdot y > z]$.

As can easily be checked, the relation now under consideration does satisfy the effectiveness condition. But there exist functions which can be defined only by an ineffective application of the minimization operation. Note that

If the effectiveness condition is satisfied, then the minimization operation leads from a relation R which is computable in the intuitive sense to a function f which is computable in the intuitive sense.

In fact, suppose that we can find out when a relation $R(x, z)$ does, and when it does not, hold for any natural numbers x and z. Then the function

$$f(z) = (\mu x)[R(x, z)]$$

is computed as follows. We consider, one by one, all the natural numbers $x = 0, 1, 2, ...$, and for a given number z we first check whether the relation $R(0, z)$ holds. If it does not, then we check whether $R(1, z)$ holds; if it also does not hold then we check whether $R(2, z)$ holds, etc. Since the effectiveness condition is satisfied, for every z there does exist a number x such that $R(x, z)$ holds. By examining one by one the relations

$$R(0, z), \; R(1, z), \; R(2, z), \; R(3, z), \; ...$$

we finally find the first of those numbers x for which, under the effectiveness condition, the relation $R(x, z)$ holds, and this number x is the value of the function $f(z)$. If the effectiveness condition is not satisfied, that is, if for some z_0 no x such that $R(x, z_0)$ exists, then we could check whether the relation $R(x, z_0)$ holds for $x = 0, 1, 2, ...$ for arbitrarily large numbers and we would find no x such that $R(x, z_0)$. Then, under the general definition of the minimization operation, the value of the function $f(z)$ would equal zero, but in order to make sure of that by successive trials we would need infinitely many such trials. Such a method would be ineffective.

Thus a relation R must satisfy the effectiveness condition if the minimization operation is to lead from a computable relation to a computable function.

III. HIERARCHY

Definition of the class of computable functions

Mathematical practice shows that all those functions for which effective methods of computation are known can be defined by the operations of substitution, simple recursion, and effective minimization, in combination, starting from some simplest mathematical operations. It has turned out that the best solution is to apply the term *computable function* in the mathematical sense to all those functions which can be defined by applications of the operations of superposition, simple recursion, and effective minimization, the starting points being the function $x+1$ as already understood and certainly computable in every sense of the word, and the constant function $0(x) = 0$.

Accordingly, the following inductive definition of the class of computable functions has been formulated:

DEFINITION 1 (Kleene [34]). *The class of computable functions* Comp *is the least class of functions defined on natural numbers and taking on natural numbers as values, and satisfying the following conditions*:

1) *The successor function* $S(x) = x+1$, *the functions equal to their arguments*: $I(x, y) = x$ *and* $I(x, y) = y$, *and the constant function* $0(x) = 0$ *are in* Comp.

2) *If two functions, f and g, are in* Comp, *then the function h, obtained by superposition*

$$h(..., x, ..., y, ...) = f(..., x, ..., g(..., y, ...), ...)$$

also is in Comp.

3) *If two functions, f and g, are in* Comp, *then the function h, obtained by induction under the schema of simple recursion*

$$h(0, x) = f(x),$$
$$h(S(n), x) = g(x, n, h(n, x))$$

is also in Comp.

4) *If a function f is in* Comp *and if the effectiveness condition* $\bigwedge u \bigvee x f(u, x) = 0$ *is satisfied, then the function h, defined by the minimization operation,*

$$h(u) = (\mu x) [f(u, x) = 0]$$

also is in Comp.

410

By way of abbreviation the above definition can be formulated in a single sentence:

The class Comp *is the least class of functions which contains zero and the successor function and is closed under the operations of superposition, simple recursion, and effective minimization.*

Condition 1) is the initial condition of inductive definition, and conditions 2)–4) are induction conditions.

We consider that definition by identification of arguments (cf. Definition 1 above) falls under superposition of functions. Moreover, the two functions each of which equals one of its arguments are adopted as initial functions. This is done because of the special form of the schema of simple recursion, as adopted here. Schema 3) of simple recursion leads only from a function $f(x)$ of one argument and a function $g(x, n, y)$ of three arguments to a function $h(n, x)$ of two arguments. The function $I(x, y)$ makes it possible to define various functions, for instance, addition, in a way which fully complies with that schema.

Now that we have defined computable functions we define computable relations as those which have a characteristic computable function:

DEFINITION 2. *A relation R is computable* $(R \in \text{Comp})$ *if and only if there exists a function $f \in$* Comp *such that*

$$\bigwedge x_1, \ldots, x_n \in \mathcal{N}\big(R(x_1, \ldots, x_n) \equiv f(x_1, \ldots, x_n) = 0\big).$$

The set of natural numbers is computable, since it is determined by a computable ralation of one argument.

It is said in general that a relation or a set is in a special class X if there exists for it a characteristic function in that special class X.

All the functions and relations so far defined in this section are computable in the sense of the definition given above. Addition, multiplication, predecessor, subtraction and exponentiation have been defined by inductive formulae (5)–(10), the starting point being the successor function and zero. The remaining ones have been defined by reference to effective minimization.

Consider now a precise proof of the fact that addition is in the class Comp. Since the functions S and $I(x, y)$ both are in Comp, hence, by condition 2) of Definition 1 so are the functions S_1 and S_2, where

$$S_1(n, y) = I\big(n, S(y)\big),$$
$$S_2(x, n, y) = I\big(x, S_1(n, y)\big).$$

Hence, by condition 3), addition is in Comp, since it satisfies the formulae

$$0+x = I(x),$$
$$S(n)+n = S_2(x, n, n+x),$$

which comply exactly with the schema of simple recursion.

The above definition of computability is conventional in nature. As we shall see, however, it is equivalent to certain other definitions which perhaps penetrate deeper into the essence of computability, but are much more difficult. For the time being we shall examine certain properties of computability already defined. These properties will be formulated in a more general way, that formulation being applicable to any classes closed under certain operations.

THEOREM 1. *Every class X of functions which contains the functions* $x+1, x \div y, x^y$ *and is closed under the operations of*

a. *superposition,*

b. *effective minimization,*

s also closed under the operation of induction falling under the schema of simple recursion.

PROOF. Starting from the functions $x+1$, $x \div y$, and x^y we shall first define several functions and relations which are in X and, hence, are computable, since the class Comp satisfies the conditions imposed upon the class X in the theorem now under consideration.

$$\begin{aligned}
& 0 = x \div x, \\
& x \leqslant y \equiv x \div y = 0, \\
& x = y \equiv (x \div y)^{(1 \div (y \div x))} = 0, \\
\text{(17)} \quad & 2 = ((x \div x)+1)+1, \\
& x \cdot y = (\mu z) [2^z = (2^x)^y], \\
& x+y = (\mu z) [2^z = 2^x \cdot 2^y].
\end{aligned}$$

By referring to the ordinary laws of the multiplication and addition of exponents we can check easily that the last two definitions do in fact define the operations of multiplication and addition, respectively.

Note next that if there are two relations defined by the formulae

$$\begin{aligned}
\text{(18)} \quad & R(x, y) \equiv f(x, y) = 0, \\
& S(x, y) \equiv g(x, y) = 0,
\end{aligned}$$

412

then the relation which defines the fact that both relations R and S hold concurrently can be defined as follows:

(19) $\quad$ $^{\prime}$ $R(x, y) \wedge S(x, y) \equiv \left(f(x, y) + g(x, y)\right) = 0,$

while the relation which defines the fact that either R or S holds can be defined as follows:

(20) $\quad$ $R(x, y) \vee S(x, y) \equiv \left(f(x, y) \cdot g(x, y)\right) = 0.$

The relation which defines the fact that R does not hold is defined thus:

(21) $\quad$ $\sim R(x, y) \equiv \left(1 \div f(x, y)\right) = 0.$

Thus both the conjunction and the disjunction of two relations from the class X yields a relation from the class X. Likewise, the negation of a relation from the class X yields a relation from the class X.

Note also a fact which we will often refer to below: The effective minimization operation makes it possible to formulate the logical operation of the universal bounded quantifier as computable, because of the equivalence

$$\left(\bigwedge x\left(x < n \rightarrow R(..., x ...)\right) \wedge \sim R(..., n, ...)\right) \equiv$$
$$n = (\mu x) [\sim R(..., x, ...)].$$

This property will be referred to below when formula (24) is used to define the set of prime numbers.

(22) $\quad$ $[x/y] = (\mu z) [y \cdot (z+1) > x \vee y = 0].$

(23) $\quad$ $x|y \equiv x \cdot [y/x] = y$

($x|y$ means: x divides y without remainder, that is, x is a divisor of y).

If x divides y without remainder, then the fraction y/x is an integer, and as such it is equal to its integral part: $y/x = [y/x]$, which means in turn that the relation $y = x \cdot [y/x]$ holds between the integers x and y.

(24) $\quad$ $\text{pr}(x) \equiv \left(x \neq 0 \wedge x \neq 1 \wedge x = (\mu z) [x = 0 \vee x = 1 \vee \right.$
$$(z > 1 \wedge z|x)])$$

(where $\text{pr}(x)$ means: x is a prime number). The above definition states that $\text{pr}(x)$ if $0 \neq x \neq 1$ and x is the least number greater than 1 that divides x without remainder. Hence, x is, obviously, the only such number greater than 1 which divides x without remainder. Thus, x is divisible without remainder only by 1 and by itself, which means exactly

413

that it is a prime number. Hence, the set of prime numbers is computable in the strict sense of the term. Likewise, the enumeration of the prime numbers, that is, the function p_n, is computable. This function can easily be defined by a combination of induction with the effective minimization operation:

$$p_0 = 2,$$
$$p_{n+1} = (\mu z) [z > p_n \wedge \text{pr}(z)].$$

Yet it is not assumed about the class X that it is closed under the schema of recursion; this is exactly that property which is to be proved. This is why the function p_n must be defined otherwise, by referring to superposition and effective minimization only. This will be somewhat more difficult and more "artificial". We first define two auxiliary functions:

(25) $\qquad \exp(x, y) = (\mu z) [y = 1 \vee y = 0 \vee \sim (y^{z+1}|x)]$

(exp(x, y) stands for the greatest exponent of a power such that if y is raised to that power it is a divisor of x).

The above definition corresponds to the meaning given to the symbol, since if z is the least number such that y^{z+1} is not a divisor of x, then this means that y^z still is a divisor of x and that it is the greatest of those powers of y which are divisors of x.

$$\text{pri}(z) = (\mu y) \Big[\sim \text{pr}(z) \vee \big\{ y > 1 \wedge$$
$$\exp(y, 2) = 2 \wedge z = (\mu v) \Big[\text{pr}(v) \wedge \big(y < 2 \vee$$
$$\exp(y, (\mu u) [\text{pr}(u) \wedge u > v]) \neq \exp(y, v) + 1 \big) \Big] \big\} \Big].$$

An informal definition of the auxiliary function pri(z) might be as follows:

$$\text{pri}(z) = \begin{cases} 0 & \text{if } z \text{ is not a prime number,} \\ 2^2 3^3 5^4 7^5 \ldots z^{n+2} & \text{if } z \text{ is the } n\text{-th prime number} \\ & (z = p_n). \end{cases}$$

For if, by its definition, z is the least prime number v whose power, which occurs in the expansion of the number y: $\exp(y, v)$, augmented by 1, differs from the power of that prime number which is the next after v:

$$\exp(y, (\mu u) [\text{pr}(u) \wedge u > v]),$$

then it must be true for all prime numbers less than z that the next prime number occurs in the expansion of y in a power augmented by 1. This, in fact, occurs if and only if

$$y = 2^2 \cdot 3^3 \cdot 5^4 \cdot 7^5 \cdot \ldots \cdot p_n^{n+2}, \quad \text{where } p_n = z.$$

We can now define p_n in accordance with that property of the auxiliary function pri:

(26) $\quad p_n = (\mu z)\,[\mathrm{pr}(z) \wedge \exp(\mathrm{pri}(z), z) = n+2].$

In fact, since $\mathrm{pri}(z) = p_0^2 p_1^3 p_2^4 p_3^5 \ldots p_n^{n+2}$, where $p_n = z$, then

$$\exp(\mathrm{pri}(z), z) = n+2.$$

The sequence of prime numbers plays an important role in the theory of computable functions. Every natural number m can be represented as a product of powers of prime numbers:

$$m = 2^{x_0} \cdot 3^{x_1} \cdot 5^{x_2} \cdot \ldots \cdot p_n^{x_n},$$

where

$$x_i = \exp(m, p_i) \quad \text{for} \quad i = 0, 1, \ldots, n.$$

Such a presentation of a number m is called its *decomposition into prime factors*. Now instead of saying that there exists a finite sequence of numbers $x_0, \ldots, x_n$ which satisfy certain conditions we may say that there exists a single number m such that the exponents $\exp(m, 2), \ldots, \exp(m, p_n)$ satisfy those conditions.

We also often augment the numbers $x_0, \ldots, x_n$ by 1 each in order to make a distinction between the numbers which are in a given sequence and may be zeros and the zero exponents whose number in the expansion of any number is infinite. Thus the sequence $x_0, \ldots, x_n$ can also be represented by the number

$$m = 2^{x_0+1} \cdot 3^{x_1+1} \cdot \ldots \cdot p_n^{x_n+1};$$

in such a case, of course, $x_i = \exp(m, p_i) \,\dot{-}\, 1$.

Precisely this procedure is required when we want to write down an inductive definition the minimization operation. This is the last and the most important part of the proof. If a function h is defined by a schema of simple recursion in which functions f and g occur:

(27) $\quad \begin{aligned} h(0, x) &= f(x), \\ h(n+1, x) &= g(n, x, h(n, x)), \end{aligned}$

then we may say that

$u = h(n, x) \equiv$ *there exists a number m such that*
1) $\exp(m, 2) = f(x) + 1$,
2) $\exp(m, p_{i+1}) = g(i, x, \exp(m, p_i) \div 1) + 1$, *for all $i < n$*,
3) $\exp(m, p_n) = u + 1$,

because

$$m = 2^{h(0,x)+1} \cdot 3^{h(1,x)+1} \cdot 5^{h(2,x)+1} \cdot \ldots \cdot p_n^{h(n,x)+1}$$

is such a number. Thus $h(n, x) + 1 = \exp(m, p_n)$, where m is, for instance, the least of those numbers m which satisfy conditions 1)–3). Hence, if the functions f, g, and h satisfy condition (27), then the function h can be defined by the effective minimization operation as follows:

$$h(n, x) = \exp\big((\mu m) \,[\exp(m, 2) = f(x) + 1 \wedge$$
$$n = (\mu i) \,[\exp(m, p_{i+1}) \neq$$
$$g(i, x, \exp(m, p_i) \div 1) + 1]], \, p_n\big) \div 1.$$

In fact, the condition $n = (\mu i) \,[\exp(m, p_{i+1}) \neq g(i, x, \exp(m, p_i) \div 1) + 1]$ implies condition 2) from the equivalence above. For if n is the least i for which the above inequality holds, then the equation specified in 2) holds for all i less than n.

Thus, whatever can be defined by the schema of recursion can also be defined by effective minimization and superposition alone. This completes the proof of Theorem 1.

THEOREM 2 (Kleene [34], J. Robinson [75]). *The class* Comp *is the least class of functions which contains the functions $x \div y$, $S(x)$, and x^y, and is closed under the operations of*

1. *superposition*,

2. *effective minimization*.

The proof follows directly from Definition 1 and Theorem 1.

Conditional definitions were analysed earlier in this section. It can easily be noted that

THEOREM 3. *If a class X is closed under the operation of superposition and contains the functions $x \div y$, $x + y$, and $x \cdot y$, then it is closed under the use of definitions by cases.*

416

2. COMPUTABLE FUNCTIONS

PROOF. If a function f is defined by cases in the following way:

$$f(x) = \begin{cases} g(x) & \text{if} \quad j(x) = 0, \\ h(x) & \text{if} \quad j(x) \neq 0, \end{cases}$$

then the function f can be defined by superposition as follows:

$$f(x) = \Big(g(x) \cdot (1 \div j(x)) \Big) + \Big(h(x) \cdot \Big(1 \div \big(1 \div (j(x)) \big) \Big) \Big).$$

Thus, the class of computable functions, as given by Definition 1, is closed under all those operations of defining new functions which do not lead outside functions computable in the intuitive sense of the term. Many other properties of computable functions will be discussed later.

1. Define the functions $[x/2]$, $[x/3]$, the remainder upon division of x by y, $[\sqrt{x}]$ and $[x/y]$ by using simple recursion and superposition.

2. Define the functions $[\sqrt[x]{y}]$ and $\binom{n}{m}$ by using effective minimization.

3. Prove rigorously that multiplication, exponentiation, and subtraction are in Comp, as has been proved for addition on pp. 411, 412.

2. SOME PROPERTIES OF COMPUTABLE FUNCTIONS

Elementary recursive operations

The present section will be concerned with certain formal properties of computable functions. Some of these properties will be formulated in a more general manner, as in the case of Theorem 1.

A new operation of *bounded minimization* will be introduced:

$$(\mu x \leqslant z)\,[R(..., x, ...)] = \textit{the least } x, \textit{ less than or equal to } z,$$
$$\textit{for which the condition } R(..., x, ...)$$
$$\textit{holds.}$$

This operation can fully be defined by means of ordinary minimization, as introduced in the preceding section:

(28) $(\mu x \leqslant z)\,[R(..., x, ...)] = (\mu x)\,[x \leqslant z \wedge R(..., x, ...)].$

The operation of bounded minimization does not lead outside functions computable in the intuitive sense of the term, because, in accordance

417

with (28), if we want to determine the value of $(\mu x \leqslant z)\,[R(..., x, ...)]$, it suffices to examine only those numbers which are less than or equal to z, and to find out whether they include any which have the property R. If they do, then the earliest of them is the value in question, and if they do not, the value in question is 0 on the strength of the definition of the general minimization operation. Thus, the computation of the value sought requires at most a z-fold check whether the relation R does or does not hold.

THEOREM 4. *If X is a class of functions which contains the functions $S(x)$, $x \div y$, $x+y$, and $x \cdot y$, and is closed under the operation of simple recursion, then X is also closed under the operation of bounded minimization.*

PROOF. Assume that $g \in X$ and a function f is defined as follows:

(29) $f(x, t) = (\mu z \leqslant x)\,[g(z, t) = 0]$.

The function f can be defined by simple recursion as follows:

$$f(0, t) = 0,$$

(30) $f(n+1, t) = \begin{cases} n+1 & \text{if } f(n, t) = 0 \wedge g(0, t) \neq 0 \wedge \\ & g(n+1, t) = 0, \\ f(n, t) & \text{if } f(n, t) \neq 0 \vee g(0, t) = 0 \vee \\ & g(n+1, t) \neq 0. \end{cases}$

In fact, if $x = 0$, then every $z \leqslant x$ has the value zero, and hence $(\mu z \leqslant x)\,[g(z, t) = 0] = 0$.

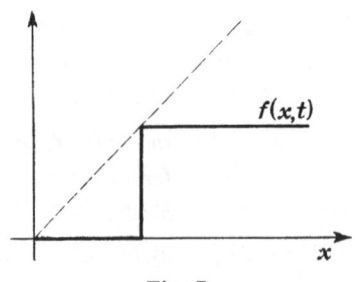

Fig. 7

In order to comprehend the induction condition (30) consider the graph of the function $f(x, t)$ as defined by formula (29). (Cf. Figure 7.) The graph of $f(x, t)$ as a function of the variable x for a fixed parameter

t is as follows: If $g(0, t) = 0$ or $g(x, t) \neq 0$ for every x, then $f(x, t) = 0$ for every x. On the other hand, if there exist numbers $x \neq 0$ such that $g(x, t) = 0$, then the graph of the function f has one jump, namely $f(x, t) = 0$ for $x < (\mu z) [g(z, t) = 0]$, and beginning with $x = (\mu z)$ $[g(z, t) = 0]$ the value of the function f always equals the number $(\mu z) [g(z, t) = 0]$.

The function f is, thus, constant, that is, it satisfies the equation $f(n+1, t) = f(n, t)$ for all n except for $n+1 = (\mu z) [g(z, t) = 0]$. For $n+1 = (\mu z) [g(z, t) = 0]$ the function f satisfies the equation $f(n+1, t) = n+1$. This is why a distinction between these two cases is made in formula (30). The correctness of formula (30) can easily be proved in a rigorous manner by induction.

If for an n

(31) $f(n, t) = (\mu z \leqslant n) [g(z, t) = 0]$

and also $f(n, t) = 0$, then either 0 is the least z such that $g(z, t) = 0$, or there are no $z \leqslant n$ such that $g(z, t) = 0$. Hence, if $f(n, t) = 0$ and $g(0, t) \neq 0$, then the latter case holds, so that $g(z, t) \neq 0$ for all $z \leqslant n$. If in such a case also $g(n+1, t) = 0$, then $n+1$ is the least natural number z for which $g(z, t) = 0$, and, thus, it also is the least number among those which are less than or equal to $n+1$. Hence, if the upper condition of formula (30) is satisfied, and if $f(n+1, t)$ is defined as $n+1$, then

$$f(n+1, t) = (\mu z \leqslant n+1) [g(z, t) = 0].$$

Consider now the lower condition of formula (30). If $g(n+1, t) \neq 0$, then $n+1$ is not the least z such for which $g(z, t) = 0$. The least $z \leqslant n+1$ such that $g(z, t) = 0$, if it exists, is less than or equal to n, and hence $(\mu z \leqslant n+1) [g(z, t) = 0] = (\mu z \leqslant n) [g(z, t) = 0] = f(n, t)$, on the assumption of formula (31).

If $g(0, t) = 0$, then $0 = (\mu z) [g(z, t) = 0]$ and, by (28), for any m, $0 = (\mu z \leqslant m) [g(z, t) = 0]$, and, hence,

$$f(n, t) = 0 = (\mu z \leqslant n+1) [g(z, t) = 0].$$

If $f(n, t) \neq 0$, then, on the assumption of (31) and by (28), there is a least $z \leqslant n$ such that $z \neq 0$ and $g(z, t) = 0$. That z is, obviously, less than $n+1$, and, thus, it equals $(\mu z \leqslant n+1) [g(z, t) = 0]$. Then, by (31),

$$f(n, t) = (\mu z \leqslant n+1) [g(z, t) = 0].$$

419

Thus, if the lower condition of formula (30) is satisfied, then

$$f(n+1, t) = f(n, t) = (\mu z \leqslant n+1) [g(z, t) = 0].$$

Hence, if either condition, the upper or the lower, of formula (30) is satisfied, then condition (31) implies the condition

(32) $\qquad f(n+1, t) = (\mu z \leqslant n+1) [g(z, t) = 0].$

Formula (32) is obtained by the substitution in (31) of $n+1$ for n. Formula (31) is thus satisfied for 0 and, as we have seen, if it is satisfied for n, then it is also satisfied for $n+1$. Thus, by induction we have that formula (31) is satisfied for every $n \in \mathcal{N}$.

By reference to formulae (19), (20), (21) and Theorem 3 we can easily write down formula (30) as a single equation, with superpositions of the functions $g, S, +, \cdot, \div$.

It follows immediately from Theorem 4 and Definition 1 that the class of computable functions is closed under the operation of bounded minimization. The operation of bounded minimization, if applicable, is more convenient than that of effective minimization, since it does not require satisfaction of any additional conditions of effectiveness and always leads from computable functions to computable functions.

THEOREM 5. *If X is a class of functions which contains S(x) and x $\div$ y and is closed under bounded minimization, then the relations in X are closed under the logical operations of sentential connectives and bounded quantifiers.*

PROOF. As noted in the preceding section (formulae (19), (20), (21)), the connectives $\vee$, $\wedge$, and $\sim$ do not lead outside the relations of a class which contains $+, \cdot, \div$. Now note only that subtraction $\div$ and the successor function suffice to define the sentential connectives. The following equivalences can easily be checked

(33) $\qquad \begin{aligned} &(a = 0 \rightarrow b = 0) \equiv \big((1 \div a) \div (1 \div b)\big) = 0, \\ &\sim (a = 0) \equiv (1 \div a) = 0. \end{aligned}$

Since all other connectives can be defined by implication and negation, the relations in X are closed under the sentential connectives.

In arithmetic the bounded quantifiers are the formulae:

$$\bigwedge x \leqslant z \dots \text{ means: } \textit{for every } x \leqslant z, \dots,$$

$$\bigvee x \leqslant z \dots \text{ means: } \textit{there exists an } x \leqslant z \textit{ such that } \dots$$

2. COMPUTABLE FUNCTIONS

The meanings of these formulae can accordingly be defined by reference to ordinary quantifiers:

$$(34) \quad \begin{aligned} \bigwedge x \leqslant zA &\equiv \bigwedge x(x \leqslant z \to A), \\ \bigvee x \leqslant zA &\equiv \bigvee x(x \leqslant z \wedge A). \end{aligned}$$

If we confine ourselves to natural numbers, the quantifiers can be defined by the minimization operation, since the equivalence

$$(35) \quad \bigvee xA(..., x, ...) \equiv A\big(..., (\mu x)\,[A\,(..., x, ...)], ...\big)$$

is true: if there exists an x such that $A(..., x, ...)$, then there obviously exists a least x, which has the property A; conversely, if $A\big(..., (\mu x)\,[A(..., x, ...)], ...\big)$, then the element $(\mu x)\,[A(..., x, ...)]$ is an element which has the property A.

The universal quantifier can in turn be defined by reference to the existential quantifier using De Morgan's laws.

Equivalences (35) and (34) also yield the equivalence

$$(36) \quad \bigvee x \leqslant zA(..., x, ...) \equiv A\,(..., (\mu x \leqslant z)\,[A(..., x, ...)], ...)$$

which shows that the bounded existential quantifier can be defined by the bounded minimization operation. The bounded universal quantifier can be defined by the bounded existential quantifier and negation:

$$\bigwedge x \leqslant zA(..., x, ...) \equiv \sim \bigvee x \leqslant z \sim A(..., x, ...).$$

It follows from Theorems 4 and 5 and Definition 1 that the class of computable relations is closed under the operations of the sentential calculus and the bounded quantifiers.

It is also obvious that if X is a class of functions closed under superposition, then by substituting into a relation in X a function in X we obtain a relation in X. In fact, if R is a relation in X, then R has the form $R(x, y) \equiv f(x, y) = 0$, where $f \in X$. If also $g \in X$, then the relation $R\big(x, g(y)\big)$ has the form

$$R\big(x, g(y)\big) \equiv f\big(x, g(y)\big) = 0,$$

and, hence, is in X.

The operations defined so far, which do not lead outside computable functions are not equivalent to one another as to the possibility of defining new functions. The effective minimization operation is the strongest in that respect, while the simple recursion operation is weaker, and the restricted minimization operation is weaker still. Yet functions definable by those weaker operations suffice for many applications. Two classes of functions, narrower than the class Comp, will now be introduced.

DEFINITION 3. *The class* Prec *of primitive recursive functions is the least class of functions which contains the functions* $S(x), 0(x), I(x, y) = x, I(x, y) = y$, *and is closed under the operations of*

a. *superposition,*

b. *simple recursion.*

DEFINITION 4. *The class* Erec *of elementary recursive functions is the least class of functions which contains the functions* $S(x), x \doteq y, x^y$, *and is closed under the operations of*

a. *superposition,*

b. *bounded minimization.*

THEOREM 6.

(a) Erec $\subset$ Prec $\subset$ Comp.

(b) *Relations in each of the classes specified above are closed under the operations of the sentential calculus and the bounded quantifiers.*

The proof follows directly from Theorems 4 and 5 and Definitions 1–4.

On the contrary, the converse inclusions are not true.

We shall now discuss the two classes defined above which are narrower than Comp. Note first that all the computable functions defined so far are elementary recursive, since in formulae (13)–(26) effective minimization can be replaced everywhere by bounded minimization if the variables bound by the operator μ are adequately restricted. Such a restriction cannot, of course, be found for a general formula of the function $h(n, x)$ as defined by schema (27). In some cases, when constructing the restricting functions we have to avail ourselves of the fact that the classes Erec, Prec, and Comp are also closed under the following operations definable by the bounded minimization operation.

The *minimum value operation* $\operatorname*{Min}_{x \leqslant z} f(x)$ is defined as:

$$(37) \qquad \operatorname*{Min}_{x \leqslant z} f(x) = f((\mu x \leqslant z) [\bigwedge y \leqslant z \, f(x) \leqslant f(y)]).$$

2. COMPUTABLE FUNCTIONS

The relation under the minimization symbol is defined by the bounded quantifier. The same holds for the following operation:

The *maximum value operation* $\operatorname*{Max}_{x \leqslant z} f(x)$:

(38) $\qquad \operatorname*{Max}_{x \leqslant z} f(x) = f((\mu x \leqslant z) [\bigwedge y \leqslant z f(x) \geqslant f(y)])$.

The operations of finite sums and products. These operations will be defined rigorously in the proof of the next theorem. For the time being their sense will be described by informal formulae:

$$\sum_{x \leqslant z} f(x, u) = f(0, u) + f(1, u) + \ldots + f(z, u),$$
(39)
$$\prod_{x \leqslant z} f(x, u) = f(0, u) \cdot f(1, u) \cdot \ldots \cdot f(z, u).$$

The above formulae, even though they explain sufficiently the meaning of the operations $\sum$ and $\prod$, are not precise because of the dots "..." which occur in them and the meaning of which cannot in genera be described with precision.

The operations of bounded induction. A class X is closed under the bounded induction operation if it satisfies the following condition: If $g, h, j \in X$, then the function f, which satisfies the formulae:

(40)
$\qquad$ (a) $\;\; f(0, u) = g(u),$
$\qquad$ (b) $\;\; f(n+1, u) = h(u, n, f(n, u)),$
$\qquad$ (c) $\;\; f(n, u) \leqslant j(n, u),$

also is in X.

The function f is, of course, determined unambiguously by conditions (a) and (b), which coincide with the schema of simple recursion. Condition (c) merely restricts the application of that schema to functions with values less than the values of a function in X.

THEOREM 7. *If a class X contains the functions $x \div y$, S, x^y, and is closed under bounded minimization and superposition, then it is also closed under the operations*: $\operatorname*{Min}_{x \leqslant z}$, $\operatorname*{Max}_{x \leqslant z}$, $\sum_{x \leqslant z}$, $\prod_{x \leqslant z}$, *and bounded induction.*

PROOF. For Min and Max, Theorem 7 follows directly from formulae (37) and (38). For induction, condition (40) (c) makes it possible to estimate the upper value of the function f by reference to a known function j, so that the operation of bounded minimization can be used.

For instance, we can first define a relation $R(n, u, y)$, which has the sense of $y = f(n, u)$, and next the function f itself:

$$R(n, u, y) \equiv \bigvee m \leqslant p_n^{n\operatorname{Max}_{x \leqslant n} j(x, u)} \Big(\exp(m, 2) = g(u) \wedge y =$$
$$\exp(m, p_n) \wedge \bigwedge x < n \big(\exp(m, p_{x+1}) =$$
$$h\big(u, x, \exp(m, p_x)\big)\big)\Big),$$

$$f(n, u) = \big(\mu y \leqslant j(n, u)\big)\, [R(n, u, y)].$$

In defining the relation R we describe a number m, whose expansion is

$$m = 2^{f(0, u)} \cdot 3^{f(1, u)} \cdot \ldots \cdot p_n^{f(n, u)},$$

in a way similar to that in the proof of Theorem 1, by referring to the fact that the class X is closed under the bounded quantifiers.

Bounded induction can, in turn, be used to define finite sums and products: If $g(z, u) = \sum_{x \leqslant z} f(x, u)$ and $h(z, u) = \prod_{x \leqslant z} f(x, u)$, then the functions g and h can also be defined as follows:

$$g(0, u) = f(0, u),$$
$$g(n+1, u) = g(n, u) + f(n+1, u),$$
$$g(n, u) \leqslant (n+1) \cdot \operatorname*{Max}_{x \leqslant n} f(x, u),$$
$$h(0, u) = f(0, u),$$
$$h(n+1, u) = h(n, u) \cdot f(n+1, u),$$
$$h(n, u) \leqslant \big(\operatorname*{Max}_{x \leqslant n} f(x, u)\big)^{n+1}.$$

The operations specified in Theorem 7 are called *elementary recursive operations*. Each of them together with superposition can be used to define all the others.

Pairing functions and graphs

A *pairing function* is any function of two arguments $f(x, y)$ which establishes a one-to-one assignment of every pair of natural numbers to a natural number, which takes as its value every natural number. In this way a function f establishes a one-to-one correspondence between the entire set $\mathcal{N} \times \mathcal{N}$ of pairs of natural numbers and the entire set $\mathcal{N}$. An example is provided by the function

$$K(x, y) = 2^x(2y+1) \doteq 1.$$

2. COMPUTABLE FUNCTIONS

Each such function has two inverse functions which give the first and the second element of the pair. In the case of the function K the inverse functions are:

$$K_1^{-1}(z) = \exp(z+1, 2).$$

$$K_2^{-1}(z) = \left[\frac{\left[\dfrac{z+1}{2^{\exp(z+1,2)}} \right] \div 1}{2} \right].$$

It can easily be checked that these functions satisfy the formulae:

$$K\big(_1^{-1}K(x, y)\big) = x,$$
(41) $$K_2^{-1}\big(K(x, y)\big) = y,$$
$$K\big(K_1^{-1}(z), K_2^{-1}(z)\big) = z.$$

In view of the above, every number may be considered a pair of two numbers. Hence, every number may also be considered a triple, a quadruple, or any n-tuple of natural numbers.

If we are not concerned with a one-to-one covering of the entire set of natural numbers, we may construct the following mapping using prime numbers:

(42) $$\langle x_0, ..., x_n \rangle = 2^{x_0} \cdot 3^{x_1} \cdot \ ... \ \cdot p_n^{x_n}.$$

This mapping will be used most frequently. The inverse functions are defined by the formulae

(43) $$(z)_i = \exp(z, p_i).$$

Obviously,

$$(\langle x_0, ..., x_i, ..., x_n \rangle)_i = x_i.$$

The function

$$L(x, y) = (x+y)^2 + y$$

is a fairly simple pairing function, which, again, does not cover the entire set of natural numbers. Its inverse functions are:

$$L_1^{-1}(z) = z \div [\sqrt{z}]^2,$$
$$L_2^{-1}(z) = [\sqrt{z}] \div L_1^{-1}(z).$$

Pairing functions make it possible to analyse many properties for functions of one argument only. For, if a class X contains a pairing function and is closed under superposition, then every function of two arguments $h(x, y)$ has its corresponding function of one, argument $h'(z)$, which is defined as follows:

(44) $h'(z) = h(K_1^{-1}(z), K_2^{-1}(z))$.

It is self-evident that the function h can conversely be defined by reference to the function h':

(45) $h(x, y) = h'(K(x, y))$,

since (45) follows from (41) and (44).

Prime numbers also make it possible to represent the history of the behaviour of a function f up to a point by means of a single number, which will be denoted by $\tilde{f}(n)$:

$$\tilde{f}(n) = \langle f(0), \ldots, f(n) \rangle.$$

This operation on a function will be called the *diagram* (or *course-of-values*) *operation for a function f up to a number n*. The operation is rigorously defined by the formula

(46) $\tilde{f}(n) = \prod_{x \leqslant n} p_x^{f(x)}$.

It follows from formulae (43) and (46) that

(47)   *For $i \leqslant n$, $f(i) = (\tilde{f}(n))_i$.*

The concept of diagram thus defined applies to functions of one argument. In the case of a function of many arguments $f(x, \ldots, y, z)$ the argument with respect to which a diagram is made will be designated by the fact that the remaining arguments will be written as parameters next to the function symbol. Thus, when treating $f(x, \ldots, y, z)$ as a function of the variable z we shall use the notation:

$$f_{x, \ldots, y}(z) = f(x, \ldots, y, z).$$

Then the following interpretation of the diagram is natural:

$$\tilde{f}_{x, \ldots, y}(z) = \prod_{t \leqslant z} p_t^{f_{x, \ldots, y}(t)} =$$
$$\langle f(x, \ldots, y, 0), f(x, \ldots, y, 1), \ldots, f(x, \ldots, y, z) \rangle.$$

Pairing functions and diagrams will often be used below. The following property is an important application of the pairing function: *If a class*

2. COMPUTABLE FUNCTIONS

X constains pairing functions, for instance the functions $(x)_0$, $(x)_1$, and if a relation $R(x, y, a)$ is in X, then the class X also contains a relation $R'(z, a)$ such that the following equivalences hold:

$$\bigvee x \bigvee y\, R(x, y, a) \equiv \bigvee z R'(z, a),$$
$$\bigwedge x \bigwedge y R(x, y, a) \equiv \bigwedge z R'(z, a).$$

In other words, *pairing functions make it possible to combine similar quantifiers into one quantifier of the same kind.*

In fact, a relation $R\big((z)_0, (z)_1, a\big)$ is a relation R' which has the said property, since the equivalence

$$\bigvee x \bigvee y R(x, y, a) \equiv \bigvee z R\big((z_0), (z)_1, a\big)$$

holds. The same applies, in a parallel way, to the universal quantifier.

Universal functions

A function of two arguments $h(n, x)$ is called *universal in the broader sense* for a class X if and only if for every function of one argument $f \in X$ there exists an n such that

(48) $\bigwedge x h(n, x) = f(x).$

A universal function in the broader sense for a class X will be called *universal in the narrower sense* if for every $n \in \mathcal{N}$ there exists an $f \in X$ such that condition (48) holds. In most cases universal functions in the broader sense will suffice; they will often be called simply *universal*.

If a class X is closed under superposition and contains a pairing function K, then a universal function can yield all the functions of X of any arguments, since it follows from (45) and (48) that for every function $f \in X$ of two arguments there exists an $n \in \mathcal{N}$ such that

$$\bigwedge x, y\, h\big(n, K(x, y)\big) = f(x, y).$$

Since we shall be almost exclusively concerned with classes X containing pairing functions and closed under superposition, we shall confine ourselves to universal functions of two arguments.

THEOREM 8. *If a class X contains the function $x+1$ and is closed under superposition, then no function universal (in the broader sense) for the class X is in the class X.*

PROOF. Suppose that h is a universal function for a class X and is in X. The operation of superposition covers the operation of identifica-

427

tion of arguments. Hence the function $h(x, x)$ is in X. The successor function may be superposed on the function h, so that $h(x, x)+1$ is in X. Since h is a universal function, the function $h(x, x)+1$ must have its number n such that, by (48),

$$\bigwedge xh(n, x) = h(x, x)+1.$$

This formula holds for every $n \in \mathcal{N}$, and, hence, in particular, for the number n which is the number of the function $h(x, x)+1$. But, on substituting n for x in the formula above, we arrive at a contradiction

$$h(n, n) = h(n, n)+1.$$

Hence, the function h cannot be in the class X, if it is universal for X.

Defining a universal function for a given class X is, thus, one of the methods of obtaining a function which is not in X. Another method, equally general in nature, is to define a function which increases more rapidly than do all the functions which are members of the class X.

Primitive recursive functions

Those mathematicians who were investigating effective methods of defining functions did not immediately come upon the full concept of computable function. It was originally supposed that the schema of simple recursion sufficed to yield all the functions computable in the intuitive sense. This was why primitive recursive functions were originally called just recursive, and it was believed that they comprised all the computable functions. But it was soon discovered that there are other methods of computing functions, methods which also are effective but cannot be reduced to simple recursion. Consequently, the narrower class, previously under consideration, came to be called the *class of primitive recursive functions*, and the class of functions computable by any effective formulae came to be called the *class of general recursive functions*.

The primitive recursive functions, even though they have lost in significance, have a number of important applications, and, hence, deserve some comment. First of all, the schema of simple recursion was an object of many investigations (Peter [60]–[62], R. M. Robinson [76]). Attention was drawn to the fact that there are many recursion schemata which are reducible to the schema of simple recursion, and that there are also schemata of the type called multiple recursion, not reducible to simple recursion. Several types of recursion, reducible to simple recursion and convenient in applications, will now be discussed.

1. CONDITIONAL RECURSIVE DEFINITIONS. A recursive definition is called *conditional* if it falls under the following schema:

    a. $f(0, u) = g(u)$,

(49)   b. $f(x+1, u) = \begin{cases} h(x, u, f(x, u)) & \text{if } R_1(x, u), \\ j(x, u, f(x, u)) & \text{if } R_2(x, u), \\ \dots\dots\dots\dots\dots\dots \\ k(x, u, f(x, u)) & \text{if } R_m(x, u), \end{cases}$

where the relations $R_1(x, u)$, $R_2(x, u)$, ..., $R_m(x, u)$ are mutually exclusive and cover all the possible cases.

Since, as has been noted in the preceding section, conditional definitions reduce to superposition, conditional recursion reduces to simple recursion. Let, for instance, g_i be a characteristic function of the relation R_i; then the function f can be defined by simple recursion as follows:

$$f(0, u) = g(u),$$
$$f(x+1, u) = h(x, u, f(x, u)) \cdot 0^{g_1(x, u)} +$$
$$j(x, u, f(x, u)) \cdot 0^{g_2(x, u)} + \dots +$$
$$k(x, u, f(x, u)) \cdot 0^{g_m(x, u)}.$$

2. RECURSION WITH k INITIAL FIXED VALUES. This term applics to inductive definitions which fall under the schema:

(50)   a. $\begin{cases} f(0, u) = g_0(u), \\ f(1, u) = g_1(u), \\ \dots\dots\dots\dots \\ f(k, u) = g_k(u), \end{cases}$

    b. *For* $x \geqslant k, f(x+1, u) = h(x, u, f(x, u))$.

Recursion which complies with schema (50) can easily be reduced to conditional recursion (49). The function f can be defined as follows:

    a. $f(0, u) = g_0(u)$,

    b. $f(x+1, u) = \begin{cases} g_1(u) & \text{if } x = 0, \\ \dots\dots\dots\dots\dots\dots \\ g_k(u) & \text{if } x = k-1, \\ h(x, u, f(x, u)) & \text{if } x \geqslant k. \end{cases}$

3. RECURSION WITH RESPECT TO A COURSE OF VALUES. When constructing inductive definitions we sometimes make the value of a function for the next number depend not on its value for the preceding number, but on its still earlier values, for example,

$$f(0, u) = g(u),$$
$$f(x+1, u) = h(x, u, f(x \dotdiv 1, u), f(x \dotdiv 2, u)).$$

Generally speaking, we may make the value of a function $f(x+1, u)$ depend on the value of a function $f(j(x), u)$ if $j(x) \leqslant x$. The operation of recursion with respect to a course of values states, in its general form, that if $j_1, ..., j_k$ are functions in Prec which satisfy the inequality $j_i(x) < x$, and if g and h also are in Prec, then a function f, defined by the conditions

(51)
a. $f(0, u) = g(u),$

b. $f(x+1, u) = h\big(x, u, f(j_1(x), u), ..., f(j_k(x), u)\big)$

also is in Prec.

Recursion with respect to a course of values reduces to simple recursion: for the function f, as defined by schema (51), we can easily define its diagram $\tilde{f}$ by using simple recursion, since the diagram $\tilde{f}$ satisfies the conditions

$$\tilde{f}(0, u) = 2^{g(u)},$$
$$\tilde{f}(x+1, u) = \tilde{f}(x, u) \cdot p_{x+1}^{h(x, u, \exp(\tilde{f}(x,u), p_{j_1}(x)), ..., \exp(\tilde{f}(x,u)p_{j_k}(x)))}.$$

Next the function f itself is defined:

$$f(x, u) = (\tilde{f}(x, u))_x.$$

4. RECURSION PROGRESSING EVERY $k+1$ NUMBERS. This term applies to inductive definitions which comply with the schema:

(52)   a.
$$\begin{cases} f(0, u) = g_0(u), \\ f(1, u) = g_1(u), \\ \dots\dots\dots\dots \\ f(k, u) = g_k(u), \end{cases}$$

b. $f(x+k+1, u) =$
$\quad h(x, u, f(x, u), f(x+1, u), ..., f(x+k, u)).$

Recursion which complies with schema (52) reduces to schema (50) of recursion with k initial fixed values, combined with recursion with respect to a course of values, since condition b of (52) can be written as follows:

For $x \geqslant k$,

$$f(x+1, u) =$$

$$h\big(x, u, f(x \div k, u), f((x \div k)+1, u), ..., f(x, u)\big).$$

The functions $x \div k$, $(x \div k)+1$, $(x \div k)+2$, ..., $(x \div k)+n$ for $n \leqslant k$ of course satisfy the condition $(x \div k)+n \leqslant x$.

5. SIMULTANEOUS RECURSION. Sometimes an induction schema is used to define several functions simultaneously. For instance, two functions, f and F, are defined by simultaneous recursion by means of functions g, h, G, H, if the following conditions are satisfied:

(53)

a. $\begin{cases} f(0, u) = g(u), \\ F(0, u) = G(u), \end{cases}$

b. $\begin{cases} f(n+1, u) = h\big(n, u, f(n, u), F(n, u)\big), \\ F(n+1, u) = H\big(n, u, f(n, u), F(n, u)\big). \end{cases}$

The functions f and F can then easily be defined if a function $j(x, y)$, which is the pairing of the functions f and F, is defined by simple recursion:

$$j(0, u) = 2^{g(u)} \cdot 3^{G(u)},$$

$$j(n+1, u) = 2^{h\big(n, u, (j(n, u))_0, (j(n, u))_1\big)} \cdot 3^{H\big(n, u, (j(n, u))_0, (j(n, u))_1\big)}.$$

Next the functions f and F are defined as follows:

$$f(n, u) = \big(j(n, u)\big)_0,$$

$$F(n, u) = \big(j(n, u)\big)_1.$$

There are other, still more complicated, recursion schemata, for example, *nested recursion*, which also reduce to simple recursion (Peter [60]–[62]). Other results obtained take on the form of theorems stating that the *general schema of simple recursion is equivalent to simpler schemata*, (if based on properly selected initial functions and the operation of superposition). The simplest schema is that of *iteration* (R. M. Robinson [76]). It applies to functions of one argument only and makes it possible to form, of a function g, a function f which is defined by induction in the following manner:

(54)

$$f(0) = 0,$$

$$f(x+1) = g\big(f(x)\big).$$

In an informal notation we may say that

$$f(x) = \underbrace{g \ldots g(0)}_{n} = g^n(0).$$

Schema (54) is a narrower schema of recursion, but nevertheless, together with superposition and the appropriate initial functions, it can yield all the functions of the class Prec. It is also known that the class Prec is essentially richer than the class Erec. These results, which are mentioned here without proofs, will not be referred to below. But a detailed proof will be given of the fact that the class Prec is essentially narrower than the class Comp. It will be demonstrated that the effective minimization operation makes it possible to define a universal function for the class of primitive recursive functions. In order to define a universal function for Prec we shall first prove the following theorem.

THEOREM 9. *The class* Prec_2 *of primitive recursive functions of at most two arguments is the least class of functions which contains the functions* 0, $I(x) = x$, $S(x)$, p_y^x, $(x)_y$, $x \cdot y$, $x+y$, *and is closed under the following operations*:

a. *superposition (restricted to substitutions which do not lead outside functions of at most two arguments)*;

b. *simple recursion: if* g, $h \in \text{Prec}_2$, *and*

$$(55) \quad \begin{aligned} f(0, y) &= g(y), \\ f(n+1, y) &= h(\langle n, y \rangle, f(n, y)), \end{aligned}$$

then $f \in \text{Prec}_2$.

PROOF. Assume Theorem 9 to be an inductive definition of the class which has conventionally been termed Prec_2. It will be demonstrated that the class Prec_2 thus defined is identical with the class of primitive recursive functions of at most two arguments. The fact that all functions in Prec_2 are primitive recursive follows directly from the conditions imposed upon Prec_2. To prove the converse inclusion we have to demonstrate that for every primitive recursive function $f(x_0, \ldots, x_n)$ of any number of arguments there exists a function $f' \in \text{Prec}_2$ such that

$$(56) \quad f(x_0, \ldots, x_n) = f'(\langle x_0, \ldots, x_n \rangle).$$

This condition is self-evident for the initial functions of Prec, since $\langle x \rangle = 2^x = p_0^x$. It is only to be checked that it is inherited under superposition and simple recursion.

If for functions f and h there exists functions f' and h' such that condition (56) holds and

$$h(y_0, \ldots, y_m) = h'(\langle y_0, \ldots, y_m \rangle),$$

then a function α, defined by superposition

$$\alpha(x_0, \ldots, x_{i-1}, y_0, \ldots, y_m, x_{i+1}, \ldots, x_n) =$$
$$f(x_0, \ldots, x_{i-1}, h(y_0, \ldots, y_m), x_{i+1}, \ldots, x_n)$$

has its counterpart α' in Prec_2, which is defined as follows:

$$(57) \quad \alpha'(z) = f'\left(\langle (z)_0, (z)_1, \ldots, (z)_{i-1} \rangle \cdot p_i^{h'(\langle (z)_i, (z)_{i+1}, \ldots, (z)_{i+m} \rangle)} \times \right.$$
$$\left. \times p_{i+1}^{(z)_{i+m+1}} \cdot p_{i+2}^{(z)_{i+m+2}} \cdot \ldots \cdot p_n^{(z)_{m+n}} \right).$$

The function α' which satisfies formula (57) can easily be defined by the functions $f', h', (z)_i, p_i^n, \cdot, +, S(x)$ and 0 without going outside the operations permitted in the class Prec_2. On replacing z in (57) by the complex $\langle x_0, \ldots, x_{i-1}, y_0, \ldots, y_m, x_{i+1}, \ldots, x_n \rangle$ we obtain the analogue of (56) for the functions α and α'.

Likewise, if the function f is obtained from the functions g and h by the application of schema (4) of simple recursion, then its analogue f' can easily be defined by schema (55) by means of g' and h'. We first define

$$H(x, y) = h'\left(\langle (x)_0, (x)_1, y \rangle\right),$$

next

$$F(0, y) = g'(\langle y \rangle),$$
$$F(n+1, y) = H\left(\langle n, y \rangle, F(n, y)\right),$$

and finally

$$f'(u) = F\left((u)_0, (u)_1\right).$$

If $h'(\langle x, y, z \rangle) = h(x, y, z)$ and $g'(\langle y \rangle) = g(y)$, then we can demonstrate easily that $f'(\langle x, y \rangle) = f(x, y)$.

We have, thus, proved by induction that for every primitive recursive function f there exists an analogue f' in Prec_2.

Suppose now that f is a primitive recursive function of two arguments. As such it has its analogue $f' \in \mathrm{Prec}_2$ such that

$$(58) \quad f(x, y) = f'(\langle x, y \rangle).$$

As multiplication and p_y^x are in Prec_2, Prec_2 also contains the function $\langle x, y \rangle = 2^x \cdot 3^y$, and, hence, also the function f, which is obtained

433

from f' and $\langle x, y \rangle$ by superposition in accordance with (58). The class Prec_2, defined by induction in Theorem 9, essentially coincides with the class of primitive recursive functions of at most two arguments.

THEOREM 10. *There exists a function $G_n(x, y)$ which is universal for the functions* Prec_2 *and computable.*

PROOF. As the functions Prec_2 form a class defined by induction, we can enumerate all these functions, beginning with the initial ones. Such an operation is described, for instance, by the following formulae:

1) I n i t i a l   c o n d i t i o n s:

$$G_0(x, y) = 0,$$
$$G_1(x, y) = x,$$
$$G_2(x, v) = x+1,$$
$$G_3(x, y) = p_y^x,$$
$$G_4(x, y) = (x)_y,$$
$$G_5(x, y) = x \cdot y,$$
$$G_6(x, y) = x+y.$$

2) I n d u c t i o n   c o n d i t i o n s: For all $n \geqslant 6$
A. if $(n)_2 < 3$:

$$G_{n+1}(x, y) = \begin{cases} G_{(n)_0}(x, y) & \text{if } (n)_2 = 0, \\ G_{(n)_0}(x, x) & \text{if } (n)_2 = 1, \\ G_{(n)_0}(x, G_{(n)_1}(x, y)) & \text{if } (n)_2 = 2; \end{cases}$$

B. if $(n)_2 \geqslant 3$:

$$G_{n+1}(0, y) = G_{(n)_0}(y, y),$$
$$G_{n+1}(x+1, y) = G_{(n)_1}(\langle x, y \rangle, G_{n+1}(x, y)).$$

The initial functions of the class Prec_2 are obviously included in the enumeration. If the enumeration includes functions G_k and G_l, then, by formula A of the induction condition, the superposition $G_k(x, G_l(x, y))$ is a function G_m with the number $m = 2^k \cdot 3^l \cdot 5^n+1$ for every $n \geqslant 3$. The remaining formulae of condition A define functions obtained by identification of arguments and reversal of the order of variables. It can easily be seen that these operations make it possible to number all the possible cases of superposition in the class Prec_2. The function obtained from G_k and G_l by the application of the schema of simple recursion is given the numbers $2^k \cdot 3^l \cdot 5^n+1$ all $n \geqslant 3$. It can easily be inferred that not only functions obtained by recursion, but all $Prec_2$ functions in general recur in that enumeration infinitely many times.

Since for every number $n \geqslant 6$ the functions $(n)_0$, $(n)_1$, $(n)_2$ are defined and take on values which are less than n, the function G_{n+1} is, under induction conditions A and B, determined unambiguously by means of functions bearing earlier numbers. Hence, also $G_n(x, y)$, as a function of three arguments, is unambiguously defined by induction for all n, x, y, namely by the definition of the sequence $\{G_n\}$ as written above. As it numbers all the functions in $Prec_2$, it is a universal function for the class Prec.

The function $G_n(x, y)$ is computable in the intuitive sense of the term: we know how to compute the initial functions of the sequence $\{G_n\}$, and the computation of the next functions reduces to the computation of certain precisely indicated previous functions in the sequence $\{G_n\}$.

By Theorem 8, the function G, being a universal function for Prec, is not in Prec. It is, however, in Comp. A rigorous proof of the computability of the function G consists in defining that function properly by the effective minimization operation. The above inductive definition of the function G cannot, of course, be subsumed under the schema of simple recursion. If it could be, G would be in Prec, contrary to Theorem 8. The inductive schema which defines G is a schema of double recursion: in condition B recursion goes not only from n to $n+1$, but simultaneously from x to $x+1$. Definitions of that kind, except for very simple cases, usually are not reducible to single recursion; they are, however, always reducible to the effective minimization operation, and that in a way which resembles very much the reduction of simple recursion to the effective minimization operation. In the proof of Theorem 1 recursion was reduced to the minimum operation by describing a number m, which was the diagram $m = \tilde{f}(n)$ of a function f defined by recursion; in the present case, too, we can describe a number m, which is, as it were, a diagram of the function $G_n(x, y)$. Since G is a function of three arguments, the number m must describe an entire three-dimensional table. Of course, a single number m is used to describe only a table which is finite in every dimension.

For a fixed y, the cross-section of that table is a two-dimensional table of the form

$$G_0(0, y), G_0(1, y), G_0(2, y), ..., G_0(u_0, y),$$
$$G_1(0, y), G_1(1, y), G_1(2, y), ..., G_1(u_1, y),$$
$$G_2(0, y), G_2(1, y), G_2(2, y), ..., G_2(u_2, y),$$
$$\cdots\cdots\cdots\cdots\cdots\cdots\cdots\cdots\cdots\cdots\cdots$$
$$G_n(0, y), G_n(1, y), G_n(2, y), ..., G_n(u_n, y).$$

A number m will be said to *represent* a table of a function G if for every triple $\langle n, x, y \rangle$ of co-ordinates from that table the formula

(59) $(m)_{\langle n, x, y \rangle} = G_n(x, y)+1$

holds. (The number 1 is added to the value of $G_n(x, y)$ in order to distinguish between those co-ordinates for which the value of G occurs in the table and equals zero, and those for which the value of G does not occur in the table represented by m.)

In order to compute the value of $G_n(x, y)$ on the strength of its inductive definition for a given triple $\langle n, x, y \rangle$ we have to know the value of G for many other triples of numbers. Inductive formulae make the value of $G_n(x, y)$ dependent on certain values of earlier n's, or on value of G for the same n, but for earlier x's. A table will be termed *closed* if for every triple $\langle n, x, y \rangle$ of co-ordinates from that table all those values of G which are necessary for computing the value of $G_n(x, y)$ from inductive formulae occur in that table. The property

$W(m) \equiv m$ represents a closed table of a function G

can be described by means of bounded quantifiers as a primitive recursive relation:

$$W(m) \equiv \bigwedge n, x, y \leqslant m\Big((m)\langle n, x, y\rangle \neq 0 \rightarrow$$
$$\Big((n = 0 \rightarrow (m)\langle n, x, y\rangle = 1)\wedge$$
$$(n = 1 \rightarrow (m)\langle n, x, y\rangle = x+1)\wedge$$
$$(n = 2 \rightarrow (m)\langle n, x, y\rangle = x+2)\wedge$$
$$(n = 3 \rightarrow (m)\langle n, x, y\rangle = p_y^x+1)\wedge$$
$$(n = 4 \rightarrow (m)\langle n, x, y\rangle = (x)_y+1)\wedge$$
$$(n = 5 \rightarrow (m)\langle n, x, y\rangle = x \cdot y+1)\wedge$$
$$(n = 6 \rightarrow (m)\langle n, x, y\rangle = x+y+1)\Big)\Big)\wedge$$
$$\bigwedge x, y, n \leqslant m\Big((m)\langle n+1, x, y\rangle \neq 0 \rightarrow$$
$$\Big((n \geqslant 6 \wedge (n)_2 < 3) \rightarrow$$

$$\Big(\big((n)_2 = 0 \to (m)\langle n+1, x, y\rangle = (m)\langle (n)_0, y, x\rangle\big) \wedge$$
$$\big((n)_2 = 1 \to (m)\langle n+1, x, y\rangle = (m)\langle (n)_0, x, x\rangle\big) \wedge$$
$$\big((n)_2 = 2 \to ((m)\langle (n)_1, x, y\rangle \neq 0 \wedge$$
$$(m)\langle n+1, x, y\rangle =$$
$$(m)\langle (n)_0, x, (m)\langle (n)_1, x, y\rangle \dotminus 1\rangle)\big)\Big)\Big) \wedge$$

$$\bigwedge x, y, n \leqslant m \Big(\big((n \geqslant 6 \wedge (n)_2 \geqslant 3\big) \to$$
$$\Big(\big((m)\langle n+1, 0, y\rangle \neq 0 \to$$
$$(m)\langle n+1, 0, y\rangle = (m)\langle (n)_0, y, y\rangle\big) \wedge$$
$$\big((m)\langle n+1, x+1, y\rangle \neq 0 \to ((m)\langle n+1, x, y\rangle \neq 0 \wedge$$
$$(m)\langle n+1, x+1, y\rangle = (m)\langle (n)_1, \langle x, y\rangle,$$
$$(m)\langle n+1, x, y\rangle \dotminus 1\rangle)\big)\Big)\Big)\Big).$$

In this formula we write, for typographical reasons, $(m)\langle n, x, y\rangle$ instead of $(m)_{\langle n, x, y\rangle}$. Note that $(m)_{\langle n, x, y\rangle} = \exp(m, p_{\langle n, x, y\rangle})$.

As can easily be seen, the above definition records the conditions which inductively define the function G, the only difference being that instead of the formula $G_n(x, y)$, which occurs in the inductive definition of G, the definition of the relation W has the formula $(m)\langle n, x, y\rangle$, which describes a point m in the table which has the co-ordinates n, x, y. An inductive proof of the implication

(60) $\big(W(m) \wedge (m)_{\langle n, x, y\rangle} \neq 0\big) \to (m)_{\langle n, x, y\rangle} = G_n(x, y)+1$

is self-evident in the light of the explanations given above. Since it follows from the properties of a function $(x)_i$ that

(61) $(m)_{\langle n, x, y\rangle} \neq 0 \to n, x, y \leqslant m$,

for $n \leqslant 6$ conditional equation (60) follows self-evidently from the definitions of the function G and the relation W.

Assume now that (60) is true for all $k \leqslant n$ ($n \geqslant 6$) and for all x, y; we shall see that it must also be true for $n+1$. By the definition of the function G for $n \geqslant 6$, the function G_{n+1} is obtained, either by superposition, if $(n)_2 < 3$, or by simple recursion, if $(n)_2 \geqslant 3$, from the functions $G(n)_0$ and $G(n)_1$. Since $(n)_0 \leqslant n$ and $(n)_1 \leqslant n$, (60) is true for $(n)_0$ and $(n)_1$ by the induction assumption. This, (61), the analysis of the various conditions involved, and the comparison of the definitions of the function G and the relation W, taken together, clearly yield equation (60) for $n+1$.

Since only a finite number of values of earlier functions $G_{(n)_1}$, $G_{(n)_2}$ is required to compute a value of $G_{n+1}(x, y)$, for every triple of the numbers n, x, y there exists a number m such that $W(m)$ and $(m)_{\langle n, x, y \rangle} \neq 0$. This statement is a condition of effectiveness

$$(62) \qquad \bigwedge n, x, y \bigvee m\big(W(m) \wedge (m)_{\langle n, x, y \rangle} \neq 0\big).$$

A more precise proof is by induction. Note first that the following relationship holds:

$$(63) \qquad \begin{aligned} &\big(W(m_1) \wedge (m_1)\langle n, x, y \rangle \neq 0 \wedge W(m_2) \wedge (m_2)\langle n, x, y \rangle \neq 0\big) \to \\ &(m_1)\langle n, x, y \rangle = (m_2)\langle n, x, y \rangle. \end{aligned}$$

It follows directly from formula (60), for if for any co-ordinates n, x, y from the table m_1: $(m_1)\langle n, x, y \rangle = g_n(x, y)+1$, and analogically for the table m_2: $(m_2)\langle n, x, y \rangle = g_n(x, y)+1$, then obviously $(m_1)\langle n, x, y \rangle = (m_2)\langle n, x, y \rangle$.

Relationship (63) states that two tables coincide as to those points which simultaneously belong to both tables. Hence, the least common multiple $[m_1, m_2]$ of two tables is a table which is the least common extension of m_1 and m_2. For if $([m_1, m_2])\langle n, x, y \rangle \neq 0$, then either $([m_1, m_2])\langle n, x, y \rangle = (m_1)\langle n, x, y \rangle$, or $([m_1, m_2])\langle n, x, y \rangle = (m_2)\langle n, x, y \rangle$. Hence, by the definition of the property W, if $W(m_1)$ and $W(m_2)$, then $W([m_1, m_2])$.

Formula (62) is proved by induction with respect to n. For $n \leqslant 6$, the following values are adopted for the number m:

$$\text{for } n = 0, \; m = p^1_{\langle 0, x, y \rangle};$$
$$\text{for } n = 1, \; m = p^{x+1}_{\langle 1, x, y \rangle};$$
$$\text{for } n = 2, \; m = p^{x+2}_{\langle 2, x, y \rangle}; \; ...;$$
$$\text{for } n = 6, \; m = p^{x+y+1}_{\langle 6, x, y \rangle}.$$

Assume now that for all $k \leqslant n$ $(n \geqslant 6)$ and for all x, y there exist m such that $W(m)$ and $(m)_{\langle k, x, y \rangle} \neq 0$. Let the least m which satisfies the conditions specified above be denoted by $\mu(k, x, y)$. For $n+1$ and for given x and y the number m is constructed as follows. If $(n)_2 = 0$, then let $m_1 = \mu\big((n)_0, x, x\big)$; if so, then for that m which satisfies the conditions $W(m)$ and $(m)\langle n+1, x, y \rangle \neq 0$ we take

$$(64) \qquad m = m_1 \cdot p^{(m_1)\langle (n)_0, y, x \rangle}_{\langle n+1, x, y \rangle}.$$

If $(n)_2 = 1$, then as the appropriate m we take

$$m = m_1 \cdot p_{\langle n+1, x, y \rangle}^{(m_1)\langle (n)_0, x, x \rangle},$$

where $m_1 = \mu((n)_0, x, x)$. If $(n)_2 = 2$, then we take

$$m = m_1 \cdot p_{\langle n+1, x, y \rangle}^{(m_1)\langle (n)_0, x, (m_1)\langle (n)_1, x, y \rangle \dot- 1 \rangle},$$

where $m_1 = \left\{ \mu((n)_1, x, y), \mu\big[(n)_0, x, \big(\mu((n)_1, x, y)\big) \langle (n)_1, x, y \rangle \dot- 1\big] \right\}$.
If $(n)_2 \geqslant 3$, then for $x = 0$ we take

$$m = m_1 \cdot p_{\langle n+1, 0, y \rangle}^{(m_1)\langle (n)_0, y, y \rangle},$$

where $m_1 = \mu((n)_0, y, y)$. For $x = z + 1$ we put

(65) $$m = m_1 \cdot p_{\langle n+1, z+1, y \rangle}^{(m_1)\langle (n)_1; \langle z, y \rangle, (m_1)\langle n+1, z, y \rangle \dot- 1 \rangle},$$

where $m_1 = [\mu(n+1, z, y), \mu((n)_1, \langle z, y \rangle, \mu(n+1, z, y))(\langle n+1, z, y \rangle \dot- 1)]$.

In each of the cases specified above it can easily be checked that the number m, as constructed in the formulae above, satisfies the conditions: $W(m)$ and $(m)\langle n, x, y \rangle \neq 0$.

By formula (60), the function G can be defined by reference to effective minimization as follows:

(66) $$G_n(x, y) = \big((\mu m) \left[W(m) \wedge (m)_{\langle n, x, y \rangle} \neq 0 \right]\big)_{\langle n, x, y \rangle} \dot- 1.$$

The function G was initially defined by double recursion. It is easy to give examples of triple recursion, which leads outside double recursion. There exists an infinite sequence of increasingly strong operations of multiple recursion (Peter [61], R. M. Robinson [77]). Every function defined by such operations is computable and can be defined by a single application of the effective minimization operation, in a way similar to that in which the function G was defined above. There are also well-known cases of computable functions which cannot be defined by any schema of multiple recursion.

Theorem on normal form

The computable function G proved definable by a single application of the operation of effective minimization. It will be proved now that this is a general phenomenon.

THEOREM 11 (on normal form, Kleene [34]). *For every function* $f \in$ Comp *there exist relations* $R, R', R'' \in$ Erec *such that*

(67) $f(\mathfrak{a}) = n \equiv \bigvee x R'(\mathfrak{a}, x, n) \equiv \bigwedge x R''(\mathfrak{a}, x, n),$

where $\mathfrak{a}$ *stands for a complex of variables* $a_1, \ldots, a_k,$ *and*

(68) $f(\mathfrak{a}) = ((\mu y) [R(\mathfrak{a}, y)])_0,$

for which the effectiveness condition

(69) $\bigwedge \mathfrak{a} \bigvee y R(\mathfrak{a}, y)$

is satisfied.

PROOF.

a. Representability of (67) will first be proved by induction.

1. For the initial functions of the class Comp condition (67) is satisfied trivially. For instance, for raising to a power

$$a^b = n \equiv \bigvee x a^b = n \equiv \bigwedge x a^b = n.$$

2. Suppose now that condition (67) is satisfied for two functions, f and g, so that in addition to formula (67) the formula

(70) $g(n) = b \equiv \bigvee y S'(n, y, b) \equiv \bigwedge y S''(n, y, b)$

is true, too. It will be proved that the superposition $g(f(\mathfrak{a})) = b$ also is representable in the desired way. That superposition can be described as follows:

(71) $g(f(\mathfrak{a})) = b \equiv \bigvee n (f(\mathfrak{a}) = n \wedge g(n) = b),$

(72) $g(f(\mathfrak{a})) = b \equiv \bigwedge n (f(\mathfrak{a}) = n \rightarrow g(n) = b).$

If the superposition in question is to be shown in its existential form, we make use of formula (71) and, in accordance with (67), we present the relations $f(\mathfrak{a}) = n$ and $g(n) = b$ as an existential sentences, which yields the formula

(73) $g(f(\mathfrak{a})) = b \equiv \bigvee n, x, y (R'(\mathfrak{a}, x, n) \wedge S'(n, y, b)).$

By making use of the pairing function we can, of course, replace the variables n, x, y by a single variable:

(74) $g(f(\mathfrak{a})) = b \equiv \bigvee z \big(R'(\mathfrak{a}, (z)_1, (z)_0) \wedge S'((z)_0, (z)_2, b) \big).$

If we want to present the superposition as a sentence which begins with the universal quantifier, we make use of (72) and select, from conditions (67) and (70), the appropriate substitutions for the relations $f(\mathfrak{a}) = n$ and $g(n) = b$:

$$(75) \qquad g(f(\mathfrak{a})) = b \equiv \bigwedge n(\bigvee xR'(\mathfrak{a}, x, n) \rightarrow \bigwedge yS''(n, y, b)).$$

In formula (75) all the quantifiers may be transferred to the beginning of the consequent as universal quantifiers, and then replaced by a single quantifier by making use of the pairing function again. This yields the equivalence

$$(76) \qquad g(f(\mathfrak{a})) = b \equiv \bigwedge z \left(R'\left(\mathfrak{a}, (z)_1, (z)_0\right) \rightarrow S''\left((z)_0, (z)_2, b\right)\right).$$

Formulae (74) and (76) show that the result of superposition can be presented in the desired form, for if R', R'', S', $S'' \in$ Erec, then Erec also contains the relations that follow the quantifier in formulae (74) and (76).

3. It remains to demonstrate that the effective minimization operation also does not lead outside the functions representable in the form (67). To do so assume that a function f is representable in the desired form, that is, that

$$(77) \qquad f(\mathfrak{a}, z) = b \equiv \bigvee xR'(\mathfrak{a}, z, x, b) \equiv \bigwedge xR''(\mathfrak{a}, z, x, b),$$

and that the effectiveness condition

$$(78) \qquad \bigwedge \mathfrak{a} \bigvee zf(\mathfrak{a}, z) = 0$$

is satisfied. It will be demonstrated that a function g, defined as follows:

$$(79) \qquad g(\mathfrak{a}) = (\mu z) [f(\mathfrak{a}, z) = 0],$$

also is representable in the form of (67). In fact, it follows from (78) and (79) that

$$(80) \qquad g(\mathfrak{a}) = n \equiv (f(\mathfrak{a}, n) = 0 \wedge \bigwedge z < n f(\mathfrak{a}, z) \neq 0).$$

Formulae (77) and (80) imply the formula

$$(81) \qquad g(\mathfrak{a}) = n \equiv (\bigwedge xR''(\mathfrak{a}, n, x, 0) \wedge$$
$$\bigwedge z < n \bigwedge x \sim R'(\mathfrak{a}, z, x, 0)).$$

All the universal quantifiers can, of course, be transferred to the beginning of the representing formula, and can next be replaced by one universal quantifier by the application of the pairing function. In this way the relation $g(\mathfrak{a}) = n$ is represented in the form with the universal quantifier at the beginning. If we want to have a representation by means of the existential quantifier, we deduce, from (80) and (77), the formula

$$(82) \qquad g(\mathfrak{a}) = n \equiv \left(\bigvee x R'(\mathfrak{a}, n, x, 0) \wedge \right.$$
$$\left. \bigwedge z < n \bigvee x \sim R''(\mathfrak{a}, z, x, 0) \right).$$

To this formula we apply the following generally valid equivalence:

$$(83) \qquad \bigwedge z < n \bigvee x R(z, x) \equiv \bigvee y \bigwedge z < n R(z, (y)_z).$$

In order to prove (83) we select as y the number $y = \langle x_0, \ldots, x_{n-1} \rangle$, and we assume for $z < n$

$$x_z = (\mu x) [R(z, x)].$$

The converse implication is obvious. By applying (83) to (82) we have

$$(84) \qquad g(\mathfrak{a}) = n \equiv \left(\bigvee x R'(\mathfrak{a}, n, x, 0) \wedge \right.$$
$$\left. \bigvee y \bigwedge z < n \sim R''(\mathfrak{a}, z, (y)_z, 0) \right).$$

The existential quantifiers $\bigvee x$ and $\bigvee y$ can be replaced by one, which thus yields a representation of the relation $g(\mathfrak{a}) = n$ in a form in which the existential quantifier occurs at the beginning. Since the sentential calculus operations and bounded quantifiers do not lead outside the relations of the class Erec, if R', $R'' \in$ Erec, then also the relations which follow the existential quantifier in the expression resulting from (84) and the universal quantifier in (81) are in Erec. The property of representability in the form (67) is, thus, an attribute of all computable functions.

b. Representability in the form of (68) is obtained directly from (67). Since the function f is defined for all natural numbers, formula (67) implies

$$(85) \qquad \bigwedge \mathfrak{a} \bigvee n, x R'(\mathfrak{a}, x, n),$$

where, for a given $\mathfrak{a}$, the number n is uniquely determined, since that number is the value of the function $f(\mathfrak{a})$. Formula (85) is transformed into an effectiveness condition

$$(86) \qquad \bigwedge \mathfrak{a} \bigvee y R'(\mathfrak{a}, (y)_1, (y)_0),$$

where, for a given $\mathfrak{a}$, the element $(y)_0$ is determined uniquely, since by (67) it equals the value of $f(\mathfrak{a})$. Hence, the function f satisfies the equation

(87) $f(\mathfrak{a}) = \big((\mu y) \, [R'(\mathfrak{a}, (y)_1, (y)_0)]\big)_0$,

which was to be proved.

The form (68) is called the *normal* (or *canonical*) *form of a computable function*.

The theorem on normal form makes it possible to easily construct a universal function for computable functions.

THEOREM 12. *A function* $F_n(x)$, *defined as follows*:

(88) $F_n(x) = \big((\mu y) \, [G_n(x, y) = 0]\big)_0$,

where G *is a universal function for the class* Prec *or for the class* Erec, *is a universal function for the class* Comp.

PROOF. By Theorem 11, for every $f \in$ Comp there exists a relation $R \in$ Erec such that $f(x) = \big((\mu y) \, [R(x, y)]\big)_0$. For the relation R there exists a characteristic function $g \in$ Erec such that $\bigwedge x, y \, (R(x, y) \equiv g(x, y) = 0)$. For the function g there is a number n such that $\bigwedge x, y \, g(x, y) = G_n(x, y)$, where G_n is the universal function for primitive recursive functions, as defined in Theorem 10. Hence, by substituting $G_n(x, y) = 0$ in (67) for $R(x, y)$, we arrive at the result that the function $f(x)$ can, in accordance with (88), be represented as $F_n(x)$.

The function $F_n(x)$, being universal for the class Comp, obviously is not in Comp, as follows from Theorem 8. This provides a simple example of a non-computable function. Since it is defined by reference to a minimization, that minimization must not be effective. The function G is computable, as is also the function $(x)_0$. Hence, if that minimization were effective, F would have to be computable. Thus, the effectiveness condition must not be satisfied. There must even be infinitely many n's and x's for which there does not exist any y such that $G_n(x, y) = 0$. Thus, the function F is universal for the class Comp in the broader sense of the term, but is not universal in the narrower sense.

Comments on the classification of non-computable relations

Now that we are outside the class Comp we face the problem of adopting a classification of non-computable sets and functions. We have arrived at the concept of computability by looking for those functions and sets which are given in the most effective way, that is, those which come closest to empirical mathematical data. When handling non-computable constructs we wish to be able to classify them from the same point of view, that is, as to the effectiveness of construction. Thus, the classification will begin with computable constructs, and its further steps will consist of decreasingly effective constructs of the arithmetic of natural numbers.

As can be seen in the case of the function F, universal for the class Comp, the transition from computable to non-computable functions takes place at the application of the non-effective minimization operation. The minimization operation for functions has its analogue in the field of relations in the quantifier operation. Thus, the relations which follow the computable relations will be those which are obtained by the application of the universal or existential quantifier operation to computable relations. When outside the sphere of computability we are usually more interested in relations and sets, than in functions. Hence the further steps of the classification will be described for sets. Of course, we can always make use of the pairing function in order to pass from sets to relations of arbitrarily many arguments.

Let R, R', R'' be any computable relations holding between natural numbers. The following predicates from the arithmetic of natural numbers define certain sets which belong to increasingly high classes of the hierarchy which we have in mind:

$$A'(x) \equiv \bigvee y \in \mathcal{N} \; R(y, x),$$

$$A''(x) \equiv \bigwedge z \in \mathcal{N} \bigvee y \in \mathcal{N} \; R'(y, z, x),$$

$$A'''(x) \equiv \bigvee v \in \mathcal{N} \bigwedge z \in \mathcal{N} \bigvee y \in \mathcal{N} R''(y, z, v, x),$$

$$\cdots\cdots\cdots\cdots\cdots\cdots\cdots\cdots\cdots\cdots\cdots\cdots\cdots\cdots\cdots\cdots$$

$$B'(x) \equiv \bigwedge y \in \mathcal{N} R(y, x),$$

$$B''(x) \equiv \bigvee z \in \mathcal{N} \bigwedge y \in \mathcal{N} R'(y, z, x),$$

$$B'''(x) \equiv \bigwedge v \in \mathcal{N} \bigvee z \in \mathcal{N} \bigwedge y \in \mathcal{N} R''(y, z, v, x),$$

$$\cdots\cdots\cdots\cdots\cdots\cdots\cdots\cdots\cdots\cdots\cdots\cdots\cdots\cdots\cdots\cdots$$

As can be seen, the hierarchy under discussion branches into two according to whether the first quantifier applied to a computable relation is universal or existential. In both branches the sets of the next class are formed from the preceding relation by the alternate application of the operations of the universal and the existential quantifier.

It is easy to comprehend intuitively that the quantifier operations are those which lead from more effective to less effective concepts. For instance, if we want to check whether $B'(x)$ holds for a given number x we must, by proceeding in the most natural way, examine all the natural numbers $y = 0, 1, \ldots$ and to see whether $R(y, x)$ holds for each of them. Thus, it is not possible to make sure in any finite period of time whether the relation $B'(x)$ holds. The definition of $B'(x)$ is fairly simple, but when it comes, for example, to $B'''(x)$, we must, for every $v \in \mathcal{N}$, look for a $z \in \mathcal{N}$ such that for it in turn $R(y, z, v, x)$ should hold for every $y \in \mathcal{N}$. Thus, the set $\mathcal{N}$ of natural numbers must be examined an infinite number of times. A single checking of a property for all natural numbers is, of course, a non-effective operation, and each quantifier necessitates the repetition of the entire procedure an infinite number of times. This is why the number of the quantifiers used in a definition of a concept is essentially a measure of the non-effectiveness of that concept.

This classification applies, obviously, only to concepts definable by arithmetical formulae in the domain of natural numbers. But it may be assumed that all concepts of classical mathematics originate from arithmetic, and so this classification covers all mathematical constructs. It will be studied in greater detail in Section 6.

We have referred above to: concepts definable in a domain by means of certain formulae. It is worth while noting that this is a different concept of definability than that considered in Chapter II, Section 6. The general definition of the concept now used is as follows:

A formula $A(x_1, \ldots, x_n)$ defines, in a domain $\mathfrak{M} = \langle X, q_1, \ldots, q_m \rangle$, a relation r of n arguments, for the interpretation of the terms $Q_1, \ldots, Q_m$ which occur in the formula A as names of the relations $q_1, \ldots, q_m$, if and only if for all $t_1, \ldots, t_n \in X$ the equivalence

(i) $\qquad r(t_1, \ldots, t_n) \equiv \{t_1, \ldots, t_n\}$ *satisfies A in $\mathfrak{M} = \langle X, q_1, \ldots, q_m \rangle$*

holds for the said interpretation of terms.

The relation R is then called definable in the domain $\mathfrak{M}$ by the formula A.

For instance, the *less than* relation in the domain of natural numbers is definable by the formula

$$\ulcorner \bigvee z(x_1 + S(z) = x_2) \urcorner,$$

if the terms S, $+$ are interpreted as names of the successor function and addition in the domain of natural numbers, since for any t_1, $t_2 \in \mathcal{N}$ the equivalence

$$t_1 < t_2 \equiv \{t_1, t_2\} \; satisfy \; \ulcorner \bigvee z(x_1 + S(z) = x_2) \urcorner \; in \; \mathfrak{N}_0$$

holds. This equivalence is a special case of equivalence (i), which occurs in the definition above.

This concept of definability will be used in Theorems 23 and 51, and also on many other occasions. The following equivalent formulation will also be used:

A formula A defines r in a domain $\mathfrak{M}$ if and only if

$$\ulcorner \bigwedge x_1, ..., x_n (R(x_1, ..., x_n) \equiv A) \urcorner \in E(\mathfrak{M})$$

for the stated interpretation of terms.

Thus, a relation r is definable by a formula A if that formula can be the definiens in a true equivalence which defines the relation r.

Those relations which are definable by formulae all of whose quantifiers are restricted to the set of natural numbers are called elementary definable, arithmetically definable, or, briefly, arithmetical relations. Thus, the classification outlined above, to which we shall return later, is a classification of arithmetical relations.

Recursively enumerable sets

We shall now analyse in greater detail the first class of non-computable relations, or sets, namely, those which are obtained from computable relations by a single application of the existential quantifier operation. Thus the sets are sets defined by predicates of the type A'. They are called *recursively enumerable*.

DEFINITION 5. *A relation $P(x_0, ..., x_n)$ is recursively enumerable if and only if there exists a computable relation $R(x_0, ..., x_n, y)$ such that for any $x_0, ..., x_n \in \mathcal{N}$ the equivalence*

(89) $P(x_0, ..., x_n) \equiv \bigvee y \in \mathcal{N} R(x_0, ..., x_n, y)$

holds.

A set Z is recursively enumerable if the one-argument relation $x \in Z$ is recursively enumerable, that is, if

$$\bigwedge x \in \mathcal{N}(x \in Z \equiv \bigvee y \in \mathcal{N} R(x, y)),$$

where R is a computable relation.

446

From the intuitive point of view it may be noted that if x is an element of a recursively enumerable set Z, then this fact can be checked in a finite number of steps, by testing the relation $R(x, y)$ for $y = 0, 1, 2, \ldots$. If, on the contrary, x is not in Z, then no finite number of tests of the relation $R(x, y)$ for $y = 0, 1, 2, \ldots$ will prove decisive, since we may always assume that $R(x, y)$ holds for some y greater than those which have already been tested. Thus, a finite number of tests can either yield an answer in the affirmative, or no answer at all, to the question whether x is in Z. Thus, from the intuitive point of view recursively enumerable sets differ from computable ones. In the case of a computable set Z for every x there exists a method which makes it possible to demonstrate, in a finite number of steps, that $x \in Z$, if it is, or that $x \notin Z$, if in fact $x \notin Z$.

As will soon be demonstrated, recursively enumerable sets form a class which is essentially broader than is the class of computable sets. First note the following properties of recursively enumerable sets:

THEOREM 13.

a. *The class of recursively enumerable sets forms a ring of sets.*

b. *If A and $\mathcal{N} - A$ are recursively enumerable sets, then they are computable.*

c. *If the union of two recursively enumerable disjoint sets T and S is a computable set, then the sets T and S are also computable.*

PROOF.

a. A class K of sets is called a *ring of sets* if the class K together with any two sets A and B contains as its elements the union of those sets and their intersection. If A and B are recursively enumerable, then, by Definition 5, they can be represented as

$$x \in A \equiv \bigvee y \in \mathcal{N} R(x, y),$$

$$x \in B \equiv \bigvee y \in \mathcal{N} R'(x, y),$$

where $R, R' \in$ Comp. Hence, the union, under the laws of logic, can be represented as

$$x \in A \cup B \equiv \left(\bigvee y \in \mathcal{N}\ R(x, y) \vee \bigvee y \in \mathcal{N} R'(x, y) \right) \equiv$$
$$\bigvee y \in \mathcal{N} \left(R(x, y) \vee R'(x, y) \right).$$

The intersection of A and B is represented in the form of (89) by making use of functions inverse to the pairing function:

$$x \in A \cap B \equiv \left(\bigvee y R(x, y) \wedge \bigvee y' R'(x, y')\right) \equiv$$
$$\bigvee z \left(R\left(x, (z)_0\right) \wedge R'\left(x, (z)_1\right)\right).$$

The numbers y and y' have been combined into the pair $\langle y, y' \rangle = z$.

b. Let A and $\mathcal{N} - A$ be representable by computable relations R and R':

(90)
$$x \in A \equiv \bigvee y R(x, y),$$
$$x \in \mathcal{N} - A \equiv \bigvee y R'(x, y).$$

Since $A \cup (\mathcal{N} - A) = \mathcal{N}$, it follows from formulae (90) that

(91) $\qquad \bigwedge x \in \mathcal{N} \bigvee y \in \mathcal{N} \left(R(x, y) \vee R'(x, y)\right).$

Condition (91) is an effectiveness condition, hence, the following function is computable:

(92) $\qquad f(x) = (\mu y) [R(x, y) \vee R'(x, y)].$

It can now easily be checked that the set A satisfies the equivalence

(93) $\qquad x \in A \equiv R(x, f(x)).$

In fact, if $x \in A$, then, by (90) there exists an $y \in \mathcal{N}$ such that $R(x, y)$ and there exists a least y_0 such that $R(x, y_0)$. Since x cannot be both in A and in its complement $\mathcal{N} - A$, if $x \in A$, then $R'(x, y)$ does not hold for any y; consequently, y_0 is the least y which satisfies the disjunction $R(x, y) \vee R'(x, y)$, so that, by formula (92), $y_0 = f(x)$. Thus, it is true that $R(x, f(x))$. Conversely, if $R(x, f(x))$, then $f(x)$ is a number y for which $R(x, y)$, and, hence, $x \in A$, in accordance with (90).

Thus, A is a computable set. Since negation does not lead outside computable relations, $\mathcal{N} - A$ also is a computable set.

c. Property c is a direct consequence of property b. In fact, if the union $T \cup S \in \text{Comp}$, then $\mathcal{N} - (T \cup S) \in \text{Comp}$, and, thus $\mathcal{N} - (T \cup S)$ is recursively enumerable. The union of two recursively enumerable sets is a recursively enumerable set; thus, $\mathcal{N} - T$ is recursively enumerable, since $\mathcal{N} - T = (\mathcal{N} - (T \cup S)) \cup S$ in view of the fact that $T \cap S = \Lambda$ and since S is recursively enumerable by assumption. Since T and $\mathcal{N} - T$ are recursively enumerable, then, by b, $T \in \text{Comp}$. Similarly, $S \in \text{Comp}$.

2. COMPUTABLE FUNCTIONS

Unlike computable sets, recursively enumerable sets do not form a field of sets, since, as we shall soon see, there exist sets X which are recursively enumerable, but not computable. Their complements, therefore, cannot be recursively enumerable, since then, by Theorem 13b, the sets X would have to be computable.

THEOREM 14. *Every recursively enumerable set A can be presented in the form*

$$(94) \qquad x \in A \equiv \bigvee y R(x, y),$$

where R is a relation of the class Erec.

PROOF. By Definition 5, if A is recursively enumerable, then it can be presented in the form of (94), where R is computable. It is to be proved that the relation R can be replaced by another relation in Erec. We refer to Theorem 11 on normal form. Let $f(x, y)$ be a characteristic function of the computable relation R:

$$(95) \qquad f(x, y) = 0 \equiv R(x, y).$$

The function f is presented in normal form (67):

$$(96) \qquad f(x, y) = n \equiv \bigvee u R'(x, y, n, u),$$

where $R' \in$ Erec. It follows from formulae (94)–(96) that the set A is representable in the form

$$x \in A \equiv \bigvee y \bigvee u R'(x, y, 0, u).$$

By combining the variables y and u into a pair we obtain the desired form

$$x \in A \equiv \bigvee z R'\big(x, (z)_0, 0, (z)_1\big),$$

where $R' \in$ Erec.

THEOREM 15. *Every recursively enumerable non-empty set A can be presented as the set of values of a function $f \in$ Erec*

$$x \in A \equiv \bigvee n \, x = f(n).$$

This theorem justifies the term: recursively enumerable sets, that is, sets which are enumerable by recursive functions, or, in other words, are sets of values of recursive functions. In Theorems 14 and 15 the class Erec can, of course, be replaced by the more comprehensive class Prec.

PROOF. By Theorem 14, a recursively enumerable set A can be presented in the form of (94), where $R \in \text{Erec}$. If A is not empty, then there exist pairs $\langle x, y \rangle$ such that $R(x, y)$. Let a be the least pair $a = \langle (a)_0, (a)_1 \rangle$ such that $R((a)_0, (a)_1)$. We first define a function $g \in \text{Erec}$, which enumerates all pairs $x = \langle (x)_0, (x)_1 \rangle$ such that $R((x)_0, (x)_1)$:

$$g(n) = (\mu x \leqslant a+n) \Big[R((x)_0, (x)_1) \wedge$$
$$\bigwedge z \leqslant a+n \big(z > x \to\ \sim R((z)_0, (z)_1) \big) \Big].$$

By this definition, $g(n)$ equals the greatest number x, less than $a+n$, for which it is true that $R((x)_0, (x)_1)$. Of course, as n increases, $g(n)$ ranges over all the numbers x such that $R((x)_0, (x)_1)$. This yields immediately that

$$x \in A \equiv \bigwedge n x = (g(n))_0.$$

Thus the desired function $f \in \text{Erec}$ is the function $f(n) = (g(n))_0$.

THEOREM 16. *There exists a recursively enumerable set of pairs, which is universal for all recursively enumerable sets. This universal set is not computable.*

PROOF. Let $G_n(x, y)$ be a computable function, universal for the class Prec, as defined in Theorem 10. By Theorem 14, every recursively enumerable set X_n is presentable in the form

(97) $\qquad x \in X_n \equiv \bigvee y G_n(x, y) = 0.$

By putting

$$\text{Rp}(x, n) \equiv \langle x, n \rangle \in \text{Rp} \equiv \bigvee y G_n(x, y) = 0$$

we obtain a set Rp, which, by definition, is a recursively enumerable set of pairs. As follows from (97), the set Rp is universal for recursively enumerable sets in the sense that

(98) $\qquad x \in X_n \equiv \langle x, n \rangle \in \text{Rp}.$

The set Rp is not computable. Nor is the set Rp_0, obtained by the identification of x with n:

(99) $\qquad x \in \text{Rp}_0 \equiv \langle x, x \rangle \in \text{Rp}.$

If Rp were computable, so would Rp_0 be. Hence, it suffices to demonstrate that Rp_0 is not a computable set. If Rp_0 were computable, the set

$\mathscr{N} - \mathrm{Rp}_0$ would, in the numbering (97), have its number k. Hence, by formulae (97)–(99), it would be true that, for every x,

$$\sim \langle x, x \rangle \in \mathrm{Rp} \equiv x \notin \mathrm{Rp}_0 \equiv x \in \mathscr{N} - \mathrm{Rp}_0 \equiv x \in X_k \equiv \langle x, k \rangle \in \mathrm{Rp}.$$

By replacing x in the above equivalences by k, we obtain a contradiction, namely,

$$\sim \langle k, k \rangle \in \mathrm{Rp} \equiv \langle k, k \rangle \in \mathrm{Rp}.$$

Every computable set is, of course, recursively enumerable, because, if X is a computable set, then X is presentable as

$$x \in X \equiv \bigvee y(y = y \wedge x \in X).$$

If X is computable, then, obviously, the relation $(y = y \wedge x \in X)$ is computable, too.

On the contrary, not every recursively enumerable set is computable; Rp_0 is recursively enumerable, but not computable. Its complement is not computable and, by Theorem 13b, not recursively enumerable. That complement can be presented by reference to the universal quantifier:

$$x \in \mathscr{N} - \mathrm{Rp}_0 \equiv \bigwedge y \, G_n(x, y) \neq 0.$$

Since it is not recursively enumerable, the complement cannot be defined by means of the existential quantifier followed by a computable relation. Likewise, we can construct sets, defined by more than one quantifier, for example, the universal followed by the existential, or the existential followed by the universal, which are not representable by means of a single quantifier followed by a computable relation. It can also be proved in a general manner that in the classification outlined in the preceding section there are always sets in the higher classes which are not sets in any of the lower classes (Kleene [34], Mostowski [52]). There are many special studies concerned with the various kinds of recursively enumerable sets (Post [64], Rogers [70], Shoenfield [84], and many others). Those sets in fact reveal a wealth of varieties.

EXERCISES

1. Define, by means of bounded minimization, superposition and the functions $x \doteq y$, $S(x)$, x^y, the functions defined by formulae (13)–(26). Find appropriate bounding functions.

2. Prove that the class of elementary recursive functions is the least class of functions which contains $S(x)$, $x \div y$, $x+y$, $x \cdot y$, and x^y, and is closed under superposition, and under any of the elementary recursive operations listed in Theorem 7.

3. Prove that restricted minimum can also be defined as follows:

$$\mu x \leqslant z(F(x,t) = 0) = \sum_{i \leqslant z} \mathrm{sgn}\left(F(i,t) \cdot \prod_{j < i} F(j,t)\right),$$

where $\mathrm{sgn}(0) = 0$ and $\mathrm{sgn}(n+1) = 1$.

4. Prove that for every class of functions X which contains the pairing function and the functions inverse to it, it is true that if the class X is the least class which contains a finite number of initial functions and is closed under superposition and other operations, then a class X_1 of functions of one argument belonging to X also can be defined by induction in a similar manner.

5. Define a primitive recursive function which is universal for elementary recursive functions.

6. Prove that every infinite recursively enumerable set can be enumerated without repetitions by a computable function. Demonstrate in that connection that such a function need not always be primitive recursive.

7. Prove that if an infinite recursively enumerable set is enumerable by a computable increasing function, then it is computable.

8. Prove that the set of values of any non-computable function f which is representable as

$$f(n) = ((\mu y)[R(n,y)])_0,$$

where $R \in \mathrm{Comp}$ and the minimum operation need not be effective, is recursively enumerable. The functions f representable in the above are called *partially computable*.

9. Prove that the class of elementary recursive relations is the least class of relations which contains the relations $x < y$, $x = y^z$, $x = y+1$, $x = y+z$, $x = y \cdot z$, and is closed under the logical operations of sentential connectives and bounded quantifiers.

3. EFFECTIVENESS OF METHODS OF PROOF

Computability of the set of well-formed formulae

We shall at first return to the intuitive concept of computability and apply it to the set of inscriptions. In so doing we will be interested only in the inscriptions belonging to the elementary arithmetic of natural numbers. That theory will, for convenience, be denoted by Ar, but its axioms will not, for the time being, be formulated precisely. Let 0, S, $<$, $+$, $\cdot$, P, $=$ be the primitive terms of the theory Ar. Thus, every well-formed formula of that theory can be written exclusively with the use of the following symbols: ( ) $\bigvee$ $\bigwedge$ $\sim$ $\to$ $\vee$ $\wedge$ $\equiv$ 0 S $<$ $+$ $\cdot$ P $=$, x_I. The last two symbols make it possible to construct infinitely many variables: $x_\mathrm{I}, x_\mathrm{II}, x_\mathrm{III}, \ldots$, while the remaining symbols stand for concepts or serve as auxiliary signs (the parentheses and the comma).

As in the case of natural numbers, a set Z of inscriptions will be called *computable* (*general recursive* or *decidable*), if there exists an effective method which makes it possible to decide for any inscription whether it is, or is not in Z. Likewise, a relation R among inscriptions will be called *computable* if there exists an effective method of deciding for any pair of inscriptions a, b whether the relation R does or does not hold between them.

Since the shape of a formula is that characteristic which is the easiest to recognize, an effective method of stating something about inscriptions must take into account the visually recognizable characteristics of the form of inscriptions. Thus, such a method is, in a sense, visual, since it consists in inspecting an inscription carefully and in transforming it graphically. In this sense, the set of c.l.c. tautologies is a computable set of inscriptions, since very simple graphic transformations (the zero-one-verification method) make it possible to decide for any formula whether it is or is not a c.l.c. tautology.

Out of the sets of inscriptions constructed from the 19 initial symbols listed above some sets are computable in the intuitive sense of the term in an obvious manner. For instance, the set of well-formed formulae is computable. If we want to know if a formula A is well-formed it suffices to inspect it visually. If all the symbols are in their proper places, that formula is meaningful, that is, well-formed. It is easy to formulate a rigorous rule which makes it possible to state, in a finite number of steps, whether a formula is a string of chaotically arranged symbols, or whether each symbol is used correctly. The correct use of each symbol can easily be described by a rigorous rule which provides a method of verification. For instance, the symbol $=$ is used correctly if it links two formulae which have been verified as well-formed term formulae. The symbol $\to$ is used correctly if it links two formulae which have been verified as well-formed sentential formulae, etc. In this way, by beginning with the shortest formulae, we can easily test formulae of increasing length. The shortest well-formed formulae are the term formulae 0 and the variables x_I, x_{II}, x_{III}, etc. As examples of well-formed sentential formulae in Ar we may quote, for instance: $x_I + x_{II}$; $\bigwedge x_I(x_I + x_{II} = P(x_I, x_{II}))$; on the other hand, the expressions $x_I + +(= x$; $S(x, x)$; $P(0(x_I)) = x_I$ are not well-formed.

The use of all mathematical terms occurring in any theory can be described in a rigorous manner. The rigorous formal rules of constructing well-formed formulae are called the *syntax* of the language of a given theory. Such rules can be described so that every person, upon inspecting closely any combination of the basic symbols of a given theory, can easily decide, by reference to those rules, whether that combination of symbols is or is not a well-formed formula. Hence, in accordance with the intuitive sense of the concept of computability the set of well-formed formulae of any mathematical theory is computable. Likewise, we can easily decide about any phrase or sentence used in everyday language whether it is a correct phrase or sentence in our mother tongue, or whether it contains some grammatical errors. If there were no effective method of testing the correctness (well-formedness) of formulae (expressions) of a theory, we would be unable to even formulate problems in that theory, not to speak of solving such problems. The set of all well-formed formulae is often termed the *language* of a given theory.

Computability of the set of axioms

When constructing a mathematical theory we want to know precisely not only how to formulate problems in that theory, but also, which items of information are to be treated in that theory as reliable and unquestionable. If we know which problems belong to a given theory and which items of information are assumed to be reliable, then the task is clear: we have to take the items of information adopted as reliable and to deduce from them answers to the largest possible number of interesting problems. Creative mathematical work consists exactly in solving such problems.

Those items of information which are adopted as reliable in a given theory are formulated in sentences which are called *axioms* or *assumptions* of that theory. They are all those sentences which are adopted in that theory without proof, and from which all other theorems of that theory are deduced. In all mathematical theories usually encountered a computable set of axioms is assumed. We do so, because if we have an effective method of deciding about every sentence whether it is an axiom or not, then we can say that we in fact know which items of information in that theory are adopted as reliable, and which require substantiation.

If there were no effective method of testing whether a formula is an axiom of a given theory or not then such an axiom system would, in a sense, be useless in proofs, since it would not be possible to decide whether an axiom has or has not been used in a given proof. Proofs would become quite arbitrary, and mathematics would cease to be an intersubjective science.

Quite often the number of axioms is just finite. They are then all listed in the introduction to the theory in question, and then it is very easy to decide whether a given formula is an axiom, or not. The testing consists in comparing the form (shape) of the formula under consideration with all the axioms listed at the outset. The formula in question is an axiom if and only if it is equiform with one of the axioms. Such is the case, for instance, in the theory of the *less than* relation, which has only eight extra-logical axioms.

Many theories, however, have an infinite number of axioms, but these axioms are usually subsumed under certain schemata which can be described rigorously. For instance, the axiom of induction in the elementary arithmetic of natural numbers, as described in Chapter II, Section 5, is a schema. Nevertheless, it is easy to test whether a sentence falls under the schema of induction. Thus, the set of axioms of the theory Ar together with the axiom of induction also is computable, even though it is infinite.

We shall hereafter apply the term *normal mathematical theory* to any theory $T = \text{Cn}(X)$ for which the set X of axioms is computable. When defining the general concept of theory in Chapter I, we did not make any restriction as to whether or not the set of those sentences which form a theory must have a recursive axiom system. Thus, the set of sentences true in any domain is a theory, since it is closed under logical consequence, but the set of sentences true in $\mathfrak{N}_0$, for instance, is not a normal mathematical theory. As follows from the investigations to be discussed below, there is no recursive axiom system for the set of all arithmetical sentences true in $\mathfrak{N}_0$.

The set of axioms of the c.l.c. is another typical example of an infinite but computable set of axioms. Each c.l.c. axiom listed in Chapter I is a schema, but a simple inspection of any definite formula makes it possible to decide whether it does or does not fall under any of the initial schemata of the c.l.c.

Effectiveness of the operation of detachment and computability of the set of proofs

The relation formulated as:

> *a sentence A is obtained by detachment from sentences B and C*

is also computable. Its computability is self-evident. A is obtained by detachment if it is the consequent of an implication. Thus, A is obtained by detachment from B and C if either $B = \ulcorner C \to A \urcorner$ or $C = \ulcorner B \to A \urcorner$. It is so because one of the premisses must be an implication with A as its consequent and the other premiss as its antecedent.

Hence the set of all correct proofs in any normal mathematical theory $T = Cn(X)$ is computable. In Chapter I a proof has been defined as a finite sequence of sentences

$$D_1, ..., D_n,$$

such that every sentence either is an axiom of T or is obtained from earlier elements of that sequence by detachment. A finite sequence of sentences may always be treated as a single inscription consisting of sentences $D_1, ..., D_n$ which follow one another. If the theory in question is normal, that is, if the set of its axioms is computable, then we can decide about any sentence from the sequence $D_1, ..., D_n$ whether it is an axiom or not. Since the detachment operation is computable, we can also easily decide about each of the sentences $D_1, ..., D_n$ whether it is obtained from earlier ones by detachment. Hence, we can easily decide about any finite sequence of sentences whether it is or is not a correct proof by detachment.

It was noted in Chapter I that all proofs can be reduced to proofs by detachment, based on axioms of logic and the specific axioms of a given theory. The axioms of logic form a computable set. Hence, if a theory is normal, then the set of correct proofs is also computable.

Besides the set of proofs we shall consider the relation

(100)   *D is a proof of a sentence A.*

This relation is computable, too. D is a proof of a sentence A if D is a correct proof consisting of a finite number of sentences $D_1, ..., D_n$ such that A is the last element of $D: A = D_n$.

The set of proofs is computable, and, hence, we are able to decide whether D is a proof or not. If D is a proof, then it is easy to see whether the last element of D is or is not equiform with the sentence A. Thus, there exists a method of testing, for any two formulae D and A, whether relation (100) does or does not hold between them.

Computability of proofs is the foundation of the intersubjective nature of mathematics, since a proof carried out by one mathematician can, thus, be checked by another mathematician. Of course, in practice, proofs are constructed in abbreviated forms. Writing out complete proofs by detachment would bring about a standstill in mathematics, since even simple proofs would swell into bulky volumes. It is only because of their abbreviated forms that following mathematical proofs is not a purely mechanical operation, but sometimes requires creative mental effort.

Recursive enumerability of the set of theorems

Since the concept of computability has been applied to sets of inscriptions, other notions of the logical classification of concepts can be similarly applied to sets of inscriptions. Thus, a set X of inscriptions will be called *recursively enumerable* if and only if there exists a computable relation $R(A, B)$ between inscriptions such that the set X can be defined as follows:

(101) $A \in X \equiv \bigvee B R(A, B)$.

Thus, it is assumed that the existential quantifier operation on a variable that ranges over inscriptions leads from computable to recursively enumerable relations, as in the case of the domain of natural numbers.

It is easy to see that in this sense the set of theorems of a normal mathematical theory always is recursively enumerable. A sentence is a theorem of a theory T if it has a correct proof:

(102) $A \in T \equiv \bigvee D \, D$ *is a proof of the sentence* A.

The relation "D is a proof of the sentence A" is computable, as has been stated above. The set of theorems of T is representable in the form of (101), and, hence, is recursively enumerable.

The relations which hold between sets of inscriptions are the same as those holding between sets of numbers. Hence, every computable set is recursively enumerable, and there are sets of inscriptions which are recursively enumerable, but not computable. There are also many theories T such that the set of theorems of the theory T is recursively enumerable, but not computable. Such theories are termed *undecidable*.

DEFINITION 6. *A theory T is decidable if and only if the set of its theorems is computable.*

A theory T is undecidable if and only if the set of its theorems is not computable.

As has been stated above, the classical sentential calculus is decidable. On the contrary, arithmetic Ar is, as we will see, undecidable. The theory of the *less than* relation is decidable, but set theory is undecidable.

Arithmetization of language

Even though the concept of computability, as applied to sets of inscriptions, has a very clear intuitive meaning, nevertheless it seems important to be able to reduce the concept of computability of a set of inscriptions to that of computability of a set of numbers. (We could also possibly attempt to perform the converse operation, that is, to reduce computability as applied to sets of numbers to the intuitively clear concept of computability of sets of inscriptions.) Such a reduction of computability of sets of inscriptions to that of sets of numbers is achieved by means of what is termed the arithmetization of language. *Arithmetization* consists in an effective enumeration of all formulae of the language of the theory under investigation by means of natural numbers. Such a numbering will be described in its application to the language of the theory Ar, which will now be analysed in greater detail.

If we want a convenient assignment of a number to every expression, we first assign certain numbers to the basic signs of a theory, and then to expressions, that is, to finite sequences of basic signs, we assign numbers by means of decomposition of numbers into prime factors. Thus, the 19 basic signs of the theory Ar are assigned the first 19 numbers:

$$(\ ) \ \lor \ \land \ \sim \ \to \ \lor \ \land \ \equiv \ 0 \ S \ < \ + \ \cdot \ P \ = \ x \ \iota \ ,$$
$$1 \ 2 \ 3 \ 4 \ 5 \ 6 \ 7 \ 8 \ 9 \ 10 \ 11 \ 12 \ 13 \ 14 \ 15 \ 16 \ 17 \ 18 \ 19$$

3. EFFECTIVENESS IN PROOFS

Now, if a is one of the 19 basic signs written above, then $\text{nr}(a)$ stands for the number assigned to it and written directly over it. Hence, $\text{nr}(\rightarrow) = 6$, $\text{nr}(\wedge) = 8$. To every expression A which is a sequence of the basic signs:

$$A = a_0 a_1 \ldots a_n$$

we assign a number $\text{Nr}(A)$ of the form

(103) $\text{Nr}(A) = 2^{\text{nr}(a_0)} \cdot 3^{\text{nr}(a_1)} \cdot \ldots \cdot p_n^{\text{nr}(a_n)}$.

Thus, the number of the formula $x_1 = 0$ is

$$\text{Nr}(\ulcorner x = 0 \urcorner) = 2^{17} \cdot 3^{18} \cdot 5^{16} \cdot 7^{10} .$$

Likewise,

$$\text{Nr}(\ulcorner P(x, x) > 0 \urcorner) = 2^{15} \cdot 3^{1} \cdot 5^{17} \cdot 7^{19} \cdot 11^{17} \cdot 13^{2} \cdot 17^{12} \cdot 19^{10}.$$

Thus, in the case of every expression A which consists of the basic symbols, its number $\text{Nr}(A)$ is easy to compute. The function Nr is computable in the intuitive sense of the term, and it establishes a one-to-one relation between expressions and a subset of natural numbers. Hence, instead of speaking about the computability of a set of expressions we may speak about the computability of the set of their numbers. We can even define with precision the computability of a set of expressions: a set X of expressions is computable if and only if the set of the numbers of the elements of the set X, that is, the image of the set X in the set $\mathcal{N}$ by the function Nr, is computable. It is in this sense that the concept of computability will be used below.

Now that we have introduced a rigorous concept of computability for sets of inscriptions we can check that the comments formulated earlier in this section concerned with the computability of certain sets of expressions remain valid for the new concept of computability.

Checking the computability of the sets specified above is laborious in detail, but the idea is easy to grasp. We have to define first a number of operations on numbers which correspond to certain simple graphic operations on expressions. Then it will become self-evident that every intuitively computable relation among expressions can be defined in a computable manner. The most important operation on expressions consists in combining two expressions into one, and is called *concatenation*. Concatenation turns two expressions, A and B, into a third one, which consists of A followed directly by B.

III. HIERARCHY

The numerical function which stands for concatenation will be denoted by two figures or numerical variables separated by an asterisk $*$.

The *length* of a formula A is the number of symbols of which A consists. The corresponding function is denoted by $\ln g(a)$.

(104) $\quad \ln g(a) = (\mu x \leqslant a)[(a)_x = 0]$.

The basic symbols of the theory Ar have been assigned numbers 1–19, greater than zero. Hence, the earliest prime number occurring to the zero power in the expansion of a number a indicates the end of the formula to which the number a is assigned.

Concatenation:

(105) $\quad a * b = a \cdot \prod_{i \leqslant \ln g(b)} p_{\ln g(a)+i}^{(b)_i}$.

Thus, in the decomposition of a number $a * b$, the $(\ln g(a)+i)$-th prime number occurs to the power to which the i-th prime number occurred in the expansion of the number b: $(a * b)_{\ln g(a)+i} = (b)_i$.

Concatenation is associative: $a * (b * c) = (a * b) * c$. The parentheses may, accordingly, be omitted. The lengths of formulae add up: $\ln g(a * b) = \ln g(a) + \ln g(b)$.

In accordance with the definitions of functions $\langle m_0, \ldots, m_k \rangle$ the following formulae hold:

$$k+1 = \ln g(\langle m_0, \ldots, m_k \rangle),$$
$$\langle m_0, \ldots, m_k \rangle * \langle m_{k+1}, \ldots, m_{k+i+1} \rangle = \langle m_0, \ldots, m_{k+i+1} \rangle.$$

Numerical analogues of logical operations can easily be described by reference to concatenation. Thus, if $a = \text{Nr}(A)$ and $b = \text{Nr}(B)$, then

$$\text{Nr}(\ulcorner (A \to B) \urcorner) = 2^1 * a * 2^6 * b * 2^2,$$
$$\text{Nr}(\ulcorner \sim (A) \urcorner) = 2^5 * 2^1 * a * 2^2, \text{ etc.}$$

Of course, $2^5 * 2^1 = 2^5 \cdot 2^1$.

Let the general convention be that if a is one of the 19 basic symbols, then

(106) $\quad \bar{a} = 2^{\text{nr}(a)}$.

Hence,

$$\text{Nr}(\ulcorner (A \to B) \urcorner) = (\overline{* a *} \overline{\to} * b \overline{*}),$$
$$\text{Nr}(\ulcorner (a < b) \urcorner) = (\overline{* a *} \overline{<} * b \overline{*}).$$

The numbers which correspond to variables will be denoted by $\bar{x}_n$. The definition is by induction:

$$\bar{x}_0 = \bar{x} = 2^{17},$$
$$\bar{x}_{n+1} = (\overline{x_n * 1}) = \bar{x} \cdot p_{\lg(\bar{x}_n)}^{18}.$$

We shall now strive to define the set of numbers for numerical formulae and the set for sentential formulae. These formulae will be divided into orders. The set of numerical formulae of each order is finite, since the variables $\bar{x}_n$ and $\bar{x}_m$ for $n \neq m$ will be classed in different orders. A finite set $m_0, \ldots, m_k$ of numbers will be represented by a single number $m = \langle m_0, \ldots, m_k \rangle$. Hence, the definition of an order as a finite set might be as follows:

$$\text{Ornf}(0) = 2^{\bar{0}} = \langle \bar{0} \rangle,$$
$$\text{Ornf}(n+1) = \text{Ornf}(n) * 2^{\bar{x}_n} * \mathop{E}_{i<\lg(\text{Ornf}(n))} \bar{S} * \overline{\left[* (\text{Ornf}(n))_i * \right]} *$$

$$\mathop{E}_{i,j<\lg(\text{Ornf}(n))} \overline{\left[* (\text{Ornf}(n))_i * \overline{+} * (\text{Ornf}(n))_j * \right]} *$$

$$\mathop{E}_{i,j<\lg(\text{Ornf}(n))} \overline{\left[*(\text{Ornf}(n))_i * \overline{\cdot} * (\text{Ornf}(n))_j * \right]} *$$

$$\mathop{E}_{i,j<\lg(\text{Ornf}(n))} P * \overline{\left[* (\text{Ornf}(n))_i * \overline{,} * (\text{Ornf}(n))_j * \right]}.$$

This definition can be interpreted as follows: The set of numerical formulae of order 0 consists of only one numerical formula 0. The set of numerical formulae of the order $n+1$ consists of: 1. the numerical formulae of the order n; 2. the variable x_n; 3. the formulae of the form $S(\varphi)$ for all formulae φ of the order n; 4. the formulae $\varphi+\psi$ for all formulae φ and ψ of the order n; 5. the formulae $\varphi \cdot \psi$ for all formulae φ and ψ of the order n; 6. the formulae $P(\varphi, \psi)$ for all formulae φ and ψ of the order n. The convention is also adopted that

$$\mathop{E}_{j,i<k} f(n, i, j) = \langle f(n, 0, 0), f(n, 0, 1), \ldots, f(n, 0, k),$$
$$f(n, 1, 0), \ldots, f(n, 1, k), \ldots, f(n, k, k) \rangle.$$

An order is thus a sequence $\langle m_0, \ldots, m_k \rangle$, where $m_0, \ldots, m_k$ are numbers of those numerical formulae which belong to that order. A formula a is of an order n if, for some $i < \lg(\mathrm{Ornf}(n))$, $a = (\mathrm{Ornf}(n))_i$. Earlier orders are contained in later orders. Hence, the lengths of orders increase: $\lg(\mathrm{Ornf}(n+1)) > \lg(\mathrm{Ornf}(n))$, because the length of an order denotes the number of formulae of that order. A numerical formula is exactly of an order n if and only if it is of the order n and is not of the order $n-1$, or if it is the formula 0 in the case when $n = 0$. The fact that a formula a is of an order n and is not of an order $n-1$ can be written in the following manner:

$$\bigvee i \Big(a = (\mathrm{Ornf}(n))_i \wedge i < \lg(\mathrm{Ornf}(n)) \wedge$$

$$\bigwedge j \leqslant \lg(\mathrm{Ornf}(n)) \big((\mathrm{Ornf}(n))_i \neq (\mathrm{Ornf}(n-1))_j \big) \Big).$$

In view of the above it can be proved that

(107)   *If a is exactly of an order n, then $n < a$.*

This relationship, which will be referred to below, can easily be checked by induction. For zero: $0 < 2^{10} = \overline{0} = (\mathrm{Ornf}(0))_0$. Note also that $n+1 < 2^{17} \cdot 3^{18} \cdot 5^{18} \cdot \ldots \cdot p_{n+1}^{18} = \overline{x}_n$. Suppose now that this formula holds for n; it will be proved that it is also valid for $n+1$. A formula which is in the sequence $\mathrm{Ornf}(n+1)$ and is not in $\mathrm{Ornf}(n)$ must, by the definition of Ornf, be either a variable $\overline{x}_n$, or the successor of a formula of exactly the order $\mathrm{Ornf}(n)$, or the result of adding, multiplying or raising to a power of two formulae of exactly the order $\mathrm{Ornf}(n)$. It has been noted above that the inequality holds for the variable. For the remaining operations it is also self-evident that if, under the induction assumption, the inequality $n < (\mathrm{Ornf}(n))_i$ holds, then $n+1 < \ldots *$ $(\mathrm{Ornf}(n))_i * \ldots$ The operation of concatenation increases more rapidly than does that of successor, regardless of with what a given formula is concatenated. Hence $n+1 < (\mathrm{Ornf}(n+1))_i$, but only for those numbers $(\mathrm{Ornf}(n+1))_i$ which are not identical with any $(\mathrm{Ornf}(n))_j$, since those formulae only which are exactly of the order $n+1$ are obtained from formulae of exactly the order n.

This estimation makes it possible to restrict quantifiers when defining the set of numerical formulae, since every numerical formula is a numerical formulae of a certain order

$$a \in \mathrm{nf} \equiv \bigvee n < a \bigvee i < \lg(\mathrm{Ornf}(n)) a = (\mathrm{Ornf}(n))_i.$$

3. EFFECTIVENESS IN PROOFS

The set of numerical formulae is, thus, computable. The set of sentential formulae will be defined in a similar way. We first define orders of sentential formulae as finite sequences of such formulae:

$$\text{Orsf}(0) = \langle \overline{(* \overline{0} * \stackrel{=}{} * 0 *)} \rangle,$$

$$\text{Orsf}(n+1) = \text{Orsf}(n) * \mathop{E}_{i < \ln g(\text{Orst}(n))} \overline{\left(* \sim * (\text{Orsf}(n))_i *\right)} *$$

$$\mathop{E}_{i,j < \ln g(\text{Orst}(n))} \overline{\left(* (\text{Orsf}(n))_i * \Rightarrow (\text{Orsf}(n))_j *\right)} *$$

$$\mathop{E}_{i,j < \ln g(\text{Orst}(n))} \overline{\left(* (\text{Orsf}(n))_i * \vee * (\text{Orsf}(n))_j *\right)} *$$

$$\mathop{E}_{i,j < \ln g(\text{Orst}(n))} \overline{\left(* (\text{Orsf}(n))_i * \wedge * (\text{Orsf}(n))_j *\right)} *$$

$$\mathop{E}_{i,j < \ln g(\text{Orst}(n))} \overline{\left(* (\text{Orsf}(n))_i * \equiv * (\text{Orsf}(n))_j *\right)} *$$

$$\mathop{E}_{i,j < \ln g(\text{Orst}(n))} \overline{\left(* (\text{Orsf}(n))_i * = * (\text{Orsf}(n))_j *\right)} *$$

$$\mathop{E}_{i,j < \ln g(\text{Orst}(n))} \overline{\left(* (\text{Orsf}(n))_i * < * (\text{Orsf}(n))_j *\right)} *$$

$$\mathop{E}_{i,j < \ln g(\text{Orst}(n))} \bigvee * \overline{x}_i * (\text{Orsf}(n))_j *$$

$$\mathop{E}_{i,j < \ln g(\text{Orst}(n))} \bigwedge * \overline{x}_i * (\text{Orsf}(n))_j.$$

The above definition can be interpreted as follows: The set of sentential formulae of the order 0 consists of only one sentential formula: $(0 = 0)$; the set of sentential formulae of the order $n+1$ consists of: 1. the sentential formulae of the order n; 2. the negations of sentential formulae of the order n; 3. the implications formed of sentential formulae of the order n; 4. the disjunctions; 5. the conjunctions, 6. the equivalences, all formed of sentential formulae of the order n; 7. the equalities ($\varphi = \psi$) formed of numerical formulae φ and ψ of the order n; 8. the *less than* relations ($\varphi < \psi$) for numerical formulae φ and ψ of the order n; 9. the quantification formulae $\bigvee x_i \alpha$; and 10. $\bigwedge x_i \alpha$ for all sentential formulae α of the order n, with i restricted for convenience by the number of elements of a given order.

In a manner similar to the case of the numerical formulae, we can easily prove by induction that for every n

(108)   *If a is a sentential formula of exactly the order n, then $n < a$.*

Hence, in defining the set of sentential formulae we may use bounded quantifiers:

$$a \in \text{sf} \equiv \bigvee n < a \bigvee i < \text{lng}(\text{Orsf}(n)), \; a = (\text{Orsf}(n))_i.$$

Since the definition of the function $\text{Orsf}(n)$ covers all the possible ways of constructing compound formulae out of simple ones, every sentential formula is a sentential formula of a certain order $\text{Orsf}(n)$; consequently, sf is in fact the set of numbers of well-formed sentential formulae. That set is, accordingly, computable.

The set of numbers of closed formulae, that is, sentences: s, is computable, too. As is already known, a *sentence* is a formula in which every variable occurs bound. This idea is symbolized as follows:

$$a \in s \equiv a \in \text{sf} \wedge \bigwedge n, q < a \Big(\bigwedge i \leqslant n+1 ((a)_{q+1} = (\bar{x}_n)_i) \to$$
$$\bigvee b, c, r < a \big((b = \overline{\bigvee} * \bar{x}_n * c \vee b = \overline{\bigwedge} * \bar{x}_n * c) \wedge$$
$$b \in \text{sf} \wedge r < q < r + \text{lng}(b) \wedge \bigwedge i < \text{lng}(b)((a)_{r+1} =$$
$$(b)_i)\big)\Big).$$

The meaning of the above definition is easy to grasp. The definition states that if a variable x_n occurs in a formula a in the q-th place, then it is contained in a formula b, contained in a, which begins with an existential or a universal quantifier that binds the variable x_n.

The set of logical axioms is computable. We shall confine ourselves to defining rigorously only the set of axioms covered by schema 1 for the distribution of the universal quantifier with respect to implication:

$$\bigwedge \dots \big(\bigwedge x_n(\alpha \to \beta) \to (\bigwedge x_n \alpha \to \bigwedge x_n \beta)\big),$$

$$a \in \text{Ax} 1 \equiv \bigvee a', b, c, n \leqslant a\Big(a', b, c \in \text{sf} \wedge a \in s \wedge$$

$$a' = \overline{\big(* \overline{\bigwedge} * \bar{x}_n * \overline{(* b * \overline{\to} * c *)} * \overline{\to} *}$$

$$\overline{(* \overline{\bigwedge} * \bar{x}_n * b * \overline{\to} * \overline{\bigwedge} * \bar{x}_n * c *)*\big)} \wedge$$

$$\big(a = a' \vee \bigvee d \leqslant a(a = d * a' \wedge$$

$$\bigwedge i < \text{lng}(d)((d)_i = 4 \vee (d)_i = 17 \vee (d)_i = 18))\big)\Big)\Big).$$

3. EFFECTIVENESS IN PROOFS

The above definition describes all the meaningful substitutions for the schema of the distribution of the universal quantifier with respect to implication. By taking it as a model we can describe all the logical axioms, and the general definition of a logical axiom can take on the form of the following disjunction:

$$a \in \mathrm{Ax}L \equiv a \in \mathrm{Ax}I \lor a \in \mathrm{Ax}1 \lor \ldots \lor a \in \mathrm{Ax}0.$$

We define in a similar way the numbers of the specific axioms of the theory Ar, that is, the set AxAr.

The relation of proof is described as follows:

$$(x\,\mathrm{proof}\,y \equiv x \text{ is a proof of a sentence } y),$$

$$(109) \quad x\,\mathrm{proof}\,y \equiv \Big(y = (x)_{\mathrm{lng}(x)-1} \land$$

$$\bigwedge i \leqslant \mathrm{lng}(x) - 1 \big((x)_i \in \mathrm{Ax}L \lor (x)_i \in \mathrm{Ax}\mathrm{Ar} \lor$$

$$\bigvee j, k < i\big((x)_j = \overline{(* (x)_k * \overset{\Rightarrow}{} * (x)_i *)}\big)\big)\Big).$$

Thus a proof is a finite sequence $x = \langle(x)_0, \ldots, (x)_{\mathrm{lng}(x)-1}\rangle$ such that the last element is identical with y and every element of the sequence either is an axiom or is derived from earlier elements by detachment.

As the above definitions show, all those sets of formulae which are computable in the intuitive sense have, following the arithmetization of syntax, proved computable in the rigorous, mathematical sense of the term. On the other hand, the set Ar of numbers of arithmetical theorems is obviously recursively enumerable:

$$(110) \quad y \in \mathrm{Ar} \equiv \bigvee x\; x\,\mathrm{proof}\,y.$$

As a result of the analyses carried out in this section we can record the following theorem.

THEOREM 17.

a. *In the theory* Ar, *the relation of proof* ($x\,\mathrm{proof}\,y$) *is computable.*

b. *The set of numbers of the theory* Ar *is recursively enumerable.*

In addition, we intuitively do not see any computable method of getting a bounded existential quantifier in place of the one which occurs in (110). In fact, no such method exists, because, as will be shown soon, the set Ar is not computable. Likewise, the set of theorems of any richer mathematical theory which contains arithmetic is not computable either.

We in fact usually ascribe to the set of theorems those properties which we have accepted to be characteristic of recursively enumerable sets (cf. p. 446). If a sentence is a theorem, then it has a proof, which we can find out by mechanically writing out all the proofs, beginning with the shortest ones and then passing to proofs of increasing length. But if a sentence is independent, we cannot find out if it is a theorem even after writing out any arbitrary but finite number of proofs. If, in the process of mechanically writing out proofs of increasing length, we do not come across a proof of a sentence A, we cannot be sure whether the sentence A is or is not a theorem, and whether it can or cannot be arrived at in a longer proof.

EXERCISES

1. Deduce the formulae:

$$(a * b)_{\ln g(a)+i} = (b)_i, \quad a * (b * c) = (a * b) * c,$$
$$\ln g(a * b) = \ln g(a) + \ln g(b).$$

2. Deduce formula (108) in a rigorous manner (by induction).

3. Define rigorously the set of numbers of $Ax LI$ (generalizations of substitutions for the laws of the sentential calculus), and also the remaining logical axioms.

4. REPRESENTABILITY OF COMPUTABLE RELATIONS IN ARITHMETIC

We now return to computable functions and relations defined on natural numbers. There are certain theorems on computable functions from which the theorem on the undecidability of the arithmetic of natural numbers is deduced. Those theorems form one of the definitions of computability. That definition was even one of the earliest definitions of that concept. The intuitive concepts which lead to that definition will soon be described in considerable detail. That property of the computable functions which emerges from those analyses will be termed *representability*. The meaning of this term will be explained soon.

Denoting numbers by numerals

We shall first adopt a convention on the notation to be used. The symbol *n* will be used to denote the numeral (figure) which corresponds to the number n. Thus, 5 is a number, and **5** is a figure (numeral) which is a name of the number 5. The distinction between numbers and numerals falls under the distinction between an object and its name, made universally in logic. Numbers are abstract objects which are denoted by numerals. Figures are certain inscriptions which are names of numbers. Thus, n is a metalogical function ranging over the names of all numbers. If that function is superimposed upon another mathematical function $f(n)$, then the result of that superposition will be denoted by **f(n)**. Thus, **f(n)** is a numeral which corresponds to the number $f(n)$, that is, a name of the number $f(n)$.

We shall adopt below a certain manner of recording numerals (figures), namely that which is the simplest to use in the language of the theory Ar. It consists in recording figures by means of the successor notation, so that

The inscription "0"   is a name of the number 0,

The inscription "$S(0)$"   is a name of the number 1,

The inscription "$SS(0)$"   is a name of the number 2,

. . . .

The inscription "$\underbrace{SSS \ldots S(0)}_{n}$" is a name of the number n,

Thus

(111) $n = \underbrace{SS \ldots S(0)}_{n}$.

In this notation, **4+2** is a name of a numeral, namely $SSSSSS(0)$, as are also **6** and **2·3**. On the other hand, the formula **4+2** is a formula of the form

$$SSSS(0) + SS(0),$$

that is, quite different than formula **6**. The equality

$$4+2 = 4+2$$

is not a logical tautology, but is a non-trivial theorem of the form

$$SSSS(0) + SS(0) = SSSSSS(0).$$

In general, **f(n)** is thus a numeral in the form

$\underbrace{S \ldots S(0)}_{f(n)}$.

In accordance with the numbering Nr of formulae, adopted in the preceding section, the function Nr(n) may be defined as follows:

(112)
$$\mathrm{Nr}(0) = \overline{0},$$
$$\mathrm{Nr}(n+1) = \overline{S}_{*}\left(_{*}\mathrm{Nr}(n)_{*}\right)^{-}.$$

The concept of representability of relations

In mathematics, a proof is the principal, and in a sense the only method of making sure of something. This trivial statement may lead us to an important definition of the concept of computability. If a function f is computable, we ought to be able to obtain its value for every given natural number n. For instance, if for $n = 10$ we have $f(10) = 57$, we ought to be able to prove that it is so in fact, that is, we ought to be able to prove, in our theory, a theorem stating that $f(10) = 57$, and any other similar theorem stating that $f(n) = k$, where n and k are definite numbers.

This intuitively convincing reasoning can be extended so as to cover all computable relations. Thus, for every computable relation $R(x, y)$, there should be an arithmetical formula $A(x_1, x_2)$ such that it is interpreted as an expression in arithmetical language of the relation R and such that if the relation $R(n, m)$ holds for the numbers n and m, then we ought to be able to prove that it is so in fact; this means that, since the predicate $A(x_1, x_2)$ is an expression of the relation R in arithmetic, then the formula $A(n, m)$ ought to be an arithmetical theorem.

The formula $A(n, m)$ is obtained by the substitution in the formula $A(x_1, x_2)$ of the numeral n for the variable x_1 and the numeral m for the variable x_2. Hence the term $A(n, m)$ has the same meaning as $A(x_1/n, x_2/m)$, and they will both be used alternately. Since negation does not lead outside computable relations, the relation $\sim R(n, m)$ is computable, too; thus, if the relation R does not hold for the numbers n and m, that is, if it is true that $\sim R(n, m)$, then this fact should also be provable in a reasonable arithmetic: accordingly, the formula $\ulcorner \sim A(n, m) \urcorner$ should be an arithmetical theorem.

By making this reasoning more precise we can write it down in a different form. Every computable relation R holding between natural numbers $\mathcal{N}$ in the domain $\mathfrak{N}_0$ ought to have the following two properties:

1) R is definable in the set $\mathcal{N}$ of the domain $\mathfrak{N}_0$ by a formula A of

468

a sufficiently rich arithmetical system T, true in $\mathfrak{N}_0$ (for the natural interpretation of the primitive concepts of arithmetic).

2) A formula $A(x_1, x_2)$, which defines the relation R, ought to be such that, for any numbers $n, m \in \mathcal{N}$, the following implications are true:

$$\text{(113)} \quad \begin{array}{l} R(n, m) \to \ulcorner A(n, m) \urcorner \in T, \\ \sim R(n, m) \to \ulcorner \sim A(n, m) \urcorner \in T. \end{array}$$

If this is so, then the fact that the relation R does, or does not, hold can be established by mathematical methods, namely by proofs in T. Obviously, the theory T must be a normal mathematical theory in the sense defined in the previous section.

Conversely, if conditions 1) and 2) are satisfied, then the relation R is computable in the intuitive sense of the word. The way in which we can establish whether it does or does not hold can be described as follows.

All the expressions of the system of arithmetic can be arranged in a sequence by the following method: we first write all the expressions of one symbol each, next those of two symbols each, then those of three symbols each, etc. Since the number of basic symbols in the theory is finite, in each group of expressions with a given number of symbols the number of expressions will be finite. By writing out, one by one, all those groups of expressions in a purely mechanical manner we can find out which of the expressions thus written out are correct mathematical proofs in the theory. This is possible, because the set of proofs is computable. Concerning an expression which is a proof it is also easy to find out which theorem it proves. Hence, we can find out whether the last element of the proof is perhaps the formula $A(n, m)$ or $\sim A(n, m)$. Since, under the law of excluded middle, for any $n, m \in \mathcal{N}$

$$R(n, m) \vee \sim R(n, m),$$

hence conditions (113) yield that

$$\text{(114)} \quad \ulcorner A(n, m) \urcorner \in T \vee \ulcorner \sim A(n, m) \urcorner \in T.$$

Thus, by writing out mechanically all the proofs in T one by one, beginning with the shortest ones and passing on to those of increasing length, at some time we must come across a proof of the formula $A(n, m)$ or a proof of the formula $\sim A(n, m)$, since either one or the other of these two formulae has a proof. Since it has been assumed that the theory T is true in the model $\mathfrak{N}_0$, if $\ulcorner A(n, m)\urcorner \in T$, then it is true in $\mathfrak{N}_0$, and thus, by 1), the relation $R(n, m)$ holds. And if $\ulcorner \sim A(n, m)\urcorner \in T$, then, by 1), it must be true that $\sim R(n, m)$. Thus, if we come across a proof of the formula $A(n, m)$, then we are sure that the relation R holds between n and m, and if we come across a proof of $\sim A(n, m)$, then we are sure that R does not hold between n and m.

Conditions 1) and 2), as stated above, explicitly treat the concept of computability as semantic in nature, defined in metamathematical research. Note that condition 1) is self-evident for functions and relations computable in the sense of Definition 1, because the class of computable functions has been defined as the least class of functions definable by certain simple methods. All computable functions and relations are then definable (in the sense of the definition given in Section 2, p. 445) by means of certain simple formulae. The only problem is whether there exists an arithmetical system which satisfies conditions (113). It will be shown soon that there are very many such systems. A fairly simple example is provided by the following system Ar, with which we shall now be concerned.

The theory Ar has the following specifically arithmetical axioms: The axioms on zero, the *less than* relation, and the successor function:

1. $\sim (x < 0)$,
2. $x < y \rightarrow \sim (y < x)$,
3. $x < y \lor y < x \lor x = y$,
4. $S(x) = S(y) \rightarrow x = y$,
5. $0 < S(x)$,
6. $x < y \rightarrow S(x) < S(y)$,
7. $x < S(y) \rightarrow (x < y \lor x = y)$.

The axioms of induction for $+$, $\cdot$, P:

8. $x+0 = x$,
9. $x+S(y) = S(x+y)$,
10. $x \cdot 0 = 0$,
11. $x \cdot S(y) = x \cdot y+x$,
12. $P(x, 0) = S(0)$,
13. $P\big(x, S(y)\big) = P(x, y) \cdot x$.

4. REPRESENTABILITY

It can easily be seen that if we consider a theory T true in the model $\mathfrak{N}_0$, then condition (113) implies that the relation R is definable by a formula $A(x_1, x_2)$, for two numbers n and m satisfy the formula $A(x_1, x_2)$ in $\mathfrak{N}_0$ if and only if $R(n, m)$. In fact, if $R(n, m)$, then, by (113), $\ulcorner A(n, m) \urcorner \in T$, and, hence, $\ulcorner R(n, m) \urcorner \in E(\mathfrak{N}_0)$ for the ordinary interpretation of the terms; n and m are names of the numbers n and m, hence n, m satisfy $\ulcorner A(x_1, x_2) \urcorner$ in $\mathfrak{N}_0$. Likewise, if $\sim R(n, m)$, then $\ulcorner \sim A(n, m) \urcorner \in T$, and hence n, m satisfy $\ulcorner \sim A(x_1, x_2 \urcorner$ in $\mathfrak{N}_0$, that is, n, m do not satisfy $\ulcorner A(x_1, x_2) \urcorner$ in $\mathfrak{N}_0$. Thus, in accordance with the definition given on p. 445, the relation R is definable by the formula $A(x_1, x_2)$.

The foregoing analyses show that it is essential to consider consistent theories which satisfy condition (113). Thus, it is relevant to introduce the following concept:

DEFINITION 7. *A relation $R(n, m)$ between natural numbers is strongly represented by a formula $A(x_1, x_2)$ in a set T of sentences if and only if, for all $n, m \in \mathcal{N}$,*

$$
(115) \quad
\begin{aligned}
R(n, m) &\equiv \ulcorner A(n, m) \urcorner \in T, \\
\sim R(n, m) &\equiv \ulcorner \sim A(n, m) \urcorner \in T.
\end{aligned}
$$

This concept can, of course, be extended so that it cover relations of three or more arguments, and also relations of one argument, that is, sets: *A set X of natural numbers is strongly represented in a set T of sentences by a formula $A(x_1)$ if and only if, for all $n \in \mathcal{N}$,*

$$(116) \quad n \in X \equiv \ulcorner A(n) \urcorner \in T,$$

$$(117) \quad n \notin X \equiv \ulcorner \sim A(n) \urcorner \in T.$$

In arithmetic, as we shall soon see, strong representability coincides with computability. In addition to the above concept of strong representability we shall introduce now the concept of weak representability, also to be used below.

DEFINITION 8. *A set X (a relation R) is weakly representable in a set T of sentences by a formula $A(x_1), (A(x_1, x_2))$ if and only if, for every $n \in \mathcal{N}$ ($n, m \in \mathcal{N}$), equivalence (116) (the first of equivalences (115)) holds.*

It follows immediately from this definition that if X is strongly representable in T, then both X and its complement $-X$ are also weakly representable in T. Strong representabiliy implies weak representability, but not conversely. Condition (115) of strong representability is equivalent to condition (113), as formulated above, combined with the condition that the theory T is consistent. Since, under the law of consistency, R and $\sim R$ cannot both hold, equivalences (115) exclude the possibility of A and $\ulcorner \sim A \urcorner$ being simultaneously theorems in T. Conversely, if T is consistent, then the implications of (113) can be strengthened so as to become the equivalences of (115).

Note also that if a relation R is strongly representable in a theory T, then it is strongly representable in every consistent extension of the theory T.

A theory T' is called an *extension* of a theory T, if $T \subset T'$, that is *all the theorems of T are theorems of T'*. Hence, if $\ulcorner A(n) \urcorner \in T$, then $\ulcorner A(n) \urcorner \in T'$, and, likewise, if $\ulcorner \sim A(n) \urcorner \in T$, then $\ulcorner \sim A(n) \urcorner \in T'$. Thus, the implications of (113) for T entail the implications of (113) for T'. Since T' is consistent, those implications can be strengthened to equivalences.

It will be proved soon that computable relations are those and only those relations which are strongly representable in arithmetic. That theorem will be proved for the arithmetic Ar, but it is also valid for very many weaker and stronger systems which contain an arithmetical part. That theorem is precisely the metamathematical definition of computability.

Examples of representation of relations and functions in arithmetic

The *less than* relation and the identity relation can serve as easy examples of relations strongly representable in Ar. Since the arithmetic Ar is true in the model $\mathfrak{N}_0$, if the symbol $<$ is interpreted as a name of the *less than* relation in the domain $\mathfrak{N}_0$, hence, we suppose that the *less than* relation is strongly represented in Ar by the predicate $x_1 < x_2$. In addition, we suppose that, the identity relation is represented by the predicate $x_1 = x_2$. By Definition 7, it ought to be true that, for all $n, m \in \mathcal{N}$,

(118)
$$n < m \equiv \ulcorner n < m \urcorner \in \text{Ar},$$
$$\sim (n < m) \equiv \ulcorner \sim (n < m) \urcorner \in \text{Ar}.$$

For instance, since $0 < 1$ and $3 < 5$, the sentences $\mathbf{0 < 1}$ and $\mathbf{3 < 5}$, that is, the sentences

$$0 < S(0),$$
$$SSS(0) < SSSSS(0)$$

ought to be theorems in Ar, because these sentences state in the language of Ar that $0 < 1$ and $3 < 5$. On the other hand, since 1 is not less than 0 and 3 is not less than 2, by (118), the sentences

$$\sim S(0) < 0,$$
$$\sim SSS(0) < SS(0)$$

ought to be theorems in Ar.

The *less than* relation is, in fact, strongly represented in Ar by the predicate $x_1 < x_2$. All the above sentences are, in fact, theorems in Ar, which can be proved by induction. First, it is easy to see that formulae (118) are satisfied for $n = 0$: by substituting in Axiom 5 any numeral m for x we obtain the sentences $0 < m$ as theorems in Ar for any $m > 0$. Next, it is easy to see that, if $\ulcorner n < m \urcorner \in \text{Ar}$, then, by Axiom 6, also $\ulcorner n+1 < m+1 \urcorner \in \text{Ar}$. Hence, by induction on the metalogical level, we have that if $\ulcorner n < m \urcorner \in \text{Ar}$, then $\ulcorner n+k < m+k \urcorner \in \text{Ar}$ for any k. If $n < m$ for any $n, m \in \mathcal{N}$, then $0 < m-n$, and, hence, $\ulcorner 0 < m-n \urcorner \in \text{Ar}$. This and the previous conclusion yield that

$$\ulcorner 0 + n < (m-n) + n \urcorner \in \text{Ar}, \quad \text{that is,} \quad \ulcorner n < m \urcorner \in \text{Ar}.$$

If, on the contrary, $\sim (n < m)$, then either $n = m$, or $m < n$. If $n = m$, then Axiom 2 yields that $\ulcorner \sim (m < m) \urcorner \in \text{Ar}$. If $m < n$, then, by what has been proved above, $\ulcorner m < n \urcorner \in \text{Ar}$, and, hence, by Axiom 2, $\ulcorner \sim (n < m) \urcorner \in \text{Ar}$. Since the arithmetic Ar is true in the model $\mathfrak{N}_0$, it is consistent, and the implications obtained above can be strengthened to become equivalences of (118).

If $n = m$, then $\ulcorner n = m \urcorner$ follows in Ar directly from the logical axiom $x = x$. If, on the contrary, $n \neq m$, then either $n < m$, or $n > m$. Hence, by the fact proved above that the *less than* relation is strongly representable, we have either $\ulcorner n < m \urcorner \in \text{Ar}$, or $\ulcorner m < n \urcorner \in \text{Ar}$. In either case, $\ulcorner \sim (n = m) \urcorner$ is obtained in Ar from Axiom 2.

Similarly, for other primitive terms of Ar it is true that the relations denoted by these terms are strongly representable in arithmetic. (This fact will soon be proved in Theorem 19.) The remaining arithmetical terms are function terms, and the relations which they represent are functions. If a function formula represents a function, then the relation of representability can be recorded in a simpler manner. This will be demonstrated on the example of addition. Since $2+3 = 5$, the formula

$$SS(0) + SSS(0) = SSSSS(0),$$

which can be written as $\mathbf{2+3 = 5}$, ought to be a theorem in Ar. Likewise, for any numbers m and $n \in \mathcal{N}$, the formula

$$\boldsymbol{m+n = m+n}$$

ought to be a theorem in Ar.

Under the convention adopted above, the formula $\boldsymbol{m+n}$ denotes a numeral in the form of

$$\underbrace{S \dots S(0)}_{m+n}.$$

Hence, the entire formula $\boldsymbol{m+n = m+n}$ denotes a formula whose schema is

$$\underbrace{S \dots S(0)}_{m} + \underbrace{S \dots S(0)}_{n} = \underbrace{S \dots S(0)}_{m+n}$$

Likewise, if exponentiation is strongly representable by a function formula $P(x_1, x_2)$, then the formulae:

$$P\big(SS(0), S(0)\big) = SS(0),$$
$$P\big(SS(0), SS(0)\big) = SSSS(0),$$

must be theorems in Ar, since $2^1 = 2$ and $2^2 = 4$. In general, all the sentences of the form

$$P(\boldsymbol{n, m}) = \boldsymbol{n^m},$$

that is, sentences of the form:

$$P(\underbrace{S \dots S(0)}_{n}, \underbrace{S \dots S(0)}_{m}) = \underbrace{S \dots S(0)}_{n^m}$$

should be in Ar.

In view of the above we can adopt the following definition, which is an application of the general definition of representability to the representation of functions by term formulae:

DEFINITION 9. *A function $f(x)$ is strongly represented in a set T of sentences by a term formula $\Phi(x)$ if and only if, for any number $n \in \mathcal{N}$,*

(119) $\ulcorner \Phi(x/n) = f(n) \urcorner \in T.$

A function of two arguments $f(n, m)$ is, of course, represented by a formula of two variables $\Phi(x_1, x_2)$ if and only if

(120) $\ulcorner \Phi(x_1/n, x_2/m) = f(n, m) \urcorner \in T.$

THEOREM 18. *If T is a consistent theory, based on a logical system with identity and such that identity is strongly represented in T by the predicate $x = y$, and if a formula $A(x, y)$ satisfies in T the uniqueness condition for every $n \in \mathcal{N}$:*

(121) $\ulcorner \bigwedge y, z \big(A(n, y) \wedge A(n, z) \to y = z \big) \urcorner \in T,$

then a relation $f(m) = n$ is strongly represented in T by the formula $A(x, y)$ if and only if, for every m, $\ulcorner A \big(m, f(m) \big) \urcorner \in T.$

In particular, *a function $f(n)$ is strongly represented in T by a term formula $\Phi(x)$, in the sense of Definition 9, if and only if the relation $f(n) = m$ is strongly represented in T by a sentential formula $\Phi(x_1) = x_2$, in the sense of Definition 7.*

This theorem obviously also applies to functions of more arguments: If a formula $A(x, y, z)$ satisfies the uniqueness condition in T, then it represents a relation $g(n, m) = k$ if and only if $\ulcorner A \big(n, m, g(n, m) \big) \urcorner \in T$ for every n and m.

PROOF. If $A(x, y)$ strongly represents a relation $f(m) = n$, then, for any m and n, it is true that

(122) $f(m) = n \to \ulcorner A(m, n) \urcorner \in T.$

Hence, in particular, by substituting $n/f(m)$ we have

$\ulcorner A \big(m, f(m) \big) \urcorner \in T.$

Conversely, if

(123) $\ulcorner A \big(m, f(m) \big) \urcorner \in T,$

then both implication (122) and the second of the implications (113), which characterize strong representation, are true. Namely, if $f(m) = n$, then formula (123) means the same as the formula $\ulcorner A(m, n) \urcorner \in T.$

Implication (122) is thus self-evident. Hence, it remains only to demonstrate that formula (123) implies the implication

(124) $f(m) \neq n \rightarrow \ulcorner \sim A(m, n) \urcorner \in T.$

In fact, since identity is strongly represented in T, hence if $f(m) \neq n$, then

(125) $\ulcorner \sim (f(m) = n) \urcorner \in T.$

Formula (121), on the substitutions $y/f(n)$ and z/n, and formulae (123) and (125) immediately yield that $\ulcorner \sim A(m, n) \urcorner \in T$ in view of ordinary logical laws. Thus, implication (124) in fact follows from (123).

In particular, if $\Phi(x)$ is a term formula, then under the laws of logic alone the sentential formula $\Phi(x) = y$ satisfies in T the uniqueness condition (121), and consequently, by the first part of Theorem 18 and Definition 9 (formula (119)), the strong representability of a relation $f(m) = n$ by the sentential formula $\Phi(x) = y$ is equivalent to the strong representability of the function $f(m)$ by the term formula $\Phi(x)$.

It can be proved for many computable functions that they are strongly representable in the theory Ar by certain term formulae in the sense of Definition 9, but this does not apply to all such functions. On the other hand, it will be proved about all computable relations that they are strongly representable, in the sense of Definition 7, by certain sentential formulae. The difference is due to inessential causes. To have every computable function represented by a term formula we would have to increase the set of term formulae by adding to Ar an operator, for instance, the minimization operator, which would make it possible to construct term formulae that would correspond to any functions definable as relations. Concerning that operator we would have to assume infinitely many axioms falling under the schemata

$$\bigvee x A(x) \rightarrow A\big((\mu x)[A(x)]\big),$$

$$\bigwedge x \big(x < (\mu x)[A(x)] \rightarrow \sim (A(x))\big),$$

$$\bigwedge x \sim A(x) \rightarrow (\mu x)[A(x)] = 0.$$

In the system Ar, with the above schemata added to it, every computable function is represented by a term formula. Such a system would, perhaps, be quite natural for serving mathematical considerations. But from the point of view of metamathematical research, with which we are now concerned, it would be inconvenient to have infinitely many axioms specifically mathematical in nature. On the contrary, the system Ar, as it is presented so far, has a finite number of axioms. None of these axioms is a schema. We shall avail ourselves in an essential way of this finite axiomatizability of the theory Ar.

For instance, when it comes to the subtraction function $n \div m$ there is in Ar no term formula which would represent it strongly. But there exists a compound sentential formula which strongly represents the relation $n \div m = k$. As we shall see, this is the formula $A(x, y, z)$ in the form

$$x = y + z \lor (y > x \land z = 0).$$

Those functions which have their symbols in the theory Ar in the form of function terms are, as can easily be seen, strongly representable in Ar in the sense of Definition 9.

THEOREM 19. *The functions $x+1$, $x+y$, $x \cdot y$, x^y are strongly represented in Ar by the term formulae $S(x)$, $x_1 + x_2$, $x_1 \cdot x_2$, $P(x_1, x_2)$. The less than relation is strongly represented in Ar by the sentential formula $x_1 < x_2$, and the identity relation is strongly represented in Ar by the sentential formula $x_1 = x_2$.*

PROOF. For the function $x+1$ formula (119) takes on the form

(126) $\ulcorner S(n) = n+1 \urcorner \in \text{Ar}.$

In fact, for every n the theorem of the form $S(n) = n+1$ is obtained from the logical axiom $\bigwedge x(x = x)$ by the substitution of the numeral $n+1$ for x.

For the function $x+y$ formula (120) takes on the form

(127) $\ulcorner m+n = m+n \urcorner \in \text{Ar},$

that is,

$$\ulcorner \underbrace{S \ldots S(0)}_{m} + \underbrace{S \ldots S(0)}_{n} = \underbrace{S \ldots S(0)}_{m+n} \urcorner \in \text{Ar}$$

for any n, $m \in \mathcal{N}$.

This formula can easily be obtained by induction with respect to the metalogical variable n. For $n = 0$, formula (127) takes on the form

(128) $\ulcorner m+0 = m \urcorner \in \text{Ar}$

(where (128) is a consequence of Axiom 8 for x/m). It will now be proved that

(129) $\quad \ulcorner m+n = m+n \urcorner \in \mathrm{Ar} \to \ulcorner m+n+1 = m+n+1 \urcorner \in \mathrm{Ar}.$

In fact, by substituting, in Axiom 9, x/m and y/n we obtain the theorem $m+S(n) = S(m+n)$. Since $\ulcorner m+n = m+n \urcorner \in \mathrm{Ar}$, also $m+S(n) = S(m+n)$ is a theorem in Ar. By (126) that theorem is identical with the theorem $m+n+1 = m+n+1$. (128) and (129) by induction yield (127).

For the function $x \cdot y$ formula (120) takes on the form

(130) $\quad \ulcorner m \cdot n = m \cdot n \urcorner \in \mathrm{Ar},$

that is,

$$\ulcorner \underbrace{S \ldots S(0)}_{m} \cdot \underbrace{S \ldots S(0)}_{n} = \underbrace{S \ldots S(0)}_{m \cdot n} \urcorner \in \mathrm{Ar}$$

for any $n,\ m \in \mathcal{N}$.

It is easy to see that

(131) $\quad \ulcorner m \cdot 0 = m \cdot 0 \urcorner \in \mathrm{Ar},$

since $m \cdot 0 = 0$. Thus, (131) is derived directly from Axiom 10 by the substitution x/m. Next it is proved that

(132) $\quad \ulcorner m \cdot n = m \cdot n \urcorner \in \mathrm{Ar} \to \ulcorner m \cdot S(n) = m \cdot (n+1) \urcorner \in \mathrm{Ar}.$

In fact, from Axiom 11 we derive the theorem

(133) $\quad m \cdot S(n) = (m \cdot n) + m.$

Thus, if $\ulcorner m \cdot n = m \cdot n \urcorner \in \mathrm{Ar}$, then also $\ulcorner m \cdot S(n) = m \cdot n + m \urcorner \in \mathrm{Ar}$. This and Theorem (127) yield that $\ulcorner m \cdot S(n) = m \cdot n + m \urcorner \in \mathrm{Ar}$, that is exactly $\ulcorner m \cdot S(n) = m \cdot (n+1) \urcorner \in \mathrm{Ar}$. By (131), formula (130) is thus true for $n = 0$, and, by (132), if it is true for a number n, then it is also true for $n+1$. Thus, we obtain by induction that formula (130) is true for every $n \in \mathcal{N}$.

For exponentiation formula (120) takes on the form

(134) $\quad \ulcorner P(m, n) = m^n \urcorner \in \mathrm{Ar}.$

For $n = 0$ formula (134) yields the theorem

(135) $\quad P(m, 0) = S(0),$

which is obtained from Axiom 12 by the substitution x/m.

It can easily be seen that

(136) $\quad \ulcorner P(m, n) = m^n \urcorner \in \mathrm{Ar} \rightarrow \ulcorner P(m, S(n)) = m^{n+1} \urcorner \in \mathrm{Ar}.$

In fact, the theorem

(137) $\quad P(m, S(n)) = P(m, n) \cdot m$

is obtained from Axiom 13 by substitution.

Thus, if $P(m, n) = m^n \in \mathrm{Ar}$, then, by (137), also $\ulcorner P(m, S(n)) = m^n \cdot m \urcorner \in \mathrm{Ar}$. This, by (130), yields that

$$\ulcorner P(m, S(n)) = m^n \cdot m \urcorner \in \mathrm{Ar},$$

that is,

$$\ulcorner P(m, S(n)) = m^{n+1} \urcorner \in \mathrm{Ar}.$$

Thus, by (135), formula (134) is true for $n = 0$, and, if it is true for a number n, then, by (136), it is also true for $n+1$. Hence, we infer by induction that (134) is true for every $n \in \mathcal{N}$.

Thus, all the functions specified above are strongly representable by primitive terms of the theory Ar. To this theorem we add the strong representability of the *less than* relation and the strong representability of the identity relation, both proved earlier by way of example.

Theorem on representability in a weak form

It will now be proved that for every computable relation R there exists a formula which strongly represents R in Ar. We shall later strengthen that theorem by proving that there exist simple formulae, for all practical purposes with one quantifier each, which have the same property.

THEOREM 20. *Every computable relation $f(m) = n$ is strongly representable in Ar by a formula $A(x, y)$ which satisfies in Ar the uniqueness condition for every $m \in \mathcal{N}$:*

(138) $\quad \ulcorner \bigwedge y, z \big((A(m, y) \wedge A(m, z)) \rightarrow y = z \big) \urcorner \in \mathrm{Ar}.$

PROOF BY INDUCTION. We consider relations of the form $f(m) = n$ for any functions $f \in \mathrm{Comp}$. We refer to the inductive definition of the class Comp as given in Theorem 2. For those functions f which are initial for the class Comp the existence of the appropriate formulae which strongly represent those functions has been proved in Theorem 19 except for the function of subtraction.

Concerning subtraction it will be demonstrated that the relation $n_1 \div n_2 = n_3$ is strongly representable by a formula $A(x_1, x_2, x_3)$ in the form

(139) $\qquad x_1 = x_3 + x_2 \vee (x_2 > x_1 \wedge x_3 = 0).$

This formula defines the relation $n_1 \div n_2 = n_3$ in $\mathfrak{N}_0$, so it is to be assumed that it also strongly represents that relation in arithmetic. In fact, if $n_1 \div n_2 = n_3$, then either $n_1 > n_2$ and then $\ulcorner n_1 = n_3 + n_2 \urcorner \in \mathrm{Ar}$ by Theorem 19, or $n_2 > n_1$ and $n_3 = 0$, and then $\ulcorner (n_2 > n_1 \wedge n_3 = 0) \urcorner \in \mathrm{Ar}$, since the relations $<$ and $=$ are strongly representable in Ar by the formulae $x_1 < x_2$ and $x_1 = x_2$. In both cases the disjunction

$$n_1 = n_3 + n_2 \vee (n_2 > n_1 \wedge n_3 = 0),$$

which is obtained from (139) by substitution, is a theorem in Ar. Thus, the implication $n_1 \div n_2 = n_3 \to \ulcorner A(n_1, n_2, n_3) \urcorner \in \mathrm{Ar}$ is true. Hence, for any $n_1, n_2 \in \mathcal{N}$,

$$\ulcorner A(n_1, n_2, n_1 \div n_2) \urcorner \in \mathrm{Ar}.$$

Thus, under Theorem 18, it remains to prove the uniqueness condition for (139). We shall first prove the following lemma:

(140) $\qquad \ulcorner \bigwedge y(n = y + k \to y = n \div k) \urcorner \in \mathrm{Ar}.$

The proof is by induction in metalogic with respect to the variable k. For $k = 0$, formula (140) becomes $n = y + 0 \to y = n$ and is derived from Axiom 8. Suppose now that we have proved (140) for a k and for any n. We will prove it for $k + 1$. If $n = y + k + 1$, that is, $n = y + S(k)$, then, by Axiom 9, $n = S(y + k)$. It follows from Axioms 2 and 5 that $0 \neq S(x)$. Hence, for $n = 0$ the equation $n = S(y + k)$ is in contradiction with the axioms, and, thus, implies any sentence. If $n \neq 0$, that is, $n = S(n \div 1)$, then we have $S(n \div 1) = n = S(y + k)$. Thus Axiom 4 yields that $n \div 1 = y + k$. Hence, from the induction assumption (140) we derive that $y = (n \div 1) \div k$, which, since $n \neq 0$, means the same as $y = n \div (k + 1)$, which was to be proved.

The uniqueness condition for (139), which remains to be proved, has the form

$$\bigwedge n, k \in \mathcal{N} \ulcorner \bigwedge y, z \big((A(n, k, y) \wedge A(n, k, z)) \to y = z \big) \urcorner \in \mathrm{Ar}.$$

To prove it we reason as follows. By (139), the antecedent of the theorem to be proved is a disjunction of four cases:

a. $n = y+k \wedge n = z+k,$

b. $n = y+k \wedge k > n \wedge z = 0,$

c. $k > n \wedge y = 0 \wedge n = z+k,$

d. $k > n \wedge y = 0 \wedge k > n \wedge z = 0.$

In case a, it follows from Lemma (140) that $y = n \dotminus k$ and $z = n \dotminus k$, and, hence, $y = z$. In case d, $y = 0 = z$. In cases b and c, since $k > n$, by the representability of the *less than* relation, $k > n$, and, thus, $n \dotminus k = 0$. In the case b, it follows from Lemma (140) that $y = n \dotminus k = 0 = z$. In the case c, it follows from Lemma (140) that $z = n \dotminus k = 0 = y$. Thus, in all the cases $z = y$.

It will be demonstrated now that this property is inherited under the operations of superposition and effective minimization.

a. S u p e r p o s i t i o n. Suppose that $A(x, y)$ strongly represents in Ar the relation $f(m) = n$ and satisfies condition (138), and that $B(y, z)$ strongly represents in Ar the relation $g(n) = k$ and sastisfies a similar uniqueness condition

(141) $\ulcorner \bigwedge z, u\big((B(n, z) \wedge B(n, u)) \to z = u\big)\urcorner \in \mathrm{Ar}.$

It will be shown that the relation $g(f(m)) = k$ is strongly representable by the formula $\bigvee y(A(x, y) \wedge B(y, z))$, and that this formula satisfies the uniqueness condition.

Since $A(x, y)$ by assumption strongly represents the relation $f(m) = n$ and $B(y, z)$ strongly represents the relation $g(n) = k$, it is true that, for any m and n,

(142) $\ulcorner A(m, f(m))\urcorner \in \mathrm{Ar},$

(143) $\ulcorner B(n, g(n))\urcorner \in \mathrm{Ar}.$

On substituting $n/f(m)$ in (143) we obtain

(144) $\ulcorner B\big((f(m), g(f(m)))\big)\urcorner \in \mathrm{Ar}.$

It follows from (142) and (144) that

(145) $\ulcorner \bigvee y\big(A(m, y) \wedge B(y, g(f(m)))\big)\urcorner \in \mathrm{Ar}.$

By Theorem 18, it remains to prove in Ar the following uniqueness condition:

(146) $\bigwedge z_1, z_2 \Big(\big((\bigvee y_1 (A(m, y_1) \wedge B(y_1, z_1)) \wedge$
$\qquad \bigvee y_2 (A(m, y_2) \wedge B(y_2, z_2))\big) \to z_1 = z_2\Big).$

Theorem (146) is obtained in Ar in the following way. The assumed uniqueness condition (138) and formula (143) yield that if $A(m, y_1)$ and $A(m, y_2)$, then $y_1 = y_2 = f(m)$. Hence, if also $B(y_1, z_1)$ and $B(y_2, z_2)$, then $B(f(m), z_1)$ and $B(f(m), z_2)$. Hence, by (141), $z_1 = z_2$.

By Theorem 18, formulae (145) and (146) mean that the formula $\bigvee y(A(x, y) \wedge B(y, z))$ strongly represents the relation $g(f(m)) = k$.

b. E f f e c t i v e   m i n i m i z a t i o n. Let $g(n, m)$ be a function of two arguments such that the relation $g(n, m) = k$ is strongly represented by the formula $A(x, y, z)$, which satisfies the uniqueness condition. Hence, for any n and m,

(147)
$$\ulcorner A(n, m, g(n, m)) \urcorner \in \text{Ar},$$
$$g(n, m) \neq 0 \rightarrow \ulcorner \sim A(n, m, 0) \urcorner \in \text{Ar}.$$

Suppose that a function f is defined by effective minimization as follows:

(148) $f(n) = (\mu m) [g(n, m) = 0],$

and that the effectiveness condition

(149) $\bigwedge n \bigvee m g(n, m) = 0$

is satisfied.

It will be proved that the relation $f(n) = m$ is strongly represented by the formula

(150) $A(x, y, 0) \wedge \bigwedge z\left(z < y \rightarrow \sim \left(A(x, z, 0)\right)\right).$

Note first that formula (150) satisfies the uniqueness condition in Ar. In fact, if (150) holds and if also

$$A(x, y_1, 0) \wedge \bigwedge z\left(z < y_1 \rightarrow \sim \left(A(x, z, 0)\right)\right),$$

then under the laws of logic it cannot be true that $y_1 < y$ or that $y < y_1$, so that, by Axiom 3, it must be true that $y_1 = y$. Thus, the uniqueness condition for (150) is a theorem in Ar. To prove strong representability, with which we are concerned, it suffices to demonstrate that, for every n,

(151) $\ulcorner A(n, f(n), 0) \wedge \bigwedge z\left(z < f(n) \rightarrow \sim \left(A(n, z, 0)\right)\right) \urcorner \in \text{Ar}.$

482

The first element of the above conjunction can be obtained very easily. It follows from (148) and (149) that $g(n, f(n)) = 0$; this and formula (147), on the substitution $m = f(n)$, yield that

(152) $\quad \bigwedge n \ulcorner A(n, f(n), 0) \urcorner \in \text{Ar}.$

It also follows from (148) that

$$\bigwedge m < f(n) g(n, m) \neq 0.$$

This and the second formula of (147) yield that

(153) $\quad \bigwedge m < f(n) \ulcorner \sim A(n, m, 0) \urcorner \in \text{Ar}.$

Note now that it follows from the arithmetical Axioms 1 and 7 (on the substitution $y/0$) that

$$\ulcorner \bigwedge x (x < S(0) \to x = 0) \urcorner \in \text{Ar}.$$

This theorem and Axiom 7 (with the substitution $y/SS(0)$) yield that

$$\ulcorner \bigwedge x \big(x < SS(0) \to (x = 0 \lor x = S(0)) \big) \urcorner \in \text{Ar}.$$

In this way, by applying Axiom 7 k times, we obtain:

$$\ulcorner \bigwedge x \big(x < S(k) \to (x = 0 \lor x = S(0) \lor \ldots \lor x = k) \big) \urcorner \in \text{Ar}.$$

Since it follows from the fact that the *less than* relation is strongly representable that $\ulcorner n < S(k) \urcorner \in \text{Ar}$ for any $n \leqslant k$, the above implication can be strengthened into the equivalence:

(154) $\quad \ulcorner \bigwedge x \big(x < S(k) \equiv (x = 0 \lor x = S(0) \lor \ldots \lor x = k) \big) \urcorner \in \text{Ar}.$

In particular,

(155) $\quad \ulcorner \bigwedge x \big(x < f(n) \equiv (x = 0 \lor x = 1 \lor \ldots \lor x = f(n) \dot{-} 1) \big) \urcorner \in \text{Ar}.$

It clearly follows from (153) and (155) that

(156) $\quad \ulcorner \bigwedge x (x < f(n) \to \sim A(n, x, 0)) \urcorner \in \text{Ar}.$

Formulae (156) and (152) yield (151). Thus, the effective minimization operation does not lead outside functions strongly representable in Ar. Consequently, all computable relations in the form $f(n) = m$ are strongly representable in Ar, and, hence, all computable relations, since they are obtained by setting computable functions equal to zero, are strongly representable in Ar.

The converse theorem, stating that every relation which is strongly representable in Ar is computable, is much easier to prove. An outline of such a proof has been contained in the introductory remarks in this section. The idea of the proof is as follows. If a relation R is strongly representable in Ar by a formula $A(x)$, the following equivalences hold:

$$(157) \quad \begin{aligned} R(n) &\equiv \ulcorner A(n) \urcorner \in \text{Ar}, \\ \sim R(n) &\equiv \ulcorner \sim A(n) \urcorner \in \text{Ar}. \end{aligned}$$

Since, under the law of excluded middle, it is true for every n that $R(n)$ or $\sim R(n)$, hence equivalences (157) yield the disjunction

$$\ulcorner A(n) \urcorner \in \text{Ar} \lor \ulcorner \sim A(n) \urcorner \in \text{Ar}.$$

Since the theory Ar is consistent, these two possibilities are mutually exclusive. The set Ar is recursively enumerable (Theorem 17). Hence, in enumerating all the theorems of Ar (for instance, by writing out all the proofs one by one) we must at some time come across either a theorem of the form $A(n)$ or a theorem of the form $\sim A(n)$. When we come across one of these theorems we know, in accordance with equivalences (157), whether it is $R(n)$ or $\sim R(n)$ which holds. Thus, we can make sure in a finite number of steps whether the relation R does or does not hold.

This yields the following

THEOREM 21. *If a relation is strongly representable in* Ar, *then it is computable.*

In the process of further, more detailed analyses to be carried out in this section we shall give another, more precise version of the proof of Theorem 21, in which concepts to be introduced later will be used. Theorems 20 and 21 show that representability in Ar is identical with recursiveness.

Representation of elementary recursive relations

This part of the present section and the next two can be omitted in the first reading, if the reader is interested in the weaker form of Theorem 30 on the undecidability of arithmetic. Cf. the next section.

Possibly the simplest proof of Theorem 20 has just been given above. Theorem 20 alone suffices to deduce Theorem 30 on the undecidability of the arithmetic Ar, but the form of this theorem would be rather weak.

4. REPRESENTABILITY

To obtain a stronger version of the theorem on the undecidability of arithmetic, namely a version which yields the form of the independent sentences, we have to strengthen the theorem on representability. Such exactly is the role of the next few theorems.

THEOREM 22. *If* $f \in$ Erec, *then there exists a formula* $A(x_1, ..., x_k)$ *which strongly represents in* Ar *the relation* $n_1 = f(n_2, ..., n_k)$ *and which contains only bounded quantifiers and satisfies in* Ar, *for any* $n_2, ..., n_k \in \mathcal{N}$, *the following condition of uniqueness*:

$$\ulcorner \bigwedge x_1, x \Big(\big(A(x_1, n_2, ..., n_k) \wedge A(x, n_2, ..., n_k) \big) \to x_1 = x \Big) \urcorner \in \text{Ar}.$$

The language of the theory Ar has no special symbols for bounded quantifiers. Hence this term is always used in the sense that the existential quantifiers $\bigvee x$ is called *bounded* if it is followed by a conjunction whose first element has the form $x < A$, where A is a term formula in which the variable x does not occur. Similarly, the universal quantifier $\bigwedge \bar{x}$ is called *bounded*, if it is followed by an implication whose antecedent has the form $x < A$, where A is a term formula in which the variable x does not occur.

The proof is by induction and refers to inductive Definition 4 of the class Erec. As often happens, we have to prove by induction a condition which is somewhat stronger than that stated in the theorem to be proved. It will be demonstrated that every function $f \in$ Erec has the following property W:

$$W(f) \equiv \text{ there exists a formula } A(x_1, ..., x_k) \text{ such that}$$

1. *the relation* $n_1 = f(n_2, ..., n_k)$ *is strongly representable in* Ar *by the formula* A;

2. *if the formula* A *contains a quantifier, then that quantifier is bounded in the sense specified above*;

3. *there exists a term formula* $B(x_2, ..., x_k)$ *such that* B *strongly represents in* Ar *a function* b *which is increasing with respect to all its arguments*;

4. *the function* b *forms the upper limit of the function* f; *for any* $n_2, ..., n_k \in \mathcal{N}$

(158) $f(n_2, ..., n_k) < b(n_2, ..., n_k)$;

485

III. HIERARCHY

5. *sentences which express the uniqueness of the formula A*

$$\ulcorner \bigwedge x, x_1 \big(\big(A(x, n_2, ..., n_k) \wedge A(x_1, n_2, ..., n_k) \big) \rightarrow$$
$$x = x_1 \big) \urcorner \in \mathrm{Ar}$$

are theorems in Ar.

Note that, in view of strong representability, condition (158) by 1 and 3 implies that the sentences

$$A(n_1, ..., n_k) \rightarrow n_1 < B(n_2, ..., n_k)$$

are theorems in Ar.

It will be demonstrated first that the property W is an attribute of the initial functions of the Erec class, and next, that it is inherited under the operations of superposition and bounded minimization, under which the class Erec is closed.

I. By Definition 4, the initial functions of the class Erec are: $n+1$, $n \dot- m$, n^m. By Theorems 18 and 19, the relations $n = S(m)$ and $n_1 = n_2^{m_3}$ are strongly represented in Ar by the formulae $x_1 = S(x_2)$ and $x_1 = P(x_2, x_3)$. The formulae

$$\big(n_1 = S(n_2) \rightarrow n_1 < S\big(S(n_2)\big)\big), \quad \cdot$$
$$\big(n_1 = P(n_2, n_3) \rightarrow n_1 < S\big(P\big(S(S(n_2)), S(n_3)\big)\big)\big)$$

also are theorems in Ar.

This shows that the relation $<$ is strongly represented by $\ulcorner x < y \urcorner$, and that $n+1$ and n^m are strongly represented by $S(x)$ and $P(x, y)$. Formula (139), which strongly represents subtraction, also satisfies the condition W, and the successor of the minuend plays the role of the function b that sets the upper limit for the difference:

$$n \dot- m < S(n).$$

II. S u p e r p o s i t i o n. The fact that superposition of functions does not lead outside the functions which have the property under consideration will be checked for functions of one argument. Let f and g be functions which satisfy the conditions W:

(159) $\ulcorner A\big(n, f(n)\big) \urcorner \in \mathrm{Ar}$,

(160) $\ulcorner C\big(n, g(n)\big) \urcorner \in \mathrm{Ar}$.

486

Next, there exist term formulae B and D such that they strongly represent certain increasing functions b and d which set the upper limits for the functions f and g, respectively:

(161) $f(n) < b(n)$,

(162) $g(n) < d(n)$;

as has been noted above, this means that, for any $n, k \in \mathcal{N}$,

(163) $\ulcorner (C(n, k) \rightarrow k < D(n)) \urcorner \in \mathrm{Ar}$.

Moreover, the uniqueness conditions are satisfied:

(164) $\ulcorner \bigwedge y, z\big((C(n, y) \wedge C(n, z)) \rightarrow y = z\big) \urcorner \in \mathrm{Ar}$,

(165) $\ulcorner \bigwedge y, z\big((A(n, y) \wedge A(n, z)) \rightarrow y = z\big) \urcorner \in \mathrm{Ar}$.

Further, all the quantifiers occurring in A and C are bounded. It is easy to guess that the superposition $m = f(g(n))$ will be strongly represented in Ar by a formula $A^*(x, y)$ in the form

(166) $\bigvee z < D(x)\, (C(x, z) \wedge A(z, y))$,

in view of the fact that

$$m = f(g(n)) \equiv \bigvee z\big(z = g(n) \wedge m = f(z)\big).$$

By (163), the variable z may be bounded in the manner shown by formula (166).

It will be proved that formula (166) has the property W.

1. A^* strongly represents the relation $m = f(g(n))$ in Ar. In fact, it follows from (159) that

(167) $\ulcorner A\big(g(n), f(g(n))\big) \urcorner \in \mathrm{Ar}$.

Formulae (160) and (163) yield

(168) $\ulcorner g(n) < D(n) \urcorner \in \mathrm{Ar}$.

It follows from (160), (167) and (168) that

(169) $\ulcorner \bigvee z < D(n)\big(C(n, z) \wedge A\big(z, f(g(n))\big)\big) \urcorner \in \mathrm{Ar}$.

The uniqueness condition for (166) follows from (164), (160), (165) and (167), hence, by Theorem 18, formula (169) implies that formula (166) strongly represents the relation $m = f(g(n))$.

2. All the quantifiers occurring in A^* are bounded.

3. The formula $B\big(D(x)\big)$ strongly represents the superposition of functions $b\big(d(n)\big)$. The function $b\big(d(n)\big)$, being a superposition of two increasing functions, is itself an increasing function. Next, we obtain from (161) that $f\big(g(n)\big) < b\big(g(n)\big)$. Since b is an increasing function, we obtain from (162) that $b\big(g(n)\big) < b\big(d(n)\big)$. Thus, finally we have

4. $f\big(g(n)\big) < b\big(d(n)\big)$.

III. B o u n d e d m i n i m i z a t i o n. Suppose that a function g satisfies the conditions resulting from the property W

$$(170) \quad \begin{aligned} g(n, m) = k &\equiv \ulcorner A(n, m, k) \urcorner \in \mathrm{Ar}, \\ g(n, m) \neq k &\equiv \ulcorner \sim A(n, m, k) \urcorner \in \mathrm{Ar}, \end{aligned}$$

and that all the quantifiers occurring in A are bounded. The other conditions resulting from W will not be used in this step of the proof.

If a function f is defined by the bounded minimization operation:

$$(171) \quad f(n, k) = (\mu m \leqslant k)\,[g(n, m) = 0],$$

then the relation $m = f(n, k)$ is strongly represented in Ar by the following predicate $A^*(x, y, w)$:

$$(172) \quad \begin{aligned} &\big(w < S(y) \wedge A(x, w, 0) \wedge \bigwedge z < w\big(\sim A(x, z, 0)\big)\big) \vee \\ &\big(w = 0 \wedge \bigwedge z < S(y)\big(\sim A(x, z, 0)\big)\big). \end{aligned}$$

Note first that this predicate satisfies the uniqueness condition. We can carry out the following reasoning in Ar.

If $A^*(n, k, w)$, and $A^*(n, k, v)$, then, as A^* is a disjunction consisting of two elements, we have to examine four cases. But since the formula $\bigwedge z < S(k)\big(\sim A(n, z, 0)\big)$ is in contradiction both with the formula $w < S(k) \wedge A(n, w, 0)$ and with the formula $v < S(k) \wedge A(n, v, 0)$, only the following two cases remain to be examined:

$$(173) \quad \begin{aligned} (\alpha)&w < S(k) \wedge (\beta)A(n, w, 0) \wedge (\gamma) \bigwedge z < w\big(\sim A(n, z, 0)\big) \wedge \\ (\delta)&v < S(k) \wedge (\varepsilon)A(n, v, 0) \wedge (\eta)\bigwedge z < v\big(\sim A(n, z, 0)\big); \end{aligned}$$

$$(174) \quad w = 0 \wedge \bigwedge z < S(k)\big(\sim A(n, z, 0)\big) \wedge v = 0.$$

In the case of (174) we obviously have $w = 0 = v$. In the case of (173) we refer to equivalence (154). It follows from it and from the strong representability of the *less than* relation that, since $w < S(k)$ and $v < S(k)$, then $v < w$, or $w < v$, or $w = v$. In this way we can omit

488

Axiom 3, which proves to be unnecessary in the proof of this theorem. If $w < v$ held, then (η) would yield the negation of (β). If $v < w$ held, then (γ) would yield the negation of (ε). Hence, the only remaining possibility is that $w = v$. Thus, we can prove in Ar the theorem

(175) $\big(A^*(n, k, w) \wedge A^*(n, k, v)\big) \to w = v.$

It will be proved now that, for any n, $k \in \mathcal{N}$, the sentence $A^*(n, k, f(n, k))$ is a theorem in Ar. Two cases are to be distinguished in this connection.

In accordance with (171) and with the meaning of the bounded minimization operation either there are no numbers $m < S(k)$ for which $g(n, m) = 0$, and then $f(n, k) = 0$, or else such numbers do exist, and then $f(n, k)$ is the least of them. We shall consider each of the two cases separately.

(i) If $\bigwedge m < S(k) \, g(n, m) \neq 0$, and, thus, $f(n, k) = 0$, then by (170),

(176) $\bigwedge m < S(k) \ulcorner \sim A(n, m, 0) \urcorner \in \text{Ar}.$

By (154), for every $k \in \mathcal{N}$, the sentence

(177) $\bigwedge z \big(z < S(k) \equiv (z = 0 \vee z = S(0) \vee z = SS(0) \vee$

    $\ldots \vee z = k)\big),$

stating that a number which is less than $S(k)$ equals one of the numbers $m < S(k)$, is a theorem in Ar.

Theorems (177) and (176) yield

(178) $\bigwedge z < S(k) \, (\sim A(n, z, 0)),$

as a theorem in Ar.

Theorem (178) may be joined with the theorem $0 = 0$ by the conjunction symbol.

(ii) If there exist numbers $m < S(k)$ for which $g(n, m) = 0$, and if $f(n, k)$ is the least of them, then it is true that $f(n, k) < S(k)$, and $g(n, f(n, k)) = 0$, and $\bigwedge m < f(n, k) \, (g(n, m) \neq 0)$. Since the relations $j < i$ and $g(n, m) = k$ are strongly representable in Ar by the predicates $x_1 < x_2$ and $A(x, y, z)$, it follows from the above that

(179) $\ulcorner f(n, k) < S(k) \urcorner \in \text{Ar}, \quad \ulcorner A\big(n, f(n, k), 0\big) \urcorner \in \text{Ar},$

(180) $\bigwedge m < f(n, k) \ulcorner \sim A(n, m, 0) \urcorner \in \text{Ar}.$

As is the case of (177), the sentence

(181) $\bigwedge z\big(z < f(n, k) \equiv \big(z = 0 \vee z = S(0) \vee \ldots \vee z = f(n, k) \dot{-} 1\big)\big)$

is a theorem in Ar. Now, (180) and (181) yield that

(182) $\ulcorner \bigwedge z < f(n, k) \sim A(n, z, 0)\urcorner \in \text{Ar}.$

(179) and (182) yield that the conjuction of sentences (179) and (182) is a theorem in Ar. By combining the two cases, (i) and (ii), we obtain that in the first case $((178) \wedge 0 = 0)$ is a theorem, and in the second, the conjunction of the theorems stated in (179) and (182) is a theorem. Thus the disjunction

$((178) \wedge 0 = 0) \vee (\textit{the conjunction of the theorems from formulae} (179) \textit{ and } (182))$

always is a theorem.

The above disjunction is obtained from formula (172) by the substitutions $x/n, y/k, z/f(n, k)$. Thus, for any $n, k \in \mathcal{N}$,

(183) $\ulcorner A^*\big(n, k, f(n, k)\big)\urcorner \in \text{Ar}.$

By Theorem 18, the formula $A^*(x, y, z)$ strongly represents the relation $m = f(n, k)$.

All the quantifiers occurring in A^* are bounded. By (171), the function $k+1$ is that increasing function which sets the upper limit for $f(n, k)$:

$f(n, k) < k+1.$

Thus, the property W is inherited under those operations under which the class Erec is closed.

Note that Axiom 3 of the system of arithmetic is not needed to prove Theorem 22. Axiom 3 is used only in connection with the proof of representability of non-bounded effective minimization in the proofs of Theorems 20 and 20*.

It follows immediately from Theorem 22 that every elementary recursive relation R is strongly representable in Ar by a formula in which all quantifiers occur bounded. The converse theorem can also be proved:

THEOREM 23. *Every sentential formula $A(x_1, \ldots, x_k)$ in which all quantifiers occur bounded strongly represents in Ar an elementary recursive relation $R(n_1, \ldots, n_k)$, namely that relation which is defined by that formula in $\mathfrak{N}_0$ for the natural interpretation of terms.*

490

PROOF BY INDUCTION.

1. As is known from Theorem 19 and from arguments carried out by way of example, the primitive formulae of the language of the theory Ar strongly represent those elementary recursive functions and relations which they define for the natural interpretation of terms.

2. If term formulae $a(x)$ and $b(y)$ strongly represent functions $f(n)$ and $g(n)$, respectively, then the superposition of these formulae $a(b(x))$ strongly represents the superposition of these functions $f(g(n))$. In fact, if, for every n,

$$(184) \quad \begin{array}{l} \ulcorner a(n) = f(n) \urcorner \in \text{Ar}, \\ \ulcorner b(n) = g(n) \urcorner \in \text{Ar}, \end{array}$$

then $\ulcorner a(b(n)) = f(g(n)) \urcorner \in \text{Ar}$. This fact has referred to in the previous proof.

3. If $A(x)$ strongly represents a relation $R(n)$, and if a term formula $a(x)$ strongly represents a function $f(n)$, then the superposition $A(a(x))$ strongly represents the relation $R(f(n))$.

In fact, if in addition to (184) the following conditions are satisfied:

$$(185) \quad \begin{array}{l} R(n) \equiv \ulcorner A(n) \urcorner \in \text{Ar}, \\ \sim R(n) \equiv \ulcorner \sim A(n) \urcorner \in \text{Ar}, \end{array}$$

then the following conditions also hold:

$$\begin{array}{l} R(f(n)) \equiv \ulcorner A(a(n)) \urcorner \in \text{Ar}, \\ \sim R(f(n)) \equiv \ulcorner \sim A(a(n)) \urcorner \in \text{Ar}. \end{array}$$

4. If $A(x)$ and $B(x)$ strongly represent relations $R(n)$ and $R'(n)$, respectively, then $\ulcorner A(x) \land B(x) \urcorner$ strongly represents the relation $R(n) \land R'(n)$; the same holds in a parallel way for the other sentential connectives. In fact, if, besides (185), we have

$$(186) \quad \begin{array}{l} R'(n) \equiv \ulcorner B(n) \urcorner \in \text{Ar}, \\ \sim R'(n) \equiv \ulcorner \sim B(n) \urcorner \in \text{Ar} \end{array}$$

then the following equivalences are also true:

$$(187) \quad \begin{array}{l} R(n) \land R'(n) \equiv \ulcorner A(n) \urcorner \in \text{Ar} \land \ulcorner B(n) \urcorner \in \text{Ar} \equiv \\ \ulcorner A(n) \land B(n) \urcorner \in \text{Ar}, \end{array}$$

as well as the implications:

$$\sim \big(R(n) \wedge R'(n)\big) \rightarrow \big(\sim R(n) \vee \sim R'(n)\big) \rightarrow$$
$$\ulcorner \sim A(n) \urcorner \in \mathrm{Ar} \vee \ulcorner \sim B(n) \urcorner \in \mathrm{Ar} \rightarrow$$
$$\ulcorner \sim A(n) \vee \sim B(n) \urcorner \in \mathrm{Ar} \rightarrow \ulcorner \sim \big(A(n) \wedge B(n)\big) \urcorner \in \mathrm{Ar}.$$

Conversely, by referring to the consistency of Ar and to equivalences (187), we obtain

$$\ulcorner \sim \big(A(n) \wedge B(n)\big) \urcorner \in \mathrm{Ar} \rightarrow \sim \big(\ulcorner A(n) \urcorner \in \mathrm{Ar} \wedge \ulcorner B(n) \urcorner \in \mathrm{Ar}\big) \rightarrow$$
$$\sim \big(R(n) \wedge R'(n)\big).$$

For negation, the parallel property follows directly from the definition of representability. For the remaining sentential connectives, the property in question holds as consequence of the fact that it holds for negation and conjunction.

5. If $A(x)$ strongly represents $R(n)$, then the formula $\bigwedge x < S(y) A(x)$ strongly represents the relation $\bigwedge n \leqslant k R(n)$. In fact, if $\bigwedge n \leqslant k R(n)$, then, by (185),

(188) $\bigwedge n \leqslant k \ulcorner A(n) \urcorner \in \mathrm{Ar}.$

This, by (177), yields that

(189) $\ulcorner \bigwedge x < S(k) \big(A(x)\big) \urcorner \in \mathrm{Ar}.$

Conversely, if (189) holds, then, on substituting the figures n for x, we have (188) for $n \leqslant k$, so that, in accordance with (185), it is true $\bigwedge n \leqslant k R(n)$. Likewise, if $\sim \big(\bigwedge n \leqslant k R(n)\big)$, that is, $\bigvee n \leqslant k \sim R(n)$, then

(190) $\bigvee n \leqslant k \ulcorner \sim A(n) \urcorner \in \mathrm{Ar}.$

If $n \leqslant k$, then $\ulcorner n < S(k) \urcorner \in \mathrm{Ar}$, and hence, by (190), for some $n \leqslant k$,

(191) $\ulcorner n < S(k) \wedge \sim A(n) \urcorner \in \mathrm{Ar}.$

Now (191), under the quantifier laws, yields that

(192) $\ulcorner \bigvee x < S(k) \big(\sim A(x)\big) \urcorner \in \mathrm{Ar},$

and hence, under De Morgan's laws,

(193) $\ulcorner \sim \big(\bigwedge x < S(k) \big(A(x)\big)\big) \urcorner \in \mathrm{Ar}.$

Conversely, if (193), then (192). If (192), then, by (177), there must exist an $n \leqslant k$ for which $\ulcorner \sim A(n) \urcorner \in \mathrm{Ar}$, so that, by (185), $\bigvee n \leqslant k \sim R(n)$, that is, $\sim (\bigwedge n \leqslant k R(n))$.

6. If $A(x)$ strongly represents $R(n)$, then the formula $\bigvee x < S(y) (A(x))$ strongly represents the relation $\bigvee n \leqslant k R(n)$. This relationship is obtained directly from 4 and 5 above, since a bounded existential quantifier can be expressed by a bounded universal quantifier and negation.

Since definition by means of superposition, sentential connectives and bounded quantifiers does not lead outside elementary recursive relations, as characterized by Definition 4 and Theorem 6, every relation which is definable by a sentential formula in which all quantifiers are bounded is an elementary recursive relation. Furthermore, parts 1 to 6 of the proof show by induction that a sentential formula in which all quantifiers are bounded strongly represents that relation which it defines for the natural interpretation of terms. Thus, every sentential formula in which all quantifiers are bounded strongly represents an elementary recursive relation, namely that which it defines for the natural interpretation of terms.

Representability of computable relations and weak representability of recursively denumerable sets

THEOREM 20*. *Every computable relation $R(n_1, \ldots, n_k)$ is strongly representable in* Ar *and in every consistent extension of* Ar *by a formula $A(x_1, \ldots, x_k)$ which begins with a single existential quantifier followed by a formula A' which represents an elementary recursive relation such that all the quantifiers in A' are bounded. If the relation R is a function, then the corresponding uniqueness condition for A is a theorem in* Ar.

PROOF. We shall begin with a proof for the functions $f \in \mathrm{Comp}$, and for clarity shall confine ourselves to functions of one argument. Then we can avail ourselves of the fact that every computable relation R' can be represented by setting a function $f \in \mathrm{Comp}$ equal to zero.

By Theorem 11 (on the normal form), the function f can be represented as follows:

(194) $f(n) = (f'(n))_0$,

(195) $f'(n) = (\mu m) [R(n, m)]$,

493

where R is an elementary recursive relation and the effectiveness condition

(196) $\bigwedge n \bigvee m R(n, m)$

is satisfied.

By Theorem 22, the relation $R(n, m)$ is strongly representable in Ar by a formula $B(x_1, x_2)$ in which all the quantifiers are bounded:

(197)
$$R(n, m) \equiv \ulcorner B(n, m) \urcorner \in \text{Ar},$$
$$\sim R(n, m) \equiv \ulcorner \sim B(n, m) \urcorner \in \text{Ar}.$$

Since

$$(z)_0 = \exp(z, 2) = (\mu u \leqslant z) [\sim (2^{u+1}|z)],$$

where $x|y \equiv \bigvee u \leqslant y \, x \cdot u = y$, the function $(z)_0$ is in Erec, too. Thus, by Theorem 22, there exists a formula $C(x, y)$ in which all the quantifiers are bounded such that for all k

(198) $\ulcorner C(k, (k)_0) \urcorner \in \text{Ar}$

and

(199) $\ulcorner \bigwedge y, z \big((C(k, y) \wedge C(k, z)) \to y = z \big) \urcorner \in \text{Ar}.$

It will be proved that the relation $f(n) = k$ is strongly represented by the following formula $A(x, y)$:

(200) $\bigvee z (C(z, y) \wedge B(x, z) \wedge \bigwedge z_1 < z \sim B(x, z_1)).$

Note first that it follows from (196) and (195) that, for any $n \in \mathcal{N}$,

(201) $R(n, f'(n))$

and

(202) $\bigwedge m < f'(n) \, (\sim R(n, m)).$

From (201) and (197) we obtain that

(203) $\ulcorner B(n, f'(n)) \urcorner \in \text{Ar}.$

From (202) and (197) we obtain that

(204) $\bigwedge m < f'(n) \ulcorner \sim B(n, m) \urcorner \in \text{Ar}.$

Since Ar includes Theorem (155) in the form

$$\bigwedge x \big(x < f'(n) \equiv (x = 0 \vee x = 1 \vee \ldots \vee x = f'(n) \dot- 1) \big),$$

hence, as in previous proofs, we obtain from (204) that

(205) $\ulcorner \bigwedge z_1 < f'(n)\,(\sim B(n, z_1))\urcorner \in \mathrm{Ar}.$

Finally, from (194) and (198) we obtain that

(206) $\ulcorner C(f'(n), f(n))\urcorner \in \mathrm{Ar}.$

By the quantifier laws and by (200) we obtain from (203), (205) and (206) that

$$\ulcorner A(n, f(n))\urcorner \in \mathrm{Ar}.$$

Thus, to prove representability it suffices to prove uniqueness:

(207) $\ulcorner \bigwedge y, u(A(n, y) \wedge A(n, u) \to y = u)\urcorner \in \mathrm{Ar}.$

To do so we can carry out the following reasoning in Ar. If $A(n, y)$ and $A(n, u)$, then, by (200), there exist z and v such that

$$(\alpha)\,C(z, y) \wedge (\beta)\,B(n, z) \wedge (\gamma)\bigwedge z_1 < z\,(\sim B(n, z_1)) \wedge$$
$$(\delta)\,C(v, u) \wedge (\varepsilon)\,B(n, v) \wedge (\eta)\bigwedge z_1 < v\,(\sim B(n, z_1)).$$

If $z < f'(n)$, then (β) and (205) would yield a contradiction. If $f'(n) < z$, then (203) and (γ) would yield a contradiction. Thus, by Axiom 3, the only possibility is that $z = f'(n)$. Likewise, the assumptions that $v < f'(n)$ and $f'(n) < v$ result in contradictions by (203), (205), (ε) and (η). Hence, by Axiom 3, we have to conclude that $v = f'(n)$. Consequently, we obtain from (α) and (δ) that $C(f'(n), y)$ and $C(f'(n), u)$. Hence, by (199), we obtain that $y = u$. Formula (207) has, thus, been proved.

Thus, in accordance with Theorem 18, the formula $A(x, y)$ strongly represents the relation $f(n) = k$.

Consequently, the relation $f(n) = k$ is strongly represented by a formula A which begins with a single existential quantifier, followed by a formula which strongly represents an elementary recursive relation $\varrho(n, z, k)$ such that

$$\varrho(n, z, k) \equiv \big((z)_0 = k \wedge R(n, z) \wedge \bigwedge z_1 < z \sim R(n, z_1)\big).$$

The theorem converse to 20* is obtained in a rigorous manner by arithmetizing the language of the theory Ar and making use of the following concept.

DEFINITION 10. *A set X is recursively reducible to a set Y if and only if there exists a primitive recursive function f such that the following equivalence holds*:

$$(208) \qquad n \in X \equiv f(n) \in Y \quad (\text{that is, } X = O_f^{-1}(Y)).$$

The term is intuitive. Deciding whether a number n is an element of X reduces to deciding whether the value $f(n)$ is an element of Y. Hence, the computability of Y obviously implies the computability of X.

The above definition leads directly to some self-evident properties of recursive reducibility. If X is recursively reducible to Y, then

$$\text{(a) } Y \in \text{Comp} \rightarrow X \in \text{Comp},$$

(209)   (b) $Y \in$ recursively denumerable $\rightarrow$
                 $X \in$ recursively denumerable.

In the case of (209) (b) the reasoning is as follows. If Y is recursively denumerable, that is, is defined thus:

$$n \in Y \equiv \bigvee m R(n, m),$$

where $R \in \text{Comp}$, then, by (208), X also can be represented in a recursively denumerable form

$$n \in X \equiv \bigvee m R(f(n), m).$$

To formulate the point of the above examples in a general manner: if Y can be represented as a quantified formula, then X can be presented in the same form.

THEOREM 24. *Every set (relation) which is weakly representable in a set of sentences T is recursively reducible to the set of numbers of the sentences of T.*

PROOF. If a set X is weakly representable in a set of sentences T, then for some formula $A(x)$ the following equivalence holds:

$$n \in X \equiv \ulcorner A(n) \urcorner \in T.$$

As we pass to arithmetized syntax, T will be used to denote the set of numbers of the sentences in the theory T. Thus, the above equivalence becomes

$$(210) \qquad n \in X \equiv \text{Nr}(\ulcorner A(n) \urcorner) \in T.$$

By Definition 10 (of recursive reducibility), it suffices to demonstrate that the function

$$f(n) = \mathrm{Nr}(\ulcorner A(\boldsymbol{n}) \urcorner)$$

is in Prec. In order to be able to speak about numbers of sentences of the type $A(x_k/\boldsymbol{n})$ we must add to the definitions introduced in Section 3 a definition of substitution of numerical constants for numerical variables. Let $\mathrm{Sb}(a, x_k/b)$ denote the formula obtained from a formula a by the substitution in it of b for a variable x_k. Of course, a, x_k, and b are interpreted as numbers. The definition of the function Sb can, for our purposes, be greatly simplified. Since the laws on quantifiers make it possible to replace variables by other variables, every relation represented in Ar can always be represented by a formula A in which a free variable x_k does not simultaneously occur in any other place as a bound variable. This simplification can go still further: we can confine ourselves to formulae $A(x_k)$ in which the free variable x_k occurs only once. If the free variable x_k occurs in $B(x_k)$ more than once, we can replace $B(x_k)$ by the equivalent formula

$$\bigvee z\big(x_k = z \wedge B(z)\big),$$

in which the free variable x_k occurs only once. The only restriction is that we must select as the variable z a variable which before the replacement did not occur in $B(x_k)$. The formula

$$\bigwedge x_k\big(B(x_k) \equiv \bigvee z(x_k = z \wedge B(z))\big)$$

is, obviously, a logical theorem. For those formulae $A(x_k)$ in which the free variable occurs only once the definition of substitution is very simple:

(211) $\quad \mathrm{Sb}(a, x_k, b) = (\mu c)\,\big[(c = a \wedge \bigwedge d, e < a\ a \neq d * \bar{x}_k * e) \vee$

$\qquad\qquad (\bigvee d, e < a\ (a = d * \bar{x}_k * e \wedge c = d * b * e))\big].$

Thus, the result of the substitution Sb either equals a if the variable x_k does not occur in a, or, if x_k occurs in a, it differs from a only by the replacement of x_k with b, while those parts of the formula a which precede and follow x_k remain the same.

Thus the function Sb is defined for all numbers a, k, b. The minimization is effective, since for every a, x_k, b there exists a c such that either

$c = a$, or $c = d * b * e$, and c satisfies the condition specified by the right side of (211). This minimization can even be bounded. It can easily be checked that $c < p_{\text{lng}(a*h)}^{(a \cdot b)^2}$ always holds, so that Sb is a primitive recursive function.

Now the proof of the fact that f is a primitive recursive function does not present any difficulties. If X is representable in T, then we may assume that it is representable by a formula $A(x_1)$, in which the variable x_1 occurs only once. The formula $A(x_1)$ has its constant number $\text{Nr}(\ulcorner A(x_1)\urcorner)$, in accordance with the numbering adopted in Section 3. The substitution $A(n)$ of the formula $A(x_1)$ has, by (193) and (112), the following number which is a function of the number n:

$$f(n) = \text{Nr}(\ulcorner A(n)\urcorner) = \text{Sb}\big(\text{Nr}(\ulcorner A(x_1)\urcorner), x_1, \text{Nr}(n)\big),$$

so that $f \in \text{Prec}$, and, by (208) and (210), the set X is recursively reducible to the set T.

THEOREM 25.

(a) *Every set (relation) which is weakly representable in a recursively enumerable set T of sentences is recursively enumerable.*

(b) *Every set (relation) which is strongly representable in a recursively enumerable set T of sentences is computable.*

PROOF.

(a) A set X which is weakly representable in T is, by Theorem 24, recursively reducible to T. Hence, by (209), if T is recursively enumerable, then X is recursively enumerable, too.

(b) If X is strongly representable in T, then it follows directly from Definitions 7 and 8 that both X and its complement $-X$ are weakly representable in T. Thus, if T is recursively enumerable, then X and $-X$ are. This and Theorem 13 b yield that X is computable.

Theorem 25 yields another, more precise version of the proof of Theorem 21.

PROOF OF THEOREM 21. The set of theorems of the theory Ar is recursively enumerable (Theorem 17); hence, by Theorem 25, every relation which is strongly representable in Ar in computable.

Theorems 20 and 21 were, historically, the earliest general definition of computability. The theory Ar can here by replaced by any more comprehensive theory, provided that the theory is consistent and recursively enumerable, because all computable relations are strongly representable in every consistent extension of Ar. The converse implication follows from Theorem 25 (b).

Since some recursively enumerable relations are non-computable, it is not true that all recursively enumerable relations are strongly represented in Ar. On the other hand, they can be characterized by weak representability.

THEOREM 26. *Every recursively denumerable relation R is weakly representable in Ar by a formula A which begins with a single existential quantifier, followed by a formula A' (which represents a relation in the class* Erec) *such that all the quantifiers which occur in it are bounded.*

PROOF. For clarity we shall confine ourselves to representation in Ar of relations of one argument, that is, sets. If X is recursively denumerable, then there exists a relation $R \in$ Comp such that

(212) $n \in X \equiv \bigvee m R(n, m)$.

Since $R \in$ Comp, by Theorem 22 there exists a formula $A(x_1, x_2)$ such that

(213)
$$R(n, m) \equiv \ulcorner A(n, m) \urcorner \in \text{Ar},$$
$$\sim R(n, m) \equiv \ulcorner \sim A(n, m) \urcorner \in \text{Ar},$$

and the formula A has the form

(214) $\bigvee x A'(x, x_1, x_2)$,

where all the quantifiers in A' occur bounded.

It can easily be seen that the set X is weakly represented in Ar by the formula

(215) $\bigvee x_2 \bigvee x A'(x, x_1, x_2)$.

In fact, if $n \in X$, then, by (212), there exists an m such that $R(n, m)$, and hence, by (213), $\ulcorner A(n, m) \urcorner \in$ Ar, so that, by (214), $\ulcorner \bigvee x A'(x, n, m) \urcorner \in$ Ar. By the laws on quantifiers it can also be proved that $\ulcorner \bigvee x_2 \bigvee x A'(x, n, x_2) \urcorner \in$ Ar.

Conversely, if $\ulcorner \bigvee x_2 \bigvee x A'(x, n, x_2) \urcorner \in$ Ar, then, in view of the fact that the theory Ar is ω-consistent by being true in the model $\mathfrak{N}_0$, it cannot be true that $\bigwedge m \ulcorner \sim \bigvee x A'(x, n, m) \urcorner \in$ Ar. Hence there exists an m such that $\ulcorner \sim \bigvee x A'(x, n, m) \urcorner \notin$ Ar, so that, by (214),

$\ulcorner \sim A(\boldsymbol{n}, \boldsymbol{m}) \urcorner \notin \text{Ar}$. This and the second of the formulae (213) yield that $R(n, m)$, and, thus, by (212), $n \in X$.

We have thus proved the truth of the equivalence

(216) $n \in X \equiv \ulcorner \bigvee x_2 \bigvee x A'(x, \boldsymbol{n}, x_2) \urcorner \in \text{Ar}.$

The set X is, thus, weakly represented in Ar by formula (215), which can easily be given the desired form: the variables x_2 and x may be restricted by a third variable, z. The equivalence

(217) $\bigwedge x_1 \left(\bigvee x_2 \bigvee x A'(x, x_1, x_2) \equiv \right.$

$\left. \bigvee z \bigvee x_2 < z \bigvee x < z A'(x, x_1, x_2) \right)$

is a theorem in Ar, since the successor of the greater of the two numbers, x_2 and x, may always be taken to be the z in question. Formulae (216) and (217) yield that the set X is weakly represented in Ar by the right side of equivalence (217), which has the desired form.

It can also be proved that every recursively denumerable set is weakly representable in every consistent extension of the arithmetic Ar, but the proof is somewhat more complicate (see [14]).

Since every recursively denumerable set is weakly representable in Ar, by Theorem 24 every recursively denumerable set is recursively reducible to the set Ar of numbers of theorems. The set Ar is said to be *universal* (with respect to reducibility) for the class of recursively denumerable sets. Hence, it follows that the set Ar is non-computable. If it were computable, then, by formula (209) (a), every set reducible to it would also have to be computable, and, thus, every recursively denumerable set would have to be computable. Since, however, by Theorem 16, there exist recursively denumerable sets which are not computable, hence the set Ar cannot be computable.

As will be seen below in Theorem 31, for computable sets, in contrast to recursively denumerable sets, there is no set which is universal with respect to recursive reducibility.

By referring to Theorem 23 we can easily prove a theorem converse to Theorem 26. If a formula A begins with a single existential quantifier, followed by a formula in which all the quantifiers occur bounded, then A weakly represents in Ar a recursively denumerable set, namely that set X which is defined in $\mathfrak{N}_0$ by the formula A for the natural interpretation of the primitive terms of the theory Ar. For formulae A which begin with a single existential quantifier followed by a formula with only bounded quantifiers, definability thus coincides with representability in Ar:

(218) X is definable in $\mathfrak{N}_0$ by $A \equiv$
    X is weakly representable in Ar by A.

Representability of functions Comp *in the arithmetic* Ar *without exponentiation*

To conclude the discussion on representability it is worth noting that the theory Ar is not the weakest theory in which all computable functions can be strongly represented. Ar has been selected for heuristic purposes. But it is possible to reduce both the number of axioms and primitive terms without invalidating Theorems 17 to 26. Exponentiation is definable by the operations of multiplication and addition and the successor function. But to understand the definition of exponentiation expressed by means of the concepts specified above, one has to possess some knowledge of the arithmetic of natural numbers, in particular, the *Chinese Remainder Theorem*. That theorem, in a form adapted to our needs, can be formulated as follows:

(219)   *For every finite sequence of natural numbers* $a_0, \ldots, a_k$ *there exist numbers* $u, v \in \mathcal{N}$ *such that, for every* $i \leqslant k$, $a_i =$ *the remainder of the division of the number* u *by the number* $1 + v(i+1)$.

This theorem, specifically arithmetical in nature, will not be proved here. We note only that it makes it possible for any finite sequence of natural numbers $a_0, \ldots, a_k$ to be represented by a pair of numbers $\langle u, v \rangle$. Every number a_i can be obtained from that pair by the formula occurring in Theorem (219). Theorem (219) can, thus, play the same role as the theorem stating that any natural number can be decomposed into prime numbers. It makes it possible to write inductive definitions in a manner similar to that used in Section 1 (Theorem 1), by using prime numbers. Since the concept of exponentiation need not be referred to in a definition of the concept of remainder, the concept of pair can be defined by reference to multiplication and addition alone. Thus, addition and multiplication suffice to write down in a computable manner inductive definitions using the operations of superposition and minimization. In particular, the inductive definition of exponentiation can be reformulated as an ordinary definition. This can be done by reference to the concept of remainder in the following manner.

We first introduce auxiliary definitions:

$$[x/y] = (\mu z) [y \cdot S(z) > x \vee y = 0],$$
$$x \dot{-} y = (\mu z) [x = y + z \vee (z = 0 \wedge y > x)],$$
$$\varrho(x, y) = x \dot{-} ([x/y] \cdot y)$$

(the remainder on dividing x by y),

$$[\sqrt{x}] = (\mu z) [S(z) \cdot S(z) > x],$$
$$G_1^{-1}(z) = z \dot{-} ([\sqrt{z}] \cdot [\sqrt{z}]),$$
$$G_2^{-1}(z) = [\sqrt{z}] \dot{-} G_1^{-1}(z),$$
$$G(x, y) = (x + y)^2 + x.$$

G is the pair function, and G_1^{-1}, G_2^{-1} are functions converse to G.

$$T(i, w) = \varrho\big(G_1^{-1}(w), S(G_2^{-1}(w) \cdot S(i))\big).$$

By Theorem (219), for every sequence $a_0, \dots, a_k$ there exists a pair w such that, for $i \leqslant k$,

$$a_i = T(i, w).$$

Thus, the function T yields the i-th element of the sequence obtained from the number w. Hence,

$$x^n = y \equiv \bigvee w\big(T(0, w) = 1 \wedge T(n, w) = y \wedge$$
$$\bigwedge i < n T(S(i), w) = T(i, w) \cdot x\big).$$

Thus, the function of exponentiation, can be defined by effective minimization:

$$x^n = T\big(n, (\mu w) \big[T(0, w) = 1 \wedge n = (\mu i) [T(S(i), w) \neq$$
$$T(i, w) \cdot x]\big]\big).$$

Theorem (219) ensures the effectiveness of the minimization operation. Thus, the above formulae show that exponentiation is definable in $\mathfrak{N}_0$ by means of the terms S, $<$, $\cdot$, $+$ and the operations of superposition and effective minimization. Hence, also the definition of the class of general recursive functions could have been presented in a simpler form, since the class GR is now seen to be the least class which contains the initial functions 0, S, $\cdot$, $+$, and is closed under the operations of superposition and effective minimization.

In a number of papers concerned with simplifying reasoning it has been found out that the following theory, with the terms $0, S, +, \cdot$, and based on seven axioms:

1. $0 \neq S(x)$,
2. $S(x) = S(y) \rightarrow x = y$,
3. $\sim (x = 0) \rightarrow \bigvee y\, x = S(y)$,
4. $x + 0 = x$,
5. $x + S(y) = S(x + y)$,
6. $x \cdot 0 = 0$,
7. $x \cdot S(y) = (x \cdot y) + x$,

is the simplest of all those arithmetical systems in which all relations are strongly representable. This theory will be denoted by Ar*. The following theorem

THEOREM 27 (Tarski, Mostowski, Robinson [57] in [99]). *A relation is computable if and only if it is strongly representable in* Ar*,

will be recorded, without proof, as the result of the analyses carried out in this part of the present section.

1. Define a class Erec⁰ (narrower than Erec) of functions which contains the functions S, $\cdot$, $+$, $\dot{-}$, and is closed under superposition and bounded minimization. Prove that in Theorems 11, 14, and 15 in Section 2 the class Erec can be replaced by the class Erec⁰.

2. Add to the theory Ar* the term $<$ and the axioms on the *less than* relation and prove for the theory Ar** thus obtained (with the terms $0, S, \cdot, +, <$) Theorems 20 to 26, replacing Ar by Ar** and Erec by Erec⁰.

H i n t: Refer to the results obtained in Exercise 1.

3. Prove that in the theory Ar with added schemata (121) every computable function is strongly representable by a term formula in the sense of Definition 9.

4. Deduce formula (218).

5. PROBLEMS OF DECIDABILITY

Essential undecidability of arithmetic

As stated in Section 3, a theory is called decidable if the set of its theorems is computable. Otherwise, it is called undecidable. The arithmetic Ar is undecidable. The fact that the set of its theorems is non-computable has been seen in connection with Theorem 26; for if all recursively enumerable sets are recursively reducible to the set Ar and if there exist recursively enumerable sets which are non-computable, then the set Ar cannot be computable.

But an even stronger theorem can be proved about arithmetic: Not only is the theory Ar undecidable, but no consistent theory which contains Ar is decidable either. Theories which have that property are called *essentially undecidable*.

DEFINITION 11.

a. *A set of sentences Z is undecidable if and only if the set of numbers of those sentences is non-computable.*

b. (Tarski [99]). *A set Z of sentences is essentially undecidable if and only if it is undecidable and if every consistent set that contains Z is undecidable.*

In all applications of Definition 11.b the consistency of the set of sentences may be interpreted so that set does not contain any pair of con_ tradictory sentences, A and $\sim A$. In the earlier chapters the concept of consistency was somewhat different: it was stated that no pair of contradictory sentences can be a logical consequence of the set of sentences.

The theorem on the essential undecidability of arithmetic can also be obtained directly from Theorem 20 (27), which states that all computable sets are strongly representable in Ar (Ar*). Namely the following theorem can be proved in a general way.

THEOREM 28. *If all computable sets are strongly representable in a set of sentences Z, then the set Z is undecidable.*

The proof is a typical example of the diagonal method. All sentential formulae $A(x_1)$ of one free variable x_1 are numbered in an effective manner. This can be done, for instance, by arranging all sentential formulae by length, the alphabetical order being adopted for the formulae of the same length. Let

$$A_0(x_1), A_1(x_1), A_2(x_1), \dots$$

be that sequence of sentential formulae.

Suppose now that the set Z is decidable. Hence for every sentence there is a method of finding out whether that sentence is or is not in Z. In particular, for the sentences of the form $A_k(n)$, and also for their negations $\sim A_k(n)$, such a method exists. Hence, there must also be a method of finding out whether a sentence of the form $\sim A_k(k)$ is or is not a theorem in Z. That method can be described as follows. For any number $k \in \mathcal{N}$ we find the k-th sentential formula of one free variable, that is, the formula $A_k(x_1)$. (The operation is effective, since the formulae have been numbered in an effective manner.) Next we precede it with the negation symbol and replace the free variable x_1 by the numeral k. As is known, these operations are effective. The sentence $\sim A_k(k)$ thus obtained is next checked by the decision method for the set Z.

The above reasoning can be summarized in the following implication: *If Z is decidable, then a set X, defined as follows*:

(220) $\quad k \in X \equiv \ulcorner \sim A_k(k) \urcorner \in Z$

is computable.

But should the set X be computable, then, by assumption, it would have to be strongly representable in Z; so there would have to exist a sentential formula $A_{k_0}(x_1)$ of one free variable such that

(221) $\quad \begin{aligned} &n \in X \equiv \ulcorner A_{k_0}(n) \urcorner \in Z, \\ &n \notin X \equiv \ulcorner \sim A_{k_0}(n) \urcorner \in Z. \end{aligned}$

Consider now the number k_0. It follows from (220) that

(222) $\quad k_0 \in X \equiv \ulcorner \sim A_{k_0}(k_0) \urcorner \in Z.$

But, on the other hand, it follows from (221)

(223) $\quad k_0 \notin X \equiv \ulcorner \sim A_{k_0}(k_0) \urcorner \in Z.$

Formulae (222) and (223) yield the contradiction $k_0 \in X \equiv k_0 \notin X$. The set X accordingly cannot be computable, and, hence, Z cannot be decidable.

THEOREM 29. *If all computable sets are strongly representable in a set of sentences Z, then the set Z is essentially undecidable.*

PROOF. If $Z \subset Z'$, and Z' is consistent, then the equivalences $n \in X \equiv \ulcorner A(n) \urcorner \in Z$ and $n \notin X \equiv \ulcorner \sim A(n) \urcorner \in Z$ imply similar equivalences for Z'. Hence, if all computable sets are strongly representable in Z, then they also are strongly representable in any superset Z' which does not contain any pair of contradictory sentences. It follows from the above and from Theorem 28 that Z' is undecidable. By Definition 11, the set Z is, thus, essentially undecidable.

THEOREM 30 (Gödel [17], Rosser [79]). *The arithmetic* Ar (Ar*) *is essentially undecidable.*

PROOF. By Theorem 20 (27), all computable sets are strongly representable in Ar (Ar*), hence, by Theorem 29, the arithmetic Ar (Ar*) is essentially undecidable.

Computable inseparability of the set of theorems of Ar *and the set of negations of theorems of* Ar

The theorem on the essential undecidability of Ar can be made somewhat stronger: it can be proved that not only does a computable set of sentences which would be an extension of the set of theorems of Ar not exist, but furthermore there exists no computable set of numbers which includes the set of numbers of theorems of Ar and is disjoint from the set of numbers of negations of theorems of Ar. In other words, the sets of numbers of theorems of Ar and of negations of theorems of Ar are computably inseparable, that is, inseparable by means of computable sets.

DEFINITION 12. *Sets X and Y are computably inseparable if and only if there does not exist any computable set Z which includes X and is disjoint with Y.*

The following auxiliary theorem will be proved:

THEOREM 31. *There does not exist any computable set Y, universal (with respect to recursive reducibility) for computable sets, that is such that, for every $X \in$ Comp and for some $f \in$ Prec,*

(224) $X = O_f^{-1}(Y)$.

PROOF. Suppose that a set $Y \in \text{Comp}$ is universal, with respect to recursive reducibility, for computable sets. It will be seen that this assumption results in a contradiction, because we can, with reference to the set Y, easily define a set $X \in \text{Comp}$ which does not satisfy condition (224) for any function $f \in \text{Prec}$. The set X is defined as follows:

(225) $n \in X \equiv G_n(n) \notin Y$,

where $G_n(k)$ is a computable function, universal for the functions Prec of one argument.

The set X is, obviously, computable. Suppose that $X = O_f^{-1}(Y)$ for some $f \in \text{Prec}$. By the definition of counter-image (Introduction, formula (17)), this would mean that

(226) $n \in X \equiv f(n) \in Y$.

The function $f \in \text{Prec}$ has its number k in the numbering G_n: $f = G_k$. Formula (226), thus, implies that, for some k and for any n, the formula

(227) $n \in X \equiv G_k(n) \in Y$

holds.

Now formulae (225) and (227) yield, for some k and for any n, the equivalence

(228) $G_n(n) \notin Y \equiv G_k(n) \in Y$.

On substituting k for n in (228) we obtain the contradiction $G_k(k) \notin Y \equiv G_k(k) \in Y$. Hence X must differ from the counter-image of Y, and accordingly formula (224) cannot be satisfied.

Sets of sentences will hereafter be identified with sets of their numbers. If Z is a set of numbers of sentences, then $\text{neg} Z$ will be used to denote the set of numbers of those sentences whose negations are in Z:

(229) $n \in \text{neg} Z \equiv$ *the negation of the sentence bearing the number* n *is in* Z.

THEOREM 32. *If every computable set is strongly representable in a set of sentences* Z, *then the sets* Z *and* $\text{neg} Z$ *are computably inseparable.*

PROOF. If every set $X \in \text{Comp}$ is strongly representable in a set Z, then for every $X \in \text{Comp}$ there exists a formula $A(x_1)$ of one free variable x_1 which occurs only once in $A(x_1)$, such that

(230) $n \in X \equiv \text{Nr}(\ulcorner A(x_1/n) \urcorner) \in Z$,

(231) $n \notin X \equiv \text{Nr}(\ulcorner \sim A(x_1/n) \urcorner) \in Z$.

By (229), equivalence (231) can be written in the following form:

(232) $n \in \mathcal{N} - X \equiv \mathrm{Nr}(\ulcorner A(x_1/n)\urcorner) \in \mathrm{neg}\,Z$.

By the definition of the operation of substitution (Section 4, formula (211)), equivalences (230) and (232) can be written thus:

(233)
$$n \in X \equiv \mathrm{Sb}\big(\mathrm{Nr}(\ulcorner A(x_1)\urcorner), x_1, \mathrm{Nr}(n)\big) \in Z,$$
$$n \in \mathcal{N} - X \equiv \mathrm{Sb}\big(\mathrm{Nr}(\ulcorner A(x_1)\urcorner), x_1, \mathrm{Nr}(n)\big) \in \mathrm{neg}\,Z.$$

It follows from the proof of Theorem 24 that the function

(234) $f(n) = \mathrm{Sb}\big(\mathrm{Nr}(\ulcorner A(x_1)\urcorner), x_1, \mathrm{Nr}(n)\big)$

is primitive recursive. The sets X and $\mathcal{N} - X$ are thus counter-images, respectively, of the sets Z and $\mathrm{neg}\,Z$, obtained by the function f. Formulae (233) and (234) yield, namely, that

(235) $n \in X \equiv f(n) \in Z$,

(236) $n \in \mathcal{N} - X \equiv f(n) \in \mathrm{neg}\,Z$.

Suppose now that there is a set $Y \in \mathrm{Comp}$, which separates the sets Z and $\mathrm{neg}\,Z$. It would have to satisfy the formulae

(237) $Z \subset Y$

and

(238) $\mathrm{neg}\,Z \subset \mathcal{N} - Y$.

It will be demonstrated that such a set Y, separating Z from $\mathrm{neg}\,Z$, would be universal for all computable sets. In fact, it would follow from (235) and (237) that, for any $X \in \mathrm{Comp}$,

(239) $n \in X \rightarrow f(n) \in Y$.

On the other hand, (236) and (238) would yield that

(240) $n \in \mathcal{N} - X \rightarrow f(n) \in \mathcal{N} - Y$.

On combining (226) and (240) into an equivalence we obtain the formula

$$n \in X \equiv f(n) \in Y,$$

which states that $X = O_f^{-1}(Y)$. Thus, any computable set X would be recursively reducible to the computable set Y, which would, thus, be a computable set universal for computable sets, which, by Theorem 31, is not possible. Hence a computable set separating the sets Z and $\text{neg}\,Z$ does not exist (Mostowski [19]).

Theorem 32 also immediately yields a proof of Theorem 29.

PROOF OF THEOREM 29. By Theorem 32, the sets Z and $\text{neg}\,Z$ are computably inseparable. If there existed a consistent and decidable set T which was an extension of the set Z, then T would separate Z and $\text{neg}\,Z$. Should T be an extension of Z, this would yield $Z \subset T$. If T also were consistent, then it could not contain those sentences whose negations are in Z. All those sentences whose negations are in Z would then have to be in the complement of T: $\text{neg}\,Z \subset \mathcal{N} - T$. Hence T would separate Z and $\text{neg}\,Z$.

Incompleteness of arithmetic

The undecidability of arithmetic is closely connected with its incompleteness. That connection is quite general in nature. We can formulate it in the theorems that follow.

THEOREM 33. *Every set of sentences which is consistent, recursively enumerable and complete is decidable.*

PROOF. The idea of the proof as referring to a theory T can be given the following intuitive formulation. If T is complete, then for every sentence A, formulated in terms of T,

(241) $A \in T \vee \ulcorner \sim A \urcorner \in T.$

By writing out, one by one, all the proofs, beginning with the shortest one, we at some time come across either a proof of the sentence A or a proof of the sentence $\sim A$, and in this way we find out whether A is or is not a theorem. This gives an effective method of deciding whether $A \in T$, if T is consistent.

By referring to the arithmetization of language we can reason with greater precision as follows. If T is consistent, then, in addition to formula (241), the following formula is true, too:

(242) $\sim (A \in T \wedge \ulcorner \sim A \urcorner \in T).$

Formulae (241) and (242), then together, state that the computable set Z of all sentences splits into two disjoint recursively enumerable sets: the set of theorems and the set of those sentences whose negations are theorems:

$$Z = T \cup \operatorname{neg} T.$$

The set T is recursively enumerable by assumption. On the other hand, the set $\operatorname{neg} T$ is recursively reducible to the set T. By Definition (229), the set $\operatorname{neg} T$ can be described as follows by means of the arithmetization of language:

$$n \in \operatorname{neg} T \equiv \; \approx *\overline{(*n\,\overline{*})} \in T.$$

Since the function $*$ is in Prec, the set $\operatorname{neg} T$ is recursively reducible to the set T, and, thus, by formula (209) (b) (see the preceding section), $\operatorname{neg} T$ is recursively enumerable.

Both T and $\operatorname{neg} T$ are recursively enumerable and their union, Z, is a computable set; hence, by Theorem 13 c, T and $\operatorname{neg} T$ are computable sets, which means that T is decidable.

Thus a theory which is recursively enumerable and undecidable is thereby incomplete, and, hence, must contain independent sentences. In particular, arithmetic, being both recursively enumerable (Theorem 17) and undecidable (Theorem 30), is incomplete. The same may be said about its extensions.

THEOREM 34. *Every extension of the arithmetic* Ar (Ar*) *which is consistent and recursively enumerable is incomplete.*

The proof follows directly from Theorems 30 and 33.

The following theorem may be considered converse to Theorem 33.

THEOREM 35. *Every decidable and incomplete theory has an extension which is decidable, consistent, and complete.*

PROOF. If a theory T is incomplete, then it is also consistent. T can easily be extended to a theory S which is complete, consistent, and recursively enumerable. That theory S will be a sum of an increasing and computable sequence $\{T_n\}$ of decidable theories which are obtained by the successive addition to T of the shortest independent sentences:

$$T_0 = T,$$

$$(243) \qquad n \in T_{k+1} \equiv \Big(n \in T_k \vee \overline{(} * (\mu m \leqslant k) \, [m \in \mathrm{s} \wedge m \notin T_k \wedge$$
$$\approx *\overline{(*m\,\overline{*})} \notin T_k]* \overset{\rightarrow}{} * n\,\overline{*} \Big) \in T_k \Big).$$

510

5. PROBLEMS OF DECIDABILITY

By (243), T_{k+1} contains all the theorems of T_k and all those sentences which are implied in T_k by the least sentence independent of T_k. This definition shows that all the theories in the sequence T_k are consistent and decidable, because the addition of an independent sentence and all its consequences leads from a consistent theory to a consistent theory (Chapter II, Theorem 15).

The minimization which occurs in (243) is bounded, and, hence, that equivalence defines the set T_{k+1} in a computable manner. Consequently, the following implication is true:

(244)   *If T_k is decidable and consistent, then T_{k+1} is decidable and consistent.*

The entire relation $n \in T_k$ is computable, too, since its characteristic function $g(n, k)$ can be defined by induction combined with the bounded minimization operation:

$$g(n, 0) = \begin{cases} 0 & \text{if } n \in T, \\ 1 & \text{if } n \notin T; \end{cases}$$

$$g(n, k+1) = g(n, k) \cdot g\Big(\!\!\big(\!\ast(\mu m \leqslant k)\,[m \in s \wedge g(m, k) \neq 0 \,\wedge \\ g(\approx \ast \overline{(\ast m \ast)}, k) \neq 0] \ast \overrightarrow{\to} \ast n \ast\big], k\Big).$$

Of course,

$$n \in T_k \equiv g(n, k) = 0.$$

Hence, the sum S of the sequence $\{T_k\}$ is a recursively enumerable set which can be written down as follows:

(245) $n \in S \equiv \bigvee k\, g(n, k) = 0.$

The sum of an increasing sequence of consistent theories is a consistent theory (cf. Chapter II, Section 4). Thus, the theory S is consistent. It is also a complete theory, for should there be a sentence m independent of S, then by the definition of S that sentence m would also be independent of every theory T_k. Since there are only finitely many sentences earlier than m (with numbers less than that of m), in the process of the construction of the successive theories T_k the sentence m would, at some moment, have to become the earliest sentence independent of the theories already defined. Hence, by (243), it had to be added to one of the theories T_{k+1}, for $k \geqslant m$, contrary to the assumption that it is independent of the theory S.

Since S is a theory which is recursively enumerable, complete, and consistent, then, by Theorem 32, S is decidable as well.

Theorems 33 and 35 can be combined into one, which yields a characteristic of essentially undecidable theories.

A theory T is essentially undecidable if and only if it is consistent and every consistent and recursively enumerable extension of it is incomplete (Tarski [99]).

Form of independent sentences

The foregoing analyses make it possible not only to prove the existence of independent sentences in every recursively enumerable and consistent extension of arithmetic, but also to describe the form of such sentences.

THEOREM 36. *In every consistent and recursively enumerable extension T of the theory* Ar *there are independent sentences in the form of*

$$(246) \qquad \bigvee x A(x),$$

in which all the quantifiers in the formula $A(x)$ are bounded.

PROOF. Let Z_0 be the set of all well-formed sentences in the form of (246) and their negations, where all the quantifiers occurring in $A(x)$ are bounded. The set Z_0 is computable, because the set of sentences is computable, and it can easily be decided about every sentence whether its form is that of (246) or whether it is a negation of a formula in the form of (246), and whether all its quantifiers, except the initial one, are bounded or not.

Consider the set $Z_0 \cap$ Ar, that is, the set of those theorems of Ar which are in the form of (246). By Theorem 20*, every computable relation (set) is strongly representable in the set $Z_0 \cap$ Ar. Hence, by Theorem 29, we have that the set $Z_0 \cap$ Ar of sentences is essentially undecidable, so that, by Definition 11 b, the set $Z_0 \cap$ Ar satisfies the condition:

(247)   *Every set which contains $Z_0 \cap$ Ar and is consistent is undecidable.*

If the theory T is a consistent and recursively enumerable extension of the theory Ar: Ar $\subset T$, then also $Z_0 \cap$ Ar $\subset Z_0 \cap T$. The set $Z_0 \cap T$, being part of a consistent set, is itself consistent. Since it contains the set $Z_0 \cap \mathrm{Ar}$, by (247) it is undecidable, and, hence, non-computable. We can now make use of an argument similar to that used in the proof of Theorem 33. If there were no sentences in Z_0 which are independent of T, then Z_0 would be the union of two disjoint sets

(248) $Z_0 = (Z_0 \cap T) \cup (Z_0 \cap \mathrm{neg}\, T)$.

By (241), the set $\mathrm{neg}\, T$ is recursively reducible to the set T. T is recursively enumerable; hence, by (209) (b) (see the preceding section), the set $\mathrm{neg}\, T$ is recursively enumerable, too. The common part of two recursively enumerable sets is itself a recursively enumerable set, and, hence, if there were no sentences in Z_0 independent of T, then, by (248), the computable set Z_0 would be the union of two disjoint recursively enumerable sets. Hence, by Theorem 13c, the sets $Z_0 \cap T$ and $Z_0 \cap \mathrm{neg}\, T$ would have to be computable, too. But we have proved $Z_0 \cap T$ to be non-computable. Consequently, there must be sentences in Z_0, that is, sentences in the form stated in Theorem 36, which are independent of the theory T.

THEOREM 37. *If a formula $A(x)$ contains only bounded quantifiers, then the existential sentence $\bigvee x A(x)$ satisfies the equivalence*

$$\ulcorner \bigvee x A(x) \urcorner \in \mathrm{Ar} \equiv \ulcorner \bigvee x A(x) \urcorner \in E(\mathfrak{N}_0),$$

for the ordinary interpretation of arithmetical terms.

PROOF. The implication $\rightarrow$ is self-evident. As shown in Chapter II, the axioms of Ar are true in $\mathfrak{N}_0$, so that all the theorems of Ar are true in $\mathfrak{N}_0$.

The implication $\leftarrow$ follows from Theorem 23. In fact, if all the quantifiers in $A(x)$ occur bounded, then, by Theorem 23, $A(x)$ strongly represents an Erec relation $R(n)$, which is defined by $A(x)$ in $\mathfrak{N}_0$. Hence, if $\ulcorner \bigvee x A(x) \urcorner \in E(\mathfrak{N}_0)$, then, by the definitions of the concept of truth and definability, there exists an n such that n satisfies $A(x)$, which means that $R(n)$ holds; thus $\ulcorner A(n) \urcorner \in \mathrm{Ar}$ by the definition of representability, and hence, under the quantifier laws, $\ulcorner \bigvee x A(x) \urcorner \in \mathrm{Ar}$.

THEOREM 38 (Rosser [79]). *In every consistent and recursively enumerable extension T of the theory Ar there exist independent sentences in the form of*

(249) $\bigwedge x A(x)$

which are true in $\mathfrak{N}_0$, *and have the property that all definite cases*: $A(0)$, $A(1)$, $A(2)$, ... *are theorems in* Ar.

PROOF. By Theorem 36, in every consistent and recursively enumerable extention T of the theory Ar there exist independent sentences in the form of $\bigvee x A(x)$, where $A(x)$ strongly represents a relation $R \in$ Erec. By Theorem 37, sentences in that form are true in $\mathfrak{N}_0$, if and only if they are theorems in Ar; hence, only false sentences in that form are independent. Thus, their negations are true in $\mathfrak{N}_0$, are obviously independent, and by De Morgan's law are in the form of (249). Since $A(x)$ strongly represents a relation in Ar, for every $n \in \mathcal{N}$, the following

(250) $\ulcorner A(n) \urcorner \in$ Ar $\vee \ulcorner \sim A(n) \urcorner \in$ Ar

holds.

Hence, it follows that if the universal sentence (249) is independent, then $\ulcorner A(n) \urcorner \in$ Ar for every $n \in \mathcal{N}$, for if $\ulcorner A(n) \urcorner \notin$ Ar for some n, then, by (250), $\ulcorner \sim A(n) \urcorner \in$ Ar; hence, $\ulcorner \bigvee x \sim A(x) \urcorner \in$ Ar, too. Sentence (249) would, accordingly, not be independent.

The earliest proofs of undecidability (Gödel, Rosser) immediately indicated the form of independent sentences. An easy exposition of the Gödel–Rosser method is to be found in Mostowski [55].

Theorem 38 may be interpreted as an indication of a cognitive limitation of ordinary (effective) deductive methods in mathematics. Every theory based on ordinary logic and on a commonly available system of axioms is recursively enumerable. Moreover, proofs can be computable, and, hence, intersubjectively verifiable, only in a theory which is recursively enumerable. Every arithmetical theory which is based on intersubjectively verifiable proofs, thus, contains undecidable universal sentences $\bigwedge x A(x)$ such that every definite case: $A(0)$, $A(1)$, $A(2)$, ... is a theorem in that theory. Thus, no arithmetic with intersubjectively verifiable proofs is ω-complete.

Undecidability of other methematical theories

The theorem stating that the arithmetic Ar (Ar*) is essentially undecidable makes it possible to prove the undecidability of many other mathematical theories. The theorems that follow make it possible to pass from the undecidability of one theory to that of another.

THEOREM 39. *If a theory T' is a finite extension of a theory T, and if T' is undecidable, then T is undecidable, too.*

PROOF. A theory T' is called a *finite extension of a theory T* if there exists a sentence A such that

(251) $T' = \text{Cn}(T \cup \{A\})$.

Hence, by the Deduction Theorem,

(252) $B \in T' \equiv \ulcorner(A \to B)\urcorner \in T$.

If the theory T is decidable, then formula (252) provides a decision method for the theory T'. Thus, if T' is undecidable, then T is undecidable, too.

From an arithmetical point of view formula (252) implies recursive reducibility of the set T' to the set T. When arithmetized, formula (252) can be written down as follows:

$$n \in T' \equiv \overline{\left(*\text{Nr}(A)* \overset{\rightarrow}{} *n*\right)} \in T,$$

where $f(n) = \overline{\left(*\text{Nr}(A)* \overset{\rightarrow}{} *n*\right)}$ is obviously a primitive recursive function. Thus, if T is computable, then T' is also computable.

Theorem 39 applies to those theories which do not differ from one another in the set of the initial constants. The theorem that follows makes it possible to pass to theories with a different set of constants.

THEOREM 40 (Tarski [99]). *If a theory T_1 is consistent with a theory T_2 (the union $T_1 \cup T_2$ is consistent), and*

(α) *if the constants of the theory T_2 are constants of the theory T_1,*

(β) *the theory T_2 is essentially undecidable and finitely axiomatizable,*

then T_1 is undecidable, and so is every subtheory of T_1 which has the same constants as T_1.

PROOF. The theory $T = \text{Cn}(T_1 \cup T_2)$ is consistent by assumption, is undecidable because it contains the essentially undecidable theory T_2, and is a finite extension of T_1, since T_2 is finitely axiomatizable. Hence, by Theorem 39, T_1 is undecidable, too. The same reasoning can be repeated for every subtheory of T_1.

Theorem 40 has numerous applications. It has been used to prove undecidability of many mathematical theories, such as group theory, the arithmetic of rational numbers, and, thus, also the theories of number fields and ordered fields. The same method has been used to prove the undecidability of the theory of modular lattices, abstract projective geometry, the theory of Brouwerian lattices, closure algebra, many versions of set theory, and many other theories. We have also the following easy theorem.

THEOREM 41. *If a sentence A contains no functions, no individual constants, and nor the atomic predicate $P(x)$, and if a sentence A' is derived from A by relativization of quantifiers to the predicate $P(x)$, and if the set $Cn(\{A\})$ is essentially undecidable, then the set $Cn(\{A'\})$ is essentially undecidable, too.*

PROOF. Consider only sets of sentences without functions and without individual constants. The assumption of non-emptiness $\bigvee x(\alpha \to \alpha)$ may be treated as one which follows from A. A theory without existential assumptions cannot be essentially undecidable, since it is true in the empty domain, and the set of sentences which are true in the empty domain is decidable. Hence, as the concept of consequence we may take consequences based on those logical theorems which are true in the empty domain. Let this be the meaning of the concept of Cn as used in this proof.

Let X be a consistent extension of the set $Cn(\{A'\})$. Suppose that the set X is computable. It will be demonstrated that under this assumption there would also exist a consistent and computable extension of the set $Cn(\{A\})$. The set of sentences Z with all quantifiers relativized to the predicate $P(x)$ is computable, since it can easily be checked whether in a given sentence all quantifiers are or are not restricted to the predicate in question. Hence, if the set X is computable, then the set $X \cap Z$ of those sentences in X which are relativized to the predicate $P(x)$ is computable, too. Let $f(n)$ be a function which changes a number n of a sentence into a number $f(n)$ of the sentence which is derived from the sentence with the number n by relativizing all its quantifiers to the predicate $P(x)$. Consider the set $Y = O_f^{-1}(X \cap Z)$. The following relation holds:

(253) $n \in Y \equiv f(n) \in (X \cap Z)$.

The function $f(n)$ is clearly primitive recursive. Hence, the set Y is recursively reducible to the set $X \cap Z$. If $(X \cap Z) \in$ Comp, then $Y \in$ Comp. The set Y is at the same time an extension of the set $Cn(\{A\})$. The relativization function $f(n)$ preserves the concept of logical consequence based on a logic true in the empty domain, because, by Theorem 13 (Chapter I, Section 9) on the relativization of theorems of a logic true in the empty domain, the following formula holds:

(254) $\quad B \in \mathrm{Cn}(\{A\}) \rightarrow f(B) \in \mathrm{Cn}(\{f(A)\}),$

where, obviously, $f(A) = A'$. Formulae (253) and (254) make it possible to conclude that

(255) $\quad B \in \mathrm{Cn}(\{A\}) \rightarrow f(B) \in \mathrm{Cn}(\{A'\}) \cap Z \rightarrow$
$\qquad f(B) \in (X \cap Z) \rightarrow B \in Y.$

Formula (255) proves the inclusion $\mathrm{Cn}(\{A\}) \subset Y$. Thus Y in fact is an extension of the theory $\mathrm{Cn}(\{A\})$. Since negation as a sentential connective does not change when quantifiers are relativized, if the set Y were inconsistent, then, by (253), the set $X \cap Z$ would have to be inconsistent, too. Now, if X is a consistent and decidable extension of the set $\mathrm{Cn}(\{A'\})$, then Y, too, is a consistent and decidable extension of the set $\mathrm{Cn}(\{A\})$. That, however, is impossible if the theory $\mathrm{Cn}(\{A\})$ is essentially undecidable.

The application of Theorems 40 and 41 will be illustrated by a very simple example of a proof of the undecidability of the arithmetic of integers. Let T be a theory with the constants $<, +, \cdot$, which stand, respectively, for the *less than* relation holding between integers, and the functions of addition and multiplication of integers. Let any sentences true in the domain of integers be axioms of T. In the theory T, we can define the set $\mathcal{N}$ of natural numbers and we can restrict the operations $+$ and $\cdot$ to the set of natural numbers.

(256) $\quad \mathcal{N}(x) = \bigvee y(y+y = y \wedge (x = y \vee x > y)).$

Formula (256) is a definition of the predicate "natural number" in the arithmetic of integers. This definition is true in the domain of integers. The axioms of the arithmetic Ar* of natural numbers relativized to the predicate $\mathcal{N}(x)$ are obviously true in the domain of integers. Let T_1 be the theory T augmented by Definition (256) and the term $\mathcal{N}$ which (256) defines: $T_1 = \mathrm{Cn}(T \cup \{(256)\})$. Let T_2 be the set $f(\mathrm{Ar}^*)$ of the axioms of Ar* restricted to the predicate $\mathcal{N}(x)$. By Theorem 41, T_2 is a theory which is essentially undecidable and finitely axiomatizable. The constants of T_2 are the constants of T_1. Conditions (α) and (β) of Theorem 40 are, thus, satisfied. The theories T_1 and T_2 also have a common model, namely the domain of integers. The union $T_1 \cup T_2$ is, thus, consistent. By Theorem 40, the theory T_1 is undecidable. By the theorem on the elimination of definable terms, the following equivalence:

$$(257) \qquad A' \in T \equiv A' \in \mathrm{Cn}\big(T \cup \{(256)\}\big) \equiv A' \in T_1$$

holds for the sentences A' in which the predicate $\mathscr{N}(x)$ does not occur; hence, the theory T is also undecidable.

We have, thus, proved that every system of sentences which contain the terms $<$, $+$, and $\cdot$, and are true in the domain of integers is undecidable. In particular, the set of axioms of the theory of ordered rings is undecidable.

It also follows from the above reasoning that there exists a finitely axiomatizable and essentially undecidable arithmetic of integers. The only care is to so select axioms for integers so that the axioms of the system Ar* for natural numbers, relativized to the predicate $\mathscr{N}(x)$, can be deduced from them.

We can, in a similar manner, show that set theory is essentially undecidable, because in set theory it is possible to define the predicate for natural numbers $\mathscr{N}(x)$ and to deduce theorems which are Ar axioms restricted to that predicate. There are even very weak systems of set theory which are undecidable.

Another conclusion easily derivable from the previous theorems is the following

THEOREM 42. *The set of c.l.c. theorems is undecidable.*

PROOF. The terms of Ar may be considered to be terms in the c.l.c., about which we obviously made no assumptions whatever in c.l.c. The classical logical calculus is true in every domain, and, hence, also in the domain $\mathfrak{N}_0$, in which the theory Ar is true. On setting $T_1 = \text{c.l.c.}$ and $T_2 = \text{Ar}$ we note that all the conditions specified in Theorem 40 are satisfied. Thus, the set of theorems of the classical logical calculus is undecidable, in contrast to the case of the set of theorems of the classical sentential calculus, for which the zero-one verification method is a decision procedure.

Proofs of decidability of mathematical theories are usually more complicated. A method of proving decidability via a proof of completeness is suggested by Theorem 32. Theorem 39 (Chapter II) on categoricity in power can sometimes be used in proofs of completeness. In this way the decidability of the elementary theory of the *less than* relation can be proved directly from Theorem 33 and Theorem 40 (Chapter II). A great deal of information on the various proofs of decidability and undecidability can be found in: Tarski [99], Shoenfield [83], Rogers [70].

EXERCISES

1. For a given predicate $P(x)$ define an arithmetical function $f(n)$ which associates with the number n of a formula A the number $f(n)$ of the formula obtained from A by restriction of all its quantifiers to the predicate $P(x)$.

2. Prove the undecidability of the theory of addition and multiplication in the domain of integers.

H i n t. The predicate $\mathcal{N}(x)$ can be defined by reference to addition and multiplication on the strength of the theorem stating that every natural number is a sum of four squares of natural numbers.

3. Formulate an axiom system of an essentially undecidable arithmetric of integers.

4. Prove the undecidability of set theory by referring to the possibility of defining in it the set of natural numbers.

5. Prove the following general lemma:

(A) *If a theory T is consistent and recursively enumerable, and if there exists a formula $A(x)$ which represents in T a non-computable set, then that theory is incomplete and undecidable.*

6. LOGICAL HIERARCHY OF ARITHMETIC CONCEPTS

It was indicated in Section 2 that analyses carried out in that section result in a logical hierarchy of mathematical concepts. The lowest level in that classification consists of computable concepts, which are the simplest of all, and upper levels are obtained by imposing quantifier operations stepwise upon lower levels. The numbers of quantifiers necessary for defining a concept is the measure of the logical complexity of that concept and it also determines that concept's place in the hierarchy mentioned. Topological classifications, Borelian and projective, may be treated as special cases of the logical classification.

Computable functionals and computable relations of higher orders

In order to present the logical classification mentioned in sufficient detail and with a sufficient degree of generality we have to introduce the concepts of computable functional of higher orders and of computable

519

relation whose arguments are mathematical entities of higher orders, such as natural functions, sets of natural numbers, relations between natural numbers, etc. In doing so we shall confine ourselves to mathematical entities of the two lowest levels.

We shall consider the functionals

(258) $\quad f(n_1, ..., n_k, \alpha_1, ..., \alpha_l, A_1, ..., A_p, R_1, ..., R_q) = m$

with numerical values, and the relations

(259) $\quad R(n_1, ..., n_k, \alpha_1, ..., \alpha_l, A_1, ..., A_p, R_1, ..., R_q),$

where the variables $n_1, ..., n_k$ range over natural numbers, the variables $\alpha_1, ..., \alpha_l$ range over natural functions with natural arguments, the variables $A_1, ..., A_p$ range over sets of natural numbers, and the variables $R_1, ..., R_q$ range over relations among natural numbers.

In a particular case the number of arguments of a certain kind may equal zero. If a functional depends on numerical arguments only (when $l = p = q = 0$), then it is simply a numerical function.

In order to avoid such long inscriptions as (258) and (259) we shall use German letters $\mathfrak{a}$, $\mathfrak{b}$, $\mathfrak{c}$ to denote any complexes of any variables whose closer specification is inessential to a given argument. For instance, a relation in the form of (259) may be symbolized as $R(n_1, \mathfrak{a})$ if in a given argument we are concerned solely with one numerical argument, or by $R(\alpha_1, \mathfrak{a})$ if it is only the functional argument α_1 which matters in a given case.

For the functionals in the form of (258) the concept of computability is defined by induction:

DEFINITION 13. I. *Computable functionals* (Comp) *are the least class of functionals K in the form given above, such that*:

1) *the class K contains as the initial functionals*:

a. *the arithmetical functionals* $S(n)$, $n+m$, $n\dot{-}m$, $n \cdot m$, n^m,

b. *the identity functionals*:

$J_1(n, \alpha) = \alpha(n)$ *for functions of one argument*,

$J_k(n_1, ..., n_k, \alpha) = \alpha(n_1, ..., n_k)$ *for functions of k arguments*,

$J'_1(n, A) = \begin{cases} 0 \text{ if } n \in A \\ 1 \text{ if } n \notin A \end{cases}$ *for sets of numbers*,

$J'_n(n_1, ..., n_k, R) = \begin{cases} 0 \text{ if } R(n_1, ..., n_k) \\ 1 \text{ if } \sim R(n_1, ..., n_k) \end{cases} \begin{cases} \text{for relations of arbitrarily} \\ \text{many arguments}; \end{cases}$

2) *K is closed under the substitution of functionals in numerical argument places;*

3) *K is closed under the effective minimization operation.*

Thus, if *f* is in *K*, then the functional

$$g(\mathfrak{a}) = (\mu n) [f(n, \mathfrak{a}) = 0]$$

also is in *K* if the effectiveness condition

$$\bigwedge \mathfrak{a} \bigvee n \, f(n, \mathfrak{a}) = 0$$

is satisfied.

DEFINITION 13. II. *A relation R in the form of* (259) *is computable if and only if there exists a computable functional f which is characteristic of the relation R, that is such that*

$$R(n_1, ..., n_k, \alpha_1, ..., \alpha_l, A_1, ..., A_p, R_1, ..., R_q) \equiv$$
$$f(n_1, ..., n_k, \alpha_1, ..., \alpha_l, A_1, ..., A_p, R_1, ..., R_q) = 0.$$

Similarity between the definition of computability as applied to natural functions and relations only and the definition given above is so close that many theorems can be transferred automatically. They will be given without proof. Definitions parallel to the definitions of primitive recursive and elementary recursive functions will now be introduced.

DEFINITION 14. I. *The class of primitive recursive functionals* (Prec) *is the least class of functionals K of type* (258) *which contains the same initial functionals as the class of computable functionals contains* (Definition 13. I) *and is closed under the operations of*:

a. *superposition: substitution of functionals in numerical argument places;*

b. *simple recursion: if g, h are in K, then the functional f which is defined as follows:*

$$f(0, \mathfrak{a}) = h(\mathfrak{a}),$$

$$f(n+1, \mathfrak{a}) = g\big(n, \mathfrak{a}, f(n, \mathfrak{a})\big)$$

also is in K.

DEFINITION 14. II. *The class of elementary recursive functionals* (Erec) *is the least class of functionals K of type* (258) *which contains as the initial functionals the arithmetical and identity functionals which are initial for the class* Comp, *and is closed under the operations of*:

a. *superposition*: *substitution of functionals in numerical arguments*,

b. *bounded minimization*: *if f is in K, then the functional defined as follows*:

$$g(n, \mathfrak{a}) = (\mu x \leqslant n) [f(x, \mathfrak{a}) = 0]$$

also is in K.

DEFINITION 14. III. *Primitive (elementary) recursive relations are those which are obtained by setting the appropriate functionals equal to zero.*

Examples of computable functionals:

$$f_i(n, \alpha) = (\mu x) [\alpha(x) = \alpha(n)],$$
$$f_2(\alpha) = \alpha(\alpha(0)),$$
$$f_3(\alpha, \beta, n) = \alpha(\beta(n)) + \beta(\alpha(n)).$$

For all the generality of the definitions given above we shall be primarily concerned with functionals defined only on numerical and functional arguments, that is, functionals in the form of $f(x_1, ..., x_k, \alpha_1, ..., \alpha_l)$.

THEOREM 43. *Computable sets form a field of sets. Quantifier-free formulae define the relations* Erec. *Bounded operations and sentential connectives do not lead outside either computable relations or relations of the class* Prec *or* Erec.

Theorem 43 does not require a separate proof, because all those proofs which are valid for computable functions can be transferred to functionals without any essential change: it suffices to join the parameter $\mathfrak{a}$. A theorem on the normal form, analogous to Theorem 11, can also be proved for computable functionals and relations; that theorem, however, has certain elements which require a separate proof.

THEOREM 44 (on the normal form).

a. *For every computable relation* $R(\alpha, \mathfrak{a})$ *there exist elementary recursive relations* $R'(n, \mathfrak{a})$ *and* $R''(n, \mathfrak{a})$ *which instead of the functional argument* α *contain a numerical argument n and are such that the following equivalences hold*:

$$(260) \qquad R(\alpha, \mathfrak{a}) \equiv \bigvee x R'(\bar{\alpha}(x), \mathfrak{a}) \equiv \bigwedge x R''(\bar{\alpha}(x), \mathfrak{a}),$$

where the functional $\bar{\alpha}(x)$ is defined thus:

(261) $\qquad \bar{\alpha}(x) = \prod_{i \leqslant x} p_i^{\alpha(i)+1}.$

b. *For every computable functional $f(\alpha, \mathfrak{a})$ there exists a relation $R \in \text{Erec}$ which contains a numerical argument x and is such that the following formulae hold*:

$$f(\alpha, \mathfrak{a}) = \big((\mu x)\big[R\big(\bar{\alpha}(x), \mathfrak{a}\big)\big]\big)_0$$

and

$$\bigwedge \alpha, \mathfrak{a} \;\bigvee x R\big(\bar{\alpha}(x), \mathfrak{a}\big).$$

(The functional $\bar{\alpha}(x)$ differs slightly from the diagram functional $\tilde{\alpha}(x)$, as defined by formula (46) in Section 2; $\bar{\alpha}(x)$ is a diagram of the function α plus one.)

It follows from (261) that

(262) $\qquad \alpha(n) = \big(\bar{\alpha}(x)\big)_n \dot{-} 1 \quad$ for $n \leqslant x.$

a. PROOF BY INDUCTION. 1. The theorem is satisfied for the initial functionals treated as relations. The functional variable occurs only in the identity functionals J_k. The corresponding relations R', R'' can be given for those functionals: the following equivalence is true (we confine ourselves to considering the functional J_1):

$$J_1(n, \alpha) = k \equiv \alpha(n) = k \equiv \bigvee x\big(\big(\bar{\alpha}(x)\big)_n = k+1\big).$$

Hence, the following relation is the desired relation R':

$$R'(n, x, k) \equiv \big((x)_n = k+1\big).$$

The following equivalence is also true:

$$\alpha(n) = k \equiv \bigwedge x\big(n+1 = \lg\big(\bar{\alpha}(x)\big) \to \big(\bar{\alpha}(x)\big)_n \dot{-} 1 = k\big).$$

Hence, the following relation is the desired relation R'':

$$R''(n, x, k) = \big(n+1 = \lg(x) \to (x)_n \dot{-} 1 = k\big).$$

2. Suppose now that the theorem under consideration is true for functionals f and g, treated as relations. Accordingly, there exist relations R', R'', S', $S'' \in \text{Erec}$ such that the following formulae hold:

(263) $\qquad f(n, \alpha, \mathfrak{a}) = m \equiv \bigvee x R'\big(\bar{\alpha}(x), n, m, \mathfrak{a}\big) \equiv$
$$\bigwedge x_1 R''\big(\bar{\alpha}(x_1), n, m, \mathfrak{a}\big),$$

(264) $\qquad g(\alpha, \mathfrak{a}) = n \equiv \bigvee x S'\big(\bar{\alpha}(x), n, \mathfrak{a}\big) \equiv \bigwedge x S''\big(\bar{\alpha}(x), n, \mathfrak{a}\big).$

The appropriate formulae for superposition follow from (263) and (264):

(265) $\quad f(g(\alpha, \mathfrak{a}), \alpha, \mathfrak{a}) = m \equiv$
$$\bigvee n\Big(\bigvee x S'(\bar{\alpha}(x), n, \mathfrak{a}) \wedge \bigvee x_1 R'(\bar{\alpha}(x_1), n, m, \mathfrak{a})\Big)$$

(266) $\quad f(g(\alpha, \mathfrak{a}), \alpha, \mathfrak{a}) = m \equiv$
$$\bigwedge n\Big(\bigvee x S'(\bar{\alpha}(x), n, \mathfrak{a}) \rightarrow \bigwedge x_1 R''(\bar{\alpha}(x_1), n, m, \mathfrak{a})\Big).$$

In formula (265) the quantifiers $\bigvee n$, $\bigvee x$, $\bigvee x_1$ are combined into one by the pairing function, so that the right side of equivalence (265) becomes:

(267) $\quad \equiv \bigvee z\Big(S'(\bar{\alpha}((z)_2), (z)_1, \mathfrak{a}) \wedge R'(\bar{\alpha}((z)_3), (z)_1, m, \mathfrak{a})\Big).$

Since the following formulae hold:

(268a) $\quad z = \lg(\bar{\alpha}(z)) \dot{-} 1, \quad (z)_i \leqslant z$

and for $z' \leqslant z$

(268b) $\quad \alpha(z') = \prod_{i \leqslant z'} p_i^{(\alpha(z))_i},$

then $(z)_1$, $(z)_2$, $(z)_3$, $\bar{\alpha}((z)_2)$, $\bar{\alpha}((z)_3)$ can be expressed in terms of $\bar{\alpha}(z)$, so that formula (267) can be written as $\bigvee z R(\bar{\alpha}(z), m, \mathfrak{a})$.

Likewise, in formula (266), on shifting all the quantifiers to the front of the formula, we can combine them into one universal quantifier and thus obtain the formula

$$\bigwedge z\Big(S'(\bar{\alpha}((z)_2), (z)_1, \mathfrak{a}) \rightarrow R''(\bar{\alpha}((z)_3), (z)_1, m, \mathfrak{a})\Big),$$

which, by reference to formulae (268), we can reduce to

$$\bigwedge z R(\bar{\alpha}(z), m, \mathfrak{a}).$$

3. Suppose that the theorem under consideration is true for a functional f treated as a relation, in which case formulae (263) hold. Suppose further that a functional g is defined by effective minimization:

(269) $\quad \begin{aligned} &g(\alpha, \mathfrak{a}) = (\mu x)[f(x, \alpha, \mathfrak{a}) = 0], \\ &\bigwedge \alpha, \mathfrak{a} \bigvee x f(x, \alpha, \mathfrak{a}) = 0. \end{aligned}$

It will be proved that the relation $g(\alpha, \mathfrak{a}) = n$ also satisfies the conditions laid down in the theorem under consideration. It follows from (269) that

(270) $\quad g(\alpha, \mathfrak{a}) = n \equiv (f(n, \alpha, \mathfrak{a}) = 0 \wedge \bigwedge u < n f(u, \alpha, \mathfrak{a}) \neq 0).$

524

By applying equivalence (263) to (270) we can define the relation $g(\alpha, \mathfrak{a}) = n$ by either of the following formulae:

(271) $\quad \bigvee x R'(\bar{\alpha}(x), n, 0, \mathfrak{a}) \wedge \bigwedge u < n \bigvee x_1 \sim R''(\bar{\alpha}(x_1), u, 0, \mathfrak{a})$,

(272) $\quad \bigwedge x_1 R''(\bar{\alpha}(x_1), n, 0, \mathfrak{a}) \wedge \bigwedge u < n \bigwedge x \sim R'(\bar{\alpha}(x), u, 0, \mathfrak{a})$.

Formula (272) can easily be reduced to the desired form. The three universal quantifiers $\bigwedge x_1, \bigwedge u, \bigwedge x$ can be combined into one by means of the pairing function: $z = \langle x_1, u, x \rangle$. Next, by reference to (268), we reduce (272) to the form $\bigwedge z P(\bar{\alpha}(z), n, \mathfrak{a})$. If we want to combine quantifiers in formula (271), then we must first handle separately the second element of the conjunction which is formula (271). The following true equivalence can be applied to that element:

(273) $\quad \bigwedge u < n \bigvee x_1 R(u, x_1) \equiv \bigvee x_1 \bigwedge u < n R(u, (x_1)_u)$.

This makes it possible to shift the unbounded quantifier so that it precedes the bounded quantifier and thus to combine the two quantifiers $\bigvee x, \bigvee x_1$ in (271) into one. (271) is, thus, equivalent to

(274) $\quad \bigvee z \Big(R'\big(\bar{\alpha}((z)_0), n, 0, \mathfrak{a}\big) \wedge \bigwedge u < n \sim R''\big(\bar{\alpha}\big(((z)_1)_u\big),$
$\qquad u, 0, \mathfrak{a}\big)\Big)$.

Formulae (268) make it possible to reduce (274) to the form

$$\bigvee z P(\bar{\alpha}(z), n, \mathfrak{a}).$$

We have, thus, proved by induction that every computable relation in the form $f(\alpha, \mathfrak{a}) = n$ satisfies formula (260). Since every computable relation R is obtained by setting a computable functional equal to zero, hence in fact all computable relations Comp satisfy the theorem under consideration.

b. The theorem on the normal form for functionals follows directly from the theorem on the normal form for relations. Let f be a computable functional. Under part a of the theorem under consideration there exists an $R \in$ Erec such that

(275) $\quad f(\alpha, \mathfrak{a}) = n \equiv \bigvee x R(\bar{\alpha}(x), n, \mathfrak{a})$.

Since $f(\alpha, \mathfrak{a})$ has a value for any α and $\mathfrak{a}$, hence the condition

(276) $\quad \bigwedge \alpha, \mathfrak{a} \bigvee n \bigvee x R(\bar{\alpha}(x), n, \mathfrak{a})$

525

is satisfied. We combine the variables n and x into one:

(277) $\qquad \bigwedge \alpha, \mathfrak{a} \bigvee z R\big(\bar{\alpha}((z)_1), (z)_0, \mathfrak{a}\big).$

Now, (277) is an effectiveness condition. By (275)–(277), the following equation holds:

$$f(\alpha, \mathfrak{a}) = \big((\mu z)[R(\bar{\alpha}((z)_1), (z)_0, \mathfrak{a})]\big)_0.$$

By applying formulae (268) we impart to the relation R the form postulated in the theorem under consideration.

If the remaining arguments of the relation R which belong to the complex $\mathfrak{a}$ are functional arguments $\beta, \gamma, \ldots$ and numerical arguments $y_1, \ldots, y_k$, then Theorem 44 may again be applied to the relation R'; this will yield that the relation R in the form of (260) can be represented as:

$$R(\alpha, \beta, \mathfrak{b}) \equiv \bigvee x_1 \bigvee x_2 R'''(\bar{\alpha}(x_1), \bar{\beta}(x_2), \mathfrak{b}) \equiv$$
$$\bigwedge x_1 \bigwedge x_2 R''''(\bar{\alpha}(x_2), \bar{\beta}(x_2), \mathfrak{b}).$$

The variables x_1 and x_2 can, of course, be combined into a pair referring to formulae (268). By applying Theorem 44 in turn to all functional variables we obtain the following

THEOREM 45 (Kleene [37], [38]). *For every computable relation* $R(\alpha_1, \ldots$ $\ldots, \alpha_n, y_1, \ldots, y_k)$ *with n function arguments and k numerical arguments there exists relations* $R', R'' \in$ Erec *with* $n+k$ *numerical arguments (and without functional arguments) such that*

$$R(\alpha_1, \ldots, \alpha_n, y_1, \ldots, y_k) \equiv$$
$$\bigvee x R'(\bar{\alpha}_1(x), \ldots, \bar{\alpha}_n(x), y_1, \ldots, y_k) \equiv$$
$$\bigwedge x R''(\bar{\alpha}_1(x), \ldots, \bar{\alpha}_n(x), y_1, \ldots, y_k).$$

For every functional $f \in$ Comp *with n function arguments and k numerical arguments there exists a relation* $R \in$ Erec, *without function arguments, such that*

$$f(\alpha_1, \ldots, \alpha_n, y_1, \ldots, y_k) =$$
$$\big((\mu x)[R(\bar{\alpha}_1(x), \ldots, \bar{\alpha}_n(x), y_1, \ldots, y_k)]\big)_0$$

and

$$\bigwedge \alpha_1, \ldots, \alpha_n, y_1, \ldots, y_k \bigvee x R(\bar{\alpha}_1(x), \ldots, \bar{\alpha}_n(x), y_1, \ldots, y_k).$$

The functionals Comp have thus been defined for functions, sets, and relations, but in place of sets and relations we need only consider their characteristic functions. Thus, we can confine ourselves to functionals defined on functions of one argument and on natural numbers. In the text below it will be assumed that all computable functionals and relations contain only numerical and functional arguments. Theorem 45 accordingly shows that any computable functional can be represented by means of a numerical relation $\in$ Erec, the functional $\bar{a}(x)$, and a single application of the effective minimization operation. As in the case of Theorem 44, Theorem 45 will be called the *theorem on the normal form*.

The theorem on the normal form has numerous applications. We shall mention two of them, one very simple, and the other of more interest. For the simpler one we shall prove a theorem which makes it possible to make substitutions not only for numerical variables, but for function variables as well. If a variable α, which, for example, stands for a function of two arguments, occurs in a formula, then that formula remains meaningful if that variable is replaced by a constant, for instance, the constant $+$, which stands for addition, that is, for a definite function of two arguments. The same applies, obviously, to functions of one argument. A constant function may, in such a case, depend on certain parameters, not necessarily numerical in nature, it may also be a functional. In general, if there is a function of a large number of arguments, $f(n, k, \mathfrak{a})$, it is always permissible to single out a certain argument, for instance, the first numerical argument, and to consider the whole as the function $f_{k,\mathfrak{a}}(n)$ of that argument; the remaining arguments are then treated as indices attached to the function symbol. The entire symbol $f_{k,\mathfrak{a}}$ is then a symbol of a function of one numerical argument. Such a symbol may, of course, be substituted in the relevant formulae for a functional variable that stands for a function of one argument; if this is done, the meaningfulness of such a formula is not affected.

THEOREM 46. *Substitution of computable functionals, treated as functions of one numerical argument, for variables that stand for single argument functions does not lead outside computable functionals (relations).*

PROOF. By Theorem 44 the functional $f(\alpha, \mathfrak{a})$ can be represented as

$$(278) \qquad f(\alpha, \mathfrak{a}) = \left((\mu x)[R(\bar{a}(x), \mathfrak{a})]\right)_0 .$$

527

If a symbol $g_{\mathfrak{b}}$, which stands for a functional and is defined as

(279) $g_{\mathfrak{b}}(x) = g(\mathfrak{b}, x),$

is substituted for α, this, by (278) yields:

$$h(\mathfrak{a}, \mathfrak{b}) = f(g_{\mathfrak{b}}, \mathfrak{a}) = \big((\mu x)[R(\bar{g}_{\mathfrak{b}}(x), \mathfrak{a})]\big)_0.$$

Since the relation $R(z, \mathfrak{a})$ is an elementary recursive relation, because by (279) and (261)

$$\bar{g}_{\mathfrak{b}}(x) = \prod_{i \leqslant x} p_i^{g(\mathfrak{b}, i)+1},$$

and because the minimization operation involved is effective by assumption, the new functional h, obtained as a result of the substitution described above, is computable. This theorem can, of course, be carried over to relations.

The theorems proved so far make it possible to prove an interesting property of computable functionals, namely their continuity. Under the general definition of continuity, a functional will be called *continuous* if it transforms a convergent sequence of functions into a convergent sequence of values. Thus, the definition of convergence serves as the starting point:

DEFINITION 15.

a. *A sequence $\{\alpha_n\}$ of functions of one argument is convergent to a function α ($\alpha = \lim_n \{\alpha_n\}$) if and only if*

(280) $\bigwedge k \bigvee m \bigwedge n \geqslant m \bigwedge x \leqslant k\, \alpha_n(x) = \alpha(x).$

b. *A functional Φ_α which is a function of one function argument and of one numerical argument:*

(281) $\Phi_\alpha(x) = \Phi(\alpha, x)$

is continuous if and only if the following condition

(282) $\alpha = \lim_n \{\alpha_n\} \to \Phi_\alpha = \lim_n \{\Phi_{\alpha_n}\}$

is satisfied for any functions $\alpha, \alpha_n \in \mathcal{N}^{\mathcal{N}}$.

In order to be able to formulate the desired theorem in a fairly strong form we shall introduce a definition of functionals computable with respect to a function δ.

DEFINITION 16. *The class of functionals computable with respect to a function δ (Comp$_\delta$) is the least class of functionals which contains, as the initial functionals,*

  a. *the functions:* $x+1$, $x+y$, $x \cdot y$, x^y, $x \doteq y$, $\delta(x)$,
  b. *the same identity functionals which were adopted in part* b *of Definition* 13,

and which is closed under the operations of

  1. *superposition,*
  2. *effective minimization.*

The difference between Definition 13 and Definition 16 is worth emphasizing. In Definition 13, all the function symbols $\alpha, \beta, \ldots$ occur only as arguments in functionals. In Definition 16, the function symbol δ is designated: it occurs not as an argument in a functional, but as a symbol whose sense has been fixed once and for all, namely as a symbol that denotes a certain function which is treated as an initial functional. The difference becomes manifest when we consider the effective minimization operation. If $f_\delta(\alpha, x)$ is a functional in the class Comp$_\delta$ and if the effectiveness condition

$$(283) \qquad \bigwedge \alpha \bigvee x\, f_\delta(\alpha, x) = 0$$

is satisfied, then, by Definition 16, the functional

$$g_\delta(\alpha) = (\mu x)[f_\delta(\alpha, x) = 0]$$

also is in Comp$_\delta$. Now it may happen that the functional f_δ can be represented as the value of a computable functional h for one functional argument set equal to δ:

$$(284) \qquad f_\delta(\alpha, x) = h(\delta, \alpha, x)$$

whereas the functional g may prove not to be presentable in a similar manner, because the effectiveness condition (283), in view of (284), states that

$$(285) \qquad \bigwedge \alpha \bigvee x\, h(\delta, \alpha, x) = 0,$$

whereas for the functional $j(\delta, \alpha) = (\mu x)[h(\delta, \alpha, x) = 0]$ to be computable the stronger condition

$$(286) \qquad \bigwedge \alpha \bigwedge \delta \bigvee x\, h(\delta, \alpha, x) = 0$$

529

must be satisfied. Now it is easy to give examples of a function δ and a functional $h \in \text{Comp}$ such that condition (285) is satisfied, but condition (286) is not. For the functionals Comp_δ it suffices if the effectiveness condition is satisfied for the function δ in question, whereas for the unrestricted functionals Comp it must be satisfied for all the functions δ.

Those functionals in Comp_δ which are independent of function arguments, that is, are just functions, are called *functions computable with respect to* δ. Functions defined by the formulae $\delta(x + \delta(x))$, $\delta(\delta(x) \cdot \delta(x))$ are computable with respect to δ. The functionals $f(\alpha, x) = \alpha(\delta(x))$, $g(\alpha, x) = \delta(\alpha(x) + \delta(x))$ are functionals computable with respect to δ. Functionals computable with respect to a computable function are computable in the non-restricted sense.

THEOREM 47. *A functional Φ in the form of* (281) *is continuous if and only if there exists a function δ such that $\Phi \in \text{Comp}_\delta$.*

PROOF a. It will be proved first that functionals with arbitrary parameters and computable with respect to a certain function are continuous. The following property

(287) $\quad \Phi \in \text{Comp}_\delta \rightarrow \bigwedge \alpha, \mathfrak{a} \bigvee n \bigwedge \beta \big(\bigwedge x \leqslant n(\beta(x) = \alpha(x)) \rightarrow$
$\Phi(\alpha, \mathfrak{a}) = \Phi(\beta, \mathfrak{a}) \big)$

will be taken as the starting point of the proof. This property is a consequence of Theorem 44 on normal form. As can easily be seen, the proof of Theorem 44 is also valid for relations and functionals computable with respect a given function δ. Hence, if $\Phi \in \text{Comp}_\delta$, then Φ has the form of

(288) $\quad \Phi(\alpha, \mathfrak{a}) = \big((\mu x)[R(\bar{\alpha}(x), \mathfrak{a})] \big)_0$, where $R \in \text{Erec}_\delta$.

The minimization in question is effective. Hence, for any α and $\mathfrak{a}$, there exists an n, namely

(289) $\quad n = (\mu x)[R(\bar{\alpha}(x), \mathfrak{a})]$,

such that if the condition

(290) $\quad \bigwedge x \leqslant n\beta(x) = \alpha(x)$

is satisfied, then, by formula (261) $\bar{\alpha}(n) = \bar{\beta}(n)$, and, accordingly, for all $x \leqslant n$, $\bar{\alpha}(x) = \bar{\beta}(x)$. Hence, by (289), it is true that

(291) $\quad n = (\mu x)[R(\bar{\beta}(x), \mathfrak{a})]$.

On the other hand, formula (288) is also valid for the function β:

(292) $\qquad \Phi(\beta, \mathfrak{a}) = \big((\mu x)\, [R(\bar{\beta}(x), \mathfrak{a})]\big)_0 \, .$

Formulae (288), (289), (291), (292) together yield the equation $\Phi(\alpha, \mathfrak{a}) = \Phi(\beta, \mathfrak{a})$, which was to be demonstrated.

Property (287) implies continuity. Suppose that

(293) $\qquad \alpha = \lim_n \{\alpha_n\} \, .$

We want to prove that $\Phi_\alpha^{\mathfrak{a}} = \lim \{\Phi_{\alpha_n}^{\mathfrak{a}}\}$, where

(294) $\qquad \Phi_\alpha^{\mathfrak{a}}(x) = \Phi(\alpha, x, \mathfrak{a}) \, .$

Under Definition 15a we have to demonstrate that for any k we can find an m such that, for $n \geqslant m$ and for $i \leqslant k$, $\Phi(\alpha_n, i, \mathfrak{a}) = \Phi(\alpha, i, \mathfrak{a})$. By property (287), for α, $\mathfrak{a}$ and for every $i \leqslant k$ there exists an n_i such that

(295) $\qquad \bigwedge \beta \Big(\bigwedge x \leqslant n_i \big(\beta(x) = \alpha(x)\big) \to \Phi(\alpha, i, \mathfrak{a}) = \Phi(\alpha, i, \mathfrak{a})\Big).$

Let $p = \max n_i$ for $i \leqslant k$, where the various n_i satisfy condition (295). It accordingly follows from (295) that

(296) $\qquad \bigwedge \beta \Big(\bigwedge x \leqslant p \big(\beta(x) = \alpha(x)\big) \to \bigwedge i \leqslant k \big(\Phi(\alpha, i, \mathfrak{a}) =$
$\qquad \Phi(\beta, i, \mathfrak{a})\big)\Big).$

Formula (296) holds for any functions, and hence, in particular, for the functions from the sequence $\{\alpha_n\}$:

(297) $\qquad \bigwedge n \Big(\bigwedge x \leqslant p \big(\alpha_n(x) = \alpha(x)\big) \to \bigwedge i \leqslant k \big(\Phi(\alpha, i, \mathfrak{a}) =$
$\qquad \Phi(\alpha_n, i, \mathfrak{a})\big)\Big).$

We now avail ourselves of assumption (293). By Definition 15a (293) means the same as formula (280). In (280) we may substitute p for k, and we thus obtain that there exists an m such that, for $n \geqslant m$, the condition

(298) $\qquad \bigwedge x \leqslant p \big(\alpha_n(x) = \alpha(x)\big)$

is satisfied. It follows from (297) and (298) that the consequent of implication (297), that is, the sentence $\bigwedge i \leqslant k \big(\Phi(\alpha, i, \mathfrak{a}) = \Phi(\alpha_n, i, \mathfrak{a})\big)$, is true for $n > m$, which was to be demonstrated. Hence, in fact, $\Phi_\alpha^{\mathfrak{a}} = \lim_n \{\Phi_{\alpha_n}^{\mathfrak{a}}\}$.

PROOF b. In proving the converse implication we refer to another condition of continuity, equivalent to Definition 15b, namely Cauchy's condition. That condition is formulated by reference to the concept of neighbourhood. The *neighbourhood* of a function is the set of those functions which coincide with the function in question for a number of initial values. Thus, neighbourhood might be identified with the number which stands for the common part of the diagrams of all the functions that are in a given neighbourhood:

(299) $\alpha \in z$ (*α is in the neighbourhood z*) $\equiv z = \bar{a}(\lg(z) \dot{-} 1)$.

Hence, if $\alpha, \beta \in z$, then $\alpha(x) = \beta(x)$ for $x < \lg(z)$. Cauchy's condition as applied to the space of natural functions is:

(300) $\Phi_\alpha \in z \equiv \bigvee u(\alpha \in u \wedge \bigwedge \beta(\beta \in u \to \Phi_\beta \in z))$.

By referring to the formula for the neighbourhood of the value Φ_α we may, by (299) and (281), define the value $\Phi(\alpha, k)$ as follows:

(301) $\Phi(\alpha, k) = n \equiv \bigvee z(\Phi_\alpha \in z \wedge k < \lg(z) \wedge n = (z)_k \dot{-} 1)$.

Formulae (300) and (301) yield

(302) $\Phi(\alpha, k) = n \equiv \bigvee z, u(k < \lg(z) \wedge n = (z)_k \dot{-} 1 \wedge$

$\alpha \in u \wedge \bigwedge \beta(\beta \in u \to \Phi_\beta \in z))$.

(303) $\Phi(\alpha, k) = n \equiv \bigvee z, u(\alpha \in u \wedge R(k, z, u))$,

where R is a relation of three numerical arguments. Since the value of the functional Φ is defined for all α and k, (303) immediately yields the following condition:

(304) $\bigwedge \alpha, k \bigvee n, z, u(\alpha \in u \wedge R(k, z, n, u))$.

The variables n, z, u may be combined into one by the pair function: $w = \langle n, z, u \rangle$, which, thus, yields the following effectiveness condition:

(305) $\bigwedge \alpha, k \bigvee w(\alpha \in (w)_2 \wedge R(k, (w)_1, (w)_0, (w)_2))$.

In accordance with formulae (303)–(305), the functional Φ can be represented in the following form:

(306) $\Phi(\alpha, k) = ((\mu w)[\alpha \in (w)_2 \wedge R(k, (w)_1, (w)_0, (w)_2)])_0$,

because the value n of the functional is the first element $n = (w)_0$ in the complex $w = \langle n, z, u \rangle$.

6. ARITHMETICAL HIERARCHY

Let δ be a characteristic function of the relation $R(k, (w)_1, (w)_0, (w)_2)$:

(307) $\qquad \delta(k, w) = 0 \equiv R(k, (w)_1, (w)_0, (w)_2).$

By (306) and (307), the functional Φ is in the form

(308) $\qquad \Phi(\alpha, k) = \big((\mu w) [\alpha \in (w)_2 \wedge \delta(k, w) = 0]\big)_0.$

The relation $\alpha \in (w)_2$ is computable. The minimization in (308) is effective. The functional Φ is, thus, computable with respect to the function δ.

This shows that the markedly topological property of functionals, discussed above, can be formulated as a purely logical property.

Finite arithmetical hierarchy

The logical classification of concepts falls into a number of levels, each of which has infinitely many steps. The first two levels are of the greatest practical importance. The first is the arithmetical hierarchy in the broader sense of the term, that is, hyperarithmetical. Its analogue in topology is the Borelian classification (cf. Kuratowski [40], p. 250). The second is the analytical hierarchy, whose analogue in topology is the projective classification (cf. Kuratowski [40], p. 360). Each level divides into two intertwining branches. Thus, the definition of each hierarchy is dual. Duality is due to the fact that a given formula may begin either with a universal or with an existential quantifier. The first steps of the arithmetical hierarchy were shown in Section 2. The general definition is obtained by alternate repetitions of the universal and the existential quantifier operations.

We shall now introduce symbols for the classes of the arithmetical hierarchy. There happens to be two symbolisms: the two branches are distinguished from one another either by the initial or the final quantifier (used in the case of computable relations). Kleene's classification in [34] referred to the latter point of view, as outlined in Section 2 (p. 444) The symbolism which is more common now is that due to Mostowski [52]: the two branches are distinguished from one another by the initial quantifier of a given formula which defines a given set. Classes of sets (relations) in the arithmetical hierarchy will be denoted by Π_n^0 in the case of the branch of those formulae which begin with the universal quantifier, and by Σ_n^0 in the case of the branch of those formulae which begin with the existential quantifier. The index n may range over ordinal numbers, but for the time being we shall confine ourselves to defining the hierarchy for those n which are finite numbers. Arithmetical classes with finite numbers are also called *elementary definable* classes. The classes Π_n^0 and Σ_n^0 are identified with the class of computable sets. The same symbolism which has been adopted for classes and relations will be used for predicates as well. A classification of predicates will be used as the starting point.

DEFINITION 17.

a. $\Pi_0^0 = \Sigma_0^0 = $ *the set of computable predicates.*

b. *The class* Σ_{n+1}^0 *contains every predicate* $Q(\mathfrak{a})$ *definable by the formula*

$$Q(\mathfrak{a}) \equiv \bigvee n \in \mathcal{N} P(n, \mathfrak{a}),$$

where $P(n, \mathfrak{a})$ *is a predicate in the class* Π_n^0.

c. *The class* Π_{n+1}^0 *contains every predicate* $Q(\mathfrak{a})$ *definable by the formula*

$$Q(\mathfrak{a}) \equiv \bigwedge n \in \mathcal{N} P(n, \mathfrak{a}),$$

where $P(n, \mathfrak{a})$ *is a predicate in the class* Σ_n^0.

A set X of any objects (a relation holding among any objects) belongs to a class $\Pi_n^0 (\Sigma_n^0)$ if and only if that set (relation) is definable in the domain $\mathfrak{N}_0$ by a predicate from that class, for the ordinary interpretation of the primitive terms. This shows that the class Σ_1^0 is identical with that of recursively enumerable sets (and relations of arbitrarily many arguments) (cf. Definition 5 in Section 2). The only difference is that, under the original definition, only relations among numbers were termed recursively enumerable, whereas now this term covers relations holding between functions, relations, etc. The class Π_1^0 is that of complements of recursively enumerable sets (and relations).

Every set (relation) which falls under the above classification is derived from a computable relation on which the two quantifier operations, $\bigwedge x \in \mathcal{N}$, $\bigvee x \in \mathcal{N}$, restricted to the set of natural numbers, are imposed alternately.

The following theorem states that in the above classification an elementary recursive relation, defined on numerical arguments, may be adopted as a basis for classification for $n > 0$.

THEOREM 48. *Relations of n function arguments and a number of numerical arguments, which are in the classes Σ_k^0, Π_k^0, are, for $k > 0$, definable, respectively by the predicates*:

$$\bigvee_{\Sigma_1^0} x R(\bar{\alpha}_1(x), \ldots, \bar{\alpha}_n(x), y_1, \ldots, y_m),$$

$$\bigvee_{\Sigma_2^0} y_1 \bigwedge x R(\ldots), \quad \bigvee_{\Sigma_3^0} y_2 \bigwedge y_1 \bigvee x R(\ldots), \quad \ldots,$$

$$\bigwedge_{\Pi_1^0} x R(\bar{\alpha}_1(x), \ldots, \bar{\alpha}_n(x), y_1, \ldots, y_m),$$

$$\bigwedge_{\Pi_2^0} y_1 \bigvee x R(\ldots), \quad \bigwedge_{\Pi_3^0} y_2 \bigvee y_1 \bigwedge x R(\ldots), \quad \ldots,$$

where R is an elementary recursive relation holding between $n + m$ numerical arguments.

PROOF. By Theorem 45 on the normal form, any computable relation $R(\alpha_1, \ldots, \alpha_n, y_0, y_1, \ldots, y_m)$ can be presented as $\bigvee u R'(\bar{\alpha}_1(u), \ldots, \bar{\alpha}_n(u), y_0, y_1, \ldots, y_m)$, and also as $\bigwedge u R''(\bar{\alpha}_1(u), \ldots, \bar{\alpha}_n(u), y_0, y_1, \ldots, y_m)$, where $R', R'' \in$ Erec. Hence, a relation in the class Σ_1^0 can be presented as $\bigvee y_0 \bigvee u R'(\bar{\alpha}_1(u), \ldots, \bar{\alpha}_n(u), y_0, y_1, \ldots, y_m)$, and a relation in the class Π_1^0, as $\bigwedge y_0 \bigwedge u R''(\bar{\alpha}_1(u), \ldots, \bar{\alpha}_n(u), y_0, y_1, \ldots, y_m)$. By combining y_0 and u into a pair $x = \langle y_0, u \rangle$ and by applying formulae (268) we can easily represent sets in the class Σ_1^0 and Π_1^0 in the form required in the theorem. Relations in other classes are derived, under Definition 17, from class Σ_1^0 and class Π_1^0 relations as further numerical variables are bound with quantifiers, alternating the universal and the existential.

THEOREM 49.

a. *The arithmetical class of a predicate does not change if a computable functional is substituted, in that predicate, for a numerical or function variable.*

b. *Every arithmetical class is a ring of sets.*

c. *A class Σ_n^0 is identical with the class of complements of sets of a class Π_n^0, and vice versa.*

d. *The universal quantifier operation which binds a numerical variable does not lead outside a class Π_n^0 for $n > 0$.*

e. *The existential quantifier operation which binds a numerical variable does not lead outside a class Σ_n^0 for $n > 0$.*

PROOF. a. The first of the properties specified above is intuitively self-evident. Every concept falling under the classification under consideration can be defined by a predicate consisting of the computable part $R(\ldots x \ldots)$, followed by a number of quantifiers: $\bigwedge \ldots \bigvee \ldots \bigwedge \ldots$ $\ldots R(\ldots x \ldots)$. Free variables obviously do not occur bounded by any quantifier; hence, any substitution for a free variable of a computable functional means substituting for it in the computable predicate R. But such substitution as was adopted in part 2) of Definition 13 (and whose legitimacy for function variables was proved in Theorem 46) does not affect the computability of the predicate R. After the substitution the predicate preserves its form, except that instead of the computable relation R we have a likewise computable relation $R(\ldots f(\ldots) \ldots)$. Thus such a substitution does not change the class of predicates.

b. If two sets, Z and Z', are in the same class, then they are both either in Π_n^0 or in Σ_n^0. In both cases we shall consider their union and their intersection. In the former case, $\mathfrak{a} \in Z \equiv \bigwedge n \in \mathcal{N} \, U(\mathfrak{a}, n)$ and $\mathfrak{a} \in Z' \equiv \bigwedge m \in \mathcal{N} \, U'(\mathfrak{a}, m)$, and then

$$\mathfrak{a} \in Z \cup Z' \equiv \bigwedge n \in \mathcal{N} \, U(\mathfrak{a}, n) \vee \bigwedge m \in \mathcal{N} U'(\mathfrak{a}, m) \equiv$$
$$\bigwedge p \in \mathcal{N} \Big(U(\mathfrak{a}, (p)_1) \vee U'(\mathfrak{a}, (p)_2) \Big);$$

$$\mathfrak{a} \in Z \cap Z' \equiv \bigwedge n \in \mathcal{N} \, U(\mathfrak{a}, n) \wedge \bigwedge m \in \mathcal{N} \, U'(\mathfrak{a}, m) \equiv$$
$$\bigwedge m \in \mathcal{N} \big(U(\mathfrak{a}, m) \wedge U'(\mathfrak{a}, m) \big).$$

In the latter case, $\mathfrak{a} \in Z \equiv \bigvee n \, U(\mathfrak{a}, n)$ and $\mathfrak{a} \in Z' \equiv \bigvee m \, U'(\mathfrak{a}, m)$, and then

$$\mathfrak{a} \in Z \cup Z' \equiv \bigvee n \, U(\mathfrak{a}, n) \vee \bigvee m \, U'(\mathfrak{a}, m) \equiv$$
$$\bigvee n \big(U(\mathfrak{a}, n) \vee U'(\mathfrak{a}, n) \big);$$

$$\mathfrak{a} \in Z \cap Z' \equiv \bigvee n \, U(\mathfrak{a}, n) \wedge \bigvee m \, U'(\mathfrak{a}, m) \equiv$$
$$\bigvee p \big(U(\mathfrak{a}, (p)_1) \wedge U(\mathfrak{a}, (p)_2) \big).$$

In both cases we can see that if the union and the intersection do not lead outside the class which contains U and U', then they also do not lead outside the class which contains Z and Z'. Thus, if lower classes are rings of sets, then upper classes are rings of sets, too. The classes Π_0^0 and Σ_0^0 are rings of sets, since $\Pi_0^0 = \Sigma_0^0 = \text{Comp}$, hence, all the classes of the arithmetical hierarchy are rings of sets.

c. The proof of property c, is by induction. Let CK denote the class of complements of sets (relations) of a class K:

(309) $\qquad X \in CK \equiv \bigvee Y \big(Y \in K \wedge \bigwedge \mathfrak{a}(\mathfrak{a} \in X \equiv \mathfrak{a} \notin Y)\big).$

This definition yields immediately the formula

$$K = CCK;$$

hence, if $L = CK$, then $K = CL$. First, it is obvious that

$$\Sigma_0^0 = C\Pi_0^0,$$

since $\Sigma_0^0 = \Pi_0^0 =$ Comp, and the set Comp is a field of sets; thus, the complement of every computable set is a computable set. Suppose now that for some n it is true that

(310) $\qquad \Sigma_n^0 = C\Pi_n^0\,;$

it will be demonstrated that the set Σ_{n+1}^0 has the same property. By Definition 17,

(311) $\qquad Z \in \Sigma_{n+1}^0 \equiv \bigvee U\big(U \in \Pi_n^0 \wedge \bigwedge \mathfrak{a}(\mathfrak{a} \in Z \equiv \bigvee x \in \mathcal{N}\, U(x, \mathfrak{a}))\big).$

Let $-U$ denote the complement of U:

(312) $\qquad \bigwedge x, \mathfrak{a}\big(-U(x, \mathfrak{a}) \equiv\, \sim U(x, \mathfrak{a})\big).$

It follows from the induction assumption (310) that $-U \in \Sigma_n^0$. Let $-Z$ denote the complement of Z:

(313) $\qquad \bigwedge \mathfrak{a}(\mathfrak{a} \in -Z \equiv \mathfrak{a} \notin Z).$

Under De Morgan's laws it follows from (311)–(313) that

(314) $\qquad \bigwedge \mathfrak{a}\big(\mathfrak{a} \in -Z \equiv \bigwedge x \in \mathcal{N} - U(x, \mathfrak{a})\big).$

Thus $-Z$ is derived by the application of the universal quantifier operation to the relation $-U$, which is in Σ_n^0. Hence, by Definition 17, $-Z \in \Pi_{n+1}^0$. In this way we arrive at the implication $\rightarrow$ of the equivalence

$$Z \in \Sigma_{n+1}^0 \equiv -Z \in \Pi_{n+1}^0.$$

The converse implication is obtained in a similar manner. By (313) and (309) the equivalence referred to above yields the formula

$$\Sigma_{n+1}^0 = C\Pi_{n+1}^0.$$

537

d. The remaining two properties specified in the theorem under consideration follow from the possibility of combining numbers into pairs and from the properties discussed so far. If a relation $Z(n, \alpha)$ is in Π_n^0 for $n > 0$, then, by Definition 17, the relation $Z(n, \alpha)$ can be presented as a result of the universal quantifier operation:

$$Z(n, \alpha) \equiv \bigwedge m \in \mathcal{N} \, U(n, m, \alpha), \quad \text{where} \quad U \in \Sigma_{n-1}^0.$$

On binding the natural variable in $Z(n, \alpha)$ with universal quantifier we arrive at the relation $Z'(\alpha)$, which can be defined as follows:

$$Z'(\alpha) \equiv \bigwedge n \in \mathcal{N} Z(n, \alpha) \equiv \bigvee n, m \in \mathcal{N} \, U(n, m, \alpha) \equiv$$
$$\bigwedge p \in \mathcal{N} \, U((p)_1, (p)_2, \alpha).$$

By the property a of the theorem under consideration, the relation $U((p)_1, (p)_2, \alpha)$ is in the same class as the relation $U(n, m, \alpha)$. Hence, $Z'(\alpha)$ is in the same class as $Z(n, \alpha)$.

e. If a relation $Z(n, \alpha)$ is in Σ_n^0 for $n > 0$, then, by Definition 17, it can be defined as follows:

$$Z(n, \alpha) \equiv \bigvee m \in \mathcal{N} U(n, m, \alpha), \quad \text{where} \quad U \in \Pi_{n-1}^0.$$

On binding the natural variable in $Z(n, \alpha)$ with the existential quantifier we arrive at the relation $Z'(\alpha)$, which can be defined as follows:

$$Z'(\alpha) \equiv \bigvee n \in \mathcal{N} Z(n, \alpha) \equiv \bigvee n, m \in \mathcal{N} U(n, m, \alpha) \equiv$$
$$\bigvee p \in \mathcal{N} U((p)_1, (p)_2, \alpha).$$

By the property a, the relation $U((p)_1, (p)_2, \alpha)$ is in the same class as $U(n, m, \alpha)$, hence Z' is in the same class as Z.

The property c specified in Theorem 49 makes it possible to present the arithmetical classification as a single branch. The following symbolism is adopted:

$$(315) \qquad Z \in P(K) \equiv \bigvee U \Big(U \in K \wedge \bigwedge \alpha \big(Z(\alpha) \equiv \bigvee n \in \mathcal{N} U(n, \alpha) \big) \Big),$$

so that $P(K)$ is the class of all results of the existential quantifiers operations carried out on relations in K. We can now, accordingly, consider a single sequence $\{E_n\}$ of classes instead of the two branches Π_n^0 and Σ_n^0:

$$E_0 = \text{Comp},$$

$$(316) \qquad E_{n+1} = \begin{cases} P(E_n) & \text{for an even } n, \\ C(E_n) & \text{for an odd } n. \end{cases}$$

Instead of applying the two quantifier operations alternately we apply alternately the operation C and the operation of one of the two quantifiers, namely $\bigvee n \in \mathcal{N}$ (although we could equally well adopt $\bigwedge n \in \mathcal{N}$ as the primitive operation). Since $C(\text{Comp}) = \text{Comp}$, the relation between the sequence $\{E_n\}$ and the sequences $\{\Pi_n^0\}$ and $\{\Sigma_n^0\}$ can easily be expressed by the following formulae:

$$\Pi_n^0 = E_{2n},$$

$$\Sigma_n^0 = E_{2n-1}.$$

THEOREM 50. *The following inclusions are true*:

(a) $\qquad \Pi_n^0 \subset \Pi_{n+1}^0, \quad \Sigma_n^0 \subset \Sigma_{n+1}^0,$

(b) $\qquad \Pi_n^0 \subset \Sigma_{n+1}^0, \quad \Sigma_n^0 \subset \Pi_{n+1}^0.$

PROOF. (b). Concepts which are in subsequent classes are derived from concepts in previous classes by the addition of one quantifier. But to define any concept in Π_n^0 (Σ_n^0) we may always add an existential (universal) quantifier which does not affect the meaning of the concept in question, for instance, one which binds a variable that does not occur in the formula involved. The addition of such a superfluous quantifier yields an equivalent predicate, which, however, under Definition 17, is in Σ_{n+1}^0 (Π_{n+1}^0). This yields the inclusions of (b).

(a) The proof of this part is by induction. For $n = 0$ the inclusions of (a) follow from the inclusions of (b), since $\Pi_0^0 = \Sigma_0^0$. Suppose now that

(317) $\qquad \Pi_n^0 \subset \Pi_{n+1}^0 \quad \text{and} \quad \Sigma_n^0 \subset \Sigma_{n+1}^0.$

Consider an arbitrary set Z which is in Π_{n+1}^0. By Definition 17, there exists an R such that $R \in \Sigma_n^0$ and

(318) $\qquad \mathfrak{a} \in Z \equiv \bigwedge k \in \mathcal{N} R(k, \mathfrak{a}).$

If $R \in \Sigma_n^0$, then, by the induction assumption, $R \in \Sigma_{n+1}^0$ and, hence, by (318) and Definition 17, $Z \in \Pi_{n+2}^0$. This yields the inclusion $\Pi_{n+1}^0 \subset \Pi_{n+2}^0$.

Likewise, consider an arbitrary set $Z \in \Sigma_{n+1}^0$. By Definition 17, it is defined by the formula

(319) $\qquad \mathfrak{a} \in Z \equiv \bigvee k \in \mathcal{N} R(k, \mathfrak{a}),$

where $R \in \Pi_n^0$. By the induction assumption, also $R \in \Pi_{n+1}^0$ and, hence, by (319) and Definition 17, $Z \in \Sigma_{n+2}^0$. This yields the inclusion $\Sigma_{n+1}^0 \subset \Sigma_{n+2}^0$.

Thus, formulae (317) yield similar formulae for $n+1$, which completes the proof.

THEOREM 51. *Every relation which can be defined in the language of the arithmetic of natural numbers by a formula in which all quantifiers are relativized to the set of natural numbers is a relation in a class Σ_n^0 or Π_n^0 for a finite n.*

PROOF. Theorem 51 is practically a direct corollary of the theorem on the reduction to normal form with all quantifiers occurring at the beginning of a given formula. If a relation ϱ can be defined in the way described in the theorem under consideration, then the predicate which defines it can be reduced to normal form

$$\bigvee \ldots \bigwedge \ldots \bigvee \ldots R(\ldots),$$

consisting of the quantifier-free part $R(\ldots)$ preceded by a number of quantifiers which bind numerical variables. The quantifier-free part $R(\ldots)$ defines a computable relation. Similar (that is, either universal or existential) quantifiers which follow one another may be combined into one by the pairing function in accordance with the equivalences:

$$\bigvee n \bigvee m \, U(n, m, \mathfrak{a}) \equiv \bigvee p \, U((p)_1, (p)_2, \mathfrak{a}),$$
$$\bigwedge n \bigwedge m \, U(n, m, \mathfrak{a}) \equiv \bigwedge p \, U((p)_1, (p)_2, \mathfrak{a}).$$

Thus, the computable part is preceded by a finite sequence of n quantifiers, alternately universal and existential. Obvious inductive reasoning yields that ϱ is in Π_n^0 or in Σ_n^0. If the first quantifier is universal, then ϱ is in Π_n^0, if it is existential, then ϱ is in Σ_n^0 for an $n \in \mathcal{N}$.

THEOREM 52. *For every class Σ_n^0 (Π_n^0) ($n > 0$) there exists a relation which is universal for all the sets in that class and which itself is also in that class.*

PROOF (Mostowski [52]). The special case of this theorem for $n = 1$ was discussed in Section 2, when reference was made in Theorem 16 to the existence of a recursively enumerable set of pairs, which is universal for the class of recursively enumerable sets. The proof of Theorem 52 is similar to that of Theorem 16. The starting point is the existence of a computable function $G_k(x, y)$ which is universal for the class of primitive recursive functions. Obviously, the function G of three arguments, as constructed in Section 2, can yield by means of the complex-function functions of many arguments:

$$G_k(x_1, ..., x_n, y_1, ..., y_m) = G_k(\langle x_1, ..., x_n \rangle, \langle y_1, ..., y_m \rangle).$$

By setting such a function equal to zero we obtain universal relations for primitive recursive relations of arbitrarily many arguments. By replacing some arguments by formulae in the form of $\bar{\alpha}_1(x), ..., \bar{\alpha}_n(x)$, we obtain, by Theorem 48, that every predicate in the class Σ_n^0 can, for some k, be represented as

$$x \, G_k'(\bar{\alpha}_1(x), ..., \bar{\alpha}_n(x), y_1, ..., y_m) = 0.$$

Hence, a predicate F^1, which is defined thus:

(320)　$F^1(k, \alpha_1, ..., \alpha_n, y_1, ..., y_m) \equiv$
$$\bigvee x \, G_k'(\bar{\alpha}_1(x), ..., \bar{\alpha}_n(x), y_1, ..., y_m) = 0$$

is, by Theorem 48, a universal predicate for the predicates of n function arguments and m numerical arguments in the class Σ_1^0. Formula (320) also proves that the predicate F^1 is in the class Σ_1^0. We adopt the following abbreviation:

(321)　$G_k'(\bar{\alpha}_1(x), ..., \bar{\alpha}_n(x), y_1, ..., y_m) = 0 \equiv G.$

By (321) and Theorem 48, the following predicates are universal for their respective classes:

$$
\begin{array}{lll}
\bigvee x \, G & \text{for} & \Sigma_1^0, \\
\bigwedge x \, G & \text{for} & \Pi_1^0, \\
\bigvee y_1 \bigwedge x \, G & \text{for} & \Sigma_2^0, \\
\bigwedge y_1 \bigvee x \, G & \text{for} & \Pi_2^0, \\
\bigvee y_2 \bigwedge y_1 \bigvee x \, G & \text{for} & \Sigma_3^0, \\
\bigwedge y_2 \bigvee y_1 \bigwedge x \, G & \text{for} & \Pi_3^0,
\end{array}
$$

(322) applies to the group above.

. . . .

The way in which these predicates are defined indicates that they are in those classes, respectively, for which they are universal.

THEOREM 53.

a. *For every $n > 0$:*
$$\Pi_n^0 \neq \Sigma_n^0.$$

b. *For every* $n \in \mathcal{N}$:

$$\Sigma_n^0 \neq \Sigma_{n+1}^0 \quad \text{and} \quad \Pi_n^0 \neq \Pi_{n+1}^0.$$

c. *For* $n > 0$, *a relation of* (322) *which is universal for* Σ_n^0 *is in* Σ_n^0, *but it is neither in* Π_n^0 *nor in* Σ_{n-1}^0; *relations* (322) *which are universal for* Π_n^0 *have a parallel property.*

PROOF (Mostowski [52]). a. Theorem 53 is a simple consequence of theorems proved previously. It can be proved in a general way that if a relation which is universal for sets in a class K is itself in K, then the class K must satisfy the inequality $K \neq C(K)$. For suppose that $K = C(K)$ and that $F(k, x, \mathfrak{a})$ is a universal relation for K and is itself in K. The relation F then satisfies the equivalences

$$(323) \qquad X \in K \equiv \bigvee k \bigwedge x, \mathfrak{a}\big(X(x, \alpha) \equiv F(k, x, \mathfrak{a})\big).$$

Since F is in K and $K = C(K)$, the predicate F', defined thus:

$$\bigwedge x, \mathfrak{a}\big(F'(x, \mathfrak{a}) \equiv \, \sim F(x, x, \mathfrak{a})\big)$$

also is in K. By (323), the set F' must have its number k and must satisfy the equivalence

$$\bigwedge x, \mathfrak{a}\big(F'(x, \mathfrak{a}) \equiv F(k, x, \mathfrak{a})\big).$$

On substituting $x = k$ we arrive at the contradiction:

$$F'(k, \mathfrak{a}) \equiv \, \sim F(k, k, \mathfrak{a}),$$
$$F'(k, \mathfrak{a}) \equiv F(k, k, \mathfrak{a}).$$

Hence the class K must satisfy the inequality $K \neq C(K)$.

Since, by Theorem 52, all the classes Σ_n^0 for $n > 0$ satisfy the conditions laid down in the above argument, for $n > 0$: $\Sigma_n^0 \neq C\Sigma_n^0$, which in view of $\Pi_n^0 = C\Sigma_n^0$ (Theorem 45 c) yields the inequality $\Sigma_n^0 \neq \Pi_n^0$.

b. The second part of Theorem 53 follows from the first and from the inclusions referred to in Theorem 50. For suppose that $\Pi_n^0 = \Pi_{n+1}^0$; we would then also have the equality $C\Pi_n^0 = C\Pi_{n+1}^0$, that is, in accordance with Theorem 49c, $\Sigma_n^0 = \Sigma_{n+1}^0$. Let us, therefore, assume both equalities

$$(324) \qquad \begin{aligned} \Pi_n^0 &= \Pi_{n+1}^0, \\ \Sigma_n^0 &= \Sigma_{n+1}^0, \end{aligned}$$

since they imply one another. By Theorem 50 b, the following inclusions hold:

$$\Pi_n^0 \subset \Sigma_{n+1}^0,$$
$$\Sigma_n^0 \subset \Pi_{n+1}^0.$$

Formulae (324) and these inclusions together yield the equality

$$\Pi_n^0 = \Sigma_n^0,$$

which contradicts the first part of Theorem 53. Hence, equalities (324) cannot hold. (For $n = 0$ the inequality b was proved separately in Section 2.)

c. It is self-evident that a relation which is universal for Σ_n^0 can be neither in Σ_{n-1}^0 nor in Π_n^0; for if it were in a lower or in a dual class, then this would hold for all the sets in Σ_n^0, since these sets are obtained from the universal relation by the fixing of the numbering argument, that is, by the substitution of a constant, and, hence, a computable function, and such substitutions, as stated in Theorem 49a, do not lead outside a given class.

The classes Π_n^0 and Σ_n^0 have been defined for relations of arbitrarily many arguments. The proof of Theorem 53, as recorded above, substantiates the inequalities stated in Theorem 53 for Π_n^0 and Σ_n^0 treated as classes of those relations which have at least one numerical argument. The theorem can easily be extended so as to cover those relations which have no numerical arguments at all.

Example of estimation of the class of a concept definable in arithmetic[1]

The concept which we shall now try to locate in the hierarchy described above will be that of the convergence of a sequence of real numbers to a fixed limit. Convergence is usually defined thus:

$$(325) \qquad \{a_n\} \Rightarrow a \equiv \bigwedge k \bigvee m \bigwedge n \left(n > m \rightarrow |a_n - a| < \frac{1}{k+1} \right).$$

This definition of convergence contains three quantifiers; hence, the concept is in the class Π_3^0. The question arises whether the concept does not perhaps belong to a lower or a dual class. If it belongs to a lower class, that would mean that the definition can be simplified from

---

[1] A. Mostowski [56].

the logical point of view in an essential way, namely by reducing the number of the quantifiers which occur in it. It will soon be demonstrated that this is impossible.

Instead of analysing arbitrary sequences we shall confine ourselves to analysing sequences of rational numbers $\{a_n\}$ in the form $a_n = 1/\alpha(n)$, convergent to zero. Such a sequence converges to zero when the function $\alpha(n)$ approaches infinity. Hence we shall concentrate on analysing natural functions $\alpha(n)$ which approach infinity. Let Z stand for the set of such functions. We shall also be concerned with functions and relations dependent on certain parameters $\mathfrak{a}$:

$$(326) \qquad \alpha_{(\mathfrak{a})} \in Z \equiv \bigwedge k \bigvee m \bigwedge n (n > m \rightarrow \alpha_{(\mathfrak{a})}(n) > k).$$

If we can demonstrate that the number of quantifiers in the definition of the set Z cannot be reduced, this will mean that such a reduction is *a fortiori* impossible for the complete definition (325) of convergence.

Definition (326) itself shows clearly that $Z \in \Pi^0_3$. The proof that Z belongs neither to a lower nor to a dual class will be split into several lemmata.

LEMMA 1. *For every relation $R \in \text{Comp}$ there is a relation $S \in \text{Comp}$ such that the following equivalence holds*:

$$(327) \qquad \bigvee x \bigwedge y R_{(\mathfrak{a})}(x, y) \equiv \bigvee m \bigwedge n > m S_{(\mathfrak{a})}(n).$$

PROOF. We consider all the pairs $\langle x, y \rangle$. In the diagram below (cf. Figure 8) each such pair has its analogue in the form of the box with the co-ordinates x, y. The shaded boxes form the diagram of the relation R, as they correspond to those pairs $\langle x, y \rangle$ for which $R(x, y)$ holds. The condition

$$(328) \qquad \bigvee x \bigwedge y R(x, y)$$

is satisfied if and only if there exists a number x such that the whole column above the co-ordinate x is situated within the diagram of R. We shall now define a function $g \in \text{Comp}$ which enumerates certain pairs $\langle x, y \rangle$, namely those which in the diagram are connected by the directed line. We begin with the pair $\langle 0, 0 \rangle$ and then move to the right until we come across the diagram of the relation R. In the diagram of that relation we move vertically upward until we go outside the diagram of R. Then we come back to the zero row and move to the right again. If the whole vertical column remains within the diagram of R, then

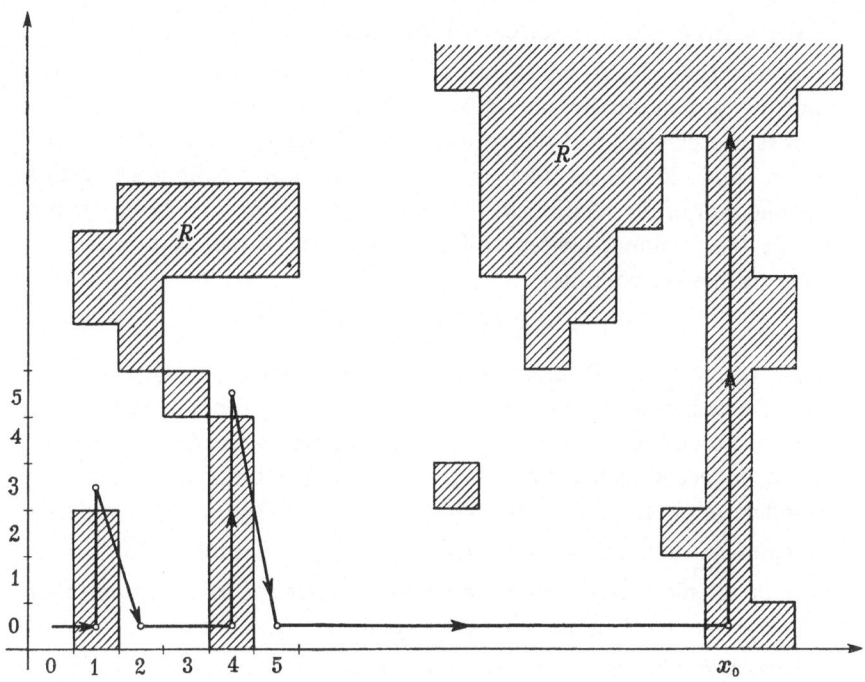

Fig. 8

the function g remains in that column. A rigorous definition of the function g can be written down as follous

$$g(0) = \langle 0, 0 \rangle,$$

$$(329) \qquad g(n+1) = \begin{cases} \langle (g(n))_0, (g(n))_1 + 1 \rangle & \text{if } R\big((g(n))_0, (g(n))_1\big), \\ \langle (g(n))_0 + 1, 0 \rangle & \text{if } \sim R\big((g(n))_0, (g(n))_1\big). \end{cases}$$

The first element $(g(n))_0$ of the pair

$$g(n) = \langle (g(n))_0, (g(n))_1 \rangle$$

stands for the number of the column. The second element, $(g(n))_1$, is the number of the row. The function g in fact behaves in accordance with its intuitive description. If the relation R is satisfied for the previous pair, then the number of the row increases by one, and the number of the column remains unchanged. If the relation R is not satisfied, then the number of the column increases by one, and the number of the row returns to zero.

545

If condition (328) is satisfied, then there exists a least x_0 for which it is true that $\bigwedge y R(x_0, y)$. On reaching the pair $g(m) = \langle x_0, 0 \rangle$ the function g continually moves along a vertical line. The co-ordinate x does not change, and the co-ordinate y increases continually. Hence, if condition (328) is satisfied, so that a certain whole column lies within the diagram of the relation R, then the function $g(n)$ for $n > m$ remains within that column, hence, within the diagram of the relation R, that is, it satisfies the condition

(330) $\qquad \bigvee m \bigwedge n > m R\big((g(n))_0, (g(n))_1\big).$

Thus, condition (328) implies condition (330). But it can easily be seen that the converse is also true. If condition (328) is not satisfied, then there is no column contained completely within the diagram of R; and hence, in accordance with the definition of the function g, there are infinitely many numbers n for which it is true that $g(n)$ lies outside the diagram of R, that is, $\sim R\big((g(n))_0, (g(n))_1\big)$. Formula (330) cannot then be satisfied. Thus, condition (328) is equivalent to condition (330), which proves the lemma under consideration.

LEMMA 2. *For every relation S the following equivalence holds*:

(331) $\qquad \bigwedge k \bigvee m \bigwedge n > m S(k, n) \equiv$
$\qquad\qquad \bigwedge k \bigvee x \bigwedge n > x (n > k \wedge S(k, n)).$

PROOF. If for every k there exists an m such that $S(k, n)$ holds for $n > m$, then for $x = \max(k, m)$ it is true that for $n > x$ both $n > k$ and $n > m$, and hence $S(k, n)$. The converse implication is self-evident.

LEMMA 3. *For every relation T the following equivalence holds*:

(332) $\qquad \bigwedge k \bigvee m \bigwedge n > m T(k, n) \equiv$
$\qquad\qquad \bigwedge k \bigvee x \bigwedge n > x \bigwedge l \leqslant k T(l, n).$

PROOF. The implication $\leftarrow$ is self-evident. The implication $\rightarrow$ will be proved:

(333) $\qquad m(l) = (\mu m) \left[\bigwedge n > m T(l, n) \right],$

(334) $\qquad x(k) = \max m(l) \quad \text{for} \quad l \leqslant k.$

If

(335) $\qquad \bigwedge k \bigvee m \bigwedge n > m T(k, n),$

then the function $m(l)$ satisfies the formula

(336) $\quad \bigwedge l \bigwedge n > m(l) T(l, n),$

for, by (335), that minimum, which is defined by formula (333), does in fact exist. It follows from (334) that

(337) $\quad (n > x(k) \wedge l \leqslant k) \rightarrow n > m(l).$

Formulae (336) and (337) imply the formula

(338) $\quad \bigwedge k, l, n\big((n > x(k) \wedge l \leqslant k) \rightarrow T(l, n)\big),$

which in turn implies the right side of equivalence (332).

LEMMA 4. *For every relation $R \in \mathrm{Comp}$ there exists a function $\alpha \in \mathrm{Comp}$ such that the following equivalence holds*:

(339) $\quad \bigwedge k \bigvee m \bigwedge n R_{(\alpha)}(k, m, n) \equiv$
$\qquad \bigwedge k \bigvee m \bigwedge n (n > m \rightarrow \alpha_{(\alpha)}(n) > k).$

PROOF. By Lemma 1, there exists a relation $S \in \mathrm{Comp}$ such that

(340) $\quad \bigwedge k \bigvee m \bigwedge n R(k, m, n) \equiv \bigwedge k \bigvee m \bigwedge n > m S(k, n).$

On assuming $T(k, n) \equiv (S(k, n) \wedge n > k)$, from Lemma 3 (formula (332)) we obtain the following equivalence

(341) $\quad \bigwedge k \bigvee m \bigwedge n > m(S(k, n) \wedge n > k) \equiv$
$\qquad \bigwedge k \bigvee x \bigwedge n > x \bigwedge l \leqslant k(S(l, n) \wedge n > l).$

Lemma 2 (formula (331)) and equivalences (340) and (341) yield the equivalence

(342) $\quad \bigwedge k \bigvee m \bigwedge n R(k, m, n) \equiv$
$\qquad \bigwedge k \bigvee x \bigwedge n > x \bigwedge l \leqslant k(S(l, n) \wedge n > l).$

By the logical laws the following equivalence is self-evident:

(343) $\quad \bigwedge l \leqslant k(S(l, n) \wedge n > l) \equiv (\mu l)[\sim S(l, n) \vee n \leqslant l] > k.$

Since $S \in \mathrm{Comp}$, the function α, defined as follows:

(344) $\quad \alpha(n) = (\mu l)[\sim S(l, n) \vee n \leqslant l] = (\mu l < n)[S(l, n)],$

is computable, too, since the minimization in (344) is bounded. Equivalences (342) and (343) and formula (344) immediately yield equivalence (339).

LEMMA 5. *The set* Z, *defined by formula* (326), *is neither in* Π_2^0 *nor in* Σ_3^0.

PROOF. By Theorem 52, there exists a relation $F_3(k, x)$, universal for the class Π_3^0, which is in the same class. Hence, by Lemma 4, there exists for it a computable function $\alpha_{k,x}(n)$ of three arguments such that the following equivalence holds:

$$(345) \qquad F_3(k, x) \equiv \bigwedge u \bigvee m \bigwedge n (n > m \rightarrow \alpha_{k,x}(n) > u).$$

By formula (326), $F_3(k, x) \equiv \alpha_{k,x} \in Z$. Should the set Z be in Π_2^0 or in Σ_3^0, then, by equivalences (345) and (326), the relation $F_3(k, x)$ would be in Π_2^0 or in Σ_3^0, which, as we know from Theorem 53, is not possible.

Thus, the set Z of functions which converge to infinity is completely contained in the class Π_3^0. Therefore, the natural definition of convergence of sequences cannot be formulated more simply in terms of the number of quantifiers. All three quantifiers which occur in it are indispensable and cannot be replaced by quantifiers of other types. Thus, an important mathematical concept has been located precisely in a certain class in the arithmetical hierarchy.

Analytical hierarchy

Let us now try to locate the concept of truth in the classification under consideration. In doing so we shall confine ourselves to sentences in the language of the arithmetic Ar and we shall consider their truth in the classical model $\mathfrak{N}_0$. Since language can be arithmetized, this concept may be considered to apply to natural numbers:

$n \in E(\mathfrak{N}_0) \equiv n$ *is a number of a sentence formulated in the language of* Ar *and true in the standard model $\mathfrak{N}_0$ for the ordinary interpretation of terms.*

The set $E(\mathfrak{N}_0)$ was defined in Chapter II.

This definition will not be analysed for the time being, since we shall first be concerned with estimating it from below. Suppose that the set $E(\mathfrak{N}_0)$ is in an arithmetical class Π_n^0. It will be proved that this assumption results in a contradiction: under this assumption it could be proved that all sets in Π_{n+k}^0 are in Π_n^0, which is in contradiction with Theorem 53. Consider accordingly an arbitrary set $X \in \Pi_{n+k}^0$. By

Definition 17, it can be defined by an arithmetical predicate $\alpha(x)$ in which one free variable occurs:

$$n \in X \equiv \alpha(n).$$

By the definition of definability (Section 2, p. 445), the set X can also be defined by the formula

$$n \in X \equiv \ulcorner\alpha(n)\urcorner \in E(\mathfrak{N}_0).$$

For a fixed predicate $\alpha(x)$, constructed in the language of the arithmetic Ar, the function $\varphi(n) = \ulcorner\alpha(n)\urcorner$, which yields the number of the sentence obtained from the predicate $\varphi(x)$ by the substitution x/n, is primitive recursive:

$$\varphi(n) = \mathrm{Sb}\Big(\mathrm{Nr}(\ulcorner\alpha(x_1)\urcorner), x_1, \mathrm{Nr}(n)\Big).$$

Hence, the definition of the set X has the form:

$$n \in X \equiv \varphi(n) \in E(\mathfrak{N}_0).$$

Thus, the set X is recursively reducible to the set $E(\mathfrak{N}_0)$. Hence the set X would be in the class Π_n^0. If accordingly the set $E(\mathfrak{N}_0)$ were in any arithmetical class Π_n^0, then all the arithmetical classes would be contained in Π_n^0, which is in contradiction with Theorem 53.

This provides an example of a concept that goes beyond a finite arithmetical hierarchy. If we wished to classify that concept in the arithmetical hierarchy, we would have to extend the arithmetical hierarchy into transfinity. That extension will not be described here, as we shall proceed directly to describing the next hierarchy, called the analytical hierarchy. As we shall see, the concept of truth is in the lowest class of the analytical hierarchy.

The analytical hierarchy has been defined as a superstructure upon the arithmetical hierarchy (Kleene [38]). It is obtained by imposing upon arithmetical predicates quantifiers which bind variables that stand for natural functions, or sets of natural numbers, or relations among natural numbers. It suffices, of course, to confine ourselves, for instance, to variables ranging over natural functions. We will do so here. The Greek letters α, β, γ, ... will stand for natural functions.

On imposing upon arithmetical predicates quantifiers which bind function variables we obtain, as in the case of the arithmetical hierarchy, two branches: one consisting of those formulae which begin with an existential quantifier, and the other consisting of those formulae which begin with universal quantifier. The classes of those branches will be denoted by Σ_n^1 and Π_n^1, respectively, as in the case of the arithmetical hierarchy, the difference being that the upper index (superscript) now is 1, and not 0, which indicates that we have passed from type 0 quantification to type 1 quantification. The definition is by induction.

DEFINITION 18. *We first define the classes Σ_n^1 and Π_n^1 for $n > 0$:*

a. *The class Σ_1^1 (Π_1^1) contains all the predicates $Q(\mathfrak{a})$ ($P(\mathfrak{a})$), defined, respectively, by the formulae:*

$$Q(\mathfrak{a}) \equiv \bigvee \alpha A(\alpha, \mathfrak{a}) \quad (P(\mathfrak{a}) \equiv \bigwedge \alpha A(\alpha, \mathfrak{a})),$$

where $A(\alpha, \mathfrak{a})$ is an arithmetical predicate (that is, is in Π_n^0 for some n).

b. *The class Σ_{n+1}^1, for $n > 0$, contains every predicate $Q(\mathfrak{a})$, defined by the formula:*

$$(346) \quad Q(\mathfrak{a}) \equiv \bigvee \alpha P(\alpha, \mathfrak{a}),$$

where $P(\alpha, \mathfrak{a})$ is a predicate in Π_n^1.

c. *The class Π_{n+1}^1, contains every predicate $Q(\mathfrak{a})$, defined by the formula:*

$$(347) \quad Q(\mathfrak{a}) \equiv \bigwedge \alpha P(\alpha, \mathfrak{a}),$$

where $P(\alpha, \mathfrak{a})$ is a predicate in Σ_n^1.

d. *Π_0^1 (Σ_0^1) denotes the common part of the classes Π_1^1 and Σ_1^1:*

$$(348) \quad \Pi_0^1 = \Sigma_0^1 = \Pi_1^1 \cap \Sigma_1^1.$$

A set X of any objects is called a set of the class Π_n^1 (Σ_n^1) if and only if it can be defined by a predicate of the class Π_n^1 (Σ_n^1).

Definition 18 is parallel to Definition 17, the only difference being that here the classes Π_0^1 (Σ_0^1) are defined in an indirect way, whereas in Definition 17 the classes Π_0^0 (Σ_0^0) were defined directly as classes of computable sets. The class Π_0^1 (Σ_0^1) could be defined directly as a class of hyperarithmetical sets, but for that purpose the arithmetical classification would have to be extended into the transfinite, whereas in Definition 17 we confined ourselves to defining the classes Π_n^0 (Σ_n^0) for finite n. The extension of the arithmetical classification into the transfinite is somewhat complicated, and is also necessary for defining the analytical hierarchy. The function-binding quantifiers which precede the number-binding quantifiers absorb all number-binding quantifiers but one, as is shown by the following theorem.

THEOREM 54. *For every set X in the class Σ_n^0 (Π_n^0) there exist recursive predicates R and R' such that the set X can be represented as*

(349) $\quad \mathfrak{a} \in X \equiv \bigvee \alpha \bigwedge x R(\alpha, x, \mathfrak{a})$,

(350) $\quad (\mathfrak{a} \in X \equiv \bigwedge \alpha \bigvee x R'(\alpha, x, \mathfrak{a}))$.

PROOF. For every relation R the equivalence

(351) $\quad \bigwedge x \bigvee y R(x, y, \mathfrak{a}) \equiv \bigvee \alpha \bigwedge x R(x, \alpha(x), \mathfrak{a})$

is true. In fact, if the left side of that equivalence is true, then the function

$$\alpha(x) = (\mu y) [R(x, y, \mathfrak{a})]$$

may be adopted as α. This function satisfies the right side of the equivalence. Conversely, if the right side of the equivalence is true, then $\alpha(x)$ is that y which satisfies the left side.

If X is in Π_{2n}^0, then X can be defined by a predicate which begins with a number-binding universal quantifier, and ends with a number-binding existential quantifier that precedes a recursive relation:

(352) $\quad \mathfrak{a} \in X \equiv \bigwedge x_0 \bigvee y_0 \bigwedge x_1 \bigvee y_2 \ldots \bigwedge x_n \bigvee y_n R(x_0, \ldots, x_n, \ldots, y_0, \ldots, y_n, \mathfrak{a})$.

By (351), the first pair of quantifiers, $\bigwedge x_0 \bigvee y_0$, may be replaced by an existential quantifier binding a function variable $\bigvee \alpha_0$ and by a universal quantifier binding a numerical variable $\bigwedge x_0$, which yields the formula

(353) $\quad \mathfrak{a} \in X \equiv \bigvee \alpha_0 \bigwedge x_0 \bigwedge x_1 \bigvee y_2 \ldots$
$\qquad \bigwedge x_n \bigvee y_n R(x_0, x_1, \ldots, x_n, \alpha(x_0), y_1, \ldots, y_n, \mathfrak{a})$.

Next we may combine the quantifiers $\bigwedge x_0 \bigwedge x_1$ by the pairing function into one universal quantifier and apply equivalence (351) again, this time to the relation which in (353) follows the quantifier $\bigvee \alpha_0$, etc. By applying (351) n times we finally transform formula (352) into

(354) $\quad \mathfrak{a} \in X \equiv \bigvee \alpha_0 \bigvee \alpha_1 \ldots \bigvee \alpha_n \bigwedge z R'(\alpha_0, \ldots, \alpha_n, z, \mathfrak{a})$,

where R' still is a recursive relation. The functions $\alpha_0, \ldots, \alpha_n$ may, of course, be combined into one by the pair function interpreted as a functional with two or more function arguments:

(355) $\quad \begin{aligned} &\langle \alpha, \beta \rangle(k) = \langle \alpha(k), \beta(k) \rangle, \\ &\langle \alpha_0, \ldots, \alpha_n \rangle(k) = \langle \alpha_0(k), \ldots, \alpha_n(k) \rangle. \end{aligned}$

The functionals

(356) $(\alpha)_i(n) = (\alpha(n))_i$

will be converse functionals. In fact, these functionals satisfy the function equations

$$(\langle \alpha, \beta \rangle)_0(n) = \alpha(n),$$
$$(\langle \alpha, \beta \rangle)_1(n) = \beta(n),$$
$$\langle (\alpha)_0, (\alpha)_1 \rangle(n) = \alpha(n).$$

Hence, formula (354) may be given the form

$$\mathfrak{a} \in X \equiv \bigvee \alpha \bigwedge z R'((\alpha)_0, \ldots, (\alpha)_n, z, \mathfrak{a}),$$

which complies with the requirements of the theorem now under consideration. If X is in any finite arithmetical class, then, by Theorem 50, it may be assumed that for a sufficiently large finite n both X and $-X$ are in Π_{2n}^0. Hence, both X and $-X$ can be represented in the form of (349). Now if $-X$ can be presented in the form of (349), then, under De Morgan's laws, X can be represented in the form of (350).

Thus, the formulae

$$\Sigma_n^0 \subset \Pi_0^1, \quad \Pi_n^0 \subset \Pi_0^1 \quad \text{for all } n \in \mathcal{N}$$

are consequences of Theorem 54.

However, it is not only the sets Σ_n^0 and Π_n^0 of the finite arithmetical hierarchy which are in Π_0^1. Π_0^1 also contains those sets which in the arithmetical hierarchy would have a transfinite number. Such sets do exist; one of them, for instance, is $E(\mathfrak{N}_0)$, from which our analyses started. All the sets in the transfinite arithmetical hierarchy are in Π_0^1, as also is the case of the finite arithmetical hierarchy. The proof of the fact that the set of true sentences is in $\Pi_1^1 \cap \Sigma_1^1$ is particularly clear. In Chapter II we defined the concept of satisfaction of a formula by a sequence. For the domain $\mathfrak{N}_0$ the sequence of objects is defined by a function. By repeating the definition of satisfaction in arithmetic we obtain an arithmetical predicate Sat(n, α):

> Sat$(n, \alpha) \equiv$ *a sequence of values of a function α satisfies a formula with a number n in the domain $\mathfrak{N}_0$ for the ordinary interpretation of terms.*

The definition of the predicate Sat(n, α) merely requires the use of quantifiers which bind variables that range over numbers (namely, numbers of formulae or numbers of finite sequences of formulae). Hence, if n is a number of a sentence, the definition of truth has the form:

$$n \in E(\mathfrak{N}_0) \equiv \bigwedge \alpha \, \mathrm{Sat}(n, \alpha).$$

But, by Theorem 4 (Chapter II, Section 1), the set $E(\mathfrak{N}_0)$ can also be defined as follows:

$$n \in E(\mathfrak{N}_0) \equiv \bigvee \alpha \, \mathrm{Sat}(n, \alpha).$$

This is obvious also in view of the fact that the set $E(\mathfrak{N}_0)$ is complete. The set $E(\mathfrak{N}_0)$ is, thus, defined dually, both as Π_1^1 and as Σ_1^1.

By Definition 18, $E(\mathfrak{N}_0) \in \Pi_0^1$.

We shall also record a theorem analogical to theorems on the arithmethical hierarchy:

THEOREM 55. *The sets in the classes Σ_n^1 and Π_n^1, for $n > 0$, can be defined, respectively, by predicates of the form*

$$
\text{(357)} \quad
\begin{array}{ll}
\underset{\Sigma_1^1}{\bigvee \alpha \bigwedge x R(\bar{\alpha}(x), \mathfrak{a})}, & \underset{\Sigma_2^1}{\bigvee \beta \bigwedge \alpha \bigvee x R(\bar{\alpha}(x), \bar{\beta}(x), \mathfrak{a})}, \quad \dots, \\[2ex]
\underset{\Pi_1^1}{\bigwedge \alpha \bigvee x R(\bar{\alpha}(x), \mathfrak{a})}, & \underset{\Pi_2^1}{\bigwedge \beta \bigvee \alpha \bigwedge x R(\bar{\alpha}(x), \bar{\beta}(x), \mathfrak{a})}, \quad \dots,
\end{array}
$$

where R is an elementary recursively relation.

PROOF. By Theorem 54, all arithmetical predicates $A(\alpha, \mathfrak{a})$ can be presented in the form of (349) and (350). Hence, by Theorem 48, Definition 18, and the properties of the pairing function, the required form is obtained.

Since the form of Definition 18 is the same as that of Definition 17, the only difference being that when Definition 17 refers to quantifiers that bind numerical variables Definition 18 refers to quantifiers that bind function variables; hence, many theorems can be transposed automatically. They will be inserted here as a matter of record:

THEOREM 56.

a. *Substitution of a computable functional does not change the class of the predicate in question.*

b. *Every analytical class is a ring of sets.*

c. $\Sigma_n^1 = C\Pi_n^1$.

d. *The operation of the universal quantifier that binds a function variable does not lead outside the class Π_n^1.*

e. *The operation of the existential quantifier that binds a function variable does not lead outside the class Σ_n^1.*

The proof is analogical to that of Theorem 49. Instead of the pairing functions we use the pairing functionals defined by formulae (355) and (356).

THEOREM 57. $\Pi_n^1 \cup \Sigma_n^1 \subset \Pi_{n+1}^1 \cap \Sigma_{n+1}^1$.

The proof is parallel to that of Theorem 50.

THEOREM 58. *Every relation which can be defined in the language of the arithmetic of natural numbers by a formula in which all quantifiers are those which bind variables that stand for natural numbers or natural functions is a relation in a class Σ_n^1 or Π_n^1 for a natural n.*

The proof is by reduction to normal form, parallel to the proof of Theorem 51, the only difference being that the following possibility must be taken into account. The formula which defines the relation under consideration after its reduction to normal form may contain a quantifier that binds a numerical variable and stands between or before those quantifiers which bind function variables. Yet such a quantifier may be transformed into one that binds a function variable on the strength of the following self-evident equivalence:

$$(358) \qquad \bigvee x R(\ldots x \ldots) \equiv \bigvee \alpha R(\ldots \alpha(0) \ldots).$$

By applying this equivalence we can reduce the formula that defines the relation under consideration to the form

$$\underbrace{\bigwedge \ldots \bigvee \ldots}_{\substack{\text{quantifiers} \\ \text{which bind} \\ \text{function} \\ \text{variables}}} \underbrace{\bigwedge \ldots \bigvee \ldots}_{\substack{\text{quantifiers} \\ \text{which bind} \\ \text{numerical} \\ \text{variables}}} R(\ldots),$$

in which the quantifier-free (computable) part $R(\ldots)$ is preceded first by quantifiers that bind numerical variables alone, and next by quantifiers that bind function variables only. By Theorem 51, that part which contains no quantifiers that bind function variables is an arithmetical predicate, and, hence, the whole is, under Definition 18, an analytical predicate.

THEOREM 59. *For every class Π_n^1 or Σ_n^1 in the analytical hierarchy, for $n > 0$, there is a relation which is universal for all the sets in that class and which itself is in that class.*

PROOF. By Theorem 55, those predicates which are in the classes of the analytical hierarchy can be represented in the form of (357), where R is an elementary recursive relation. By substituting for R a computable relation universal for the class of of primitive recursive relations we obtain predicates which are universal for a given class. Thus, for instance, the relations H_1 and H_2, defined as follows:

$$
(359) \quad
\begin{aligned}
H_1(k, z) &\equiv \bigvee \alpha \bigwedge x\, G_k\big(\bar{\alpha}(x), z\big) = 0, \\
H_2(k, z) &\equiv \bigwedge \beta \bigvee \alpha \bigwedge x\, G_k\big(\langle \bar{\alpha}(x), \bar{\beta}(x)\rangle, z\big) = 0
\end{aligned}
$$

are universal relations for those sets of numbers which are, respectively, in Σ_1^1 and Π_2^1. G is the previously encountered known computable function which is universal for the class of primitive recursive functions. The very form of the universal functions indicates that they are in the same class.

THEOREM 60. *For every n, $\Sigma_n^1 \neq \Sigma_{n+1}^1$, $\Pi_n^1 \neq \Pi_{n+1}^1$; for $n > 0$, $\Pi_n^1 \neq \Sigma_n^1$; the relation which is universal for Σ_n^1 is in Σ_n^1, and is neither in Π_n^1 nor in Σ_{n-1}^1.*

The proof is parallel to that of Theorem 53.

Estimation of analytical class and the concept of well-ordering

We shall estimate the concept of well-ordering as an example of locating a concept in the analytical hierarchy. In doing so we shall confine ourselves to orderings of natural numbers. To put it more rigorously, we shall confine ourselves to analysing orderings in a recursive set of natural numbers, namely the set of those numbers in whose decomposition into prime factors all further prime numbers occur to the zero power if a number p_i occurs to the zero power:

$$
(360) \quad x \in \mathrm{Seq} \equiv \Big\{ x > 0 \wedge \bigwedge i \leqslant x \bigwedge j \leqslant x \big(((x)_i = 0 \wedge i < j) \to (x)_j = 0 \big) \Big\}.
$$

These numbers will represent finite sequences. The length of such a sequence is well-defined by the definition of length (Section 3, formula (104)):

(361) $\lg(z) = (\mu x \leqslant z)\,[(z)_x = 0]$.

It follows immediately from formulae (360) and (361) that

(362) $(x \in \mathrm{Seq} \wedge y < \lg(x)) \to (x)_y > 0$.

DEFINITION 19. I. *A relation ϱ orders the set* Seq *if and only if*:

$$\bigwedge x, y \in \mathrm{Seq}(x\varrho y \to\; \sim (y\varrho x)) \wedge \bigwedge x, y, z \in \mathrm{Seq}((x\varrho y \wedge y\varrho z)$$
$$\to x\varrho z) \wedge \bigwedge x, y \in \mathrm{Seq}(x\varrho y \vee y\varrho x \vee x = y).$$

II. *A relation ϱ which orders the set* Seq *well-orders the set* $U \subset$ Seq (*briefly*: $\varrho\,\mathrm{B\,Ord}\,U$) *if and only if*:

ϱ *orders the set* Seq,

$U \subset$ Seq,

$$\bigwedge \alpha \Big(\bigwedge x\,(\alpha(x+1)\varrho\,\alpha(x)) \to \bigvee x\,\alpha(x) \notin U\Big).$$

The last condition of well-ordering (the most important one) states that in the set U there are no infinite sequence descending with respect to the relation ϱ, that is, there are no sequences of the form

$$\ldots \alpha(3)\varrho\,\alpha(2)\varrho\,\alpha(1)\varrho\,\alpha(0)$$

which are fully contained in the set U.

The very form in which Definition 19 is written down shows that the concept $\mathrm{B\,Ord}$ is in the class Π_1^1. It only remains to prove that it is not in Σ_1^1. To do so we shall need a selected ordering holding between the numbers of the set Seq. That ordering will be denoted by $\prec$:

(363) $n \prec m \equiv \Big\{\big(\lg(m) < \lg(n) \wedge \bigwedge i < \lg(m)\,((m)_i = (n)_i)\big) \vee$
$$\bigvee x \leqslant n\big(x < \lg(n) \wedge x < \lg(m) \wedge (n)_x < (m)_x \wedge$$
$$\bigwedge i < x\,((m)_i = (n)_i)\big)\Big\}.$$

The relation $\prec$ is not a well-ordering in Seq, but it is an ordering. Before demonstrating that let us consider the intuitive sense of the relation $\prec$. In formula (363) distinction is made between two cases. In the first, the sequence n is just a continuation of the sequence m, that is, m is an initial segment of n. In the second, the lengths of the sequences

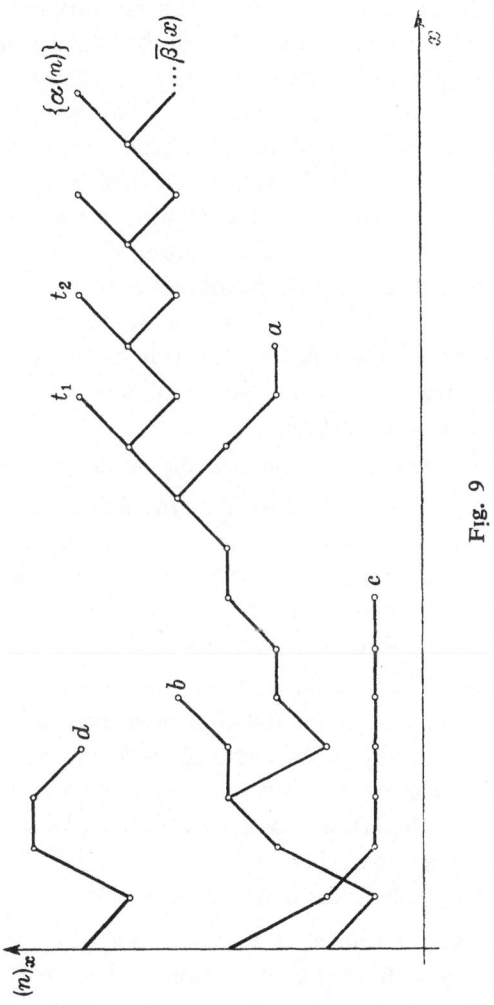

Fig. 9

may be arbitrary. The only requirement is that there is a place x such that up to it both sequences are identical, but at x the term of the sequence n is smaller than the respective term of the sequence m. To put it briefly, a sequence m is later than a sequence n if either n is a continuation of m, or at the first place at which they differ from one another the term in m is greater than the term in n. A diagram can be used in order better to visualize the relation (Fig. 9). On the horizontal co-ordinate we mark the numbers of the successive terms of the sequence. Above a number x we mark the values $(n)_x$ if they are other than zero. The successive values $(n)_x$ are connected with line segments. The polygonal line obtained in this way is the diagram of the sequence associated with the number n. The order of the sequences marked in Figure 9 with the letters a, b, c, d is: $a \prec b \prec c \prec d$. If we denote by e the common part of the sequences a and b, then it is true that $a \prec b \prec e \prec c \prec d$.

By extending a diagram to infinity we obtain an infinite diagram, that is, a diagram of a function.

LEMMA 6. *The relation $\prec$ is an ordering in the set* Seq.

PROOF. For any pair n, $m \in$ Seq it is true that either $n = m$, or one of the two sequences is a continuation of the other, or there exists an x such that $x < \lng(m)$ and $x < \lng(n)$, and x is the earliest place at which n and m differ from one another. Usually we shall have to consider those mutually exclusive cases separately.

The relation $\prec$ is antisymmetric. If $n \prec m$ and if n is a continuation of m, then m cannot be a continuation of n, nor can n and m differ from one another in any place $x < \lng(m)$, so that it is not true that $m \prec n$. If n and m differ from one another in a place $x < \lng(m)$ and if $x < \lng(m)$, $x < \lng(n)$, and x is the earliest place in which n and m differ from one another, then, if $(n)_x < (m)_x$, then it is not true that $(m)_x < (n)_x$. Thus, in both cases it is not true that $m \prec n$.

The relation $\prec$ is transitive. If $n \prec m$ and $m \prec k$, then we have to consider four cases. 1. If n is a continuation of m and m is a continuation of k, then obviously n also is a continuation of k. 2. If n is a continuation of m, then $(m)_x = (n)_x$ for some $x < \lng(m)$; hence, if $(m)_x < (k)_x$, where $x < \lng(m) < \lng(n)$, then $(n)_x < (k)_x$ and $x < \lng(n)$, and if x is the earliest place in which $(m)_x \neq (k)_x$, then x also is the earliest place in which $(n)_x \neq (k)_x$, so that $n \prec k$. 3. If, in reverse, m is a continuation of k and if x is the earliest place such that $(m)_x \neq (n)_x$ and $(n)_x < (m)_x$ and $x < \lg(n)$ and $x < \lg(m)$, then either $x < \lg(k)$ or $x \geq \lg(k)$. If $x \geq \lg(k)$, then $\lg(k) < \lg(n)$, and, for $i < \lg(k)$, $(n)_i = (m)_i = (k)_i$, so that n is a continuation of k. If $x < \lg(k)$, then, since $(m)_i = (k)_i$ for $i < \lg(k)$, x, being the earliest place in which n differs from m, is thereby the earliest place in which n differs from k and $(n)_x < (m)_x = (k)_x$, so that $n \prec k$. 4. If x is the earliest place such that $(n)_x \neq (m)_x$, and if $x < \lg(n)$, $x < \lg(m)$, and $(n)_x < (m)_x$, and if y is the earliest place at which $(m)_y \neq (k)_y$, and if $y < \lg(m)$, $y < \lg(k)$, and $(m)_y < (k)_y$, then if $x \leq y$, then $x < \lg(k)$, and x is the earliest place at which $(n)_x \neq (k)_x$, since, for $i < y$, $(m)_i = (k)_i$. Thus, if $x < y$, then $(n)_x < (m)_x = (k)_x$, and, hence, $n \prec k$. If $x = y$, then $(n)_x < (m)_x = (m)_y < (k)_x = (k)_y$, and hence also $n \prec k$. Finally, if $y < x$, then $y < \lg(n)$ and y is the earliest place at which $(n)_y \neq (k)_y$, since, for $i < x$, $(n)_i = (m)_i$. Thus, $(n)_y = (m)_y < (k)_y$, so that $n \prec k$.

The relation $\prec$ is connected in the set Seq. If m, $n \in$ Seq and m and n differ from one another, then we have to consider two cases. Either one of the two sequences, m and n, is a continuation of the other, or not. If m is a continuation of n, then $m \prec n$; if n is a continuation of m, then $n \prec m$. If neither is a continuation of the other, then there exists an x such that $x < \lg(m)$ and $x < \lg(n)$, and x is the earliest place in which m and n differ from one another, so that $(m)_x < (n)_x$ or $(n)_x < (m)_x$ and $(n)_i = (m)_i$ for $i < x$. Hence, either $m \prec n$, or $n \prec m$.

LEMMA 7. *For any function α a sequence of numbers $\{\bar\alpha(n)\}$ is an infinite descending sequence (with respect to the relation $\prec$) of the elements of the set* Seq:

$$\ldots \prec \bar\alpha(3) \prec \bar\alpha(2) \prec \bar\alpha(1) \prec \bar\alpha(0).$$

PROOF. Under the definition (261) of the functional $\bar\alpha(x)$, each of the numbers $\bar\alpha(n)$ is an element of the set Seq. Further more, by (262), $(\bar\alpha(n))_i = \alpha(i)+1$ for $i \leq n$, and $n = \lg(\bar\alpha(n)) \dotminus 1$. Hence

(364)
$$(\bar\alpha(n))_i = (\bar\alpha(n+1))_i \quad \text{for} \quad i < \lg(\bar\alpha(n)),$$
$$\lg(\bar\alpha(n)) < \lg(\bar\alpha(n+1)).$$

Hence, $\bar\alpha(n+1)$ is a continuation of $\bar\alpha(n)$, so that, by (363) it is true that $\bar\alpha(n+1) \prec \bar\alpha(n)$. The sequence $\{\bar\alpha(n)\}$ is, thus, a descending one.

LEMMA 8. *If* $\{\alpha(n)\}$ *is an infinite descending sequence* (*with respect to the relation* $\prec$) *of elements of the set* Seq, *then there exists a function* β *such that*

(365) $$\bigwedge n \bigvee k \bigwedge i < \lg\left(\bar{\beta}(n)\right)\left(1 + \beta(i) = \left(\bar{\alpha}(k)\right)_i\right)$$

(*finite sequences which are parts of the sequence* $\{\alpha(k)\}$ *are prolongations of partial diagrams of* $\bar{\beta}(n)$).

An example of such a function β is plotted in Figure 9. The finite sequences are marked there by the letters b, a, t_1, t_2, etc. All the finite sequences which are members of the sequence $\{\alpha(n)\}$ are continuations of the partial diagrams of the function β.

PROOF. The definition of the function β is arrived at in the following manner. By assumption, $\alpha(0)$ is the largest element of the sequence $\{a(n)\}$. Hence for every n it is true that

$$\left(\alpha(n)\right)_0 \leqslant \left(\alpha(0)\right)_0 ;$$

since the number of those numbers which are less than $\left(\alpha(0)\right)_0$ is finite, and there are infinitely many numbers in the sequence $\{\alpha(n)\}$, there must exist a number $k \leqslant \left(\alpha(0)\right)_0$ such that there are infinitely many elements of the sequence $\{\alpha(n)\}$ such that $\left(\alpha(n)\right)_0 = k$. The least such number k is selected as the first value of the function β', that is, as $\beta'(0)$. Next we define $\beta'(1)$ in a similar manner. Namely, we consider only those elements of the sequence $\{\alpha(m)\}$ which satisfy the formula

(366) $$\left(\alpha(m)\right)_0 = \beta'(0).$$

As we have demonstrated, their number is infinite, and their length $\geqslant 1$. But only a finite number of them can have a length which equals exactly 1, since there are finitely many elements of length 1 which precede a selected element z of length 1. Thus, there are infinitely many such elements which, next to formula (366), also satisfy the formula

(367) $$\lg\left(\alpha(m)\right) > 1.$$

Let $\alpha(m_0)$ be the first element in the sequence $\{\alpha(m)\}$ which satisfies both formulae, (366) and (367). Since the sequence $\{\alpha(m)\}$ is decreasing, it is true for infinitely many of those elements of $\{\alpha(m)\}$ which satisfy (366) and (367) that

(368) $$\alpha(m) \prec \alpha(m_0).$$

Since $\lg(\alpha(m_0)) > 1$, it follows from (368) and from the definition (363) of the relation $\prec$ that

(369) $(\alpha(m))_1 \leqslant (\alpha(m_0))_1.$

Since there are finitely many numbers x which are less that $(\alpha(m_0))_1$, and since there are infinitely many elements of $\{\alpha(m)\}$ which satisfy formulae (366)–(369), there must exist a number x such that there are infinitely many elements in $\{\alpha(m)\}$ which satisfy (366)–(369) and such that $(\alpha(m))_1 = x$. The least such x is selected as $\beta'(1)$.

The general definition of the function β is as follows:

$$\beta'(0) = (\mu x) \text{ [there are infinitely many } m \text{ such that } (\alpha(m))_0 = x],$$

(370) $$\beta'(n+1) = (\mu x) \Big[\text{there are infinitely many } m \text{ such that for } i \leqslant n \big((\alpha(m))_i = \beta'(i)\big) \text{ and } (\alpha(m))_{n+1} = x\Big],$$

$$\beta(n) = \beta'(n) \div 1.$$

It can easily be proved by induction that the minimizations which occur in Definition (370) are effective. Assume that the minimization which occurs in the definition of the number $\beta'(n)$, in accordance with formula (370), is effective; it will be proved that the same applies to $\beta'(n+1)$. By assumption there are infinitely many elements in $\{\alpha(m)\}$ such that

(371) $(\alpha(m))_i = \beta'(i)$ for $i \leqslant n$

and such that they all have a length $\geqslant n+1$. It is not possible that all but a finite number of them should have length exactly $n+1$, since there is always a finite number of those elements whose length is exactly $n+1$ and which precede a selected element of length $n+1$. Hence, there are infinitely many such elements which besides formula (371) also satisfy the formula

(372) $\lg(\alpha(m)) > n+1.$

Let $\alpha(m_0)$ be the first element in the sequence $\{\alpha(m)\}$ which satisfies both (371) and (372). Since the sequence $\{\alpha(m)\}$ is decreasing, it is true for infinitely many elements of $\{\alpha(m)\}$ which satisfy (371) and (372) that

(373) $\alpha(m) \prec \alpha(m_0).$

561

Since $\lg(\alpha(m_0)) > n+1$, it follows from (373) and the definition (363) of the relation $\preccurlyeq$ that

(374) $(\alpha(m))_{n+1} \leqslant (\alpha(m_0))_{n+1}$.

There are finitely many numbers x which are less then $(\alpha(m_0))_{n+1}$, and infinitely many elements of $\{\alpha(m)\}$ which satisfy formulae (371)–(374). There must accordingly be a number x such that there are infinitely many elements of $\{\alpha(m)\}$ which satisfy (371)–(374) and such that $(\alpha(m))_{n+1} = x$. Thus, the minimization used to define $\beta'(n+1)$ is effective.

It follows from the effectiveness of the minimizations which occur in (370) that the function β' in fact satisfies the condition imposed by the minimization operation, that is, that

(375)   *For every n there are infinitely many m such that, for $i \leqslant n$,*
$(\alpha(m))_i = \beta'(i)$.

Formula (365), which we intended to obtain, follows immediately from statement (375).

LEMMA 9. *For every relation R the set $U_a(R)$ defined as follows:*

(376) $x \in U_a(R) \equiv \big(x \in \text{Seq} \wedge \,\sim R(x, \mathfrak{a}) \wedge \bigwedge y, z < x \,((y, z \in \text{Seq} \wedge$
$x = y * z) \rightarrow \,\sim R(y, \mathfrak{a}))\big)$

satisfies the equivalence

(377) $\bigwedge \alpha \bigvee x R\big(\bar{\alpha}(x), \mathfrak{a}\big) \equiv \preccurlyeq \text{B Ord } U_a(R)$.

The function $*$ stands for concatenation (cf. formula (105) in Section 3). The formula $x = y * z$ means that the sequence x is a continuation of the sequence y by the sequence z, so that y is the initial part of x. Condition (376) states that $x \in U_a(R)$ if $\sim R(x, \mathfrak{a})$, and also that for any initial part y of a sequence x it is true that $\sim R(y, \mathfrak{a})$.

PROOF. a. Implication $\leftarrow$. If there existed a function α such that

(378) $\bigwedge x \sim R\big(\bar{\alpha}(x), \mathfrak{a}\big)$,

then, by Lemma 7, the sequence $\{\bar{\alpha}(n)\}$ would be a descending sequence of elements of the set $U_a(R)$, since it is true for every n that $\bar{\alpha}(n) \in \text{Seq}$, and, by (378), $\bar{\alpha}(n) \in U_a(R)$. This last is so because for every n the initial parts of an element $\bar{\alpha}(n)$ are only those elements which have the form $\bar{\alpha}(m)$ for $m < n$, and which, by (378), do not bear the relation

R to $\mathfrak{a}$. Thus, should formula (378) be true, the relation $\prec$ would not well-order the set $U_\mathfrak{a}(R)$.

b. Implication $\rightarrow$. Conversely, if $\prec$ does not well-order $U_\mathfrak{a}(R)$, then there exists an infinite descending sequence $\{\alpha(n)\}$ of elements of the set $U_\mathfrak{a}(R)$. By Lemma 8, there exists a function β such that formula (365) is satisfied, that is, such that every partial diagram $\bar\beta(m)$ of the function β is an initial segment of some finite sequence which is a member of the sequence $\{\alpha(n)\}$. Hence, by the definition of the set $U_\mathfrak{a}(R)$ (formula (376)), it is true for every m that

$$\sim R\big(\bar\beta(m), \mathfrak{a}\big).$$

Hence, $\sim \bigvee x R(\bar\beta(x), \mathfrak{a})$, so that the left side of equivalence (377) is not satisfied, either.

LEMMA 10 (Kleene [38]). *The predicate $\prec$ BOrd is a predicate in the class Π_1^1 and is not a predicate in the class Σ_1^1.*

PROOF. Let $G_n(x, y)$ be a computable function universal for the relations Erec. Let also

(379) $R(x, y, n) \equiv G_n(x, y) = 0.$

By Theorems 59 and 60, the relation

(380) $r(n, y) \equiv \bigwedge \alpha \bigvee x R\big(\bar\alpha(x), y, n\big)$

is a relation in Π_1^1 universal for the sets of numbers in the class Π_1^1 and is not in Σ_1^1. By Lemma 9 and formula (380), the relation r can be represented as follows:

(381) $r(n, y) \equiv \prec \mathrm{B\,Ord}\, U_{n,y}(R).$

If the predicate $\prec \mathrm{B\,Ord}\, U$ were in the class Σ_1^1, then, by (381), the relation r would be in Σ_1^1, which is not possible. In the definition of well-ordering the universal quantifier which binds function variables (or set variables), thus, cannot be eliminated (replaced by a quantifier which binds numerical variables) or be replaced by an existential quantifier which binds function variables or set variables.

EXERCISES

1. Give examples of a functional $h \in \mathrm{Comp}$ and a function δ such that formula (285) is true, whereas formula (286) is false.

2. Prove the equivalence of the definitions (282) and (300) of continuity.

3. Carry out rigorous proofs of Theorems 56, 57, 60.

4. Locate the following predicates with precision in the logical hierarchy:

a. A (natural) function α is increasing.

b. A function α has a finite number of values.

c. A function α is restricted.

d. Sat(n, α) (from p. 552).

e. A set X of natural numbers is infinite.

f. A functional Φ is continuous.

g. A relation R densely orders the set of natural functions.

h. A relation R continuously orders the set of natural functions.

Exercises a–e can be solved on the basis of certain known results. Solutions of f–h are not known to the present author and they may involve major difficulties.

5. Assume that an ordinal number α is in a class K if and only if there exists a relation in the class K which orders the set of natural numbers into the type α. Prove that the sets of ordinal numbers of the classes Erec, Prec, Comp, Π_n^0, Σ_0^1 coincide. For that purpose it suffices to note that the relation $\prec$ and the set $U_\alpha(R)$ are in Erec for $R \in$ Erec. If an ordinal number α_0 of any of the classes specified above were greater than any ordinal number of the class Erec, then, by (377), the following equivalence would hold: $\bigwedge \alpha \bigvee x R(\overline{\alpha}(x), \alpha) \equiv$ the relation $\prec$ on the set $U_\alpha(R)$ is isomorphic with a segment of α_0. The right side of this equivalence can be written down as Σ_1^1, since the existence of an isomorphism means the existence of a function. Thus, the concept of well-ordering would be of the type Σ_1^1.

Those ordinal numbers which are determined by recursive relations which well-order the set of natural numbers are called *constructive ordinal numbers*.

A HISTORICAL OUTLINE

The development of mathematical logic is closely linked with the development in our civilization of abstract and speculative thinking, and, thus, above all, with the development of mathematics. Laws of logic came to be recognized, while they were being used spontaneously in mathematics. Thinkers endowed with a sense of the analytic approach abstracted laws of logic from mathematical considerations and also from those everyday considerations which were speculative in nature, and then combined these laws into logical theories. Such an approach made the development of logic secondary in importance to and dependent upon the growth of other disciplines, especially mathematics. Logic developed mainly as an auxiliary methodological discipline, useful in proofs. The need for a particular branch of logic is recognized, when a great deal of experience with proofs is accumulated in a given field. When this happens, those investigators who are sensitive to logical problems start working on systematization, and construct a logic which provides proof techiques with clear criteria of correctness. Proofs become intersubjectively verifiable.

Such, more or less, was the nature of formal logic up to the middle of the 19th century. It was also frequently confused with epistemological considerations. About the middle of the 19th century formal logic began to move away from philosophy and to become more and more a mathematical discipline. At present, in view of the methods used in logical research and the knowledge needed to pursue such research, formal logic must certainly be classified as a mathematical science. Comparison of mathematical theories and other investigations of such theories have become the main sphere of logical research. Logic has become largely the science of the structure and various other properties of mathematical theories. It has, thus, become a science about mathematics, that is, in current terminology, it has become metamathematics.

Ancient and mediaeval logic

It was noted even in antiquity that an important role is played in speculative reasoning by certain concepts which recur in more abstract arguments, whether mathematical or philosophical.

The great accumulation of intellectual experience by philosophers:
the Sophists, Socrates (469–399), Plato (427–347) and their predecessors,
and also by Greek mathematicians and Athenian rhetoricians, made
Aristotle (384–322) feel the need for systemating the methods of proof
used in his times. Plato's dialogues, for instance, contained anylyses
of abstract concepts. Logical relationships between extensions of concepts accordingly played an important role in Plato's arguments.
Pre-Aristotelian mathematicians, for example, the Pythagoreans, also
examined relationships between certain simple sets of numbers and
between sets of geometrical figures. Those relationships were systematized
by Aristotle as a logical theory termed syllogistic, which is a calculus
of the extensions of concepts. It is confined to the formulations: *every
a is a b, some a's are b's, no a is a b,* and *some a's are not b's* and
certain simple relationships between them. Today, syllogistic is considered a very weak fragment of logic. It can be interpreted as a small
part of a calculus of terms, or a calculus of predicates, or a general
set theory. Nevertheless, it was an adequate tool for most of the
arguments which were carried out in antiquity and in the Middle Ages.
Up to the 18th century the proofs used in practice in and out of mathematics provided too few examples of other kinds of arguments to en-
courage the construction of basically new theory of terms.

The syllogistical relationships between extensions of terms were
formulated by Aristotle as implications, that is, conditional statements,
in the form *if ... then ...* The first element of an implication consisted
of two sentences combined with the sentential connective *and* and was
thus a conjunction. Implication was also used by pro-Aristotelian
Eleatics and early Megarians when they were formulating their antinomies. It was also used by ancient mathematicians. Pythagoreans proved
the irrationality of a number by *reductio ad absurdum.* In every proof
which uses *reductio ad absurdum* a conditional statement occurs
quite naturally. Later Megarians: Diodorus Cronus (4th cent. B.C.)
and his disciple, Philo of Megara (the turn of the 4th cent. B.C.), are

reputed to be first to realize, in a totally abstract manner, certain properties of conditional statements used in proofs and were said to have given, for all practical purposes, truth–table definition of implication. The Stoics: Zeno of Citium (336–264) and Chrysippus (232–205?), who were the contemporaries of the Megarians or somewhat later, felt the need to construct a general theory of implication. The investigations of the Megarians and the Stoics gave rise to what is known as the sentential calculus of the Stoics, which is a comprehensive fragment of the contemporary sentential calculus. It contains mainly theorems concerned with implication.

Sentential logic is more abstract than syllogistic, and, hence, it is not surprising that it developed later than syllogistic and perhaps partly under the influence of the former.

Modern mathematics as the main cause of the emergence of modern logical calculi

The insignificant development of Stoic sentential logic and of syllogistic in antiquity and the Middle Ages can be explained by the slow development of mathematics and other sciences at that time. Those sciences were not providing examples of any new kinds of reasoning. Modern man was also satisfied with ancient logic for a long time. Apparently only the development of modern mathematics accounts for serious progress in logic. Modern thought was more recently marked by the study of a wide range of sets of numbers and geometrical figures. The elementary calculus of sets was developed successively by G. Leibniz (1646–1716), Leonhard Euler (1707–1783), William Hamilton (1788–1856), Augustus De Morgan (1806–1878), and George Boole (1815–1864). The last named formulated an elementary calculus of sets as a complete system now called Boolean algebra. This was the last stage in the evolution of the theory of extensions of concepts, initiated in antiquity by Aristotle's syllogistic.

But the predicate calculus—the most important achievement of contemporary logic—did not emerge through successive improvements of ancient logic. It grew out of the specifically modern mathematical investigations connected with infinite operations and the concepts of continuity and limit, which are fundamental for mathematical analysis and the whole of contemporary mathematics. The origin of mathematical

567

analysis is a typical example of the fact that easier and more concrete issues are noticed earlier than are those issues which from the present-day point of view are more fundamental, but more difficult, since they are more abstract in nature. The concepts of derivative and integral, since their origin in the investigations of Newton (1642–1727) and Leibniz were understood intuitively for over 150 years, even though they lacked systematic definitions in the sense of present-day requirements. The definitions of those concepts formulated at that time seem non-systematic and vague to us. Reflections on infinitesimal quantities now appear as misleading speculations. Yet creators of those concepts comprehended their meaning well. Newton, when writing about *Ultimae rationes* in the introduction to his *Principles*, in all probability meant limits of quotients in the present-day sense of the terms, although he did not use present-day language. Comphrehension, one might say, outpaced the ability to couch ideas in formally exact terms.

The founders of the differential calculus noticed immediately the geometrical and the mechanical interpretation of the operations involved. Those operations were, in fact, discovered precisely because of their geometrical and mechanical interpretations. The investigators of the calculus also arrived at all the fundamental formulae of the differential calculus within a short period of time. The situation might seem paradoxical: a calculus was developing although rigorous definitions of the initial concepts were nonexistent. Yet, this is not as strange as it might seem. That calculus, when interpreted algebraically, is much easier to grasp than the definitions themselves. More than 150 years of intellectual training and experience in using that calculus was needed for an orderly definition of a limit, named after Cauchy (1789–1857), to be discovered. For its time that definition was very difficult to formulate, since it contains three successive quantifiers: universal, existential, and universal.

It is now difficult to discover whether Cauchy or someone else was the first to formulate the definition of a limit now ascribed to Cauchy and when during the period between 1820 and 1860 it was first formulated. Nevertheless, when it comes to systematization of mathematical analysis Cauchy did pioneering work, since he was the first to use, in his *Cours d'analyse* (1821), rigor modelled on that of Euclidean geometry: when presenting mathematical analysis he tried to define all the concepts he introduced; in formulating theorems he stated all the assumptions

568

required; he also proved theorems which his predecessors considered self-evident, for example, many elementary theorems on existence.

In some of the theorems and proofs of Cauchy's course of mathematical analysis we can find use of quantifiers, that is, formulations of the type *for every x ..., there exists an x such that ...,* or *an x can be found such that....* It is more difficult to find such formulations in definitions. Quantifiers had been used earlier and more consistently by Bolzano (1781–1848). Bolzano's formulations, even in his earlier years, seem to be more precise than those of Cauchy, and indicate better logical foundation. Yet, Bolzano's effect upon the manner of mathematical research in the 19th century seems to have been less than Cauchy's. Bolzano was active in Prague and was more creative as a philosopher than as a mathematician. For many years he was not connected with any university. Yet, even in his small book, *Betrachtung über einige Gegenstände der Elementar-geometrie* (Prague, 1804), he made some use of quantifiers, occasioned by looking for the most precise ways of formulating even very simple theorems. For instance, in that book, Part II, Section 10, p. 50, he formulated a theorem in a thoroughly modern way:

Zu einem gegebenen Puncte *a* und in einer gegebenen Richtung *R* gibt es Einen und nur Einen Punct *m*, dessen Entfernung von *a* der gegebenen des Punctes *y* von *x* gleiche.

In English: For a given point *a* and a given direction *R* there exists one and only one point *m* whose distance from *a* is equal to the distance between points *x* and *y*.

Such a formulation would do credit to any contemporary author who realized the importance of quantifiers. In Bolzano's *Rein analytischer Beweis...* (1817) quantifiers are used even more frequently and more consistently. The discovery of the usefulness of quantifiers was a natural consequence of the development of the symbolism which used variables to range over values in mathematical analysis and geometry. The use of variables arose in turn from the need for a precise and general formulation of theorems of algebra and algebraic geometry. The process began in antiquity. In most ancient and mediaeval formulae variable names were just names which could mean anything, any number in arithmetic and any set of objects in syllogistic. From the present-day

569

point of view it might be said that in antiquity and the Middle Ages the variables were free or, what amounts to the same thing, were bound by a universal quantifier at the beginning of a given theorem. Nevertheless, the use of variables, especially in arithmetical reasoning in the late Middle Age and early modern times, greatly improved and provided basic formal symbolism that made the modern analytical investigations of Fermat, Descartes, and others possible. Those investigations revealed in time the need to use variables in different roles in one and the same sentence. It no longer sufficed to use variables which were all free in the same way and stood for every thing in the same way. Some variables were free, but only in a part of a given sentence, while others depended on the earlier variables. The use of variables in new roles was particularly necessary in sentences concerning the existence of roots of certain equations or the existence of intermediate values. The root of an equation with variable co-efficients is not a definite number which we can know, so it can be denoted by any variable symbol, but its value still depends on the values of the co-efficients. For fixed co-efficients, on the other hand, the variable which stands for the root does not denote just anything, as in the case of a free variable. Bolzano's *Rein analytischer Beweis...* was concerned precisely with a rigorous presentation of a purely analytical proof of the existence of roots of an arbitrary polynomial. Thus, the development of mathematical ideas promoted the development of symbolism and brought out the usefulness of certain standard phrases in formulating ideas precisely.

Bolzano might have influenced the two German authors, Weierstrass and Frege, who contributed to advances in the use of quantifiers. In addition to the use of quantifier phrases Bolzano was also concerned with a philosophical analysis of their meanings in his philosophical book *Wissenschaftslehre* (1837), which might have influenced Frege.

From the point of view of logic, Cauchy and Bolzano opened a new epoch in the history of the foundations of mathematics, which covered the whole of the 19th and the early 20th century. They directed the interests of mathematicians toward the logical foundations of mathematics. The logical analysis of mathematical concepts found its culmination in Frege's *Grundgesetze der Arithmetik begriffsschriftlich abgeleitet* (Vol. 1, 1893), and apparently came to an end with the appearance, in 1925, of the second edition of *Principia Mathematica* by Whitehead and Russell. That period was marked by the successive reduction of higher branches of classical mathematics to the arithmetic of natural numbers, and next by the reduction of arithmetic to set theory. All the axioms required in all branches of classical mathematics were discovered. Every branch of mathematics was given a formalized, axiomatic form, and comparison of various mathematical theories became the object of research of a new branch of the foundations of mathematics, called metamathematics. At the time the second edition of *Principia Mathematica* appeared, the conviction that everything could be formalized was commonplace. The study of the various properties of such formalizations proved interesting. About 1930 the discoveries made by Gödel totally shifted the interest of logicians from the construction of formal systems to the study of such systems. The basic results obtained in the period of the logical analysis of the foundations of mathematics was summed up in the Introduction.

Cauchy's definition of limit at any rate was to be found in present-day formulation in Weierstrass's (1815–1897) formulation in his lectures held about 1860. Formulations of the type *for every $\varepsilon > 0$ there exists an n_0 such that...* came then to be fashionable owing to Weierstrass's influence as a teacher. Quantifiers, that is, phrases of the type *for every...* (universal) and *there exists an... such that...* (existential), came to be since used more and more systematically and consciously in proofs, although for many years they still had neither special names of their own, nor symbols, and their use was based only on logical intuition. Thus, the students of the foundations of mathematical analysis were the first to use quantifiers systematically, even though they did not construct any general logical theory of quantifiers. The words *every* and *exists* were, of course, used also by ancient and mediaeval thinkers, but it was only in the analyses of 19th century mathematicians that those words came to be used as operators which bind variables. They occur in such phrases as *for every x ..., for every x there exists an y such that...*, etc. Thus, a quantifier is followed by a variable which, as it is said today, is bound by it.

The development of the classical predicate calculus

The use of quantifiers was introduced in logic over a short period of time by several logicians, to a large extent independently of one another.

Most of the credit goes to C. S. Peirce (1839–1914), E. Schröder (1841–1902), and G. Frege (1848–1925). Gaining a clear historical perspective is made difficult by a lack of good historical studies of the problem. Peirce was perhaps the first to arrive at the concept of quantifier, even before 1870, but his writings appear to provide very few elements of the predicate calculus. In Frege's *Begriffsschrift, eine der arithmetisch nachgebildeten Formelsprachen des reinen Denkens* (Halle 1879) the nucleus of the predicate calculus is clearly visible. Thus, the birth of the quantifier as a logical concept falls during the period between Peirce's earliest works (1867) and the appearance of Frege's book. Church in his *Introduction to Mathematical Logic* (Vol. 1, Princeton 1956) states his opinion (cf. p. 288) that C. S. Peirce introduced quantifiers later than Frege, although he did it independently of the former. Quantifiers are, in any case, to be found in Peirce's works dating from 1885 and published in *The American Journal of Mathematics* (Vol. 5), where on p. 194 there is a reference probably to Mitchell's paper in *Studies in Logic* (1883). Thus, Peirce probably adopted Mitchell's idea and improved it slightly. At any rate it seems that the foundations of the predicate calculus developed independently both in Europe and in America.

If we disregard the chronology of the discoveries, we can give the following description of the specific approaches and achievements of the various authors. Peirce apparently arrived at the concept of quantifier by analysing the meaning of universal and existential statements. He introduced the term *quantifier* and the symbols: $\prod$ and $\sum$ for the universal and the existential quantifier, respectively. He applied those quantifiers to compound sentences in which the elementary sentences, as a rule, stated relations. Peirce used the symbolism of Boolean algebra and the algebra of relations. His works belong to the algebraic trend in logic, but he used the concept of relation in a very general sense, not restricted to relations of two arguments. Thus, Peirce's papers from the period 1880–1885, in which quantifiers occur quite frequently, may be classified as investigations in predicate logic. But he did not use the predicate calculus: theorems were given without proofs, as self-evident, and many systences were adduced merely as examples of the use of quantifiers for recording various ideas concisely.

E. Schröder also represented the algebraic trend. In his lectures on the algebra of logic he tried to prove certain theorems of the predicate calculus. His interpretation of the quantifiers was clearly algebraic: the universal quantifier was treated as an infinite Boolean product, and the existential, as an infinite Boolean sum. This emphasized the relationship between the quantifier operations and infinite multiplication and addition in arithmetic. We assume that Peirce also realized that relationship, since to denote quantifiers he chose the corresponding mathematical symbols for infinite operations. In spite of the algebraic form of his exposition Schröder did realize the sentential nature of the quantifiers. He accordingly interpreted the universal quantifier as an infinite conjunction, and the existential, as an infinite disjunction. He also emphasized the sentential nature of the quantifiers by the use of the phrases *for every...* and *for some....* The sentential interpretation of the algebra of logic, that is, Boolean algebra of finite operations, had been known from the studies by Boole himself.

Frege examined the meanings of the logical concepts in mathematics independently of the algebraic trend. His interpretation of logic was purely sentential. He axiomatized the sentential calculus completely, as it were finishing the job begun in antiquity by the Stoics. The quantifiers were analysed from the sentential point of view. For instance, in his *Begriffsschrift* Frege interpreted the traditional square of oppositions purely in terms of quantifiers. Thus, he incorporated syllogistic into the predicate calculus. In his later works he was concerned with the logical construction of the concept of number and with deducing the whole of arithmetic from logic. In his *Grundgesetze*, mentioned above, he have the first outline of a system of the logical foundations of mathematics.

Frege and G. Peano (1858–1932) were the first to use, on a large scale, logic in proofs on mathematical theorems. In their works, logic was merged with the study of the foundations of mathematics. Peano's symbolism for the sentential connectives proved more convenient than Frege's. Peano was also the first to give an axiom system of the arithmetic of natural numbers, and to prove many mathematical theorems in that field by distinctively logical methods. It was due to their work that the sentential interpretation of logic came to be fully accepted at the turn of the century. Logic, together with set theory, came to be regarded as the foundation of mathematics. Logicians attempted to treat the whole of logic as a single formalized system. The system suggested by Frege in his *opus magnum* proved inconsistent, and a system which was fairly satisfactory from the formal point of view was presented first by A. N. Whitehead (1861–1947) and B. Russell (1872–1970) in their *Principia Mathematica* (1910). The predicate calculus was treated there in approximately the current manner, and has not undergone any major modifications since. The only development has been the separation of the logical calculus, mainly the predicate calculus, from a fragment of set theory, the ramified theory of types. In *Principia* these two systems still formed a single whole, which gave rise to many formal objections. By 1930 there was a firm conviction that these two systems were two separate fields of research.

A fully modern interpretation of the predicate calculus as a logical theory independent of type theory (often called first-order functional calculus or first-order predicate calculus) was first given by Hilbert and Ackermann in their *Grundzüge der theoretischen Logik* (1928). Thus, the very formulation of the predicate calculus took exactly 50 years, but it was preceded by more than 50 years of successive introduction of quantifier phrases into mathematical practice.

The presentation of the classical logical calculus, that is, the logic of sentences and the logic of quantifiers, and of some elementary properties and applications of that calculus constituted Chapter I of the present book.

The birth of set theory

The development of set theory was linked closely with that of logic. Set theory appears to have developed more or less as logic did. Up to the middle of the 19th century syllogistic and Boolean algebra sufficed as the only general science of sets. It was only the investigations of the foundations of mathematical analysis which revealed the need to distinguish between many different sets, relations, and functions. That was why in the second half of the 19th century, owing mainly to the work of G. Cantor (1845–1918), a separate discipline was born that investigated sets, relations, and functions in the broadest sense of the terms. That discipline came to be called set theory and in the early 20th century was formulated as a rigorous system in several versions. Precision in this field was due to the general search for axiomatizations and all kinds of formal perfection, which dominated the late 19th century. As mentioned above, that trend could be traced back to Cauchy's lectures. The formalization of set theory was accelerated by discoveries at the turn of the century of numerous antinomies, which easily creep into set theory if certain measures of precaution, usually unnecessary in other branches of mathematics, are not observed. One version of the formalization of set theory is axiomatic set theory, initiated by E. Zermelo (1871–1954). Another approach is represented by type theory, due to Russell, which in its turn has several versions, the most important being that suggested by L. Chwistek (1884–1944). Axiomatic set theory is a stronger system and it complies better with the needs of mathematical practice. Type theory can be treated as a fragment of axiomatic set theory. Another variation on set theory is the theory suggested by W. V. Quine (b. 1908). Each of the theories mentioned above is interpreted differently by its author as to its philosophical import, but they all strive to satisfy the same formal requirements. A rather unique philosophical interpretation was that of S. Leśniewski (1886–1939), who interpreted simplified type theory as a *sui generis* expansion and extension into infinity of the gramatical analysis of a simple sentence, connected with the classification of words into parts of speech. That interpretation complies very well with the grammatical system of Slavic languages and might have been inspired by it.

The logical calculus of sentences and quantifiers and set theory are the two fundamental branches of the foundations of mathematics.

Metamathematical research from 1900 to 1930 as a result of trends toward formalization in mathematics

Until the 20th century formal logic was rather an auxiliary discipline. Only in the 20th century did it become an independent science with its own subject matter, namely, axiomatic systems and logical calculi. Thus, logical research in the 20th century has been concerned mainly with metalogic and metamathematics. This is why the major part of the present book (Chapters II and III) has been dedicated to discussing results of matemathematical research in the 20th century.

Such research now commands the attention of the overwhelming majority of logicians. The problems raised were born from the search, increasing in intensity at the turn of the 19th century, for the greatest possible formal precision in mathematics. In the first 30 years of the present century these problems were concerned mainly with the foundations of the formal axiomatic method. They can be summarized with the question: What cognitive values can be guaranteed by the observance of the logical requirements of formal correctness? The values in question are, above all, the consistency and completeness of mathematical systems. It seemed natural to strive to construct theories which would be: 1) complete, that is, would provide answers to all mathematical questions which can be formulated in such theories, 2) consistent, with consistency being properly guaranteed by a proof outside the theory. The search for theories which have these two formal properties was the main objective of what was called the Hilbert program, formulated around 1900 by D. Hilbert (1862–1943) and his "formalist" school. Around 1930, due mainly to the research of K. Gödel (b. 1906), logicians reached a conclusion concerning the above problem. It follows from Gödel's research that the majority of interesting mathematical theories are incomplete theories, because every mathematical theory which contains the arithmetic of natural numbers is incomplete. Most creative work done by mathematicians takes place in the field of such theories. Moreover, no mathematical theory which contains the arithmetic of natural numbers can be proved to be consistent unless the proof resorts to means which are stronger than the theory in question. In other words, every proof of the consistency of a theory T, which contains the arithmetic of natural numbers, requires a theory S, which contains the theory T and many other theorems. The Hilbert program, formulated in an *a priori* manner, has proved impossible to carry out. Nevertheless, at one time it played an important role by stimulating research. Moreover, even though the search for absolute proofs of consistency has proved sterile, reduction of the consistency of one theory to the consistency of another theory, which is called a relative proof of consistency, remains valid. It provides certainty as to the consistency of some theories if their consistency is reduced to that of a well-known theory whose consistency we believe in on the strength of centuries of experience. Relative consistency proofs also make it possible to obtain other interesting results.

Gödel's studies have also demonstrated the specific cognitive limitations of the logical axiomatic method. In every theory T which is consistent and contains the arithmetic of natural numbers there are sentences $A(x)$ with one free numerical variable such that all the sentences.

(a) $A(0), A(1), A(2), \ldots$

are theorems in T, but the general statement

(b) $A(x)$ *holds for every natural number* x

cannot be proved in the theory T, nor can its negation.

This is a paradox with profound epistemiological significance, since it refers to theories with arbitrarily rich axiom systems and arbitrarily strong rules of proof, provided that these axioms and rules do not go beyond what is required of usual theories. We require usual theories to provide formal, intersubjectively verifiable proofs. Proofs must be such that it be possible to ascertain which rules and which axioms have been used in a given proof. If a theory can be described so that all the proofs which are accepted in it satisfy these conditions, then the paradox stated above holds. It is worth noting that the rule which would make it possible to pass from type (a) statements to type (b) statements is not verifiable, because if we were to apply it we would need an infinite number of type (a) statements.

This result of Gödel's investigations was one of the main topics in Chapter III.

Issues in metamathematical research after 1930

Gödel's investigations around 1930, because of their importance, ultimately directed the attention of a majority of mathematicians to metamathematical issues. His investigations also provided solutions to problems discussed at the time and thus completed the first period of metamathematical investigations. This was followed by a greater specialization of interests. Two trends seem to be discernable in the period from 1930 to the present day. One is represented by research on the logical methods of construction of mathematical concepts. The other, by research on models of axiomatic systems. The former originated with investigations by J. Herbrand (1908–1931) and Gödel into computable functions. In that trend the concept of quantifier plays the fundamental role, since quantifiers make it possible to obtain a classification of all logical constructions, a classification which is of great importance in the field. The starting point for that classification is offered by those concepts in whose definitions quantifiers play no essential role. They are the computable concepts. Other levels of that classification are obtained by applying to lower level concepts, alternately, the universal and the existential quantifier. This classification was introduced by S. C. Kleene (b. 1909) and A. Mostowski (b. 1913).

This trend also includes the study of decidability. The mathematical concept of computability is linked with those of decidability of theories, effectiveness of methods, and inter-subjective verifiability. A function is computable if there is a method of computing its value, a method that can be formulated as a simple rule. Such a method is intersubjectively verifiable, since everyone can study the rule and check the computations. Likewise, a method of proof is effective and at the same time intersubjectively verifiable if there is a rule that makes it possible to easily check the correctness of the proof. The logical classification of mathematical concepts begins with concepts that are computable in the sense formulated above. Such concepts are, so to speak, the most tangible, and belong to the empirical date of arithmetic. By applying to computable concepts the phrases: *for every x...*, and *there exists an x such that...*, that is, quantifiers, we obtain concepts of increasing complexity, that is, higher up in the hierarchy of concepts. Gödel's theorem on the undecidability of arithmetic becomes almost a side issue in the investigations into the logical classification of mathematical concepts. All the basic issues of the logical classification of concepts were discussed in Chapter III.

The other trend originated with studies by T. Skolem (1887–1963), Löwenheim, Gödel, and A. Tarski (b. 1901) on models, the concept of satisfaction, and the completeness of the predicate calculus. While the investigations representing the first trend are concerned with the problem: what can be stated and proved about mathematical reality by logical methods, those representing the second trend go, so to speak, in the opposite direction: they start from proved statements and established theories and are concerned with the problem as to what mathematical reality is in fact described by those theories.

The studies which are included in the second trend also have an epistemological aspect which is not devoid of paradox. Every interesting system has many models and, thus, describes many domains simultaneously. The attempt to confine oneself to a single specific domain becomes unrealistic. Arithmetic, for instance, has many models, and there is no formal method of singling out "the true" natural numbers. And yet it seems that any two mathematicians concerned with arithmetic mean one and the same object by the set of natural numbers. This trend in research was discussed in Chapter II.

Modern constructivism and modern non-classical logics as a reaction against extension of logical concepts by classical logic and set theory

Constructivism, which includes intuitionism, is a separate thread in the history of logic. In early arithmetic, mathematical concepts used to be constructive, and as late as the 18th century most authors used to define mathematical functions by means of certain formulae. The 19th century saw a far-reaching extension of concepts. Any many-to-one relation came to be called a function. As it turned this extension of concepts went much further than was originally thought. When this excessive extension of concepts was recognized in the 20th century, two trends arose in the study of the foundations of mathematics: the attempt to discover the limits of constructibility of mathematical concepts, and the attempt to limit oneself to constructible concepts. The former trend gave birth to the theory of computable functions, ramified type theory, and the logical classification of mathematical concepts. The latter gave rise to intuitionistic logic, originated by L. E. T. Brouwer (1882–1967) and A. Heyting (b. 1888), which now forms the most important system competing with classical logic. Modal logic, which has been pursued by some logicians since Aristotle, is much less substantiated by mathematical practice and mathematical intuitions. No mathematician has ever consistently taken the stand of modal logic. Thus, modal logics, like the many-valued logics initiated by J. Łukasiewicz (1878–1956) and E. L. Post (1897–1954), are a product of philosophical fantasy rather than a result of essential mathematical or logical needs. Modal logics, originated in their modern form by C. I. Lewis (b. 1883) around 1912, have been most exhaustively presented in *Symbolic Logic* by C. I. Lewis and C. H. Langford (1932). Investigations in that field strive to understand the philosophical concept of possibility, which goes beyond the sphere of logic. From the point of view of classical logic the importance of modal logics lies in their attempt at using the

579

concept of possibility to narrow down the meaning of implication and to construct the concept called strict implication, governed by laws which come closer to the intuitions linked with implication as used in every-day language. There are many systems of strict implication and modal logic, suggested by different authors. The attempts of the intuitionists and the propounders of strict implication and modal logic are an interesting example of a reaction against the perhaps excessive broadness of concepts used in classical logic. These attempts, intended to narrow down the meanings of logical terms, began in the first quarter of the 20th century, as soon as the full system of classical logic came to be formulated, and they are still of current interest. They perhaps penetrate more deeply into the meaning of logical concepts than classical logic does, since the latter seems to simplify the criteria for when to use the various logical (sentential) connectives. Yet, for the majority of mathematicians the classical logical calculus has remained the principal logical instrument in their research. For this reason only a few remarks have been made here about the non-classical calculi.

BIBLIOGRAPHY

[1] J. L. Bell and A. B. Slomson, *Models and Ultraproducts*, North-Holland Publ. Co., 1969.

[2] P. Bernays, *Axiomatic Set Theory*, Amsterdam 1958.

[3] P. Bernays und D. Hilbert, *Grundlagen der Mathematik*, Berlin 1934-39.

[4] E. W. Beth, *The Foundations of Mathematics*, Amsterdam 1959.

[5] G. Birkhoff, *Lattice Theory*, New York 1948.

[6] I. M. Bocheński, *Formale Logik*, Freiburg-München 1956.

[7] G. Cantor, 'Ein Beitrag zur Mannigfaltigkeitslehre', *Journ. Reine Angew. Math.* **84** (1878) 242-258.

[8] A. Church, *Introduction to Mathematical Logic*, Princeton 1956.

[9] A. Church and S. C. Kleene, 'Formal Definitions in the Theory of Ordinal Numbers', *Fund. Math.* **28** (1937) 11-21.

[10] P. J. Cohen, *Set Theory and the Continuum Hypothesis*, New York 1966.

[11] W. Craig, 'Linear Reasoning. A New Form of the Herbrand-Gentzen Theorem', *Journal of Symbolic Logic* **22** (1957) 250-268.

[12] W. Craig, 'Three Uses of the Herbrand-Gentzen Theorem in Relating Model Theory and Proof Theory', *Journal of Symbolic Logic* **22** (1957) 269-285.

[13] A. Ehrenfeucht, 'Application of Games to Some Problems of Mathematical Logic', *Bull. Acad. Polon. Sci., Cl. III*, **5** (1957) 35-37.

[14] U. Felgner, 'Models of ZF-Set Theory', *Springer Lecture Notes* **223** (1971).

[15] G. Frege, *Die Grundlagen der Arithmetik*, Wrocław 1884.

[16] K. Gödel, 'Die Vollständigkeit der Axiome des logischen Funktionenkalküls', *Monatsh. Math. Phys.* **37** (1930) 349-360.

[17] K. Gödel, 'Über formal unentscheidbare Sätze der Principia Mathematica und verwandter Systeme I', *Monatsh. Math. Phys.* **38** (1931) 173-198.

[18] A. Grzegorczyk, 'Some Classes of Recursive Functions', *Rozprawy Matematyczne* **4** (1953) 1-45.

[19] A. Grzegorczyk, A. Mostowski and C. Ryll-Nardzewski, 'The Classical and the ω-complete Arithmetic', *Journal of Symbolic Logic* **23** (1958) 188-208.

[20] T. Hailperin, 'Quantification Theory and Empty Individual-domains', *Journal of Symbolic Logic* **18** (1953) 197-200.

[21] L. Henkin, 'A Poof of Completeness for the First Order Functional Calculus', *Jornal of Symbolic Logic* **14** (1949) 159-166.

[22] L. Henkin, 'A Generalization of the Concept of ω-completeness', *Journal of Symbolic Logic* **22** (1957) 1-41.

BIBLIOGRAPHY

[23] J. Herbrand, 'Recherches sur la théorie de la démonstration', *C. R. Soc. Sci. Varsovie, Cl. III*, 33(1930) 61.

[24] A. Heyting, *Intuitionism*, Studies in Logic, Amsterdam 1956.

[25] A. Heyting, *Constructivity in Mathematics*, Studies in Logic, Amsterdam 1959.

[26] T. J. Jech, 'Lectures in Set Theory with Particular Emphasis on the Method of Forcing', *Springer Lecture Notes* **217** (1971).

[27] R. B. Jensen, 'Modelle der Mengenlehre', *Springer Lecture Notes* **37** (1967).

[28] H. J. Keisler, *Model Theory for Infinitory Languages*, Studies in Logic, Amsterdam 1971.

[29] J. L. Kelley, *General Topology*, Princeton 1964.

[30] S. C. Kleene, *Mathematical Logic*, New York 1967.

[31] S. C. Kleene, *Introduction to Metamathematics*, Amsterdam 1967.

[32] S. C. Kleene and R. E. Veseley, *The Foundation of Intuitionistic Mathematics*, Amsterdam 1965.

[33] S. C. Kleene, 'On Notation for Ordinal Numbers', *Journal of Symbolic Logic* **3** (1939) 150–155.

[34] S. C. Kleene, 'Recursive Predicates and Quantifiers', *Trans. Amer. Math. Soc.* **53** (1943) 41–73.

[35] S. C. Kleene, 'On the Interpretation of Intuitionistic Number Theory', *Journal of Symbolic Logic* **10** (1945) 109–124.

[36] S. C. Kleene, 'Recursive Functions and Intuitionistic Mathematics', *Proceedings of the International Congress of Mathematicians*, Cambridge 1950, vol. I, 679–685.

[37] S. C. Kleene, 'Arithmetical Predicates and Function Quantifiers', *Trans. Amer. Math. Soc.* **79** (1955) 312–341.

[38] S. C. Kleene, 'Hierarchies of Number-Theoretic Predicates', *Bull. Amer. Math. Soc.* **61** (1955) 193–216.

[39] S. C. Kleene and E. Post, 'The Upper Semi-Lattice of Degrees of Recursive Unsolvability', *Ann. of Math.* **59** (1954) 379–407.

[40] C. Kuratowski, *Topology I*, Warszawa 1948.

[41] C. Kuratowski, *Introduction to Set Theory and Topology*, Warszawa 1972.

[42] C. Kuratowski and A. Mostowski, *Set Theory*, Warszawa 1971.

[43] A. Lindenbaum und A. Tarski, 'Über die Beschränktheit der Ausdrucksmittel deduktiver Theorien', *Ergebnisse eines mathematischen Kolloquiums* **7** (1936) 15–23.

[44] J. Loeckx, *Computability and Decidability. An Introduction for Students of Computer Science*, Berlin 1972.

[45] R. C. Lyndon, 'An Interpolation Theorem in the Predicate Calculus', *Pacific Journal of Mathematics* **9** (1959) 129–142.

[46] J. Łoś, 'On the Categoricity in Power of Elementary Deductive Systems and Some Related Problems', *Coll. Math.* **3** (1954) 58–62.

[47] J. Łoś, 'The Algebraic Treatment of the Methodology of Elementary Deductive Systems', *Studia Logica* **2** (1955) 151–211.

[48] J. Łukasiewicz, *Elements of Mathematical Logic*, Oxford 1963.

BIBLIOGRAPHY

[49] J. Łukasiewicz, *Selected Works*, Warszawa–Amsterdam 1970.

[50] E. Mendelson, *Introduction to Mathematical Logic*, Princeton N. J. 1964.

[51] A. P. Morse, *A Theory of Sets*, New York 1965.

[52] A. Mostowski, 'On Definable Sets of Positive Integers', *Fund. Math.* 34 (1947 81–112.

[53] A. Mostowski, *Constructible Sets with Applications*, Warszawa–Amsterdam 1969.

[54] A. Mostowski, 'On Rules of Proof in the Pure Functional Calculus of the First Order', *Journal of Symbolic Logic* 16 (1951) 107–111.

[55] A. Mostowski, *Sentences Undecidable in Formalized Arithmetic*, Amsterdan 1952.

[56] A. Mostowski, 'Examples of Sets Definable by Means of Two and Three Quantifiers', *Fund. Math.* 42 (1955) 259–270.

[57] A. Mostowski, R. M. Robinson and A. Tarski, 'Undecidability and Essential Undecidability in Arithmetic', *Undecidable theories* 55, Amsterdam 1953, 31–74.

[58] D. Nelson, 'Recursive Functions and Intuitionistic Number Theory', *Trans. Amer. Math. Soc.* 61 (1947) 307–368.

[59] P. S. Novikov, 'On the Algorithmic Unsolvability of the Word Problem in Group Theory', *Trudy Mat. Inst. im. Steklova* 44, Moscow 1955; *Amer. Mat-Soc. Trans.* 9, ser. 2.

[60] R. Peter, 'Über den Zusammenhang der verschiedenen Begriffe der rekursiven Funktion', *Math. Ann.* 110 (1935) 612–632.

[61] R. Peter, 'Über die mehrfache Rekursion', *Math. Ann.* 113 (1936) 489–527.

[62] R. Peter, *Rekursive Funktionen*, Budapest 1951.

[63] E. L. Post, 'Formal Reduction of the General Combinatorial Decision Problem', *Amer. Journal Math.* 65 (1943) 197–215.

[64] E. L. Post, 'Recursively Enumerable Sets of Positive Integers and Their Decision Problems', *Bull. Amer. Math. Soc.* 50 (1944) 284–316.

[65] E. L. Post, 'Recursive Unsolvability of a Problem of Thue, *Journal of Symbolic Logic* 12 (1947) 1–11.

[66] W. V. Quine, 'Quantification and the Empty Domain', *Journal of Symbolic Logic* 19 (1954) 177–179.

[67] W. V. Quine, *Mathematical Logic*, Combridge 1958.

[68] H. Rasiowa and R. Sikorski, *The Mathematics of Metamathematics*, Warszawa 1968.

[69] L. Rieger, 'On Countable Generalized δ-Algebras with a New Proof of Gödel's Completeness Theorem', *Czechoslovak Mathematical Journal* 1 (76) (1951) 29–40.

[70] H. Rogers Jr., *Recursive Functions and Effective Computability*, New York 1967.

[71] A. Robinson, 'Complete Theories', *Studies in Logic and the Foundation of Mathematics*, Amsterdam 1955.

[72] A. Robinson, 'A Result on Consistency and its Application to the Theory of Definition', *Indagationes Mathematicae* 18 (1956) 47–58.

[73] A. Robinson, *Introduction to Model Theory and to the Metamathematics of Algebra*, Amsterdam 1963.

BIBLIOGRAPHY

[74] J. Robinson, 'Definability and Decision Problems in Arithmetic', *Journal of Symbolic Logic* **14** (1949) 98–114.

[75] J. Robinson, 'General Recursive Functions', *Proc. Amer. Math. Soc.* **1** (1950) 703–718.

[76] R. M. Robinson, 'Primitive Recursive Functions', *Bull. Amer. Math. Soc.* **53** (1947) 925–942.

[77] R. M. Robinson, 'Recursion and Double Recursion', *Bull, Amer. Math. Soc.* **54** (1948) 987–993.

[78] P. C. Rosenbloom, *The Elements of Mathematical Logic*, New York 1950.

[79] B. Rosser, 'Extension of a Theorem of K. Gödel', *Journal of Symbolic Logic* **1** (1936) 87–91.

[80] B. Russell, *The Principles of Mathematics*, Cambridge 1903, 2nd ed. London 1956.

[81] G. E. Sacks, *Saturated Model Theory*, Reading 1972.

[82] H. Scholz, *Geschichte der Logik*, Berlin 1931.

[83] J. R. Shoenfield, *Mathematical Logic*, Reading 1967.

[84] J. R. Shoenfield, *Degrees of Unsolvability*, Amsterdam 1971.

[85] T. Skolem, 'Sur la portée du théorème de Skolem–Löwenheim', *Les Entretiens de Zürich sur les Fondements des Sciences Mathématiques*, Zurich 1941, 25–47.

[86] C. Spector, 'Recursive Well-Ordering', *Journal of Symbolic Logic* **20** (1955) 151–163.

[87] W. Szmielew, 'Elementary Properties of Abelian Groups', *Fund. Math.* **41** (1955) 203–271.

[88] W. Szmielew and A. Tarski, 'Mutual Interpretability of Some Essentially Undecidable Theories', *Proceedings of the International Congress of Mathematicians, Cambridge* 1950, vol. I, p. 734.

[89] A. Tarski, *Logic, Semantics, Metamathematics*, Oxford 1956.

[90] A. Tarski, 'Über einige fundamentale Begriffe der Metamathematik', *C. R. Soc. Sci. Varsovie, Cl. III* **23** (1930) 22–29.

[91] A. Tarski, 'Einige Betrachtungen über die Begriffe der ω-Widerspruchfreichet und ω-Vollständigkeit', *Monatsh. Math. Phys.* **40** (1933) 97–112.

[92] A. Tarski, 'Pojęcie prawdy w językach nauk dedukcyjnych', *R. C. Soc. Sci. Varsovie, Cl. III*, (1933).

[93] A. Tarski, 'Grundzüge der Systemenkalküls', *Fund. Math.* **25** (1935) 503–526 and **26** (1936) 283–301.

[94] A. Tarski, *Einführung in die mathematische Logik und in die Methodologie der Mathematik*, Wien 1937.

[95] A. Tarski, 'The Semantic Conception of Truth and the Foundation of Semantics', *Philosophy and Phenomenological Research* **4** (1944) 341–376.

[96] A. Tarski, 'Undecidability of the Theories of Lattices and Projective Geometric', *Journal of Symbolic Logic* **14** (1949) 77–78.

[97] A. Tarski, 'Some Notions and Methods on the Borderline of Algebra and Metamathematics', *Proceedings of the International Congress of Mathematicians* **1** (1950) 705–720.

BIBLIOGRAPHY

[98] A. Tarski, *A Decision Method for Elementary Algebra and Geometry*, 2nd ed, Berkeley and Los Angeles 1951.

[99] A. Tarski, *Undecidable Theories*, Studies in Logic and the Foundations of Mathematics, Amsterdam 1953.

[100] A. M. Turing, 'On Computable Numbers, with an Applications to the Entscheidungsproblem', *Proc. London Math. Soc.* **42** (1937) 230–265.

[101] R. L. Vaught, 'Applications of the Generalized Skolem–Löwenheim Theorem to Problems of Completeness and Decidability', *Bull. Amer. Math. Soc.* **59** (1953) 396–397.

[102] B. L. Van der Waerden, *Algebra*, Berlin 1955.

[103] A. N. Whitehead and B. Russell, *Principia Mathematica*, Cambridge 1910, 2nd ed., 1925, preprint 1957.

[104] P. Vopenka and P. Hájek, *The Theory of Semisets*, Prague 1972.

[105] E. Zermelo, 'Untersuchungen über die Grundlagen Axiome der Mengenlehre I', *Math. Ann.* **65** (1908) 261–281.

585

INDEX OF SYMBOLS

588

INDEX OF NAMES

INDEX OF NAMES

SUBJECT INDEX

SUBJECT INDEX

SUBJECT INDEX